PLEISTOCENE MAMMALS OF NORTH AMERICA

PLEISTOCENE MAMMALS OF NORTH AMERICA

Björn Kurtén
and
Elaine Anderson

COLUMBIA UNIVERSITY PRESS
New York   1980

Library of Congress Cataloging in Publication Data

Kurtén, Björn.
    Pleistocene mammals of North America.

    Bibliography: p.
    Includes index.
    1. Mammals, Fossil. 2. Paleontology—Pleistocene.
3. Paleontology—North America. I. Anderson, Elaine,
joint author. II. Title.
QE881.K82   569′.097   79-26679
ISBN 0-231-03733-3

Columbia University Press
New York   Guildford, Surrey

Copyright © 1980 Columbia University Press
All rights reserved
Printed in the United States of America

For
John *and* Alice Guilday

Contents

List of Figures	*ix*
List of Tables	*xii*
Introduction	*xiii*
Part 1. Mammalian Faunas of the Blancan and Pleistocene	
Key to Collections	*2*
1. Chronology	*3*
2. Blancan Faunas	*6*
3. Irvingtonian Faunas	*22*
4. Rancholabrean Faunas	*37*
5. Intercontinental Correlation and Migrations	*90*
Part 2. Mammalian Species of the Blancan and Pleistocene	
6. Introductory Notes	*99*
7. Order Marsupialia	*101*
8. Order Insectivora	*103*
9. Order Chiroptera	*116*
10. Order Edentata	*128*
11. Order Carnivora	*146*
12. Order Rodentia	*209*
13. Order Lagomorpha	*275*
14. Order Perissodactyla	*283*
15. Order Artiodactyla	*295*
16. Order Sirenia	*340*
17. Order Proboscidea	*343*
18. Order Primates	*355*
19. Extinction	*357*
References	*367*
Appendix 1. Wisconsinan Radiocarbon Dates	*403*
Appendix 2. Stratigraphic Ranges	*407*
Index to Latin Names of Organisms	*419*
Index to Common Names and Organisms	*423*
Index to Localities and Stratigraphic Terms	*433*
Index to Authors	*438*
Figure Credits	*443*

List of Figures

1.1 Late Pliocene and Pleistocene chronology in North America.
2.1 First and last occurrences of taxa in the Blancan and Irvingtonian.
2.2 Scene near Hagerman in the Blancan.
3.1 Scene near Cudahy in the Irvingtonian.
3.2 Scene near Cumberland Cave in the Irvingtonian.
4.1 Scene near Fairbanks in the late Rancholabrean.
4.2 Scene near Medicine Hat in the Rancholabrean.
4.3 Scene at Rancho La Brea in the late Rancholabrean.
4.4 Scene near Melbourne in the late Rancholabrean.
4.5 Scene near American Falls in the Rancholabrean.
4.6 Scene near Boney Spring in the late Rancholabrean.
4.7 Scene at San Josecito Cave in the Rancholabrean.
4.8 Scene near New Paris No. 4 in the late Rancholabrean.
4.9 Scene near Ingleside in the Rancholabrean.
8.1 Distribution of *Sorex cinereus*.
8.2 Distribution of *Sorex arcticus*.
8.3 Mandibles and dentitions of shrews.
8.4 Limb bones of moles.
9.1 Mandibles and dentitions of vespertilionid bats.
10.1 Scutes of armored edentates.
10.2 *Holmesina septentrionalis* carapace.
10.3 *Glyptotherium* sp. carapace.
10.4 Distribution of *Glyptotherium*.
10.5 *Megalonyx jeffersonii* skull.
10.6 *Nothrotheriops shastensis* skull and dentition.
10.7 *Glossotherium harlani* skull.
11.1 *Martes diluviana* and *Martes nobilis* skulls and mandibles.
11.2 *Mustela vison* skull and mandible.
11.3 *Gulo gulo* skull, mandible, and dentition.
11.4 *Trigonictis idahoensis* mandible and dentition.
11.5 *Taxidea taxus* skull, mandible and dentition.
11.6 *Satherium piscinaria* and *Enhydra lutris* skull, mandible, and dentitions.
11.7 *Conepatus leuconotus* and *Brachyprotoma obtusata* dentitions.
11.8 *Borophagus diversidens* skull and mandible.
11.9 *Canis dirus* and *Canis latrans* P^4–M^2.
11.10 Canid mandibles.
11.11 *Bassariscus casei* and *Procyon lotor* dentitions.
11.12 Distribution of *Tremarctos*.
11.13 Distribution of *Arctodus*.
11.14 *Arctodus simus* and *Ursus americanus* mandibles.

11.15 Ursid P^4–M^2.
11.16 *Arctodus* and *Ursus* mandibles and dentitions.
11.17 Distribution of *Smilodon fatalis* and *Homotherium serum*.
11.18 Machairodont skulls and jaws.
11.19 Felid humeri and ulnae.
11.20 Restoration of *Lynx issiodorensis*.
11.21 *Lynx rufus* and *Felis concolor* cheek teeth and mandibles.
11.22 *Chasmaporthetes ossifragus* mandible and dentition.
11.23 *Odobenus rosmarus* skull and mandible.
11.24 *Eumetopias jubata* and *Zalophus californicus* skull and mandibles.
11.25 *Erignathus barbatus* and *Cystophora cristata* skull and mandibles.
12.1 *Aplodontia rufa* skull, mandible, and cheek teeth.
12.2 Sciurid mandibles and dentitions.
12.3 Distribution of *Spermophilus tridecemlineatus*.
12.4 Geomyid skull, mandibles, and dentitions.
12.5 Heteromyid skull, mandibles, and dentitions.
12.6 Beaver skulls, mandibles, and dentitions.
12.7 Cricetine M_{1-3}.
12.8 Cricetine and microtine mandibles.
12.9 Development of dentine tracts.
12.10 Microtine M_1's.
12.11 *Microtus* M_1's.
12.12 Muskrat, lemming, and bog lemming M_1's.
12.13 Zapodid lower dentitions.
12.14 *Erethizon dorsatum* and *Hydrochoerus holmesi* cheek teeth.
13.1 Leporid skull, mandible, and teeth.
14.1 Equid upper cheek teeth.
14.2 *Equus* M_1's.
14.3 *Equus* metatarsals and pes.
14.4 *Tapirus veroensis* mandible and dentition.
15.1 Restoration of *Platygonus compressus* and *Mylohyus nasutus*.
15.2 *Platygonus vetus* and *Mylohyus nasutus* lower dentitions.
15.3 *Platygonus compressus* and *Platygonus vetus* skull and zygomatic arch.
15.4 Distribution of *Camelops*.
15.5 *Camelops hesternus* skull, mandible, and dentition.
15.6 *Camelops hesternus* forefoot.
15.7 Distribution of *Hemiauchenia* and *Palaeolama*.
15.8 Antler shapes in cervids.
15.9 *Navahoceros fricki* and *Sangamona fugitiva* mandible and dentitions.
15.10 Cervid limb bones.
15.11 Restoration of *Capromeryx minor*, *Stockoceros onusrosagris*, and *Antilocapra americana*.
15.12 *Stockoceros conklingi* skull, mandible, and dentition.
15.13 *Stockoceros conklingi* metacarpus and metatarsus.
15.14 *Oreamnos harringtoni* metacarpus and metatarsus.

15.15 Ovibovine horn shapes.
15.16 *Eucevatherium collinum* metacarpus and metatarsus.
15.17 *Symbos cavifrons* skull.
15.18 *Bison* skulls.
16.1 *Hydrodamalis gigas* skull.
17.1 Restoration of *Mammut americanum* and *Mammuthus columbi*.
17.2 Proboscidean upper molars.
17.3 Mammoth hunting scene.

List of Tables

1.1 Classical Glacial-Interglacial Sequences in North America and Europe.
2.1 Blancan Faunas in the Great Plains Area.
3.1 Irvingtonian Faunas in the Great Plains Area.
5.1 Correlation of Land Mammal Ages in North and South America.
5.2 Correlation of Plio-Pleistocene Land Mammal Faunas in Europe and North America.
5.3 Correlation of Plio-Pleistocene faunas in the USSR and China.
19.1 Blancan/Pleistocene Extinction of Mammal Species According to Stratigraphic Range.
19.2 Extinctions of Mammal Species by Size.
19.3 Blancan, Irvingtonian, and Wisconsinan Extinctions with Species Grouped by Order.
19.4 Extinction of Mammal Species by Families.
19.5 Wisconsinan/Holocene Extinctions According to Origin.
19.6 Last Appearances of Species Becoming Extinct during the Wisconsinan.

Introduction

The fossil record tells us of the organisms of the past, their relationships, evolutionary history, and migrations. It also informs us about the life communities, climates, and biogeography of former times. For the most recent geological past, the Pleistocene epoch, our information is especially complete. With present advances in methods of relative and absolute dating, temporal discrimination and correlation of local time sequences (even continents apart) have attained a high degree of sophistication.

Meanwhile, the "new systematics" (Huxley, 1940), with its emphasis on the study of populations, has replaced the old typological approach in taxonomy. As a result, the species, rather than the genus, is becoming the basic taxonomic unit in paleontology, as befits its role as the most important category in the study of evolution (Mayr, 1963). The emphasis on the species is further enhanced in the "punctuated equilibrium" theory of evolution, which envisages quantum shifts between essentially static species-stages (produced by allopatric speciation) as the most important mode of progressive evolution (Eldredge and Gould, 1972).

Although research on the Pleistocene mammals of North America dates well back to the eighteenth century and important work was carried out by Edward D. Cope and others in the nineteenth, the main efforts of North American mammalian paleontologists were long devoted to the continent's spectacular record of Tertiary mammals, almost to the exclusion of the Pleistocene. As a result, Pleistocene paleomammalogy in North America lagged behind that in Europe. In the early twentieth century, added impetus to Pleistocene work was given by the studies of J. W. Gidley, O. P. Hay, J. C. Merriam, G. G. Simpson, C. Stock, and other leading paleontologists, and in the last few decades, Pleistocene paleomammalogy has come vigorously to the fore. This is certainly in large measure due to the inspiring work of the late Claude W. Hibbard, whose trailblazing work almost tended to give the impression that Meade County, Kansas—"Hibbie's" favorite hunting ground—was the place where the Pleistocene occurred. In reality, of course, the interest is continentwide, as witness more than 250 local faunas unearthed to date in the United States, Canada, and Mexico.

The scope given the Pleistocene has varied. This reflects on the position of the Blancan Land Mammal Age, now considered to have commenced about 3.5 m.y. B.P. (million years before present) and ended about 1.8–1.9 m.y. B.P. Hibbard et al. (1965) placed the early Blancan in the Pliocene and the late Blancan in the Pleistocene. At the time of this writing, it appears probable that the Pliocene–Pleistocene boundary will be fixed in the time range 1.5–1.8 m.y. B.P. (Bandy, 1972; Haq, Berggren, and Van Couvering, 1977), so we are including all of the Blancan in the Pliocene (Zakrzewski, 1975b). In this book, we treat the Blancan as well as the truly Pleistocene mammals. Apart from the historical reasons, we feel that this will give a more coherent picture of the origin and evolution of the Pleistocene fauna. It should be noted that the faunal break between the Hemphillian (early Pliocene) and Blancan (late Pliocene) is more pronounced than that between the Blancan and the succeeding Irvingtonian (McGrew, 1948). Thus Blancan mammalian faunas are more akin to those of the Pleistocene than those of the preceding Land Mammal Ages.

Historically, the study of Pleistocene mammals in North America began with the

DISCOVERIES OF EXTINCT MAMMALIAN SPECIES (BLANCAN TO HOLOCENE)[a]

	Before 1899	1900–19	1920–39	1940–59	1960–77	Total
Micromammals	5	6	51	65	59	186
Macromammals	57	19	33	15	20	144
Total	62	25	84	80	79	

[a] Micromammals include Insectivora, Chiroptera, Rodentia, and Lagomorpha. Macromammals include Edentata, Carnivora, Artiodactyla, Perissodactyla, and Proboscidea.

large mammals. Of the extinct Blancan and Pleistocene species belonging to the orders Edentata, Carnivora, Artiodactyla, Perissodactyla, and Proboscidea (mainly large mammals), about 40 percent were described before 1900 (not counting the numerous synonyms). For orders of small mammals (Insectivora, Chiroptera, Rodentia, and Lagomorpha), the corresponding figure is less than 3 percent. In the first two decades of the twentieth century, relatively few new species were described in either category. Beginning in the 1920s', discovery of small mammal species has proceeded apace, with some 25–30 new species per decade. Discovery of large mammals reached a new peak in 1920–39, but has since abated to some extent. The data (for extinct species) are summarized in the accompanying table. In addition to these extinct taxa, a large number of extant species have been found in the fossil state.

The purpose of the present book is to summarize information on the Blancan and Pleistocene mammals of North America. The area covered is Canada, the continental United States, and northern Mexico. The work was started with the intention of producing a companion volume to the previously published book on European Pleistocene mammals (Kurtén, 1968). The main departure from the format followed in that book is the more detailed treatment of the localities and faunal succession. In Part I, we list and briefly characterize the main sites with local faunas and compare these faunas with those of the adjoining continents, South America and Eurasia.

Part II is a critical and annotated list of the Blancan and Pleistocene mammal species of North America. Many Pleistocene species are still in existence and may be studied as living entities against the background of their own fossil record, thus providing a tie-in between paleozoology and neozoology. Many species extinct in North America persist in other continents. In many cases, the origin of extinct and living species may be traced through long stretches of geologic time. In addition to the construction of such phylogenetic trees, there are additional possibilities for paleontological study of evolution. Population analysis can give information on aspects of adaptation, population dynamics, natural selection, and microevolution on the deme, subspecies, and species level. With larger assemblages of species and a longer temporal perspective, problems of species origination, survival and extinction, community evolution, climatic change, and long-range anagenetic trends may be attacked. Intercontinental comparisons should help to unravel paleobiogeographical problems on a global scale.

In the hope of furthering such studies, we have chosen not to isolate ourselves within the black box of the North American Pleistocene. Use of a provincial taxonomy hampers the study of intercontinental relationships. Thus we have deliberately tried to look across geographical borders, natural or artificial, and seek relationships with faunas in other areas. The present-day fauna of North America has many species in common with other biogeographic realms, and the same appears to have been true in the Pleistocene—to an extent we hardly realized at the outset.

In order to make the text accessible for interdisciplinary use, we have tried to avoid unnecessary technicality. However, technical terminology necessarily enters into the taxonomic and anatomical characterizations and geological and paleoecological discussions. We refer the reader to the *Dictionary of Geological Terms,* prepared by the American Geological Institute (1976), and to Romer (1966).

The project, initiated in 1970, was supported by the National Science Foundation Grant GB 31287 to Bryan Patterson, Harvard University, in October 1971. It included visits, repeated ones in many cases, to most of the major collections of North American Pleistocene mammals and to a number of sites. We wish to express our gratitude to the National Science Foundation, our hosts, and the numerous other colleagues who took so much time and trouble to show us the material in their care, guide us in the field, answer innumerable questions, and contribute a wealth of information (much of it unpublished) and who extended many other courtesies to us as well. Their names are listed below.

No one has equal expertise on all the varied groups of mammals represented in the Pleistocene faunas of North America. Our decision, therefore, was, to concentrate our first-hand research upon selected taxa, including the Carnivora and, to some extent, the Tayassuidae and Cervidae; a part of the results has been published, and the remainder will be published in due course. With regard to other taxa, as well as for the site descriptions and material relating to faunal history and correlation, we have felt the need to consult authoritative opinion. The response has been magnificent. We are deeply grateful to all the colleagues who generously took time to read early drafts of the book, correct mistakes, and add information—in many cases, again, unpublished. Many chapters passed through the hands of several reviewers, as may be seen from the listing below. Richard H. Tedford reviewed the entire manuscript. The responsibility for remaining errors and omissions, however, is ours.

At all stages of the work we have had the heartening experience of a solid backing from the entire profession. The same is true for the publisher, Columbia University Press, where the understanding, patience, and help of Editor-in-Chief John D. Moore and Associate Executive Editors Joe Ingram and Vicki Raeburn have been invaluable throughout.

We also wish to thank our original illustrator, Margaret Lambert, who cheerfully spent long working days in sketching innumerable bones, skeletons, life restorations, and landscape impressions. Brett Hendey in Capetown acted as "post office" in forwarding her drawings to us from Rhodesia. In the late stages, James D. Senior, Hubert Pepper, Erica Hansen, Riggert Munsterhjelm, and Barbro Elgert came to our assistance with additional illustrations, and Dawn B. Adams made some needed corrections. Their help was invaluable. We also want to thank Darlene Emry and Virginia Nelson for help with typing and proofreading. Finally, gratitude is due to our families, who patiently bore with our long and irregular absences and gave their loyal support during the years that it took to complete the book.

Collections Visited

American Museum of Natural History, Dept. of Vertebrate Paleontology, Dept. of Mammals; Brigham Young University, Dept. of Zoology; Carnegie Museum of Natural History, Section of Vertebrate Fossils; Central Missouri State College; Dayton Museum of Natural History; Field Museum of Natural History, Dept. of Geology; Florida Geological Survey; Florida State University, Dept. of Anthropology; Fort Hays Kansas

State College, Sternberg Memorial Museum; Harvard University, Museum of Comparative Zoology; Idaho State University Museum; Illinois State Museum; Los Angeles County Museum of Natural History, Vertebrate Paleontology Section; Midwestern University, Dept. of Biology; National Museums of Canada, Paleontology Division; National Museum of Natural History, Dept. of Vertebrate Paleontology, Bird and Mammal Laboratories; Panhandle-Plains Museum; Philadelphia Academy of Natural Sciences; Princeton University, Dept. of Geology; Royal Ontario Museum; San Diego State University, Dept. of Zoology; Southern Methodist University, Museum of Paleontology; Texas Memorial Museum, Vertebrate Paleontology Laboratory, University of Texas; University of Alaska, Dept. of Biology; University of Arizona, Dept. of Geology; University of California, Museum of Paleontology; University of Colorado Museum; University of Florida, Florida State Museum; University of Georgia, Dept. of Geology; University of Iowa, Dept. of Geology; University of Kansas, Museum of Natural History; University of Michigan, Museum of Paleontology; University of Nebraska State Museum; University of Texas at El Paso, Museum of Arid Land Biology; University of Toronto, Dept. of Zoology; University of Utah, Dept. of Geology; University of Wyoming, Dept. of Geology; U.S. Geological Survey, Menlo Park; West Texas State University, Dept. of Geology; Yale University, Peabody Museum of Natural History.

Colleagues

Daniel B. Adams; William A. Akersten; Sydney Anderson; Donald Baird; Brenda Beebe; Craig C. Black; William H. Burt; Charles S. Churcher; William A. Clemens, Jr.; Mary Courtright; A. W. Crompton; Walter W. Dalquest; Mary R. Dawson; Theodore Downs; Gordon Edmund; Howard L. Emry; Ralph E. Eshelman; David E. Fortsch; Michael Frazier; Theodore Galusha; David D. Gillette; Joseph T. Gregory; John E. Guilday; Russell D. Guthrie; Michael Hager; E. Raymond Hall; Harold Hamilton; C. R. Harington; Arthur H. Harris; Oscar Hawksley; Q. B. Hendey; Charles Hendry; Caroline A. Heppenstall; Claude W. Hibbard; Robert S. Hoffmann; David M. Hopkins; Marie L. Hopkins; J. Howard Hutchison; Antti Järvi; George T. Jefferson; Farish A. Jenkins; Eileen Johnson; J. Knox Jones, Jr.; Karl Koopman; George Krochak; Paul E. Langenwalter; Barbara Lawrence; Arnold P. Lewis; G. Edward Lewis; Jason A. Lillegraven; Everett H. Lindsay; Lewis Lipps; Ernest L. Lundelius, Jr.; Allen D. McCrady; H. Gregory McDonald; Helen McGinnis; Paul O. McGrew; Charles W. Mack; Malcolm C. McKenna; Martin E. McPike; James H. Madsen; Vincent J. Maglio; Larry D. Martin; Paul S. Martin; George J. Miller; Susanne J. Miller; Wade E. Miller; Richard Mills; Rachel H. Nichols; Mary J. Odano; Stanley J. Olsen; E. C. Olson; John H. Ostrom; John L. Paradiso; Paul W. Parmalee; David C. Parris; Bryan Patterson; Robert W. Purdy; Horace F. Quick; Clayton E. Ray; Charles A. Repenning; Richard Reynolds; Robert E. Reynolds; Horace G. Richards; King A. Richey; Peter Robinson; Alfred S. Romer; Hind Sadek-Kooros; Donald E. Savage; Gary J. Sawyer; C. Bertrand Schultz; Gerald E. Schultz; Holmes A. Semken, Jr.; Andrei Sher; Elwyn L. Simons; George G. Simpson; Morris F. Skinner; Bob H. Slaughter; James M. Soiset; A. MacS. Stalker; Lloyd G. Tanner; Beryl Taylor; Richard H. Tedford; Norman Tessman; James E. Trever; William D. Turnbull; Judith Van Couvering; Richard G. Van Gelder; Leigh Van Valen; Michael R. Voorhies, Danny N. Walker; Myrl V. Walker; S. David Webb; Lars Werdelin; Joe Ben Wheat; David P. Whistler; John A. White; Frank C. Whitmore, Jr.; John A. Wilson; John W. Wilson, III; Michael Wilson; Steve Windham; Nelda E. Wright; Richard J. Zakrzewski.

Reviewers

William A. Akersten; Philip R. Bjork; Charles S. Churcher; Walter W. Dalquest; Mary R. Dawson; Daryl P. Domning; Gordon Edmund; Michael Frazier; David D. Gillette; John E. Guilday; E. Raymond Hall; C. R. Harington; J. Howard Hutchison; George T. Jefferson; Karl Koopman; Ernest L. Lundelius, Jr.; H. Gregory McDonald; Larry D. Martin; John E. Mawby; Wade E. Miller; Ronald M. Nowak; John M. Rensberger; Charles A. Repenning; Holmes A. Semken, Jr.; Morris F. Skinner; Bob H. Slaughter; James M. Soiset; Beryl Taylor; Richard H. Tedford; S. David Webb; Joe Ben Wheat; David P. Whistler; John A. White; Frank C. Whitmore, Jr.; Richard J. Zakrzewski.

PART ONE

MAMMALIAN FAUNAS OF THE BLANCAN AND PLEISTOCENE

Key to Collections

AMNH	American Museum of Natural History
ANSP	Academy of Natural Sciences, Philadelphia
BYU	Brigham Young University
CM	Carnegie Museum of Natural History
CMSC	Charleston Museum, South Carolina
CMSU	Central Missouri State University
DOMNH	Dayton, Ohio, Museum of Natural History
F:AMNH	Frick Laboratory, American Museum of Natural History
FC	Fauna Cedazo (Mooser private collection)
FHSM	Sternberg Memorial Museum of Fort Hays
FMNH	Field Museum of Natural History
FSM	Florida State Museum
HU	Harvard University
IGM	Instituto Geológico de México
IOWA	University of Iowa
ISM	Illinois State Museum
ISUM	Idaho State University Museum
KUVP	University of Kansas, Vertebrate Paleontology
LACM	Los Angeles County Museum of Natural History
MALB	Museum of Arid Land Biology, El Paso, Texas
MCZ	Museum of Comparative Zoology
MSB	Museum of Southwestern Biology, University of New Mexico
MWU	Midwestern University
NMC	National Museums of Canada
ROM	Royal Ontario Museum
SDSM	South Dakota School of Mines
SMNH	Saskatchewan Museum of Natural History
SMUMP	Southern Methodist University, Museum of Paleontology
TMM	Texas Memorial Museum
TTM	Texas Tech Museum
TU	Tulane University
UALP	University of Arizona, Laboratory of Paleontology
UAM	University of Alaska Museum
UCM	University of Colorado Museum
UCMP	University of California, Museum of Paleontology
UGV	University of Georgia, Vertebrate Paleontology
UMMP	University of Michigan, Museum of Paleontology
UNSM	University of Nebraska State Museum
USNM	National Museum of Natural History
UTZ	University of Toronto, Zoology
UUVP	University of Utah, Vertebrate Paleontology
UW	University of Washington
UWYO	University of Wyoming
VU	Vanderbilt University
WTSU	West Texas State University
YPM	Yale Peabody Museum

CHAPTER ONE

Chronology

In the classical scheme, originating with Chamberlin (1895) and others, the Pleistocene of North America contains a succession of four glaciations with three intervening interglacials. In the Alpine region of Europe, four glaciations have likewise been distinguished; all of these except the first have their counterparts in northern Europe. The glaciations and interglacials are listed in table 1.1.

TABLE 1.1
CLASSICAL GLACIAL–INTERGLACIAL SEQUENCES IN
NORTH AMERICA AND EUROPE

	North America	Europe *Alpine*	Scandinavian
Last glaciation	Wisconsinan	Würm	Weichselian
Last interglacial	Sangamonian	Riss–Würm	Eemian
Penultimate glaciation	Illinoian	Riss	Saalian
Penultimate interglacial	Yarmouthian	Mindel–Riss	Holsteinian
Antepenultimate glaciation	Kansan	Mindel	Elsterian
Antepenultimate interglacial	Aftonian	Günz–Mindel	Cromerian
First glaciation	Nebraskan	Günz	—

A straightforward correlation between the American and European glaciations was long considered probable (e.g., Flint, 1971). Recent geochronological work, however, has thrown doubt on this correlation. In addition, the story has been shown to be considerably more complicated by the fact that, in both areas, the four classical glaciations were preceded by mountain glaciations. Furthermore, subdivision of each glaciation into a number of cold and warm phases is possible.

Interpretation of glacial–interglacial oscillation is based on sediments and their contained fossils, as well as isotopic paleotemperature determination. Deep-sea cores are particularly instructive in this respect. Ice cores from arctic and antarctic regions also yield isotopic temperature records. In addition, changes in the sea level and the groundwater table may also be studied. The result is in all cases a relative chronology.

Faunal succession in North America has been divided into a sequence of "Land Mammal Ages," of which we are concerned with the three most recent: the Blancan (Wood et al., 1941), Irvingtonian, and Rancholabrean (Savage, 1951). The most ancient of these, the Blancan, is characterized by the presence of mammal genera such as *Borophagus, Nannippus,* and *Equus* (*Dolichohippus*); the Irvingtonian by *Lepus, Ondatra,* and *Mammuthus* but not *Bison;* and the Rancholabrean by *Bison.* There are, of course, several other useful index fossils. Again, this is a relative chronology.

Among reliable "absolute" dating methods, annual rhythmite chronologies (glacial and other annual varves, tree rings) were the first to be developed. De Geer (1940 and earlier works) succeeded in dating late-glacial and postglacial time by the varve method. By analogous procedures, the lengths of the Holsteinian and Eemian interglacials in central Europe have been determined (Meyer, 1974; Müller, 1974). In all these instances (including the Holocene interglacial), interglacial durations of between 10,000 and 16,000 years have been obtained. However, chronologies for the earlier interglacials are floating ones.

According to the "astronomical theory" (Zeuner, 1952) developed originally by W. Köppen and A. Wegener and later by M. Milankovitch, changes in climate are closely related to perturbations in the earth's orbit around the sun that affect the distribution of solar radiation over the earth's surface. By correlating glaciations and interglacials with minima and maxima in the radiation curve, it was thought that a reliable chronology for the Ice Age could be established. Almost from the start, the theory has been under violent attack, but recent results on deep-sea cores indicate that there is indeed such a correspondence (Hays, Imbrie, and Shackleton, 1976). The perturbation cycles are of the same length as those deduced from the temperature indicators of the cores. Especially noteworthy is the fact that the interglacials appear as brief episodes in a massively cold-dominated history, which corroborates the rhythmite-count results. It is evident, however, that the original correlations (see Zeuner, 1952) are mostly incorrect.

The introduction of radiometric methods made "absolute" dating of geological time a reality. The most influential methods at present are radiocarbon, or C-14, dating (Libby, 1955), potassium–argon, or K–Ar, dating (Evernden et al., 1964), and fission track dating (Fleischer, Price, and Walker, 1969). Many other methods are being developed, for instance, amino acid racemization dating (Turekian and Bada, 1972), thermoluminescence dating (Hall, 1969) and the related electron-spin-resonance dating (Ikeya, 1977), and obsidian hydration dating (Friedman, Smith, and Clark, 1969).

Study of the remanent magnetism found in rocks has shown that the earth's magnetic field has repeatedly reversed its polarity and, in combination with the technique of radiometric dating, has led to the establishment of a paleomagnetic time scale. The last 4 million years of earth history may thus be divided into the following four magnetic epochs (from oldest to youngest): the Gilbert reversed, the Gauss normal, the Matuyama reversed, and the Brunhes normal. Within all of these epochs except the last, shorter reversal episodes also occurred.

The synthesis of this disparate array of data is still in its early phase, and many of the conclusions presented here are highly tentative in nature.

The chronology of the glacial–interglacial succession follows, with minor modifications, that presented by Cooke (1973; see also Haq, Berggren, and Van Couvering, 1977), bringing together material from North America, Europe, and other continents and from deep-sea cores. The Nebraskan glaciation appears to have started about 1.5 m.y. B.P. and been a correlative of the European Donau glaciation. The Kansan glaciation began at approximately 0.9 m.y. B.P. and is correlated with the Günz of Europe. An important datum line is formed by the Cromerian interglacial of Europe, dated at 0.7 m.y. B.P. (Matuyama–Brunhes paleomagnetic boundary), which is thought to be a correlative of the Yarmouthian interglacial of North America. The Illinoian glaciation started at about 0.6 m.y. B.P., and the Wisconsinan at about 0.3 m.y. B.P. The Sangamonian interglacial may correspond to the Holsteinian of Europe in this scheme, but this is a controversial point. Indeed, the chronology as a whole is highly tentative and provisional.

Fitting the Land Mammal Age succession into this framework is facilitated to some extent by a combination of magnetic and radiometric stratigraphy and biostratigraphy.

Thus the first appearance of some typical Blancan forms like *Geomys* and *Odocoileus* is dated at 3.5 m.y. B.P. and considered to mark the beginning of the Blancan (Lindsay, Johnson, and Opdyke, 1975). Irvingtonian entrants into the fossil record include *Ondatra*, *Lepus*, and *Dipodomys* (1.9 m.y. B.P.) and *Mammuthus* (ca. 1.5 m.y. B.P.) (Lindsay, Johnson, and Opdyke, 1975). The immigration of *Bison*, which, like *Mammuthus*, originated in the Old World, does not yet have a precise date. The consensus of most workers is that it occurred during the Illinoian, and the date may then be on the order of 0.5 m.y. B.P.

The Pliocene–Pleistocene boundary has been recently dated by studies of the type area in Italy, the place where the Pleistocene was first defined by Lyell, in 1834 (it was then called "newer Pliocene"). The figures range from ca. 1.9 m.y. B.P. (Bandy, 1972) to 1.6 m.y. B.P. (Haq, Berggren, and Van Couvering, 1977). There is thus a fair, though not exact, coincidence between the Blancan–Irvingtonian and Pliocene–Pleistocene boundaries.

The provisional synthesis is set forth in figure 1.1, which may be referred to for the dating of the faunas that follow and the species discussed in the second section of the book.

If the picture of faunal evolution as a whole is thought of in terms of an unfinished jigsaw puzzle, its individual pieces would correspond to the various local faunas. A local fauna may be defined as an aggregate of species from a restricted area and time interval (see Tedford, 1970, for a detailed discussion).

In the next three chapters, local faunas of the Blancan, Irvingtonian, and Rancholabrean, and the sites that have produced them, are reviewed. Each chapter has a brief introduction giving the chronological framework. The faunas have been arranged in alphabetical order by state or province, except where direct stratigraphic superposition makes it more expedient to treat successive local faunas as one topic (indexed under the name of the oldest fauna).

The number of mammalian species known from the local fauna is given with the name of the fauna, and the map reference number follows the geographical location. At the end of each site description, we have gathered references to important publications so that the reader may locate the pertinent literature. Also given are the acronyms of the collections (as far as known to us) in which important material is stored. (A key to the collections follows the Introduction.)

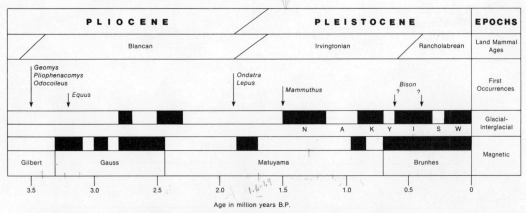

1.1 Late Pliocene and Pleistocene chronology in North America. First occurrence of certain genera indicated. Glacials (black) and interglacials: *N*, Nebraskan; *A*, Aftonian; *K*, Kansan; *Y*, Yarmouthian; *I*, Illinoian; *S*, Sangamonian; *W*, Wisconsinan. In magnetic stratigraphy, black indicates normal, white reversed polarity. For discussion, see text.

CHAPTER TWO

Blancan Faunas

The Blancan sites reviewed below (see map 1) are distributed from the states of Washington and Idaho to California and Florida, an almost continentwide coverage. The most nearly complete successions of Blancan local faunas known at present are in Idaho (Hagerman), California (Anza Borrego State Park), Arizona (San Pedro Valley), and the Great Plains (especially Meade County, Kansas). The dating of the Hagerman section is still controversial, but the Anza Borrego, San Pedro, and (in part) the Great Plains sequences have been well established by a combination of radiometric and paleomagnetic dating (Lindsay, Johnson, and Opdyke, 1975; Opdyke et al., 1977) and so may be regarded as standards for the dating of Blancan events.

At Anza Borrego, the section extends from the Hemphillian (pre-Blancan Land Mammal Age) through the Blancan and well into the Irvingtonian. Three successive faunas are recognized, the Layer Cake, Arroyo Seco, and Vallecito Creek, corre-

TABLE 2.1
BLANCAN FAUNAS IN THE GREAT PLAINS AREA[a]

Chronology[b]	Texas	Kansas	Nebraska and Colorado
Irvingtonian —1.9—			
Blancan Pre-Nebr. IGL?	Red Light Hudspeth	Borchers	
		Pearlette-"B" 2.0	
Sierran GL?	Blanco	Seger	Dixon White Rock	Donnelly Ranch
Matuyama —2.4— Gauss	Cita Canyon	Sanders Deer Park		Sand Draw Broadwater Lisco
Deadmans Pass GL?	Red Corral	(Angell Gravels) ——?——	
		Bender	
		Rexroad Keefe Canyon	
Gauss —3.3— Gilbert			
		Fox Canyon	
Blancan —3.5— Hemphillian			
		Saw Rock Canyon	
Gilbert —5.1—			

[a] Figures give dates in million years B.P.
[b] IGL, interglacial; GL, glacial.

sponding approximately to the Gilbert, Gauss, and Matuyama magnetic epochs, respectively. For most of the taxa, only preliminary studies are at hand; future work will doubtless lead to finer subdivision.

The San Pedro Valley section at Curtis Ranch extends from the early Blancan to the early Irvingtonian. In Kansas and Texas, the sequence has to be pieced together from shorter local sections by using paleomagnetic and radiometric data augmented with biostratigraphic analysis. Table 2.1 gives an interpretation of the Blancan faunas in the Great Plains, based on Lindsay, Johnson, and Opdyke (1975), with approximate dates for local faunas from Nebraska and Colorado as suggested by biostratigraphic correlation.

Stratigraphic ranges of certain genera, selected for their chronologic significance and wide geographic range, have been plotted by Lindsay, Johnson, and Opdyke (1975) for a combined Texas–Kansas–Arizona sequence (also including Irvington in northern California) and by Opdyke et al. (1977) for the Anza Borrego section. The data of these authors are summarized in figure 2.1, which gives earliest (→) and latest (+) occurrences of these genera from the early Blancan to the early Irvingtonian.

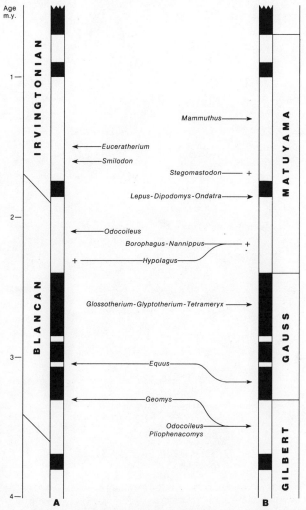

2.1 First (→) and last (+) occurrences of certain taxa in the Blancan and Irvingtonian. *A*, Anza Borrego; *B*, Great Plains region. Data from Lindsay, Johnson, and Opdyke (1975), Opdyke et al. (1977).

Map 1. Blancan Faunas

As may be noted, the Blancan is inaugurated by appearance of *Odocoileus, Pliophenacomys*, and *Geomys* in the late Gilbert magnetic epoch, about 3.5 m.y. B.P. *Equus* appears somewhat later, in the early Gauss epoch. Of these genera, both *Geomys* and *Equus* appear somewhat later in the Anza Borrego section, whereas *Odocoileus* appears at a much later date. Such dates may, of course, be modified by future discoveries. It should be noted that the only faunas definitely known to date from the earliest Blancan (late Gilbert epoch, 3.3+ m.y. B.P.) are Fox Canyon and Layer Cake, and only the former has been studied in detail.

A second burst of first occurrences comes in the late Gauss epoch, around 2.5–2.7 m.y. B.P., with the appearance of *Glossotherium, Glyptotherium,* and *Tetrameryx* (also *Tremarctos* in the Anza Borrego section, although this genus is also known from Hagerman, almost certainly earlier). This time period may well be regarded as the dividing line between the early and the late Blancan. Somewhat later, in the early part of the Matuyama epoch, are the last occurrences of such common Blancan genera as *Hypolagus, Borophagus,* and *Nannippus* (all of which were present in the late Hemphillian). Finally, with the appearance of *Lepus, Dipodomys,* and *Ondatra,* and perhaps *Euceratherium* and *Smilodon,* at about the beginning of the Olduvai magnetic event (1.9 m.y. B.P.), the Blancan–Irvingtonian boundary is reached.

As noted above, very early Blancan faunas are rare. The Fox Canyon fauna contains several species apparently ancestral to taxa in the succeeding early Gauss epoch. Of these, *Borophagus direptor,* possibly ancestral to *Borophagus diversidens,* has been tentatively identified at Asphalto, where it is associated with *Equus*. Other early Blancan faunas, most of them apparently postdating the Gilbert epoch, include Arroyo Seco (in part), Bender, Benson, Hagerman, Keefe Canyon, Rexroad, Taunton, and White Bluffs. New genera in the early Blancan, besides those already mentioned, include *Trigonictis, Ursus, Megantereon, Pliopotamys, Pratilepus,* and *Platygonus;* the sole extant mammal species known from the early Blancan is *Taxidea taxus*.

The later part of the Blancan (late Gauss and early Matuyama) is represented by the numerous sites whose descriptions follow. Changes in the fauna are mainly on the species level. Terminal Blancan is represented by the Borchers fauna of Kansas, which overlies the Pearlette "B" ash dated at 2.0 m.y. B.P.

Although very early cold-climate indicators are known from the Far North (a tillite dated at about 3.6 m.y. B.P., for instance; see Denton and Armstrong, 1969), no evidence for cold climatic conditions has been found in the early Blancan faunas. There is evidence for at least two glacial events within the Blancan, however (Dalrymple, 1972).

Map 1. Blancan Faunas*

1. Layer Cake (1)
2. Arroyo Seco (2–3)
3. Vallecito Creek (part) (4)
4. San Timoteo (B)
5. Asphalto (B)
6. Coso Mountains (2)
7. Tehama (B)
8. White Bluffs (2)
9. Taunton (2)
10. Wichman (B)
11. Grand View (4)
12. Hagerman (2)
13. Anita (4)
14. Tusker (4)
15. California Wash (4)
16. Bonanza (3)
17. Benson (2)
18. Donnelly Ranch (4)
19. Hudspeth (4)
20. Red Light (4)
21. Blanco (4)
22. Cita Canyon (3)
23. Fox Canyon (1)
24. Keefe Canyon (2)
25. Rexroad (2)
26. Bender (2)
27. Sanders (3)
28. Deer Park (3)
29. Seger (4)
30. Borchers (4)
31. Dixon (3)
32. White Rock (3)
33. Broadwater (3)
34. Lisco (3)
35. Sand Draw (3)
36. Delmont (B)
37. Santa Fe River IB (3)
38. Haile XVA (4)

*Numbers in parentheses in key represent: 1, very early Blancan; 2, early Blancan; 3, middle Blancan; 4, late Blancan; B, Blancan, exact age unknown

The oldest event is represented by the Deadman Pass Till (Curry, 1966), for which bracketing radiometric dates are 2.7 and 3.0 m.y. B.P.; the second (Sierran) is represented by the McGee Till (Cox, Doell, and Dalrymple, 1963), underlain by a basalt flow dated at 2.6 m.y. B.P. Both events appear to represent alpine glaciations and are not comparable to the great Pleistocene glaciations. The middle to late Blancan deposits and faunas of the Great Plains suggest that climatic changes occurred corresponding to the alpine glacial–interglacial cycles, but such correlations remain tenuous at present. (Cox, Doell, and Dalrymple, 1963; Curry, 1966; Dalrymple, 1972; Denton and Armstrong, 1969; Lindsay, Johnson, and Opdyke, 1975; Opdyke et al., 1977.)

ARIZONA

Anita (12 spp.), Coconino Co. (1–13)

This fauna comes from a fissure filling in the Permian Kaibab limestone in the Plateau Region of northern Arizona. Lindsay and Tessman (1974) regard Anita as an early Blancan fauna, but the association of *Hypolagus* and *Lepus* may indicate a later date. At least partially wooded surroundings are indicated by the presence of hare, woodrat, and peccary. (Hay, 1921, 1927; Lindsay and Tessman, 1974.) USNM.

Benson (23 spp.) and California Wash (11 spp.), Cochise Co. (1–17, 1–15)

A series of fossiliferous deposits in the San Pedro Valley in southeastern Arizona has yielded two major faunas, the Benson (Blancan) and Curtis Ranch (Irvingtonian). The Benson section (Mendevil, Green Saddle, Eastside Is., Post Ranch, Gray Point, and other localities) occurs in the middle member of the St. David Formation and consists of claystones and mudstones with intermittent diatomites which may have been associated with springs. It is found in normally magnetized deposits overlain by a reversed magnetozone correlated with the Mammoth magnetic event, and an underlying ash bed is dated at 3.1 m.y. B.P. (by fission tracks on zircons), which may thus be taken as the age of the fauna. The Benson fauna, along with Rexroad, Bender, and Sanders (Kansas) and Red Corral and Cita Canyon (Texas) faunas are probably correlated with the Gauss magnetic epoch.

The mammalian fauna includes a large percentage of micromammals; the presence of immigrant *Sigmodon* is significant. Large mammals include *Nannippus, Equus,* cf. *Cuvieronius, Camelops, Capromeryx,* and *Platygonus.* Amphibians and water birds (ducks, a diver) are also found at Benson. Gidley (1922) noted the presence of some articulated skeletons, which he thought suggested that the animals were bogged down in water holes at the borders of salt lakes; however, Lindsay (personal communication, 1976) believes that there was almost through drainage in the San Pedro Valley during the deposition of the St. David Formation, so that the presence of salt lakes is unlikely.

The Curtis Ranch section of the sequence has disclosed a long series of fossiliferous deposits. The oldest fossil-bearing locality (Bonanza) comes on top of the Kaena reversed magnetozone (2.8 m.y. B.P.) and is followed by other bone-producing levels in the upper Gauss and lower Matuyama zones; the Curtis Ranch faunal sites straddle the transition from lower Matuyama to Oldavai and so may be given an average age of 1.86 m.y. (early Irvingtonian). The California Wash local fauna occurs at a somewhat lower level in the Matuyama zone. At this level, *Nannippus* is already absent, and *Ondatra* is present, but *Lepus* has not yet appeared. Faunal studies on the numerous Benson and Curtis Ranch localities are in progress. (Gazin, 1942; Gidley, 1922a, 1922b, 1926; Johnson, Opdyke, and Lindsay, 1975; Lindsay, personal communication, 1976; Lindsay and Tessman, 1974; Lindsay, Johnson, and Opdyke, 1975.) UALP,5 USNM.

Tusker (25 spp.), Graham Co. (1–14)

The Tusker local fauna comes from the 111 Ranch in the Safford Valley, about 25 km SE of the city of Safford in southeastern Arizona. The fauna, which is of Blancan age, has not yet been fully studied (see Lindsay and Tessman, 1974, for a preliminary faunal list). In addition to the mammals, *Geochelone* and an unidentified bird are present. Coexistence of *Nannippus* and *Equus,* as well as *Hypolagus* and *Lepus,* is noted. Micromammals are numerous. The presence of *Glyptotherium* and *Neochoerus* indicates subtropical conditions with permanent water and lush vegetation. (Downey, 1962; Lindsay and Tessman, 1974.) UALP.

In addition to the local faunas listed above, preliminary lists for other Arizona Blancan faunas are given in Lindsay and Tessman (1974).

CALIFORNIA

Asphalto (3 spp.), Kern Co. (1–5)

This small, probably Blancan, fauna comes from an asphalt bed in the Tulare Formation near Buena Vista Lake in south-central California. The Tulare Formation overlies the San Joaquin Clay (Upper Etchegoin of some authors). *Borophagus, Ischyrosmilus,* and *Equus* are present. (Merriam, 1917; Savage, 1951; Stirton, 1936.) UCMP.

Coso Mountains (8 spp.), Inyo Co. (1–6)

The principal fossil-bearing deposits are on the western flanks of the Coso Mountains, 15 km E of Olanche, and consist of alluvial fan materials overlain by volcanic tuffs and basaltic lava beds. Additional localities up to 11 km apart have produced vertebrate remains in the same stratigraphic position. A dolichohippine horse is the most common fossil. The microtine *Cosomys* may indicate an early Blancan age. A plains paleohabitat with bodies of standing water has been inferred. (J. R. Schultz, 1937.)

Layer Cake (7 spp.) and Arroyo Seco (20 spp.), Vallecito Creek (10+ spp.),
  San Diego Co. (1–1, 1–2, 1–3)

More than 2,500 m of superposed sediments in the Vallecito–Fish Creek area of the Colorado Dessert of southern California (in the Anza Borrego Desert State Park) carry fossil vertebrates ranging in age from Hemphillian to Irvingtonian. Over 90 vertebrate taxa have been identified from about 450 localities. An extension of the Gulf of California, the Colorado Desert forms a structural basin which has been filled with sediments derived from the bordering uplands and the Colorado River. The strata consist of poorly indurated Plio–Pleistocene mudstones, siltstones, and sandstones, dipping some 24° to 26° to the south, and comprise the Palm Springs Formation. Recent paleomagnetic studies (Opdyke et al., 1977) showed that the formation had sufficient magnetic stability to delineate magnetic stratigraphy ranging from the Cochiti event at the bottom (Hemphillian) to the Matuyama reversed magnetic epoch at the top (Irvingtonian).

The Layer Cake fauna occurs about 2,100–2,500 m below the top of the Vallecito–Fish Creek section. This faunal interval approximates the late Gilbert magnetic epoch, which lasted about 0.7 m.y., and is correlated with the Fox Canyon fauna in Kansas. At the lowest vertebrate-fossil level (~3.8 m.y. B.P.), the Clarendonian-Hemphillian horse *Pliohippus* has been recognized, which indicates that the lowest level may be Hemphillian in age. Taxa include the rabbits *Hypolagus* cf. *regalis* and *Nekrolagus*.

The Arroyo Seco fauna is found in the continuation of the section up to some 1,200 m from the top. This interval approximates the Gauss magnetic epoch and lasted about 1.1 m.y. It is correlated with the Benson (Arizona), Red Corral and Cita Canyon (Texas), and Rexroad, Bender, and Sanders (Kansas) faunas. In the lower part of this section, *Geomys* makes its first appearance (~3.3 m.y. B.P.). Higher up, *cf. Equus* appears (~2.9 m.y. B.P.). In the upper part, one of the earliest known occurrences of *Tremarctos* is recorded (~2.5 B.P.); *Sylvilagus* (replaces *Hypolagus*) and true *Equus* also come in at this time. Small mammals, including shrews, a vespertilionid bat, *Anzanycteris anzensis*, *Prodipodomys*, and *Sigmodon* are also present in the Arroyo Seco fauna.

Most of the Vallecito Creek fauna occurs within the Blancan Land Mammal Age but also includes an early Irvingtonian component (see Vallecito Creek in chapter 3).

Study of these faunas is still in progress; the area may become one of the most important standards for Blancan and Irvingtonian faunal evolution. (Downs and White, 1968; Opdyke et al., 1977; White, 1970.) LACM.

San Timoteo (6 spp.), Riverside Co. (1-4)

Situated in the San Timoteo Badlands, the San Timoteo beds have been interpreted as slope-wash of a semiarid region. They rest unconformably upon the Eden Formation, which contains a Hemphillian vertebrate fauna. Presence of a dolichohippine horse indicates a Blancan age. (Frick, 1921.)

Tehama (9 spp.), Tehama Co. (1-7)

The Tehama fauna was found near the town of Tehama, about 64 km S of Redding in the northern part of the Sacramento Valley. The Tehama Formation, consisting of sandy clays with occasional intercalations of cross-bedded sands and gravels, rests unconformably upon Cretaceous beds, and is overlain disconformably by the Red Bluff Pleistocene. Its thickness is about 600 m. Fossil mammals have been taken near the base and at a horizon about 60 m higher. The fauna with a dolichohippine horse, *Borophagus*, and a few other large mammals is probably coeval with Asphalto and thus Blancan in age. (Russell and VanderHoof, 1931; VanderHoof, 1933.)

COLORADO

Donnelly Ranch (14 spp.), Las Animas Co. (1-18)

Two superposed faunas, the pre-Nebraskan Donnelly Ranch and the Rancholabrean Mesa de Maya were discovered at the Donnelly Ranch site in southeastern Colorado. The site is located on the north side of the basaltic bluff, Mesa de Maya, 116 km E of Trinidad at an elevation of 1,800 m. Fossils were found during the excavation of a large irrigation ditch; the Blancan assemblages came from locally derived sands and clays at the base of the deposit. The sediments are normally magnetized and are within the Gauss normal magnetic epoch, which gives an age of about 2.5 m.y. B.P. The fauna most closely resembles those of Broadwater, Sand Draw, and Blanco. Savanna, plains, and marsh communities are represented, and the presence of *Geochelone*, *Tapirus*, and *Sigmodon* indicates a warmer, milder climate with little annual fluctuation in precipitation. (Hager, 1975.) UWYO.

Florida

Haile XVA (20 spp.), Alachua Co. (1-38)

Numerous fossiliferous cave and fissure fillings have been exposed by limestone mining operations near the town of Haile in northern Florida. The sinkhole site Haile XVA is the first late Blancan site described from the Gulf Coastal Plain. The fossil-bearing matrix consists of alternating coarse sands and clays. About 27 m above sea level, it contains the remains of many aquatic vertebrates (sharks, brackish-water fish, turtles, birds, the otter *Satherium,* and a beaver) as well as terrestrial vertebrates including many forest forms (an Old World flying squirrel *Cryptopterus* and the peccary *Mylohyus*). The probable habitat was a springhead of a coastal stream that flowed through a tropical or subtropical forest or forest savanna. There is a close alliance between several of the mammalian taxa and Plio–Pleistocene species in South America, which strengthens the correlation of the Chapadmalalan stage in South America with the Blancan in North America. Edentates are well represented, which indicates a late Blancan influx from the south. On biostratigraphic grounds, Haile XVA probably represents a preglacial wet period during the Blancan. The site is believed to be slightly younger than Santa Fe River IB. (Robertson, 1976; Webb, 1974a.) FSM.

Santa Fe River IB (23 spp.), Gilchrist Co. (1-37)

Santa Fe River IA is a river-bottom site, about 3 m above sea level. In the adjacent bank, a localized deposit (site IB) has produced a wholly Blancan fauna. A relatively late date within the Blancan is indicated by the fauna. A dry, upland phase is suggested by the presence of grassland animals like *Platygonus* and *Chasmaporthetes;* however, *Mylohyus, Odocoileus, Tapirus,* and other woodland forms are also present. A remarkable member of the fauna is the giant flightless bird *Titanis walleri* Brodkorb, which belongs to the otherwise Neotropical family Brontornithidae.

Santa Fe River IB is regarded as the oldest Blancan site currently recognized in Florida. Elements of the fauna have been described by Brodkorb (1963), Kurtén (1965), R. A. Martin (1969), and Webb (1974a); a faunal list appears in this last reference. (Webb, personal communication, 1977.) FSM.

Idaho

Grand View (29 spp.), Owyhee Co. (1-11)

The Grand View local fauna comes from exposures on both sides of the Snake River, about 110 km W of Hagerman. Although some of the larger mammals are related to those from Hagerman, there are significant differences, such as the presence of modern genera (*Synaptomys, Ondatra, Coendou*). In addition, several species, including *Ophiomys parvus* and *Ondatra idahoensis,* are advanced compared to Hagerman representatives (*Ophiomys taylori* and *Pliopotamys minor,* respectively), which indicates a later date. Another difference is the absence or rarity of antelopes and peccaries at Grand View. The magnetic signature indicates a position in the early Matuyama epoch, with overlying sediments in the Olduvai event. The age is thus late Blancan. (Bjork, 1970; Lindsay, personal communication, 1976; Shotwell, 1970.) ISUM, LACM.

Hagerman (53 spp.), Twin Falls Co. (1-12)

The Hagerman fauna is probably the richest Blancan fauna known at present. The fossiliferous beds are located in the Glenn's Ferry Formation, along the Snake River, west of the town of Hagerman and about 48 km downstream from Twin Falls at the eastern edge of the Snake River Plain. The beds are about 100 m thick and cover an area some 17 km^2, in which about 300 localities are known. The formation consists of floodplain and stream deposits interbedded with ash beds and lava flows. The drainage was probably to the northwest, as in the case of the nearby present Snake River. Beginning in the late 1920s, parties from the U.S. National Museum, U.S. Geological Survey, University of Michigan Museum of Paleontology, and Idaho State University have worked in the area collecting both invertebrate and vertebrate fossils, mapping, and conducting stratigraphic investigations.

Absolute dates for the Glenn's Ferry Formation are conflicting. A lava flow 50 m beneath the Horse Quarry has been dated at 3.48 m.y. B.P., whereas ash beds have been dated at 3.2 and 3.3 m.y. B.P. However, a series of considerably older dates, ranging from 4.4 to 6.2 m.y. B.P., have recently been obtained for basalts in the formation (Armstrong, Leeman, and Malde, 1975). All the basalts except one (the earliest-dated lava flow in the formation) are reversely magnetized. These radiometric dates would place the beginning of the Blancan more than 2 m.y. earlier than is suggested elsewhere. Unpublished results by N.D. Opdyke (Lindsay, personal communication, 1976), however, indicate that the Hagerman Horse Quarry is in the lower Gauss epoch (less than 3.3 m.y. B.P). The fauna is thus early Blancan, slightly younger than Rexroad.

The pollen flora (e.g., *Celtis, Populus,* Pinaceae) and the fauna, with many fish, frogs, water snakes, shorebirds, and aquatic mammals, indicate warmer and more humid conditions than those existing today, with low montane vegetation in areas that are now treeless. The mammalian fauna is rich in both small and large species. A restoration of the fauna appears in figure 2.2. At Gidley's famous Horse Quarry, more than 150 skulls of *Equus (Dolichohippus) simplicidens* were recovered. Carnivores are abundant

2.2 Scene at Hagerman, Idaho, in the early Blancan. Shown are *Equus simplicidens, Trigonictis idahoensis* and its prey, *Paenemarmota barbouri.* Beaver dam in the background.

[Bjork, (1970) recognized 17 spp.], and peccaries, camelids, antelopes, and mastodonts are also present. Thanks to the work of Claude W. Hibbard and his students, many species of micromammals have been recognized and described. Most of the mammals suggest marsh and meadow habitats, but valley-slope and upland forms are also present and increase in the top part of the sequence. Other changes (mostly gradual) within the fauna are the disappearance of *Dipoides* and *Cosomys*, increase in numbers of *Castor*, *Pliopotamys*, and *Ophiomys*, and microevolution within certain rodent lineages. (Armstrong, Leeman, and Malde, 1975; Bjork, 1970; Brodkorb, 1958; Chantell, 1970; Evernden et al., 1964; Feduccia, 1967; Gazin, 1933a,b, 1934a,b, 1935, 1936, 1938; Hibbard, 1959, 1962, 1969; Hibbard and Bjork, 1971; Hibbard and Zakrzewski, 1967; Holman, 1968; Lindsay, personal communication, 1976; J. R. Macdonald, 1966; Zakrzewski, 1967b, 1969.) USNM, UMMP, ISUM.

KANSAS

Deer Park (9 spp.) and Sanders (11 spp.), Meade Co. **(1–28, 1–27)**

The Ballard Formation ("Meade Formation" of some authors), which consists of a basal sand and gravel member (Angell) grading upward into silts and clays, with a prominent buried caliche (the Missler Member), rests unconformably upon the Rexroad Formation (see the discussion of the Fox Canyon, Kansas, fauna that follows). It contains four superposed faunas, of which the most important are Deer Park, beneath the caliche, and Sanders, above it. There is also a good molluscan fauna (the Spring Creek fauna) at the base of the Missler Member.

The Deer Park fauna was recovered from a system of dendritic artesian-spring deposits with outlet in a paleobasin eroded during Ballard times. The animals were trapped in boiling quicksand. Reptiles and birds and mammals including *Nannippus*, *Equus*, *Stegomastodon*, and small rodents are represented. A warm climate is indicated by the presence of *Geochelone* and *Testudo*.

The section at Sanders is normally magnetized and presumably falls in the Gauss magnetic epoch. The fauna consists of mollusks, amphibians, reptiles, birds, and mammals including shrews, rodents, and equids. The habitat was a marshy floodplain surrounded by grassland. A continued warm climate is suggested by the mollusks and small mammals present.

In the sedimentation cycle represented by the Ballard, the Angell Member may represent a pre-Nebraskan alpine glaciation, whereas Deer Park and Sanders are interglacial and correlated with the Cita Canyon, Broadwater, and Sand Draw faunas. (Hibbard, 1956b, 1970b; Lindsay, Johnson, and Opdyke, 1975; Skinner and Hibbard, 1972; D. W. Taylor, 1966.) UMMP.

Dixon (14 spp.), Kingman Co. **(1–31)**

This fauna, in an unnamed deposit possibly equivalent in age to the Belleville or Ballard Formation and overlain by the Crooked Creek Formation, comprises small mammals (shrews and rodents), birds, reptiles, amphibians, and fishes, as well as a large assortment of mollusks. The paleohabitat included a pond or large stream with abundant, shallow-water vegetation and gallery woods. The climate was moist with mild winters. The fauna resembles that of Sand Draw, and thus had a somewhat northern aspect (e.g., presence of *Synaptomys rinkeri*), which suggests a cool pre-Nebraskan phase of the late Blancan. (Eshelman, 1975; Hibbard, 1956b, 1970b; Skinner and Hibbard, 1972; D. W. Taylor, 1966.) UMMP.

Fox Canyon (30 spp.), Keefe Canyon (25 spp.), and Rexroad (50 spp.), Meade Co. **(1-23, 1-24, 1-25)**

Four successive faunas, the Fox Canyon, Keefe Canyon, Rexroad, and Bender are known from the Rexroad Formation, which consists of stream, pond, and artesian-spring deposits with occasional lignites trapped in a local basin which owes its origin, at least in part, to solution of underlying Permian salt and gypsum beds. A still earlier fauna, the Saw Rock Canyon (18 spp., Hibbard, 1964), is regarded as terminal Hemphillian because of the complete absence of diagnostic Blancan forms. The Hemphillian/Blancan transition is thus recorded within the Rexroad Formation.

The Fox Canyon locality exhibits reverse magnetism (Lindsay, Johnson, and Opdyke, 1975) and presumably dates from a late part of the Gilbert epoch (3.3-3.5 m.y. B.P.). The vertebrates were found in a pocket of stream-deposited sandy silt at a low level in the formation, about 5 m below the massive caliche that is near the top of the formation. Small mammals are abundant (6 spp. of insectivores, 15 spp. of rodents), but large mammals very rare. Blancan elements include *Geomys, Symmetrodontomys,* and *Odocoileus.* (Hibbard, 1950, 1953a, 1967; Zakrzewski, 1967a).

The Keefe Canyon deposit was formed in an old artesian-spring basin. The animals were trapped in quicksand or mired in bogs at the edge of the spring. A vertical tube in the underlying silt beds allowed the water to escape. Some of the material, churned by the artesian water, is highly polished. The fauna is very similar to that of Rexroad but contains a somewhat larger number of large mammals, including *Megatylopus, Titanotylopus, Equus,* and *Stegomastodon* (Hibbard and Riggs, 1949). Stratigraphically, this fauna is somewhat higher than the Fox Canyon fauna and is considered approximately equivalent in age to the Rexroad fauna.

The upper part of the section, with the Rexroad and Bender faunas, is normally magnetized and is referred to the Gauss epoch, beginning about 3.3 m.y. ago. The Rexroad fauna occurs in artesian-spring and pond deposits between two buried caliches in the upper part of the formation, and the Bender fauna, not yet fully studied, comes from the top of the formation above the caliche.

Dominated by small mammals (20 spp. of rodents), the Rexroad local fauna is one of the best-known mammalian assemblages from the Blancan; in addition, numerous mollusks, fishes, amphibians, reptiles, and birds are present. Of the 50 species of mammals recognized, all but *Taxidea taxus* are extinct. The discovery and development of this site is due to the late Claude W. Hibbard, who pioneered the washing and sieving methods of collecting the small mammal remains so important in interpreting paleoenvironments and evolutionary lineages.

In Rexroad times, the area appears to have been a tallgrass savanna traversed by braided streams with gallery forests. A warm, moist, and frost-free climate prevailed (as indicated by the presence of ibis and the giant turtle *Geochelone*). Many species are typical of marsh, stream, and pond habitats; the presence of semibrowsing mammals (*Stegomastodon, Megalonyx*) suggests that abundant broad-leaved shrubs and trees existed in the area. Shortgrass prairie probably occurred in the driest upland areas only. The fauna comprises typical Blancan forms (*Equus, Trigonictis, Ogmodontomys,* etc.), as well as at least 27 species that are absent in later Blancan faunas and may be regarded as Hemphillian holdovers. The Rexroad fauna is correlated with Benson and Hagerman. (Hibbard, 1970b, and references therein.) UMMP.

Seger (4 spp.) and Borchers (25 spp.), Meade Co. **(1–29, 1–30)**

These two local faunas occur in the Crooked Creek Formation, which rests unconformably upon the Ballard Formation. The Seger faunule comes from the lower member (Stump Arroyo). The upper, Atwater Member contains an ash bed of Pearlette type "B," dated at 1.9 m.y. B.P., and the Borchers fauna overlies this bed. The deposits here are reversely magnetized, which indicates a date within the Matuyama epoch. Although the Borchers fauna, with its southern affinities, has been regarded as interglacial (Yarmouthian or Aftonian), it is evident from radiometric dates that it must be older, and most probably falls in an "interglacial" preceding the Nebraskan. Thus it should be very late Blancan in age. The small Seger fauna consists entirely of large mammals. *Nannippus* and dolichohippine horses, still common at Seger, are absent in the Borchers fauna, which instead has many modern genera, including several carnivores, *Lepus, Ondatra,* and a large horse. (Hibbard, 1941d, 1951a, 1954a, 1970b; Lindsay, Johnson, and Opdyke, 1975.) UMMP.

White Rock (43 spp.), Republic Co. **(1–32)**

This large fauna, with predominantly small mammals, occurs in the Belleville Formation, a paleovalley deposit of the ancestral Republican River in northcentral Kansas, at an elevation of about 492 m. This formation is in part a correlative of the Red Cloud sands and gravels and the Holredge, Fullerton, and Grand Island formations, all in Nebraska. In addition to mammals, the fauna comprises mollusks, ostracods, a beetle, fish, amphibians, reptiles, and birds. The fauna is regarded as a transported fossil assemblage representative of more than one contemporaneous community. The evolutionary level of the mammals indicates an age more recent than that of the Sand Draw local fauna but older than Borchers. Taxa include *Ondatra idahoensis, Synaptomys rinkeri, Zapus sandersi sandersi,* and dolichohippine and hemionine horses. The White Rock fauna appears to be most similar in age and ecology to the Dixon local fauna in Kingman County and is thus late Blancan in age.

More diverse than the modern mammalian fauna of the area, the White Rock fauna suggests a climate with less extreme temperature and moisture fluctuations than now occur in north-central Kansas. Apparently, there occurred pre-Nebraskan cooling sufficient to bring in some northern forms but not sufficient to force southward the taxa characteristic of milder climates. This resulted in such forms as *Geochelone* and *Sigmodon,* characteristic of relatively warm, mildly seasonal climates, being coexistent with boreal microtines and *Zapus.* Paleohabitats include permanent water, stream-river bank, lowland meadow, savanna valley, valley slope, and upland prairie. (Eshelman, 1975.) UMMP.

NEBRASKA

Broadwater (33 spp.), Morrill Co. **(1–33)**

Several localities in the extensive Broadwater Formation have yielded a rich fossil fauna. The formation consists of a basal, gravel member, a middle member (Lisco) containing the Broadwater local fauna, and an upper, gravel member. The main quarry, located 8.8 km E and 1.2 km N of the town of Broadwater in western Nebraska, was excavated from 1936 to 1942, and test pits were dug in 1965 and 1967; it is the largest single quarry excavation in the state. A glacial–interglacial–glacial cycle may be indicated by the alternation of the gravels with a diatomite and peat bed (in which most of the fossils were found) within the Lisco Member. The fauna would thus date from a late Blancan "interglacial," as may also be the case at Sand Draw. A considerable range of biota is represented, with permanent-stream (*Procastoroides, Satherium*), gallery-forest (*Stegomastodon, Megalonyx*), and prairie (*Equus, Hemiauchenia, Capromeryx*) animals present. Animals of the streamside habitat predominate. (Barbour and Schultz, 1937; L. Martin, personal communication, 1977; Schultz and Martin, 1970a; Schultz and Stout, 1948.) UNSM.

Lisco (12 spp.), Garden Co. **(1–34)**

The sites which have produced the Lisco fauna are in the Lisco Member of the Broadwater Formation in western Nebraska. The facies here, however, is a colluvial and loesslike silt, and the mammals suggest upland steppe or prairie. Although probably broadly correlative with the Broadwater fauna, the Lisco fauna is not necessarily precisely contemporaneous. Both the Broadwater and Lisco faunas are being restudied. (Schultz and Martin, 1970a; Schultz and Stout, 1948.) UNSM.

Sand Draw (43 spp.), Brown Co. **(1–35)**

This rich and well-studied local fauna comes from the Keim Formation, a preglacial paleovalley fill in the nonglaciated region of north-central Nebraska. It consists of fine sand, silt, and lake clay, with channel deposits. It is overlain by the gravels of the Long Pine Formation, constituting the first evidence of fluvioglacial outwash in the area. Next come interstadial sands and silts (Duffy Formation) and a second gravel outwash (Pettijohn Formation).

A pollen flora from the Keim Formation is dominated by Gramineae and *Artemisia;* among the arboreal pollen, *Pinus* is predominant. The fauna, apart from the mammals, comprises 29 gastropod, 6 bivalve, 14 ostracode, 11 fish, 4 amphibian, 16 reptile, and 10 or more bird (mostly water bird) species. Ecological analysis of the fauna suggests a lower elevation (with lakes, large winding streams, and numerous oxbows) and a more equable climate than that of the present day. The presence of the giant tortoise *Geochelone* is notable. The mammals include characteristic late Blancan species and constitute the most varied assemblage of this time, with many small and large species. They appear to represent at least five distinct communities: (1) stream and lake bank (*Dipoides, Procastoroides, Pliopotamys, Satherium,* and many other species coming only to drink); (2) marsh and semiaquatic (*Planisorex, Ogmodontomys, Pliolemmus*); (3) savanna valley (*Stegomastodon, Gigantocamelus, Geomys, Sorex, Scalopus, Hypolagus, Spermophilus, Pliophenacomys*); (4) valley slope (*Peromyscus, Neotoma*); and (5) upland (*Prodipodomys, Onychomys, Bensonomys, Canis, Equus, Hemiauchenia*). Taxonomically, the fauna is similar to that of Broadwater, and its age is late Blancan. (Skinner and Hibbard, 1972.) F:AMNH, UMMP.

NEVADA

Wichman (5 spp.), Mineral Co. **(1–10)**

This faunule is represented by waterworn bone and tooth fragments of large mammals. The remains were found in quartzitic sands deposited in pockets in rhyolite tuffs, 6 km N of the discontinued Wichman Post Office (Univ. Calif. Loc. V3914) in western Nevada. A Blancan age is indicated by the presence of a zebrine horse and an advanced mastodont. (Macdonald, 1956.) UCMP.

South Dakota

Delmont (7 spp.), Douglas Co. **(1–36)**

The Hieb Sand Pit, in unconsolidated crossbedded sands 7 km S of Delmont in southeastern South Dakota, has yielded a Blancan large-mammal fauna. The sand is overlain by late Wisconsinan till and underlain by silts resting on the Cretaceous bedrock. A horse is probably zebrine; *Procastoroides, Titanotylopus,* and *Homotherium* are also present. No precise dating within the Blancan is possible at present. (Martin and Harksen, 1974.) SDSM.

Texas

Blanco (45 spp.), Crosby Co. **(1–21)**

The Blanco local fauna, named for the regional landmark Mt. Blanco, is the type of the Blancan Land Mammal Age in North America. The Blanco Formation is exposed in a section less than 2 km long and with a maximum thickness of about 20 m; it is located on the north side of Crawfish Draw, 18 km N of Crosbyton in the Texas Panhandle. The deposits are a localized basin accumulation, and the absence of aquatic vertebrates and the varied nature of the sediments suggest a shallow-drained rather than a closed basin. Recent mineralogical, sedimentological, pollen, and diatom studies (see G. E. Schultz, 1977) indicate that the Blanco beds were deposited in a large playa during an arid to semiarid climatic interval. Water accumulation was probably seasonal.

The age of the Blanco fauna is still disputed. Two ashes overlie the fossiliferous beds. The Guaje Ash, exposed near the surface of the cover sands just north of Mt. Blanco, has been dated 1.4 ± 0.2 m.y. B.P. (Izett, Wilcox, and Borchardt, 1972) by glass shards on fission tracks and gives a minimum age for the fauna. The Blanco Ash, which lies 7.6 m below the Guaje Ash (but still above the fossiliferous zone), has been dated 2.8 ± 0.3 m.y. B.P. by the same method (G. E. Schultz, 1977), which thus suggests an age in excess of 2.8 m.y. for the fauna. Paleomagnetic dating indicates that the entire fossiliferous section is reversely magnetized, and Lindsay, Johnson, and Opdyke (1975) place it in the early Matuyama interval, i.e., less than 2.4 m.y. B.P. Most workers believe the Blanco fauna to be late Blancan in age and closely correlative with those from Sand Draw, Broadwater, Deer Park, and Cita Canyon, but Dalquest (1975) concluded that the fauna shows more resemblances to the Rexroad fauna (early Blancan) than to any other.

The first discovery of vertebrate fossils at Blanco was by W. F. Cummins, who did prospecting in the Staked Plain in 1889–90; the material was studied by Cope. Subsequently, many parties have worked in the area. The fauna is dominated by large grazing herbivores (Dalquest, 1975); microtine rodents and aquatic mammals are absent. *Megalonyx leptostomus, Borophagus diversidens, Platygonus bicalcaratus, Nannippus phlegon,* and *Equus simplicidens,* among others, were described from Blanco. In addition, one bird (*Creccoides osbornii* Shufeldt) and two land tortoises (*Geochelone* and *Gopherus*) have been found. (Dalquest, 1975; Evans and Meade, 1945; Izett, Wilcox, and Borchardt, 1972; Meade, 1945; Johnston and Savage, 1955; G. E. Schultz, 1977.) AMNH, MWU, TMM, TTM.

Cita Canyon (44 spp.), Randall Co. **(1-22)**

The Cita Canyon fauna comes from exposures around the heads of small canyons which drain southward into North Cita Canyon. The main quarry (Univ. Calif. Loc. V3721) is located about 5.6 km S and 21 km E of Canyon, in the Texas Panhandle. The fossiliferous beds are friable lake and playa sandstones or sands about 8.5 m in thickness. They rest uncomformably upon the Triassic beds and are overlain with apparent conformity by mudstones, sands, and caliche containing mollusks and scattered vertebrate fossils.

The fossils occur in several distinct horizons but are considered to belong to the same fauna, although subfaunal distinctions may reflect possible ecological, climatological, or geochronological changes. The lower part of the section shows normal polarity, whereas the upper unfossiliferous part above the quarry is reversed; this may represent the transition from the Gauss to the Matuyama magnetic epoch. The entire fossiliferous level is within the normally magnetized zone and so should be latest Gauss age, i.e., about 2.5 m.y. B.P., which probably makes it slightly older than the Blanco fauna.

Fossils were first discovered here in 1935 by Floyd V. Studer and C. Stuart Johnston, and although there have been several papers on parts of the fauna, the assemblage has never been fully described. In addition to mammals, a fish, a frog, turtles (*Geochelone, Gopherus,* and a pond turtle), snakes, lizards, and water birds have been identified. Large mammals predominate, but a small but significant microfauna was recently recovered by screen-washing 15 tons of matrix. The microfauna includes *Sorex, Hesperoscalops, Hypolagus, Spermophilus, Perognathus, Prodipodomys, Zapus, Sigmodon,* and several microtines that show resemblances to those from the Sanders and Dixon faunas in Kansas. Carnivores are well represented and include a large sample of *Canis lepophagus,* as well as *Acinonyx studeri* and *Lynx rufus* (the report of *L. issiodorensis* is incorrect). The microfauna indicates a trend toward moister conditions. Studies on this important fauna are continuing. (Johnston, 1939; Johnston and Savage, 1955; Kurtén, 1974; Lindsay, Johnson, and Opdyke, 1975; Mawby, 1965; Savage, 1955, 1960; G. E. Schultz, 1977; Skinner and Hibbard, 1972.) WTSU.

Hudspeth (18 spp.), Hudspeth Co. **(1-19)**

An assemblage closely resembling and contemporaneous with the nearby Red Light local fauna (see below) was found in the Hueco Bolson, just upriver from the Red Light locality at the southern tip of Hudspeth County. Equids dominate the fauna, and a warm climate is indicated. The age is probably late Blancan. (Strain, 1966.) TMM.

Red Light (29 spp.), Hudspeth Co. **(1-20)**

The Red Light local fauna comes from the Love Formation in the Red Light Bolson (or basin), which straddles the present-day Rio Grande. The sediments and faunas in the bolson fill reflect a gradual change from arid-climate playa to moist-climate fluvial conditions, the latter occurring during the time of the Red Light local fauna. Although climatic conditions may have been much like those of today, the drainage pattern was quite different. The fauna (numerous equids and camelids) indicates mainly open country with sandy soil, covered by grass or brush vegetation. The presence of large land tortoises shows that temperatures rarely if ever approached the freezing point. Unfortunately, the fossils are poorly preserved and often waterworn, and only generic determinations are possible in many cases. Since the fossils were collected in beds indicating moist conditions, the fauna probably lived during a pluvial period in the late Blancan. (Akersten, 1972.) TMM.

WASHINGTON

Taunton (10+ spp.), Adams Co. **(1–9)**

This rich but still largely undescribed fauna comes from fluviatile deposits resembling those of the stratotype Ringold Formation at White Bluffs, 24 km to the south (see the discussion in the next section). Fish remains are very common. Mammals include a zebrine horse, *Hypolagus, Pliopotamys, Procastoroides,* and numerous antilocaprids; peccaries and cervids are rare. A remarkable find is the lesser panda, *Parailurus.* The main habitat seems to have been savanna or open grassland, with gallery woods lining the minor Taunton tributary which entered the Columbia River at White Bluffs. The fauna is thought to be slightly younger than the White Bluffs fauna and close in age to the Hagerman local fauna. (Tedford and Gustafson, 1977.) AMNH.

White Bluffs (25 spp.), Franklin Co. **(1–8)**

The fauna comes from the upper section of the Ringold Formation in bluffs along the Columbia River, upstream from Pasco, in south-central Washington. The formation consists of nearly horizontally-bedded continental sediments and is informally divided into three parts. The fossiliferous upper section, about 153 m thick, is a heterogeneous mixture of floodplain, stream channel, and lake deposits. The fauna consists of many Blancan taxa, including *Nekrolagus, Ophiomys, Borophagus,* and *Trigonictis.* Of the larger mammals, browsers predominate, and the cervid *Bretzia pseudalces* is the most abundant species. In addition, several species of fish and reptiles have been identified. The assemblage shows close affinities to the early Blancan Hagerman local fauna in Idaho but is probably slightly older. (Fry and Gustafson, 1974; Gustafson, 1978.) UW.

CHAPTER THREE

Irvingtonian Faunas

The base of the Pleistocene is fixed typologically as the base of the Calabrian marine Stage in Italy. This has been shown to correlate either with the beginning (Bandy, 1972; Opdyke, 1972) or the end (Haq, Berggren, and Van Couvering, 1977) of the Olduvai magnetic event in the Matuyama epoch. The transition between the Blancan and the Irvingtonian falls in the same time range. Although the transition is gradual, there is a point at which the fauna becomes more characteristic of the Irvingtonian than Blancan; this occurs at about 1.9 m.y. B.P., at least in the San Pedro Valley (Opdyke et al., 1977).

In defining the Irvingtonian Land Mammal Age, Savage (1951) regarded *Mammuthus* as the characteristic guide fossil. *Mammuthus,* however, is absent in the Anza Borrego and San Pedro Valley sections and appears in the Great Plains about 1.5 m.y. B.P., although it seems more than likely that the genus was present farther north at an earlier date, judging from the presence of *Mammuthus* associated with *Borophagus* (otherwise not known in post-Blancan faunas) at Wellsch Valley. It is not improbable that *Mammuthus, Lepus, Euceratherium,* and *Panthera* (jaguar) crossed Beringia about 1.9 m.y. B.P. All are characteristic new taxa in the Irvingtonian, and all, except perhaps *Lepus,* are of Palearctic origin.

In addition to such immigrants, Irvingtonian faunas (map 2) contain new genera of Nearctic origin like *Ondatra* and *Dipodomys.* On the other hand, common Blancan taxa like *Nannippus, Equus* (*Dolichohippus*), *Borophagus,* and *Chasmaporthetes,* are absent or (in the two latter instances) survive only to the very early Irvingtonian. Also absent is *Bison,* a Rancholabrean immigrant from the Palearctic.

With the onset of continental glaciations, which dominate the history of much of the Irvingtonian and Rancholabrean, it has become customary to assign local faunas to named glacial–interglacial stages (see table 3.1). In fact, there are only a few areas where such a tie-in is possible at present, e.g., the central Great Plains (Schultz and Hillerud, 1977). A sequence of volcanic ashes, datable by radiometric methods, provides the following chronological framework: the Pearlette Ash Type "O", 0.6 m.y.B.P.; the Bishop Ash, 0.8 m.y. B.P., the Pearlette Ash Type "S", 1.2 m.y. B.P.; and, finally, the very late Blancan Pearlette Ash Type "B", 2.0 m.y. B.P. In earlier work, the various ash levels were not distinguished from each other, and this threw the stratigraphy into confusion.

Paleomagnetic stratigraphy will also serve up to the boundary between the Matuyama and Brunhes magnetic epochs (0.7 m.y. B.P.). However, paleomagnetic data are only available for a few sections at present. Included are Curtis Ranch in the San Pedro Valley, with a rich earliest Irvingtonian fauna; Vallecito Creek in the Anza Borrego section, straddling the Blancan–Irvingtonian boundary; Irvington, California, with *Mammuthus,* in the upper Matuyama; and Cudahy, Kansas, in the lower Brunhes just beneath a Type "O" ash (0.6 m.y. B.P.).

After the spate of new taxa inaugurating the Irvingtonian, faunal turnover appears to have continued at a somewhat slower rate. The important Blancan proboscidean *Stegomastodon* is still present in the early Irvingtonian and vanishes from the fossil record about 1.6 m.y. B.P. (Lindsay, Johnson, and Opdyke, 1975), and it appears that

TABLE 3.1
IRVINGTONIAN FAUNAS IN THE GREAT PLAINS AREA[a]

Glacial/interglacial	Texas–Oklahoma	Kansas	Nebraska
Early Illinoian	Doby Springs Berends	Sandahl Adams	Angus Mullen 2
Yarmouthian	Slaton	Rezabek Kanopolis	Hay Springs
	Pearlette "O" 0.6		
Kansan	Vera	Cudahy	
		Kentuck?	Pearlette "S" 1.2
			Sappa
Aftonian	Rock Creek Gilliland Holloman		
Nebraskan		Nash?	Mullen 1?

[a] Figures give dates in million years B.P.

the camelid *Titanotylopus,* another Blancan holdover, also became extinct in mid-Irvingtonian times. The grade attained by evolving lineages is also useful in subdividing the Irvingtonian. Thus the early Irvingtonian mammoths are *Mammuthus meridionalis,* whereas late Irvingtonian forms belong to *Mammuthus columbi.* The species *Smilodon fatalis* appears to be confined to the later Irvingtonian; the early Irvingtonian form is *Smilodon gracilis.*

Finally, characteristic of local faunas assigned to the late Irvingtonian is a massive appearance of extant species. This parallels the situation in Europe, where the representation of modern species increased rapidly during the early mid-Pleistocene, so that extant species make up about 50 percent of the total species known from the Cromerian–Elsterian. (Bandy, 1972; Haq, Berggren, and Van Couvering, 1977; Lindsay, Johnson, and Opdyke, 1975; Opdyke, 1972; Opdyke et al., 1977; Savage, 1951; Schultz and Hillerud, 1977.)

ALASKA

Cape Deceit (13 spp.), south shore of Kotzebue Sound **(2–50)**

The fauna comes from the Cape Deceit Formation, along an exposure of some 250 m of shoreline between Deering and Cape Deceit in western Alaska. The unit consists of silts with gravels, sands, and peat layers and is overlain unconformably by the Imachuk and Deering formations. The vertebrate fossils occur in the upper part of the Cape Deceit Formation. The earliest known North American occurrences of *Microtus, Lemmus, Cervus,* and *Rangifer* are recorded from Cape Deceit, and the presence of an ancestral lemming, *Predicrostonyx,* and a large pika, *Ochotona whartoni,* are also noted. As might be expected, the fauna shows closer affinities to Siberian faunas than to North American assemblages. The paleoenvironment was apparently a treeless lowland tun-

Map 2. Irvingtonian Faunas

dra, which indicates that the extant arctic fauna has had a long arctic-adapted evolutionary history. The existence of primitive voles and lemmings in the fauna argues for a date well back in the Irvingtonian, perhaps in the Aftonian–Kansan time range. (Guthrie and Matthews, 1971.) UAM.

ALBERTA

Medicine Hat (Units D and E, Kansan, 11 spp.; Unit G,
   Yarmouthian, 7 spp.) **(2–55, 2–56)**

Pleistocene deposits exposed at bluffs along the South Saskatchewan River, from about 13 km W to 16 km NW of Medicine Hat, display an alternation of glacial tills and floodplain deposits (grading from gravels to clays) in which a number of superposed faunas have been discovered. The oldest surficial deposit (unit D) is a coarse river gravel, with some sand beds, and directly overlies the Cretaceous beds. There are no stones originating from the Canadian Shield, which indicates that the bed was laid down before the arrival of the first Laurentide glacier. This unit intercalates with the dark gray silt and sand beds of Unit E, which appears to be a floodplain deposit. The next bone-bearing unit (G) has sharp contact with Unit E in some places, but elsewhere the contact is gradational, with the two units interfingering. It is considered to be Yarmouthian in age.

Bones have been found throughout the three units; they are not strongly waterworn, and articulated bones have been found. The fauna of Units D and E include such typical Irvingtonian forms as *Mammuthus meridionalis, Equus scotti, Equus calobatus,* and *Hemiauchenia blancoensis.* The fauna of Unit G is more advanced and includes *Mammuthus columbi* and also *Cynomys, Equus, Camelops,* and *Rangifer.* (Churcher, personal communication, 1972; Stalker and Churcher, 1970, 1972.) NMC, ROM, UTZ.

ARIZONA

Curtis Ranch (27 spp.), Cochise Co. **(2–58)**

The San Pedro Valley series of Blancan to early Irvingtonian fossiliferous deposits are included with the descriptions of the Blancan faunas (see Arizona, Benson fauna). The Curtis Ranch fauna comes from the Gidley and *Glyptotherium* localities in the

Map 2. Irvingtonian Faunas*

50. Cape Deceit (M)
51. Vallecito Creek (E–M)
52. Bautista Creek (I)
53. Irvington (M)
54. Delight (I)
55. Medicine Hat Units D and E (M)
56. Medicine Hat Unit G (L)
57. Wellsch Valley (E)
58. Curtis Ranch (E)
59. Slaton (L)
60. Rock Creek (E)
61. Gilliland (E)
62. Vera (M)
63. Holloman (E)
64. Berends (L)
65. Cudahy (L)
66. Adams (L)
67. Sandahl (L)
68. Nash (I)
69. Kentuck (L)
70. Kanopolis (L)
71. Rezabek (L)
72. Sappa (M)
73. Angus (L)
74. Hay Springs (L)
75. Rushville (L)
76. Gordon (L)
77. Mullen I (E)
78. Mullen II (L)
79. Java (M)
80. Conard Fissure (L)
81. Punta Gorda (M)
82. Pool Branch (L)
83. Coleman IIA (L)
84. Inglis IA (E)
85. Trout (L)
86. Cumberland Cave (L)
87. Port Kennedy Cave (E)

*Letters in parentheses in key represent: E, early Irvingtonian; M, middle Irvingtonian; L, late Irvingtonian; I, Irvingtonian (exact age unknown).

upper part of the middle member of the St. David Formation. Paleomagnetic investigation indicates an age of 1.8–1.9 m.y., (early Irvingtonian). Environmental conditions appear to have remained much the same as they were in early Blancan times, although water birds and amphibians are missing; the seven reptile taxa include *Geochelone*. Significant changes in the mammalian fauna have occurred, however. *Nannippus* is extinct, and *Lepus, Dipodomys,* and *Ondatra,* genera characteristic of the Blancan/Irvingtonian transition, are present. *Stegomastodon* still survives, and the largest known glyptodont exists here. Like Benson, the Curtis Ranch fauna has many small mammals, including *Spermophilus, Geomys, Peromyscus,* and *Sigmodon.*

At a higher level, in the upper Matuyama zone, there is an Irvingtonian faunule at the Prospect locality. (Gazin, 1942; Gidley, 1922b, 1926; Johnson, Opdyke, and Lindsay, 1975; Lindsay and Tessman, 1974; Lindsay, Johnson, and Opdyke, 1975.) UALP, USNM.

Arkansas

Conard Fissure (44 spp.), Newton Co. (2–80)

This fissure filling, in the limestones of the Ozark Mountains about 6 km W of Willcockson and N of Buffalo River in northern Arkansas, at an elevation of 314 m, has produced a large fauna. It was described by Barnum Brown and has been revised recently by Russell Graham. Small mammal bones, probably deposited by owls, are extremely abundant (especially *Blarina* and *Peromyscus*), but there is also a sizable large-mammal fauna, with *Mylohyus* and *Odocoileus* the most numerous. Besides mammals, remains of amphibians, reptiles, and birds have been identified. Many of the bones are rodent- or carnivore-gnawed. The age of the fauna, considered a single biostratigraphic unit (Graham, 1972), is in dispute. Reference to the Irvingtonian is suggested by the absence of *Bison* and the presence of forms like *Felis inexpectata* and *Ondatra annectens*. A Kansan or early Illinoian date is possible. The boreal affinities of some of the mammals, e.g., *Sciurus hudsonicus, Erethizon dorsatum,* and *Sorex fumeus,* suggest a glacial phase. A forest with open glades was the probable habitat. (Brown, 1908, Gidley and Gazin, 1938; Graham, 1972.) AMNH.

California

Bautista Creek (8 spp.), Riverside Co. (2–52)

Lacustrine sands and clays in the San Jacinto Mountains, 9.6 km SE of San Timoteo Canyon, have produced a small fauna with *Lepus,* an asslike horse, tapir, and fragments representing a few other taxa. The age is considered to be Irvingtonian. (Frick, 1921; Savage, 1951.) F:AMNH.

Irvington (21 spp.), Alameda Co. (2–53)

This fauna is the type of the Irvingtonian Land Mammal Age in North America (Savage, 1951). It was recovered from stream-deposited sandy clays, sands, and gravels exposed SE of Irvington; the gravel pits that yielded the fauna are now abandoned and partly overlain by Interstate Highway 680. Strata adjacent to the pits, and at the same level where the fossils were collected, are reversely magnetized and are in the Matuyama magnetic epoch (0.7+ m.y. B.P., mid-Irvingtonian or later). Most of the mammals are large, and grazing forms predominate. The presence of an advanced *Mammuthus* is significant; also present are *Eucheratherium*, *Tetrameryx*, camelids, and a large *Smilodon*. No aquatic or semiaquatic mammals have been found. (Lindsay, Johnson, and Opdyke, 1975; Savage, 1951; Stirton, 1939.) UCMP.

Vallecito Creek (50 spp.), San Diego Co. **(2–51)**

In the extensive badlands topography of the Anza Borrego Desert State Park, 80 km NE of San Diego and 56 km SW of the Salton Sea, 2,440 m of an uninterrupted sequence of fossiliferous beds have yielded early Blancan (Layer Cake and Arroyo Seco, which see) and late Blancan/early Irvingtonian (Vallecito Creek) faunas. The beds were discovered by Harley J. Garbani in 1954. Paleomagnetic studies of the Vallecito Creek section (the upper 1,220 m) show that most of it occurs within the Blancan Land Mammal Age but that it also includes the early part of the Irvingtonian Land Mammal Age and is within the Matuyama magnetic epoch. The duration of the Vallecito Creek section is about 0.8 or 1.5 m.y. (depending on the interpretation of the normal magnetic polarity event in the middle of the section). The rich fauna documents the gradual replacement of typical Blancan mammals by Irvingtonian genera (*Dipodomys* replaces *Prodipodomys*, *Sylvilagus* and *Lepus* replace *Hypolagus*) and the appearance of new forms [*Odocoileus* and *Nothrotheriops* at 2.1 m.y. B.P.; *Smilodon*, *Eucheratherium* (earliest known bovid in North America), and *Dipodomys* at 1.9 or 1.6 m.y. B.P.]. The appearance of *Lepus* is associated with the disappearance of *Stegomastodon*, as in the San Pedro Valley, Arizona. *Mammuthus*, an indicator fossil for the Irvingtonian, has not been found at Vallecito Creek. Faunal associates indicate a grasslands habitat with wooded areas at least near streams; the numerous remains of *Geomys* suggest sandy, friable soil. Precipitation was probably moderate. Vallecito Creek is correlated with Blanco and Curtis Ranch faunas. This important fauna is under study. (Downs and White, 1968; Opdyke *et al.*, 1977; White, personal communication, 1972; White and Downs, 1961.) LACM.

FLORIDA

Coleman IIA (38 spp.), Sumter Co. **(2–83)**

The rich and well-studied Coleman IIA fauna comes from a sinkhole fill in the Eocene limestones uplifted by the Ocala Arch, about 25 m above sea level in central Florida. A study of the breakage patterns of the large, long bones showed greenstick fractures, which suggests that at least the larger mammals fell into the sink. It was probably also a roosting place for bats and owls, and the latter may have deposited the remains of small mammals, which are very numerous. Rapid deposition within a short period of time is indicated. The fauna suggests a somewhat more open landscape and drier climate than today, although savanna species are not as common as at Inglis. The fauna appears to be late Irvingtonian, and a late Kansan to early Illinoian date is suggested. (R. A. Martin, 1974a.) FSM.

Inglis IA (26 spp.), Citrus Co. **(2–84)**

This site is a fossiliferous fissure filling close to present sea level, and the environment presumably was a coastal savanna. Presence of Blancan holdovers (*Chasmaporthetes*, *Platygonus bicalcaratus*), absence of *Mammuthus*, and presence of *Ondatra*, *Sylvilagus*, and *Lepus* suggest an early Irvingtonian date. Webb (1974a) has suggested that deposition occurred during transition from a glacial to an interglacial interval (Nebraskan–Aftonian?). The fauna, listed by Webb (1974a), is thought to be somewhat earlier than that from Punta Gorda. (Klein, 1971; Webb, 1974a.) FSM.

Pool Branch (10 spp.), Polk Co. **(2-82)**

This fauna, composed mainly of large mammals, has not yet been studied in detail. It has been assigned to the late Irvingtonian by Webb (1974), who has also listed the species. (Webb, 1974a.) FSM.

Punta Gorda (8 spp.), Charlotte Co. **(2-81)**

This is a coastal marsh deposit with a clayey bone bed 1 m thick, overlying marine shell marls of the Caloosahatchee Formation, at an elevation less than 2 m above sea level. The fauna, listed in Webb (1974a), includes *Equus scotti,* an early *Mammuthus* of the *M. meridionalis* type, and the pampathere *Kraglievichia paranense,* which is a Blancan survivor. As usual in Florida, the deposit seems to represent backfill from an early phase of an interglacial, and we provisionally regard it as transitional Nebraskan/Aftonian. Aquatic and forest-dwelling mammals predominate, but the mammoth may have inhabited the grassy uplands. (Webb, 1974a.) FSM.

KANSAS

Adams (10 spp.), Meade Co. **(2-66)**

A collapse basin about 1.6 km in diameter north of the Cimarron River, known as the Butler Spring basin, has produced four superposed local faunas, of which the Adams is the oldest. The fauna is found in coarse, yellow crossbedded sands and gravels laid down by the ancestral Cimarron River, a much larger stream than that existing at present. Mollusks, amphibians, reptiles, and mammals were found, with large forms predominating among the latter. Several mollusks, as well as the vole *Microtus pennsylvanicus,* are northern and indicate cooler summers. The pollen flora is dominated by *Pinus, Artemisia* and other Compositae, and Gramineae. The fauna has been considered to be Illinoian, probably late Irvingtonian in age. Recent paleomagnetic studies have shown that the sediments are normally magnetized, presumably representing an early phase of the Brunhes magnetic epoch, as is the case with Cudahy. Lindsay, Johnson, and Opdyke (1975), however, have placed the Adams fauna in the Rancholabrean Land Mammal Age. (Hibbard, 1970b; Hibbard and Taylor, 1960) Kapp, 1965; Lindsay, Johnson, and Opdyke, 1975; G. E. Schultz, 1967.) UMMP.

Cudahy (37 spp.), Meade Co. **(2-65)**

The Cudahy fauna, composed mainly of small mammals, occurs in silts immediately underlying the Cudahy ash bed, a Pearlette type "O" ash dated 0.6 m.y. B.P. The fossiliferous beds, as well as the ash, are normally magnetized, and the fauna is thus in the Brunhes magnetozone (maximum age, 0.7 m.y.) close to the Brunhes–Matuyama boundary. A very late Kansan or very early Illinoian age is borne out by the fauna, which has numerous northern forms (*Spermophilus richardsonii,* a form related to *Synaptomys borealis,* the *operarius* group of *Microtus,* and several soricids). The molluscan fauna also indicates a cool climate. The fossils accumulated in basins containing water or in slow-moving streams. Abundant plant fossils and the presence of iron concretions point to marshy conditions. The good stratigraphic control and the richness of the fauna enhance its importance in the Pleistocene history of North America. A represen-

3.1 Scene at Cudahy, Kansas, in the Irvingtonian. Shown are *Mammuthus meridionalis* and *Tetrameryx knoxensis*.

tation of the fauna appears in figure 3.1. (Hibbard, 1944, 1970b; Leonard, 1950; Lindsay, Johnson, and Opdyke, 1975; Paulson, 1961.) UMMP.

Kanopolis (34 spp.), Ellsworth Co. (2–70)

An abandoned sand and gravel pit in the town of Kanopolis, elevation 479 m, has yielded a large Yarmouthian interglacial fauna rich in both invertebrates and vertebrates. The pit is situated at the confluence of a small tributary with the main channel of the Smoky Hill River. The fish fauna (Neff, 1975) is the largest (15 spp.) known from the Great Plains. The Kanopolis mammals (Hibbard *et al.*, 1978) include both small and large forms, 18 extinct species, the northernmost record of the pampathere, *Holmesina septentrionalis*, and the earliest known North American occurrences of *Perognathus hispidus*, *Reithrodontomys humulis*, *Neotoma floridana*, and *Microtus pennsylvanicus*. The climate was probably more equable than now, with warmer winters and more precipitation. (Hibbard *et al.*, 1978; Neff, 1975; Zakrzewski, personal communication, 1976). UMMP.

Kentuck (20 spp.), McPherson Co. (2–69)

A small channel deposit that is lithologically distinct from the nearby McPherson Formation has yielded a mixed assemblage of both primitive (*Gigantocamelus*, *Ondatra annectens*) and advanced (*Ondatra nebracensis*) species and warm (*Sigmodon hispidus*) and cold-adapted (*Synaptomys*) species. The deposit filled an older cut through a thick Pearlette Ash bed and underlying deposits, which resulted in the mixing of the fossils. The locality is about 11 km E of the Sandahl local fauna in northwestern McPherson County. The ages of the Kentuck assemblage are part Kansan and part Yarmouthian interglacial or interstadial. Grassland and stream-border communities are indicated by the presence of prairie dogs and muskrats, respectively. (Semken, 1966.) UMMP.

Rezabek (9 spp.), Lincoln Co. (2–71)

The Rezabek fauna, consisting of mollusks, fish, amphibians, reptiles, a few birds, and nine species of mammals, was taken from a gravel pit in Lincoln County in central Kansas. The fossils occur in compact fine gray silts containing fine to coarse sand above the gravel. The deposits are believed to be Yarmouthian or Illinoian in age. The presence of *Castoroides* and *Neofiber* indicate a swamp and marsh habitat. (Hibbard, 1943a.) KUVP.

Sandahl (31 spp.), McPherson Co. (2–67)

Pleistocene rocks in northwestern McPherson County were formerly designated as the McPherson *Equus* beds but are now divided into the Lower Pleistocene Meade Group and the Upper Pleistocene Sanborn Group (which includes the McPherson and Loveland formations). Most of the fossil vertebrates came from the McPherson Formation, and the term "McPherson *Equus* beds" should be restricted to deposits in this formation. The Sandahl local fauna includes a number of local faunules collected from the bank of a road-cut in the McPherson Formation at an elevation of 433 m. Mollusks, fish, frogs, salamanders, snakes, turtles, and birds, as well as a large mammal fauna, have been identified. Several communities, including permanent stream, stream border, gallery forest, lowland meadow-savanna, and upland prairie, are represented. On stratigraphic, paleoecological, and evolutionary evidence, the age of the fauna is Illinoian. The climate was cooler and wetter than now, and the present-day area of sympatry of the 11 extant mammal species lies in the High Plains of southeastern Wyoming and north-central Colorado. (Semken, 1966.) UMMP.

Maryland

Cumberland Cave (49 spp.), Allegany Co. (2–86)

Cumberland Cave is undoubtedly the best-documented Irvingtonian fauna in eastern North America. It was exposed in a railroad cut at the base of an escarpment of Devonian limestone, 6.4 km NW of Cumberland in western Maryland. Excavation by parties from the U.S. National Museum began in 1912 and continued until 1915, but, unfortunately, much material was destroyed by steam shovels and dynamite before the significance of the find was known. The cave sediments are composed of unstratified clays and breccias more or less cemented together by stalactic material. Many of the bones are broken, but none is waterworn. Originally, the cave entrance was at the crest of the ridge, and the opening probably acted as a trap with about a 30 m drop. Small mammals abound, but remains of animals larger than a black bear are rare. A representation of the fauna appears in figure 3.2.

The fauna has a northern component (*Gulo, Synaptomys, Phenacomys, Clethrionomys*) and a western component (*Taxidea, Canis priscolatrans, Spermophilus tridecemlineatus*), as well as typical Appalachian elements (*Marmota, Glaucomys, Tamias*). Rodents are well

3.2 Scene near Cumberland Cave, Maryland, in the Irvingtonian. Shown is *Martes diluviana* stalking *Coendou cumberlandicus,* with *Platygonus vetus* looking on.

represented (21 spp.), as are carnivores (14 spp.). Both *Platygonus* and *Mylohyus* are present. About 40 percent of the fauna is extinct. The identification of tropical crocodile remains was erroneous; the material turned out to be immature bear canines.

The age of the Cumberland Cave fauna is still in dispute; the fauna shows similarities to Port Kennedy (older) and Conard Fissure. Zakrzewski (1975) has noted that if Cumberland Cave was located on the Great Plains, the association of *Ondatra annectens, Atopomys,* and *Neofiber* and the absence of *Microtus pennsylvanicus* would suggest a pre-Illinoian age. From his study of *Microtus* and *Pitymys* from Cumberland Cave, Van der Meulen (1978) has suggested a Middle Irvingtonian age for the fauna. Study of the cave and its fauna, by the Carnegie Museum, is continuing. (Gidley, 1913a; Gidley and Gazin, 1933, 1938, Guilday, 1971b; Van der Meulen, 1978; Zakrzewski, 1975.) CM, USNM.

Nebraska

Angus (43 spp.), Nuckolls Co., **(2–73)**

This rich, still largely unstudied fauna from south-central Nebraska is found below the Loveland Loess (Illinoian) and Sangamonian Paleosol and thus may be referred to the late Yarmouthian or early Illinoian. Descendants of most of the smaller mammals are still found in about the same area, which suggests that the climate was no more severe than it is now. There is no *Bison*. For a preliminary faunal list, see Schultz and Martin (1970b). (L. Martin, personal communication, 1971.) UNSM.

Hay Springs (19 spp.), Rushville (28 spp.), and Gordon (18 spp.), Sheridan Co. **(2–74, 2–75, 2–76)**

These three local faunas come from six localities south of the towns of Hay Springs, Rushville, and Gordon in northwestern Nebraska. They occur mainly in the Grand Island Formation, although some material comes from the overlying Upland Formation. These deposits are found in an ancestral Niobrara River Valley, cut into the Pliocene Ogallala Formation. The gravels of the Grand Island Formation and the marls and peats of the Upland Formation make up the fourth terrace fill in this area; it is mantled by the Loveland Loess. Earlier reports of a Pearlette-type ash in the area have been discounted by Schultz and Martin (1970b). The Hay Springs fauna, collected in the 1890s, has often been regarded as a mixed fauna since material from several quarries was lumped together. *Microtus pennsylvanicus,* a post-Illinoian species, was identified in the fauna but is now regarded as intrusive. Collections made after the 1920s have had better stratigraphic control, and the Terrace 4 fill containing these three faunas ranges in age from late Kansan to early Illinoian. Important elements include primitive (*Ondatra annectens*) as well as advanced muskrats (*O. nebracensis*), typically Irvingtonian species (*Canis armbrusteri, Equus calobatus, Platygonus vetus*), and advanced species (*Smilodon fatalis*). *Bison* is not present. (Hibbard, 1958b; Matthew, 1902; Nelson and Semken, 1970; Schultz and Martin, 1970b; Schultz and Stout, 1948, 1961; Schultz and Tanner, 1957; Semken, 1966.) AMNH, UNSM.

Mullen I (10 spp.) and Mullen II (35 spp.), Hooker and Cherry Cos. **(2–77, 2–78)**

The Mullen assemblage, collected from several quarries in the Terrace 4 fill along the Loup River north of Mullen, represents at least two faunas: a late Irvingtonian fauna (Mullen II) and a much earlier, reworked fauna (Mullen I). The early fauna includes Blancan survivors such as *Procastoroides, Pliopotamys,* and *Pliophenacomys,* as well as diagnostic early Irvingtonian forms such as *Ondatra nebracensis* and *Lepus.* An early Pleistocene, perhaps early Kansan, age is suggested. The much larger late Irvingtonian (early Illinoian) fauna, Mullen II, was described originally by Jakway (1962), in an unpublished thesis. The fauna resembles that of Angus. In addition, a later component (*Alces, Bison, Symbos*) seems to be present. New excavations and a restudy of the material are underway. (Jakway, 1962; L. D. Martin, 1972; C. B. Schultz, 1934; Schultz and Martin, 1970b, 1972; Schultz and Tanner, 1957.) UNSM.

Sappa (13 spp.), Harlan Co. **(2–72)**

At the type locality of the Sappa Formation in south-central Nebraksa, a fauna which is still under study has been collected immediately below type "S" Pearlette ash (1.2 m.y. B.P.); fossils have also been found above the ash bed. The fauna includes fish, amphibians, reptiles, birds, and mostly small mammals and is reportedly similar to the Cudahy fauna of Kansas but should be considerably older. (Schultz and Martin, 1970b.) UNSM.

OKLAHOMA

Berends (16 spp.), Beaver Co. **(2-64)**

This fauna from the Oklahoma Panhandle consists of mollusks and vertebrates that come from deposits overlying a Pearlette-type ash. Cool, moist conditions are indicated, and the area of sympatry of the extant mammals lies in North Dakota. Two habitats, a marsh and an upland plain, are represented. The evolutionary level of the cricetids like *Peromyscus* and *Ondatra* is consistent with an Illinoian age, and the fauna may be regarded as terminal Irvingtonian (early Illinoian). (Hibbard, 1970b; Starrett, 1956; Taylor, 1954.) UMMP.

Holloman (18 spp.), Tillman Co. **(2-63)**

The Holloman gravel terrace, about 50 m above the North Fork of the Red River, near Frederick in southwestern Oklahoma, has produced a large-mammal fauna. An early Irvingtonian age is suggested by the presence of a few Blancan holdovers such as *Titanotylopus* and *Stegomastodon*, coexisting with the earliest North American mammoth, *Mammuthus meridionalis*. A mild climate and partially wooded environment are suggested by the presence of the giant tortoise *Geochelone* and tapir, deer, stegomastodont, and several grazing forms.

A low terrace on a level with the present floodplain has produced *Bison*. Human artifacts reported from the high terrace are intrusive. Until recently, most of the Holloman fossils have been in a private collection; Dalquest (1977) restudied the fauna and reduced the number of species from 25 to 18. He considers the Holloman local fauna to be equivalent in age to, or slightly younger than, the Gilliland fauna. (Dalquest, 1977; Hay and Cook, 1930; Hibbard, 1970b; Hibbard and Dalquest, 1966; Meade, 1953; Sellards, 1932.) MWU.

PENNSYLVANIA

Port Kennedy Cave (43 spp.), Montgomery Co. **(2-87)**

This classical site, studied by Cope in the late 1800s, has produced a large Irvingtonian fauna. It was discovered by workmen quarrying limestone at Port Kennedy in southeastern Pennsylvania. The fissure no longer exists. It has been variously dated as Kansan, Yarmouthian, and Illinoian. Truly boreal elements (e.g., *Gulo*) are apparently too rare to warrant definite assignment to a glacial interval. Small-mammal remains are mostly so poorly preserved that the various species and genera erected by Cope are of uncertain status, and the affinities of some of them will probably never be known (Guilday, personal communication 1976). Thus judgment of the probable age of the fauna must be based mainly on the larger and better preserved taxa. The fauna resembles the faunas of Conard Fissure and Cumberland Cave, both of which are probably Kansan in age, yet has a decidedly more archaic stamp, as evinced by the primitive *Smilodon gracilis* and small form of the black bear, *Ursus americanus* (which resemble their Blancan predecessors, *Meganteron hesperus* and *Ursus abstrusus*, respectively). *Megalonyx wheatleyi*, described from Port Kennedy, is evolutionarily intermediate between the Blancan and Rancholabrean species. We tentatively refer the deposit to the Aftonian or early Kansan. The identification of *Bison* sp. (Hay, 1923) in the fauna is doubtful. (Cope, 1871, 1899; Gidley and Gazin, 1938; Guilday, 1971b and personal communication, 1976;

Hay, 1923; Hibbard, 1955a; Hibbard and Dalquest, 1966; Kurtén, 1963b; McDonald, 1977.) ANSP.

SASKATCHEWAN

Wellsch Valley (15 spp.) **(2–57)**

Located about 40 km N of Swift Current and 220 km E of Medicine Hat, the Wellsch Valley site contains a surficial bone-bearing deposit and at least four overlying unfossiliferous tills. Unit C, from which an Irvingtonian fauna has been recovered, is made up of alluvium, pond deposits, and slope wash that was probably laid down by streams or during flash floods. The predominantly plains fauna, which is considered to predate the oldest (Kansan?) fauna at Medicine Hat, includes the Blancan survivor *Borophagus* and the early mammoth *Mammuthus meridionalis*. The fauna shows similarities to that of Cape Deceit, Alaska, but also includes southern elements (ground sloth, peccary). An early Irvingtonian (Aftonian?) age is indicated. Recent paleomagnetic evidence (Harington and Shackleton, 1978) indicates that the fossiliferous unit was deposited during the time of the Olduvai event, i.e., about 1.7 m.y. ago. (Harington and Shackleton, 1978; Stalker and Churcher, 1972.) ROM.

SOUTH DAKOTA

Java (29 spp.), Walworth Co. **(2–79)**

A paleostream channel cut into the Cretaceous Pierre Shale in the Missouri Coteau, 35 km E of the Missouri Trench in eastern Walworth County, contains a sandy stream deposit which has produced a rich fauna, with mollusks and vertebrates. Only preliminary reports are available for most of the taxa. The deposit is overlain unconformably by late Wisconsinan till. The elevation is 610 m. Aquatic (beaver, muskrat), grassland (pocket gopher, vole), and woodland or shrubland (jumping mouse, pack rat) mammals are present. A climate cooler than that at present, at least in summer, is indicated by the presence of the boreal lemming, *Synaptomys*, and the spruce vole, *Phenacomys*. Presence of the extinct species *Zapus sandersi* and *Ondatra annectens* suggests a Kansan date. An early date is also suggested by the presence of the primitive *Microtus (Allophaiomys)*. (R. A. Martin, 1973a,b.) SDSM.

TEXAS

Gilliland (26 spp.), Baylor and Knox Cos. **(2–61)**

The Seymour Formation, consisting of fluviatile sands and gravels grading upward into silts and clays capped by caliche, has produced two superposed faunas, the Gilliland and Vera. The site is in north-central Texas at an elevation of 427 m. The Gilliland fauna closely resembles those of Holloman and Rock Creek and, similarly, includes *Mammuthus meridionalis* and *Geochelone*. Presence of the latter indicates a frost-free climate, probably maritime and humid. Thus the fauna seems to be interglacial, probably early Irvingtonian; Dalquest (1977) believes it is slightly older than Holloman and Vera. Several communities of mammals are represented: stream and semiaquatic, gallery forest along the stream, savanna valley, and upland. (Dalquest, 1977; Hibbard and Dalquest, 1962, 1966.) MWU, UMMP.

Rock Creek (14 spp.), Briscoe Co. (2–60)

In the so-called *Equus,* or Sheridan, beds, a large sample of *Equus scotti* was found at the head of Rock Creek. The main fauna, however, comes from another quarry. Only large mammals have been found, among which the rare muskox *Soergelia* may be noted. Presence of *Mammuthus meridionalis* and *Geochelone* suggests correlation with Gilliland and Holloman and an early Irvingtonian age. (Gidley, 1900; Hibbard and Dalquest, 1966; Troxell, 1915a,b.) AMNH, YPM.

Slaton (30 spp.), Lubbock Co. (2–59)

This fauna comes from a lake bed accumulated in a deep depression in the Pliocene Bridwell Formation and later dissected by Yellowhouse Creek. The exposure represents the northern margin of the lake, where fossiliferous clays accumulated in shallow, vegetation-choked water. The filling of the bed took place in more arid conditions, as indicated by a sandy facies. The water rat (*Neofiber*) is common; the presence of large alligators speaks for a good-sized lake. Although deer and armadillo remains indicate that there were trees and thickets in the vicinity, no large browsers have been found, and the abundance of grazing mammals suggests open country. The climate was mild, with frost-free winters. The absence of *Bison* in this grassland fauna is significant and indicates a pre-Rancholabrean age. Although Hibbard (1970b) concluded, based on the molluscan fauna, that the date was Sangamonian, a late Yarmouthian to earliest Illinoian age appears most probable on the basis of the mammals. (Dalquest, 1967; Hibbard, 1970b.) MWU.

Vera (13 spp.), Baylor and Knox Cos. (2–62)

Immediately below the Pearlette ash bed in the upper part of the Seymour Formation, a large molluscan fauna and vertebrates including fish, amphibians, reptiles, birds, and mammals have been found. The fauna is similar to that of Cudahy. The climate was moister than at present, with cooler summers. Dalquest (1977) dates it ca. 600,000 years B.P. (Dalquest, 1977; Hibbard and Dalquest, 1966.) MWU, UMMP.

WASHINGTON

Delight (Washtuckna Lake) (13 spp.), Adams Co. (2–54)

Fossils have been known from the sandhills of south-central Washington since the late 1880s, when Cope described three species of cervids from Whitman County. Matthew (1902) called the locality Washtuckna Lake (now designated Lake Kahlotus). The site, now called Delight, is about 14 km N of the lake and 4 km S of the town of Delight. All of the specimens are float and are fragmentary and weathered. The fauna consists of medium-sized and large mammals, of which nine species are extinct. The age is believed to be Irvingtonian or later. (Fry and Gustafson, 1974; Matthew, 1902.) AMNH.

WEST VIRGINIA

Trout Cave (41 spp.), Pendleton Co. (2–85)

This mountain cave is situated in the Ridge and Valley Province of the central Appalachians, 4.8 km SW of Franklin. The cave, a solution feature of the Lower Devonian Helderberg limestone, is formed along the 55-m terrace of the South Fork of the

Potomac River. The lower levels of the 3.6-m stratified sequence are contemporaneous with Cumberland Cave (i.e., are late Kansan), and the upper levels appear to be Wisconsinan in age. Separating them is an ill-defined flowstone level about 1.8 m below the surface. The vertebrate fauna consists mainly of isolated teeth and fragmentary skeletal remains presumably deposited by raptorial birds. Eight of the species are extinct, including *Atopomys salvelinus*, which was described from this locality. The fauna is under study. (Guilday, personal communication, 1976; Zakrzewski, 1975.) CM.

CHAPTER FOUR

Rancholabrean Faunas

The Irvingtonian–Rancholabrean transition, as defined by Savage (1951), is marked by the appearance of *Bison*, an invader from Eurasia. Rancholabrean faunas are also characterized by the presence of many recently extinct species of large mammals and many extant species, especially among the carnivores and rodents (Savage, 1951), although it should be noted that many modern species are now known from late Irvingtonian local faunas.

Presence of *Bison* in faunas here regarded as Irvingtonian has been asserted (Schultz and Frankforter, 1946; Schultz and Stout, 1948), but the absence of *Bison* in the rich faunas of the Hay Springs area in Nebraska is doubtless significant (Savage, 1951). Schultz and Hillerud (1977) have concluded that *Bison* appeared in the Great Plains in "post-Kansan to early Illinoian times."

In the compilation of stratigraphic ranges of mammalian genera by Hibbard et al. (1965), 48 genera were recorded as new in the Rancholabrean. Of these, however, no less than 19 have since been identified in the Irvingtonian or even the Blancan. The remaining genera include many chiropterans, for which chances of preservation are so uneven that their absence carries little weight, and several other genera that must be placed in the same category because of local distribution, small size, or preferred habitat. The stratigraphically most significant "new" genera are probably the Palearctic immigrants (*Rangifer, Oreamnos, Ovis, Homo,* and the mooses and muskoxen) and a few genera with Neotropical affinities.

Conversely, of the relatively few (10) genera listed in Hibbard et al. (1965) as becoming extinct during the Irvingtonian or at its end, at least four survived into Rancholabrean times (they are now regarded as synonyms of Rancholabrean or Recent genera). Altogether, the distinction between the Irvingtonian and Rancholabrean is less sharp than was originally assumed, as in the case of the Blancan–Irvingtonian transition and as occurs in general when more detailed information becomes available.

There is at present no good basis for an absolute chronology of the Rancholabrean, apart from its final phase (which lies within the range of C-14 dating). The entire age is within the Brunhes magnetic epoch. For the main part of the Rancholabrean, we thus have to rely on interrelationships of local faunas and attempts to fit the succession of faunas into a glacial–interglacial framework tentatively correlated with that of the Palearctic region.

Although correlations have not been very successful (Cooke, 1973), there can be little doubt that the Eemian interglacial of Europe has its correlative in the American Sangamonian, which presumably had its optimum at about 100,000 B.P., and that the Wisconsinan glaciation *sensu stricto* commenced about 70,000 B.P. (Flint, 1971; Hays, Imbrie, and Shackleton, 1976; Woldstedt, 1969). On the other hand, faunal evidence, as well as the somewhat meager radiometric dates at hand, suggests a correlation of the Illinoian with the European Elster–Mindel glaciation (Cooke, 1973). Thus the Rancholabrean would accommodate two full glacial–interglacial cycles preceding the Wisconsinan and corresponding to the European Elsterian–Holsteinian–Saalian–Eemian sequence. In addition, each cycle is likely to be subdivided, forming a complex with various stadials and interstadials. Indeed, there is mammalian evidence of such com-

Map 3. Late Illinoian and Sangamonian Faunas

plexity (Schultz and Martin, 1970). Graham and Semken (1976) have suggested that the climate in the Great Plains may have been more equable during certain glacial phases than in typical interglacials, as a result of which there may be local faunas with apparently incompatible elements. Such faunas are met with well back in the Irvingtonian (Kentuck, Cumberland Cave) and may also be found in the Rancholabrean (Jinglebob).

In the Butler Spring area in Kansas, a series of four superposed local faunas occurs (G. E. Schultz, 1967). The oldest (Adams) was regarded as early Illinoian, but recent paleomagnetic studies place it in the early Rancholabrean (Lindsay, Johnson, and Opdyke, 1975); the others (Butler Spring, Cragin Quarry, Robert) have been assigned to the Rancholabrean. Of these, the Cragin Quarry fauna has a "warm" aspect (the giant tortoise *Geochelone* is present) and is presumably interglacial. The Robert fauna has been radiometrically dated as late Wisconsinan. Unfortunately, large mammals are scarce in most of the faunas, and there is no record of *Bison*.

Schultz and Hillerud (1977) have shown that, at least in the Great Plains, giant bison (*Bison latifrons*) occurs only in early Rancholabrean faunas, with smaller forms of *Bison* appearing later on. Although the giant bison may survive in certain areas alongside its smaller relative, presence of different species of *Bison* makes it possible to distinguish early and late Rancholabrean local faunas.

The longevity of many mammalian species is much greater than the relatively short span of the Rancholabrean, and this makes faunal correlation difficult. However, studies of microevolutionary trends on the intraspecific level (e.g., for *Ursus americanus, Ondatra zibethicus,* and *Bison latifrons*) may be useful. Origination, extinction, and replacement of species in certain rapidly evolving genera like *Peromyscus* (several species became extinct in the early Rancholabrean, only one in the late) and *Microtus* may also give important evidence. Immigrants at various points in time, if widespread like *Bison* and *Panthera leo* (absent in earliest Rancholabrean faunas), may give datum lines. One such datum line may be provided by late Wisconsinan immigrants from Beringia (*Ursus arctos, Alces alces*).

In addition to the Pleistocene sites (see maps 3 and 4), we have reviewed a number of early Holocene localities (younger than 10,000 years B.P.; see map 5) that have yielded information on the appearance or extinction of species, former geographic ranges, or other topics. (Cooke, 1973; Flint, 1971; Graham and Semken, 1976; Hays, Imbrie, and Shackleton, 1976; Hibbard et al., 1965; Savage, 1951; Schultz and Frankforter, 1946; Schultz and Hillerud, 1977; Schultz and Martin, 1970b; Schultz and Stout, 1948; G. E. Schultz, 1967; Woldstedt, 1969.)

Map 3. Late Illinoian and Sangamonian Faunas (Rancholabrean)

Late Illinoian
100. Fairbanks I
101. Doby Springs
102. Butler Spring
103. Mt. Scott
104. Duck Creek
105. Santa Fe River IIA
106. Ashley River (part)

Sangamonian
110. Old Crow River Loc. 44
111. Newport Bay Mesa Loc. 1066

112. San Pedro
113. Silver Creek
114. American Falls
115. Medicine Hat Unit K
116. Doeden
117. Mesa de Maya
118. Cedazo
119. Tequixquiac
120. Ingleside
121. Easley Ranch
122. Coppell
123. Cragin Quarry
124. Jinglebob

125. Saskatoon area
126. Fort Qu'Appelle
127. Bradenton
128. Rock Springs
129. Withlacoochee River
130. Waccasassa River
131. Williston IIIA
132. Reddick
133. Payne's Prairie
134. Haile VIIA
135. Haile VIIIA
136. Haile XIB
137. Toronto

Aguascalientes, Mexico

Cedazo (41 spp.) (3–118)

The Cedazo local fauna occurs in the Tacubaya Formation, and has been collected a few kilometers southeast of the city of Aguascalientes, along exposures formed by two intermittent tributaries of the Rio Aguascalientes. The formation, highly varied lithologically and with an important element of volcanic ejecta, consists of waterlaid deposits. It rests unconformably upon mid-Tertiary sandstones and reaches a thickness of 8–10 m; most of the fossils were found near the base of the formation.

The rich fauna consists predominantly of large mammals, with grazing forms highly abundant and varied (equids, camelids, antilocaprids). Few browsers and no aquatic mammals have been found, and the absence of the latter may in part be due to the paucity of small-mammal remains in general. The paleohabitat is thought to have been grassland or prairie, with patches of brush and trees along the watercourses. The manner of preservation and occurrence suggests a unit fauna. The stratigraphic relationships of the formation, as well as the presence of only giant bison, suggest a pre-Wisconsinan age, and presence of *Panthera leo atrox*, unknown in deposits older than Sangamonian, indicates that the fauna is post-Illinoian. (Mooser and Dalquest, 1975b.) FC, IGM.

Alaska

Eschscholtz Bay (11 spp.) (4–150)

As early as 1816, fossils were found on the beach and in the frozen bluffs bordering Eschscholtz Bay, north of the Seward Peninsula. Several expeditions were sent to the area in the early 1900s to look for frozen carcasses of mammoths and other late Pleistocene mammals. Elephant Point and Historic Bluff are two of the important collecting localities. At the latter site, a partial skeleton of a woolly mammoth, with patches of hair, skin, tendons, and flesh still adhering, was recovered by a party from the American Museum of Natural History in 1907. Fossils are still being found there, and much work remains to be done in the area. The age of the fauna is Wisconsinan. (Harington, 1970a; Quackenbush, 1909.) AMNH.

Fairbanks I (22 spp.) and Fairbanks II (40 spp.) (3–100, 4–152)

Two faunas, one of pre-Wisconsinan, probably Illinoian, age (Fairbanks I), the other late Wisconsinan (Fairbanks II), have been recovered in the environs around Fairbanks. The fossil localities are situated on the north side of the broad Tanana River Valley, at the base of the low rounded hills that comprise part of the Yukon–Tanana River upland. Thick beds of loess cover the slopes and ridges, and the upland valleys are filled with gravel overlain by colluvial silt. The Fairbanks area, part of the vast Beringian refugium, was never glaciated, but glaciers extending out from the Alaska Range came to within 80 km of the present-day city of Fairbanks. Today, permafrost still covers much of the region. Fossils, embedded in the frozen muck, were exposed during hydraulic gold-mining operations. Because of the method of collecting, few stratigraphic data are available for the early collections, but most of the fossils apparently came from the upper part of the sections and are Wisconsinan in age. Enor-

4.1 Scene near Fairbanks, Alaska, in the late Rancholabrean. Shown are *Mammuthus primigenius*, *Arctodus simus*, and *Symbos cavifrons*.

mous numbers of fossils were collected at that time; in 1938 alone, 8,008 cataloged specimens were sent to the American Museum of Natural History in New York.

Recent geologic studies (Péwé, 1975) have revealed a rather extensive Illinoian (Fairbanks I) fauna at Cripple Creek Sump, Gold Hill, and Lower Cleary Creek. The extinct muskox *Praeovibos* may serve as an index fossil for Illinoian deposits. These deposits contain early records (perhaps the earliest) of *Bison* and other Eurasian immigrants and thus are of great importance.

The late Pleistocene (Fairbanks II) community was dominated by grazers, and remains of *Bison*, *Equus*, and *Mammuthus* outnumber all other mammals. The composition of the fauna suggests that interior Alaska, as well as other parts of the Beringian refugium, supported a heterogeneous plant community dominated by grasses and grasslike plants. *Microtus miurus*, an inhabitant of well-drained alpine meadows; was the most abundant small mammal. Many species of carnivores have been identified, and Guthrie (1968a) estimated that the average predator–prey ratio was approximately 1 wolf per 130 ungulates and 1 lion per 250 ungulates, numbers close to the predator–prey ratio in eastern Africa today. The presence of *Saiga* and *Bos* (yak) is noteworthy since they are the only Palearctic artiodactyls that reached Alaska but did not spread southward. In addition to bones, partial carcasses of *Spermophilus*, *Lynx*, *Mammuthus*, *Equus*, *Rangifer*, *Alces*, *Symbos*, and *Bison* have been found in organic-rich silt of Wisconsinan age. All the carcasses show some signs of decay, which indicates that they were not frozen instantaneously, and evidence suggests that the deaths represent natural mortality in a harsh environment. Radiocarbon dates ranging from $11,735 \pm 130$ to $40,000+$ years B.P. have been obtained on the mammal remains. Reduction of preferred habitat (a shift from grassland to forest/shrub tundra) and increased predation by man at the end of the last glaciation were probably the main factors in the extinction of many species of the Fairbanks fauna. A representation of the fauna appears in fig-

Map 4. Wisconsinan Faunas

ure 4.1. (Anderson, 1977; Guthrie, 1968a, 1968b, 1972; Harington, 1970a; Péwé, 1975.) AMNH.

Lost Chicken Creek (9 spp.) **(4–153)**

A late Pleistocene large-mammal fauna has been recovered from a placer site below the highway near Chicken, Alaska, 178 km W of Dawson, Yukon Territory. A bison specimen from this site yielded a radiocarbon date of 10,370 ± 160 years B.P., and the

Map 4. Wisconsinan Faunas (Rancholabrean)

150. Eschscholtz Bay
151. Tofty Placer District
152. Fairbanks II
153. Lost Chicken Creek
154. Old Crow River Locs. 14N and 11A
155. Upper Hunker Creek
156. Gold Run Creek
157. Newport Bay Mesa Loc. 1067
158. La Mirada
159. Rancho La Brea
160. Costeau Pit
161. Emery Borrow Pit
162. Zuma Creek
163. Schuiling Cave
164. Manix
165. Channel Islands
166. Carpinteria
167. China Lake
168. McKittrick
169. Maricopa
170. Kokoweef Cave
171. Hawver Cave
172. Potter Creek Cave
173. Samwel Cave
174. Fossil Lake
175. Smith Creek Cave
176. Tule Springs
177. Gypsum Cave
178. Glendale
179. Rampart Cave
180. Ventana Cave
181. Papago Springs Cave
182. Lehner Ranch
183. Murray Springs
184. Naco
185. Wilson Butte
186. Rainbow Beach
187. Dam
188. Wasden
189. Jaguar Cave
190. Medicine Hat Unit M
191. Medicine Hat Unit O
192. Medicine Hat Unit R
193. Empress
194. Cochrane
195. Taber Child Site
196. Hand Hills
197. Warm Springs I
198. Natural Trap Cave
199. Little Box Elder Cave
200. Bell Cave
201. Horned Owl Cave
202. Chimney Rock Animal Trap
203. Isleta Cave
204. Clovis
205. Blackwater Draw
206. Burnet Cave
207. Dry Cave
208. San Josecito Cave
209. Williams Cave
210. Lubbock Lake
211. Quitaque
212. Schulze Cave
213. Kincaid Shelter
214. Friesenhahn Cave
215. Cave Without A Name
216. Laubach Cave
217. Levi Shelter
218. Longhorn Cavern
219. Clamp Cave
220. Howard Ranch
221. Clear Creek
222. Hill-Shuler
223. Moore Pit
224. Taylor Bayou
225. Sims Bayou
226. Ben Franklin
227. Domebo
228. Afton
229. Jones
230. Robert
231. Red Willow
232. Brynjulfson Caves
233. Enon Sink
234. Boney Spring
235. Trollinger Spring
236. Zoo Cave
237. Bat Cave
238. Cherokee Cave
239. Herculaneum
240. Crankshaft Cave
241. Kimmswick
242. Peccary Cave
243. Avery Island
244. Natchez
245. Alton
246. Evansville
247. Carter
248. Big Bone Lick
249. Glass Cave
250. Welsh Cave
251. Mammoth Cave
252. Savage Cave
253. Robinson Cave
254. Saltpeter Cave
255. Big Bone Cave
256. Little Salt River Cave
257. Craighead Caverns
258. Kyle Quarry
259. Cave on Lookout Mountain
260. Whitesburg
261. Cave North of Whitesburg
262. Riverside Cave
263. Guy Wilson Cave
264. Baker Bluff Cave
265. Carrier Quarry
266. Ladds
267. Little Kettle Creek
268. Watkin's Quarry
269. Chipola River
270. Wakulla Spring
271. Aucilla River IA
272. Ichetucknee River
273. Haile XIVA
274. Arredondo
275. Kendrick
276. Sabertooth Cave
277. Seminole Field
278. North Havana Road
279. Joshua Creek
280. Jupiter Inlet
281. Vero
282. Melbourne
283. Sebastian Canal
284. Winding Stairs Cave
285. Meadowview Cave
286. Saltville
287. Gardner Cave
288. Early's Pit
289. Clark's Cave
290. Natural Chimneys
291. Organ-Hedricks
292. Cavetown
293. New Paris No. 4
294. Frankstown Cave
295. Bootlegger Sink
296. Durham Cave
297. Ottawa (Green's Creek area)
298. Roosevelt Lake
299. First American Bank Site

site may be critical in determining the last appearance of several extinct species in eastern Beringia. (Harington, 1978; Whitmore and Foster, 1967.) USNM.

Tofty Placer District (13 spp.) **(4-151)**

A Wisconsinan-early Holocene fauna was recovered from valley-bottom silt deposits at the western end of the Yukon-Tanana upland near Tofty, 150 km W of Fairbanks, at an elevation of 183 m. The rodent fauna is similar to that found at Fairbanks II. Vegetational studies suggest that, during the late Pleistocene, the timberline was at least 397 m below its present level in central Alaska and that the area was recolonized by forest vegetation more than 6,800 years ago. (Repenning, Hopkins, and Rubin, 1964.)

Pleistocene fossils have, in addition, been found on the Arctic Coastal Plain (Péwé, 1975), St. Paul Island in the Pribilofs, and Unalaska and St. Lawrence islands (Ray, 1971). Remains of marine mammals have been identified from several sites, primarily near the west coast (Repenning, personal communication, 1974), and bones of Steller's sea cow were found on Amchitka in the Aleutians (Gard, Lewis, and Whitmore, 1972).

ALBERTA

Bindloss (3 spp.), **(5-352)**

Remains of *Panthera leo atrox, Equus conversidens,* and *Mammuthus* have been recovered from a gravel pit located about 30 m above the Red Deer River at an elevation of 650 m near the town of Bindloss. A post-Wisconsinan age of more than 8,000 years B.P. is suggested for the deposit. (Churcher, 1972; Harington, 1971c.) NMC.

Cochrane (4 spp.) **(4-194)**

Four species of ungulates are known from gravel pits comprising the second terrace above the north bank of the Bow River at Cochrane. Bison bones from these deposits have yielded two radiocarbon dates averaging 11,065 years B.P. The first record of *Equus conversidens* from the Canadian prairies was reported from Cochrane. The mammal remains indicate a cool, grassland habitat in the late Wisconsinan. (Churcher, 1968, 1975.) NMC, ROM.

Empress (8 spp.) **(4-193)**

Gravel pits located at the junction of the Red Deer and South Saskatchewan rivers near Empress have yielded a late Wisconsinan/early Holocene fauna. Preservation of the bones from Empress and Bindloss gravel pits is similar, which suggests that the terraces were laid down at the same time under similar climatic conditions. (Churcher, 1972; Stalker, 1971.)

Hand Hills (7 spp.) **(4-196)**

Fossils have been found in the Hand Hills Conglomerate at the summit of the Hand Hills, which rise 213 m above the prairie. The climate was probably warmer and drier than at present, with friable soil that supported a grasslands fauna including horse and mammoth. The presence of *Cynomys ludovicianus* far to the northwest of its modern range is noted. The suggested age of the fauna is middle to late Pleistocene. (Harington, 1978; Storer, 1975.)

Medicine Hat (Unit K, Sangamonian, 29 spp.; Unit M, early Wisconsinan,
  3 spp.; Unit O, middle Wisconsinan, 17 spp.; Unit R, late Wisconsinan,
  4 spp.; Unit T, early Holocene, 5 spp.) **3–115, 4–190, 4–191, 4–192, 5–353.**

The richest known Pleistocene vertebrate assemblage from the northern Great Plains has been found near the city of Medicine Hat. Here surficial deposits ranging in age from mid-Kansan to Recent have been recovered from some 12 bluffs along the South Saskatchewan River. Major collecting localities are Surprize, Mitchell, Island, and Evil-smelling bluffs. No single bluff shows a complete section, but a composite section more than 300 m thick shows at least four separate glacial advances and several interglacial or interstadial intervals. The area containing bones is extensive, but a cover of overburden between 20 and 60 m thick has restricted collecting to places where the river has exposed the beds.

The oldest Rancholabrean deposit (Unit K) has yielded a typical Sangamonian fauna (*Panthera leo atrox, Bison latifrons*). A representation of the fauna appears in figure 4.2. Fractured cherts have been found in the "artifact band," which consists of strongly cross-bedded, poorly sorted gravel, sand, and silt lenses. Although several physical and chemical processes have been proposed to explain the presence of the chipped stones, none of them adequately does (see Stalker, 1977), and the possibility remains that man was present on the Canadian prairies in the Sangamonian. The climate was probably slightly warmer than it is today, though temperature fluctuations may have been greater, and the landscape was probably a grassland, with trees and shrubs along the river.

Wisconsinan faunas are well represented in Units M, O, and R, with *Mammuthus primigenius, Equus conversidens, Camelops hesternus,* and *Bison* predominating. There is also an early Holocene fauna (Unit T) that includes some late appearances of Pleistocene species.

The diverse fauna of Medicine Hat is under study. Many evolutionary trends are evident (in *Camelops, Mammuthus,* and *Equus*), and many ecological problems remain to be

4.2 Scene near Medicine Hat, Alberta, in the Rancholabrean. Shown are *Camelops hesternus, Hemiauchenia macrocephala,* and *Rangifer tarandus.*

Map 5. Early Holocene Faunas

worked out. The deposit is an important key to the understanding of the last million and a half years on the northern Great Plains. (Churcher, personal communication; Stalker, 1969, 1977; Stalker and Churcher, 1970, 1972; Szabo, Stalker, and Churcher, 1973.) UTZ, ROM, NMC.

Taber Child Site (1 sp.) **(4–195)**

A partial skull and fragmentary skeleton of an 18-month-old child were found at Woodpecker Bluff on the east bank of the Oldman River, 3.2 km NW of Taber. The bones are well preserved, lightly mineralized, and stained orange by limonite-hematite, with black spots of manganese; the preservation is unlike any post-Wisconsinan fossils from the same area but is similar to those of mid-Wisconsinan age at Medicine Hat. Radiocarbon dates from nearby bluffs support the view that the Taber Child dates from well back in the Wisconsinan, probably between 30,000 and 40,000 years ago. The specimen is thus the oldest direct evidence of man on the Canadian prairies before the retreat of the Wisconsinan ice. (Stalker, 1977.) NMC.

ARIZONA

Lehner Ranch (4 spp.), Cochise Co. **(4–182)**

Lehner Ranch is located in the upper San Pedro Valley, 3.2 km S of Hereford at an elevation of 1,278 m. The base formation (dated Pliocene) is a red-brown clay with soft white caliche; it is overlain by a series of beds, all of which have been reduced to remnants by erosion. Much later in time, a stream, now called Mammoth Kill Creek, cut through these beds into the basal clay. The south bank of the old channel was nearly perpendicular and formed part of a meander bow. Here bones of extinct mammals, mainly *Mammuthus,* have been found in direct association with Clovis artifacts. A minimum of nine mammoths have been recovered. Pollen taken from the same stratigraphic unit as the fossils and artifacts reveals that the environment 11,000 years ago was a desert grassland similar to that found in southeastern Arizona today. Climatic differences were not pronounced; the mean annual rainfall may have been 75–100 mm greater (with the present seasonal distribution) and the mean annual temperature 3°–4° lower than that at present. The water table may have been higher owing to continued discharge of groundwater. An average radiocarbon date of 11,260 years B.P. was obtained for the bone bed and Clovis level. (Haury, Saylor, and Wasley, 1959; P. S. Martin, 1967; Mehringer and Haynes, 1965.) UALP.

Murray Springs (14 spp.), Cochise Co. **(4–183)**

Another Clovis site in the San Pedro Valley is Murray Springs, where bone tools and flakes have been found in association with two skeletons of *Mammuthus.* Besides mammoth, remains of camelids, tapir, horse, bison, and some small mammals have been recovered. The stratigraphy of the site is correlated with that of Lehner Ranch, 16 km to

Map 5. Early Holocene Faunas

350. Moonshiner Cave	355. Centipede Cave	360. Polecat Creek
351. Middle Butte Cave	356. Damp Cave	361. Devil's Den
352. Bindloss	357. Klein Cave	362. Hornsby Springs
353. Medicine Hat Unit T	358. Miller's Cave	363. Eagle Cave
354. Casper	359. Meyer Cave	364. Hosterman's Pit

the south. A radiocarbon date of 11,230 years B.P. was obtained on charcoal of *Fraxinus* (ash). (Haynes, 1969.) UALP.

Naco (3 spp.). Cochise Co. **(4–184)**

Naco, a mammoth-kill site, is located 17 km SE of Lehner Ranch. Here mammoth and bison were found in association with Clovis artifacts. The pollen spectrum is similar to that of Lehner Ranch. A radiocarbon date of $8,980 \pm 270$ years B.P., obtained on a mammoth vertebra, is inconsistent with geological and archeological evidence and is regarded as an unacceptable date. (Haynes, 1967; Mehringer, 1967.) UALP.

Papago Springs Cave (30 spp.), Santa Cruz Co. **(4–181)**

Situated at an elevation of about 1,586 m in the Canelo Hills, near Sonoita, Papago Springs Cave was formed by groundwater seeping through the fractured limestone. During the late Pleistocene, the main room of the cave was open to the surface and apparently was used as a shelter by many species of animals. Fossils accumulated to a depth of several feet and were cemented together by mineralized seepage water. The filling period ended when large boulders blocked the entrance. *Bassariscus sonoitensis* and *Stockoceros onusrosagris* were described from Papago Springs, and the latter species is the predominant member of the fauna. (Skinner, 1942.) AMNH.

Rampart Cave (9 spp.), Mojave Co. **(4–179)**

Located in a cliff wall 151 m above the Colorado River, 83 km E of Boulder City, Nevada, Rampart Cave contains the largest and best-preserved known deposit of ground sloth dung. The surface of the 12×15-m chamber is covered with dung to a depth of more than 1.2 m. Although *Nothrotheriops shastensis*, the predominant animal in the fauna, did not inhabit the cave continuously, radiocarbon dates on the dung samples go back more than 38,300 years, with the greatest accumulation occurring between 13,000 and 11,000 years ago. The pollen record shows seasonal as well as long-term changes in the vegetation, and most of the plants are found in the area today. In the summer of 1976, vandals broke into the cave and started a fire, which, despite the efforts of Forest Service personnel, continued to smoulder for almost a year. The fire weakened the cave roof and destroyed much of the unique dung sample. (P. S. Martin, 1975; P. S. Martin, Sabels, and Shutler, 1961.) UALP, USNM.

Ventana Cave (16 spp.), Pima Co. **(4–180)**

A large rock shelter located at the southern tip of Window Mountain (La Ventana) has yielded an extinct fauna associated with Clovis artifacts. The cave is situated on the Papago Indian Reservation, about 141 km W of Tucson and 125 km S of Phoenix, in an area now dominated by paloverde trees and saguaro and other cacti. The overall length of the shelter is about 55 m, with possible living areas ranging in width from 6 to 20 m. An erosive remnant of basalt divides the cave into two areas, the lower cave, containing fill from human activities and the upper cave, where bones and artifacts have been found in the conglomerate (layer E) and volcanic debris (layer D). A spring in the cave continues to provide a reliable source of water.

The presence of *Spermophilus lateralis* is significant. Today, the nearest occurrence is 260 km N of the cave, and the species usually lives at elevations above 2,150 m (1,400 m higher than Ventana Cave). Remains of *Tapirus*, *Equus* (the most abundant mammal), and other large grazers suggest savannalike conditions with permanent water; the climate was probably slightly cooler and wetter than that at present. Charcoal from the volcanic debris level has been dated at 11,300 ± 1,200 years B.P. (Haury, 1950; Mehringer, 1967.)

See Lindsay and Tessman (1974) for other Rancholabrean localities and faunas in Arizona. Many of these are still under study, and only generic determinations are available.

Arkansas

Peccary Cave (51 spp.), Newton Co. **(4–242)**

This cave, in northwestern Arkansas, has yielded a large mammalian fauna that is still under study. During the period of deposition, the environment was probably a grassy parkland with permanent water and a climate similar to that found in southern Minnesota today. Of the 29 identified species of small mammals, 9 show northern or boreal affinities, and 8 species of the medium-sized and large mammals are extinct. The age of the fauna is Wisconsinan. (Davis, 1969; Semken, 1969.) IOWA.

California

Carpinteria (26 spp.), Santa Barbara Co. **(4–166)**

Bones were discovered in 1927 in an asphalt bed located between Ventura and Santa Barbara. Analysis of the plant remains showed a forest environment dominated by pines and containing cypress, live oak, and low shrubs, similar to the plant community found today around Monterey, 330 km N of Carpinteria. The presence of *Sciurus* and *Eutamias* in the fauna also indicates a forest habitat. Five of the mammal species are extinct, and the fauna contains a large number of young animals. (Stock, 1937; R. W. Wilson, 1933b.) LACM.

Channel Islands (4 spp.), Santa Barbara Co. **(4–165)**

Pleistocene fossils have been found on the northern Channel Islands (San Miguel, Santa Rosa, Santa Cruz, and Anacapa), a chain of small islands lying off the California coast and separated from the mainland by the 21- to 32-km-wide Santa Barbara Channel. Workers have postulated that the islands were a seaward extension of the Santa Monica Mountains, but recent studies (Johnson, 1978) show that there is no geologic, topographic, or eustatic evidence to support this view. However, these islands were connected to each other during glacial times, forming a super-island called Santarosae.

Remains of relatively small mammoths were first reported in 1873 and were later described as a distinct species, *Mammuthus exilis* (Stock and Furlong), characterized by a shoulder height of 2.4–2.7 m in adults. The taxon is now considered to be a subspecies of the Wisconsinan mammoth. Remains of this species are most numerous on Santa Rosa but have also been found on San Miguel and Santa Cruz islands. Their reduced stature is attributed to adverse environmental conditions on the islands, such as limited food supply, and to inbreeding. Johnson (1978), noting that living elephants are excellent distance swimmers, believes that the mammoths reached the islands by swimming from the mainland. Because of the glacially lowered sea level, the distance would have been less than it is today. In addition to mammoths, two extinct species of *Peromyscus* (*P. nesodytes* and *P. anyapahensis*) were described from the islands, and recently remains of *Mammut americanum* were recognized on Santa Rosa (Madden, personnal communication, 1976). Carbon-14 dates of 15,800 ± 280 and 29,650 ± 2500 years B.P. have been obtained on cypress wood and charred mammoth bone, respectively. (Johnson, 1978; Madden, personal communication, 1976; Orr, 1956; Stock, 1935; Stock and Furlong, 1928.) LACM.

China Lake (11 spp.), Kern Co. (4–167)

The China Lake fauna comes from lacustrine deposits scattered over a wide area on the grounds of the U.S. Naval Ordinance Test Station near the town of Ridgecrest. During the late Wisconsinan, China Lake (now dry) fluctuated in size, and the climate was wetter and more humid than at present. Pre-Clovis and Clovis artifacts have been recovered, along with fragmentary (broken up by the growth of evaporite mineral crystals), sandblasted remains of 11 species of birds, including 9 water birds, and 11 species of predominantly plains-living mammals. A radiocarbon date of 18,600 ± 4,500 years B.P. was obtained on *in situ* fossil-mammoth ivory. (Fortsch, 1972 and personal communication, 1976; Jefferson, personal communication, 1976.) LACM.

Costeau Pit (21 spp.), Los Angeles Co. (4–160)

This site is located in the San Joaquin Hills near the town of El Toro, at an elevation of 90 m. The fauna, which includes both small and large forms, suggests a grassland environment with a semiarid climate similar to that found in the area today. The age of the fauna (based on wood samples) is greater than 40,000 years, but is Rancholabrean since *Bison* is present. Twelve species are extinct. (W. E. Miller, 1971.) LACM.

Emery Borrow Pit (21 spp.), Los Angeles Co. (4–161)

Rancholabrean vertebrates, invertebrates, and plants have been found in the green clay beds of the La Habra Formation. The main quarry is a stream channel on the edge of a large lacustrine deposit. The presence of *Palaeolama* is noteworthy. This extensive fauna is under study. (Langenwalter, personal communication, 1976.) LACM.

Hawver Cave (26 spp.), Eldorado Co. (4–171)

Located in the foothills of the Sierra Nevada Range 8 km E of Auburn, at an elevation of 393 m, Hawver Cave was formed by the cutting of a limestone lens by the Middle Fork of the American River. The exposed limestone weathered to form fissures and chasms, and pools of water exist in the cave throughout the year. Bones are found in the cave breccia; most are broken. Although Hawver Cave is located 250 km SE of Potter Creek and Samwel caves, the faunas are similar, and *Euceratherium, Megalonyx,* and *Nothrotheriops* are found in all three caves. The Wisconsinan fauna at Hawver is a mixture of forest and plains animals, with plaindwellers predominating. (Stock, 1918.) UCMP.

Kokoweef Cave (28 spp.), San Bernardino Co. (4–170)

This nearly vertical solution cavern is located on the northwest side of Kokoweef Peak, 6.4 km S of Mountain Pass in northeastern San Bernardino County. The cavity was formed by solution in Paleozoic marine limestone and is about 91.5 m deep. Some 15 m of roughly stratified Pleistocene sediments have been exposed by mining activities. The rich late Pleistocene fauna includes gastropods, amphibians, reptiles, birds, and mammals. The mammalian fauna consists of both mountain-dwelling (*Ochotona princeps, Spermophilus lateralis*) and foothills plains species. An alpine forest, with *Pinus aristata, Pinus flexilis, Abies concolor,* and *Juniperus osteosperma,* probably existed close to the cave. Interdisciplinary studies are in progress on this well-preserved, relatively high-elevation late Rancholabrean fauna. (R. E. Reynolds, personal communication, 1976.) LACM.

La Mirada (15 spp.), Los Angeles and Orange Cos. (4-158)

This site was exposed in the banks of Coyote Creek, in the eastern part of La Mirada. Only a small part of the fauna was recovered before the banks of the creek were lined with cement. Carbon-14 dating of wood samples found in association with the fauna indicates an age of 10,690 ± 360 years B.P. The fauna is similar to the one from Rancho La Brea. Four species are extinct. (W. E. Miller, 1971.) LACM.

McKittrick (44 spp.), Kern Co. (4-168)

An oil seep, McKittrick is situated in the foothills of the Coast Range in the southwestern part of the San Joaquin Valley, at an elevation of 300 m. The summers were probably hotter and the winters colder than at Rancho La Brea, and the flora included pine, juniper, and manzanita. The fauna is similar to that of Rancho La Brea but contains more mountain-dwelling species. The dominant camelid was *Hemiauchenia,* whereas at Rancho La Brea, *Camelops* was dominant. Canids outnumber all other members of the fauna. A radiocarbon date of 38,000 ± 2,500 years B.P. has been obtained on plant material. (Berger and Libby, 1966; J. R. Schultz, 1938.) LACM.

Manix (19 spp.), San Bernardino Co. (4-164)

A Wisconsinan fauna has been recovered from lake beds at Manix, in the Mojave Desert about 40 km E of Barstow. Lake fluctuations have been correlated with stadial and interstadial conditions, and snails found at Manix also occur in the mountains. Cooler, perhaps rainy summers are indicated. Three communities are recognized: lacustrine, with waterfowl, turtles and fish; shrub–woodland, inhabited by *Nothrotheriops* and *Ovis;* and grassland, with herds of camelids, bison, and horses. Herbivores outnumber carnivores 35 to 1, and *Camelops cf. minidokae* is the most abundant member of the fauna. (Buwalda, 1914; Jefferson, personal communication, 1976.) LACM.

Maricopa (25+ spp.), Kern Co. (4-169)

Located in the southwestern corner of the San Joaquin Valley, 50 km E of McKittrick, this site was probably a late Pleistocene water hole near the foot of the mountains. The 30 to 60-cm-thick bone bed lies between shallow sag ponds developed along a fault and the base of a low, broad hill and is covered with a minimum of 60 cm of oil-impregnated sand, silt, and clay overburden. The bones are found in a mixture of clay and asphalt but, unlike the bones from Rancho La Brea, are not heavily impregnated with asphalt and are, in fact, dry and chalky. The late Pleistocene climate at Maricopa was probably cooler and wetter than at present. Ten species are extinct, and *Canis latrans* is the predominant member of the fauna. This Wisconsinan fauna is similar to that of McKittrick, and the material is currently under study. (Macdonald, 1967; Soiset, personal communication, 1975.) BYU, LACM.

Newport Bay Mesa (Loc. 1066, 19 spp.; Loc. 1067, 17 spp.), Orange Co. (3-111, 4-157)

Two vertebrate faunas are recognized in the vicinity of Upper Newport Bay. Locality 1066 contains an extensive invertebrate fauna, as well as birds and mammals. Most of the bones are permineralized, and a number are encrusted with sand and marine shells. Both terrestrial and marine (*Enhydra*, phocids, otariids) mammals are present. Radiometric dating of the correlative Palos Verdes Sand yielded an age of 100,000 to 130,000 years. Locality 1067, a small, fossil-bearing pocket in a road-cut, consists almost entirely of small mammals; it is believed to be slightly older than Rancho La Brea. (W. E. Miller, 1971.) LACM.

Potter Creek Cave (42 spp.), Shasta Co. (4–172)

Located in the foothills of the Cascade Range, 454 m above sea level, this Wisconsinan-age cave contains a rich mammalian fauna. It was formed along a fissure by percolating water, and the stratified deposits consist of pebbly clay, cave breccia, stalagmites, and volcanic ash. Bones were discovered here in 1878, and the cave was excavated in the early 1900s. Fifteen species are extinct, including *Arctodus simus*, *Nothrotheriops shastensis*, and *Thomomys microdon*, which were described from Potter Creek Cave. (Sinclair, 1903; Stock, 1918.) UCMP.

Rancho La Brea (42 spp.), Los Angeles Co. (4–159)

One of the best known Pleistocene deposits in the world, Rancho La Brea is located just off Wilshire Boulevard in downtown Los Angeles. Renowned for the extensive collections of well-preserved, Wisconsinan-age mammals and birds, as well as numerous insect and plant remains and some gastropods, Rancho La Brea is the type locality of the late Pleistocene Rancholabrean Land Mammal Age in North America. A representation of the fauna appears in figure 4.3.

The earliest description of the site was made by the Portala Expedition in 1769–70. Although mention had been made of bones in the asphalt deposits and animals becoming entrapped in the pits, it wasn't until 1905 that the scientific value of the fossils was recognized. From 1906–1915, parties from the University of California, the Los Angeles County Museum, and other institutions carried out major excavations. In 1915, G. Allen Hancock, owner of the site, gave the 23-acre tract, now called Hancock

4.3 Scene at Rancho La Brea, California in the late Rancholabrean. Shown are *Smilodon fatalis* attacking *Glossotherium harlani*, with *Canis dirus* looking on, and the giant condor, *Teratornis merriami* circling overhead.

Park, to Los Angeles County and requested that its scientific features be preserved, studied, and exhibited to the public.

Of the museum's 101 numbered excavations (pits), about 16 proved to be major deposits. Because of the craterlike dimensions of some of them, it was popularly believed that the pits represented open, conical, tar-filled pools, often camouflaged by water or wind-blown dust, where unwary animals came to drink and became enmired, their struggles attracting large numbers of predators, who were then also entrapped. Because the viscous, yet fluid asphalt continues to bubble and upwell, it was assumed that this caused mixing of the bones and consequent loss of stratigraphy. Fossils have never been found in the open asphalt lakes, however. Rather, remains have been entombed in hard, asphalt-impregnated sand and clay, frequently admixed with gravel lenses and cobbles of fluviatile origin. Also contrary to popular belief, entrapment was not a continual happening—in fact, an average of only one animal entrapped per year over the 30,000-year span represented at Rancho La Brea is more than sufficient to account for the enormous number of bones recovered (Akersten, personal communication, 1976).

Although the fauna has been extensively studied, the geology has been generally neglected. Recently, Woodard and Marcus (1973) showed that well-defined stratigraphy is present in the Rancho La Brea deposits. The asphaltic, fossiliferous deposits are confined to the upper part of Member B and the overlying Member C of the Palos Verdes Sand. The uppermost bone-bearing strata are largely fluviatile in origin, as shown by the surrounding sediments, the freshwater limestone, and molluscan fauna. Bone occurs in variously-sized pockets; the larger, more continuous ones may represent areas of asphaltic quicksand in which the animals were trapped. Other smaller deposits containing waterworn and abraded material probably represent localized fluviatile concentrations in ponds or stream channels that were later enveloped by asphalt permeating upward and sideward from active vents and fissures. The individual pits have not been continually active, nor were they all active for the same length of time—this accounts for their faunal differences. Pit 3 shows gross stratification of both the fossil deposits and the sediments; age increases with depth, and a series of carbon-14 dates ranging from $12,650 \pm 160$ to $19,300 \pm 395$ years B.P. were obtained on *Smilodon* bones with the use of advanced techniques to remove petroleum and humic acid contaminants. Five C-14 dates on plant and animal material from Pit 91 (reopened in 1961, excavation continuing) range from 25,000 to 40,000 years B.P.

The late Pleistocene climate in the Los Angeles Basin was probably cooler with more fog and greater rainfall, than today. In Pit 91, several distinct contemporaneous plant communities have been recognized: coastal pine and cypress forest (similar to that found on the Monterey Peninsula today), chaparral and foothill woodland, coastal redwood forest (southernmost occurrence), riparian woodland along streambanks, and freshwater acquatics and emergents assemblage. The mixture of cool, moist, coastal vegetation and a warm, dry, inland vegetation is accounted for by stream drift and does not indicate a dramatic climatic change (Warter, 1976).

The fauna of Rancho La Brea is dominated by carnivorous mammals and raptorial birds. Canids (mainly *Canis dirus*) make up about 57 percent of the carnivore population and raptors make up 67.5 percent of the total avifauna. Of the 133 species of birds identified, 19 are extinct, including the giant condor, *Teratornis merriami* Miller, the largest known flying bird. Insects, especially beetles, are the most abundant invertebrates. The presence of blowfly larvae, dermestids, and carrion beetles in bone cavities suggests that the decaying material was exposed to the elements for some time before burial.

Interdisciplinary studies are continuing on the fauna and sediments. The literature is extensive—Stock (1965) lists over 200 publications. (Akersten, personal communication, 1976; Berger and Libby, 1968; Ho, Marcus and Berger, 1969; Howard, 1962; Marcus, 1960; Stock, 1965; Warter, 1976; Woodard and Marcus, 1973.) LACM, UCMP.

Samwel Cave (38 spp.), Shasta Co. **(4-173)**

Located on the east bank of the McCloud River, 25 km from the mouth of the river and 455 m above sea level, this limestone cave consists of two chambers. The deep first chamber has yielded the greater number of extinct forms. Most of the bones are covered with stalactite crystals and rodent-gnawed, and many are split. Complete carnivore skeletons indicate that the predators used the cave as a den and dragged in their prey. *Euceratherium collinum* and *Martes nobilis* were described from Samwel Cave, and eight species are extinct. A local Indian legend recounts how an Indian maiden fell to her death in the first chamber. (Furlong, 1906.) UCMP.

San Pedro (17 spp.), Los Angeles Co. **(3-112)**

This fauna was apparently taken from late Pleistocene marine sands and gravels in a San Pedro lumberyard. The fauna closely resembles that of Locality 1066 at Newport Bay Mesa, and the two faunas are believed to be the same age. The marine invertebrates present suggests that water temperatures near shore were more diverse (both warmer and cooler) than at present. Nine species are extinct. (Langenwalter, 1975; W. E. Miller, 1971.) LACM.

Schuiling Cave (11 spp.), San Bernardino Co. **(4-163)**.

Twenty-nine species of reptiles, birds, and mammals are known from Schuiling Cave, a late Pleistocene deposit in the Mojave Desert. The cave is situated about 3 m above the floor of a dry canyon that cuts through the volcanics and alluvial fill. Much of the cave fill is alluvial in nature, and some of the bones were probably washed in during stream deposition or flooding. Human artifacts have been found in the upper levels, and a human skull fragment was recently found at the same depth as the fossil bones. The faunal remains, including several species of water birds, indicate a climate more equitable or cooler than that of the present day, an abundant nearby source of water, and a great amount of grass and woodland vegetation. (Downs *et al.*, 1959; Jefferson, personal communication, 1976.) LACM.

Zuma Creek (12 spp.), Los Angeles Co. **(4-162)**

A small fauna is being recovered from Pleistocene sediments laid down in a deltaic estuary environment near the present Zuma Creek. So far, two extinct species of *Tapirus* and *Equus* have been recovered; the other species are found in the area today. (Langenwalter, personal communication, 1973.) LACM.

COLORADO

Chimney Rock Animal Trap (32 spp.), Larimer Co. **(4-202)**

This site is a circular depression in the Casper Sandstone approximately 19.7 m across and 3 m deep, with an overhang of from 1.2 to 7.5 m. It is located 48 km SW of Laramie, Wyoming, at the extreme southern end of the Laramie Basin, at an elevation of 2,410 m. A preponderance of carnivores in the deposit is typical of an animal trap, where predators are attracted by trapped animals, jump in, and are unable to escape. The deposit is unstratified. The fauna is similar to that of Little Box Elder and Jaguar caves. Bone from the 1.2-m level yielded a date of 11,980 ± 180 years B.P. (Hager, 1972.) UWYO, UCM.

Mesa de Maya (14 spp.), Las Animas Co. (3-117)

A major unconformity of at least 2.0 m.y. duration is present between the pre-Nebraskan Donnelly Ranch fauna and the Sangamonian Mesa de Maya fauna. The latter assemblage was recovered from locally derived sediments from the middle to the top of the deposit. The site is located 116 km E of Trinidad, at an elevation of 1,800 m. The fauna, representing grasslands and marsh communities, closely resembles that of Cragin Quarry in Meade County, Kans. Pollen analysis indicates that the flora was similar to that found in the area today. The climate was probably more equable, with cooler summers, than that of the present day. Of the 14 species of mammals recovered, 7 are extinct. (Hager, 1975.) UWYO.

Recently, two faunas, Selby (14 spp.) and Dutton (20 spp.) have been recovered in eastern Colorado (Yuma Co.). At both sites, extinct mammals are associated with pre-Clovis artifacts, and dates on bone run about 20,000 years B.P. The faunas are under study. (Graham, personal communication, 1978.) USNM.

FLORIDA

Arredondo (32 spp.), Alachua Co. (4-274)

Fissure fillings in the Ocala Limestone (Eocene) near Arredondo, at an elevation of 25 m, have produced faunas which may be approximately contemporaneous. With two exceptions, *Synaptomys australis* and *Desmodus stocki*, the small mammals belong to extant species, whereas almost all the larger mammals are extinct. On stratigraphic grounds, i.e., position in relation to the Wicomico Terrace (thought to be Sangamonian in age), this and several other faunas have been referred to the late Illinoian or early Sangamonian. Revisions of the interglacial shorelines in Florida have shown this date to be untenable, however (Alt and Brooks, 1964).

The fauna indicates moist and forested conditions; grazers like horse, bison, and *Platygonus* are absent, and presence of the vampire bat (*Desmodus stocki*) suggests warmer winters than at present.

The fauna also comprises amphibians, reptiles, and birds. The box turtle *Terrapene carolina* (Linnaeus) gives particularly useful evidence. The large subspecies *T. carolina putnami* is thought to have inhabited coastal marsh and lowland savanna areas, like the living *T. carolina major*, and thus indicates an interglacial transgression; the smaller subspecies *T. carolina carolina* (or *bauri*) is typical of higher, more interior and forested areas. At Arredondo, the sequence from *T. carolina putnami* to *T. carolina carolina* seems to indicate a regression from an interglacial high and so suggests a late Sangamonian age. (Auffenberg, 1958, 1963, 1967; Bader, 1957; Brodkorb, 1959; Kurtén, 1965; R. A. Martin, 1967, 1968a; Webb, 1974a.) FSM.

Aucilla River IA (33 spp.), Jefferson Co. (4-271)

There are many fossiliferous sites of diverse character and age along the Aucilla River in the Florida Panhandle. Of these, the underwater Site IA has yielded a large fauna, which is currently being studied by David Gillette. The presence of grazers (horse, bison, *Platygonus*) indicates a somewhat drier climate. The bison (*B. bison antiquus*), studied by Robertson, 1974 points to a Wisconsinan age. (Gillette, personal communication; Robertson, 1974; Webb, 1974a.) FSM.

Bradenton (14 spp.), Manatee Co. (3–127)

This is a coastal deposit on the Gulf of Mexico, with white beach sands and laminae of fossiliferous dark sand with clay balls. The site is in the bank of a drainage canal (elevation ca. 3 m) in the town of Bradenton. Both small and large mammals are represented, and 10 species are extinct. The box turtle is the lowland type. Thus the age is probably Sangamonian, which is corroborated by the presence of *Bison latifrons* and *Microtus hibbardi*. (Auffenberg, 1963; Robertson, 1974; Simpson, 1930a.) AMNH, FSM.

Devil's Den (48 spp.) Levy Co. (5–361)

The fauna of this sinkhole trap lived in either the very late Wisconsinan or the early Holocene; unpublished carbon-14 dates and stratigraphic data obtained by H. K. Brooks (Martin and Webb, 1974) indicate a date of 7,000–8,000 years B.P. The site is located in a pasture just north of the town of Williston. A narrow aperture opens into the sink, from which lateral passages lead off; almost all of the fossils came from one of these passages, termed chamber 3. The fossils (except for the bats, which roosted there) accumulated by trap-fall.

The fauna includes man and domestic dog, as well as a number of extinct species (for which the suggested age would be late or terminal Wisconsinan) including the following: *Synaptomys australis, Megalonyx jeffersonii, Canis dirus, Tremarctos floridanus, Smilodon fatalis, Mammut americanum, Equus* sp., and *Platygonus compressus*. Extant northern forms are *Ondatra zibethicus, Microtus pennsylvanicus,* and *Myotis grisecens;* their presence suggests conditions somewhat cooler than those at present. The environment was mainly open woodland, but denser forest also occurred, probably closer to the sink. Fish, amphibian, reptile, and bird fossils have also been found. (Kurtén, 1966; Martin and Webb, 1974.) FSM.

Haile VIIA (11 spp.), VIIIA (18 spp.), XIB (26 spp.), XIVA (11 spp.), Alachua Co. (3–134, 3–135, 3–136, 4–273)

At an elevation of about 25 m above sea level, numerous fossiliferous fissure fillings have been discovered in the Ocala Limestone of the Haile area [Roman numerals refer to quarries, letters to individual deposits (thus there is no correlation with stratigraphy)]. The deposits are of varying ages, the oldest currently recognized is Haile XVA (Blancan, which see). Of the sites listed here, VIIA, VIIIA, and XIB are regarded as Sangamonian. The presence of *Terrapene carolina putnami* (Hay) suggests a lowland environment and hence a high interglacial stand of the sea. At Haile VIIIA, a sequence from the upland *T. carolina bauri* in the lower strata to *T. carolina putnami* in the upper bears witness to a transgressive sea and dates the deposit to the early Sangamonian. *Bison latifrons* (site VIIIA) and *Desmodus* (site XIB) appear to be typical Sangamonian forms in this area. Other species common to these Haile sites include *Dasypus bellus, Holmesina septentrionalis, Tremarctos floridanus, Panthera onca augusta,* and *Hemiauchenia macrocephala*.

Site XIVA, with the small upland *Terrapene,* has a dry-climate fauna; the evolutionary level of *Sigmodon* is consistent with an early Wisconsinan date, but a late Illinoian age is not excluded. In any case, a low sea level is indicated. The fauna consists mainly of small mammals. (Auffenberg, 1963, 1967; Brodkorb, 1963; Kurtén, 1965; Ligon, 1965; R. A. Martin, 1974b; Robertson, 1974; Webb, 1974a.) FSM.

Hornsby Springs (10 spp.), Alachua Co. (5–362)

The site is northeast of High Springs (elevation, 25 m); the spring run empties into the Santa Fe River. Most of the fauna comes from the spring bottom down to some 20 m under the surface. Eight of the mammals are Pleistocene megafauna. There is a radiocarbon date of 9,880 ± 270 years B.P. (Bader, 1957; Dolan and Allen, 1961; Webb, 1974a.) FSM.

Ichetucknee River (30 spp.), Columbia Co. (4–272)

Several sites along this river, a tributary of the Santa Fe, have yielded a rich, apparently mainly Wisconsinan fauna. At an elevation of about 10 m, the fauna includes the upland box turtle and comprises many forest forms (*Sciurus, Glaucomys, Tapirus, Mylohyus, Palaeolama, Mammut*) as well as some grazers. Twelve species are extinct. (Auffenberg, 1963; Kurtén, 1965; McCoy, 1963; R. A. Martin, 1969; Neill, 1957; Robertson, 1974; Simpson, 1930a; Webb, 1974a.) FSM.

Kendrick IA (19 spp.), Marion Co. (4–275)

Limestone quarrying at Kendrick, in north-central Florida, has disclosed a number of fissure fillings. The elevation is 25 m. The box turtle is of the upland type, which suggests a Wisconsinan date. Both micromammals and megafauna are represented, and 6 of the species are extinct. (Kurtén, 1965; Webb, 1974.) FSM.

Melbourne (49 spp.), Brevard Co. (4–282)

Early discovery of human remains (Loomis, 1924) gave the impetus to fieldwork and faunal studies. The fossils come out of the Melbourne Formation, which consists of sands with marine shells and local bone accumulations; the formation rests upon the Pleistocene Anastasia Formation and is overlain by Recent beds. The main fossil collection comes from the Golf Course locality, 5 km W of the center of Melbourne on the eastern coast of central Florida. Lying 3 m above sea level, the deposit is considered to be late Wisconsinan to sub-Recent. The rich fauna has an important element of browsers and other swamp (*Castoroides, Neofiber*), river (*Neochoerus, Tapirus, Glyptotherium*), and forest (*Mylohyus, Odocoileus*) species, as well as some plains species (*Bison, Mammuthus, Equus*). A representation of the fauna appears in figure 4.4. Twenty-three mammals are extinct. The Melbourne Formation at Merritt Island, Brevard Co. is also productive. (Gazin, 1950; Gidley, 1928; Loomis, 1924; Ray, 1958; Simpson, 1929a; Webb, 1974a.) AMNH, FSM, USNM.

Payne's Prairie (10 spp.), Alachua Co. (3–133)

The fossils in this area were mainly collected at a drainage canal 7 km S of Gainesville (elevation, 22 m). The fauna, consisting of reptiles and mammals, with lowland *Terrapene*, is suggestive of a pond and marsh habitat. Both *Hemiauchenia* and *Palaeolama* are present. A Sangamonian age has been assigned to the fauna. (Auffenberg, 1963; R. A. Martin, 1969; Webb, 1974a.) FSM.

Reddick (52 spp.), Marion Co. (3–132)

One of the richest Rancholabrean faunas in Florida comes from fissure fillings in the Ocala Limestone just south of Reddick (elevation, 25 m). A particularly prolific deposit is termed Rodent Bed IB. In addition to the mammals, 9 amphibians, 32 reptiles, and

4.4 The late Rancholabrean fauna in coastal Florida near Melbourne included *Neochoerus pinckneyi*, *Tremarctos floridanus*, and *Holmesina septentrionalis*.

64 bird species have been identified. Abundance of bats, barn owls, swallows, and vultures suggests a cave where rodent and snake bones were stockpiled by small predators. Truly aquatic forms are absent. Twenty-one species are extinct. The presence of large lowland box turtles, as well as the probably interglacial transients *Desmodus*, *Tadarida*, and *Felis pardalis*, suggests a Sangamonian age. (Auffenberg, 1963; Brodkorb, 1957; Gut, 1959; Gut and Ray, 1963; Hamon, 1964; Kurtén, 1965; Ray, Olsen, and Gut, 1963; Webb, 1974a.) FSM.

Rock Springs (17 spp.), Orange Co. **(3–128)**

This site, some 6 km S of Mt. Plymouth in north-central Florida, has yielded Pleistocene (also Miocene and Recent) vertebrates, found in the bed of the spring run about 100 m below the head. Presence of southern or southwestern forms (*Felis yagouaroundi*, *Mormoops megaphylla*) is notable. A Sangamonian age is suggested. (Auffenberg, 1963; Kurtén, 1965; Ray, Olsen, and Gut, 1963; Webb, 1974a.) FSM.

Sabertooth Cave (27 spp.), Citrus Co. **(4–276)**

Also referred to as Allen, or Lecanto, Cave, this site is in the Ocala Limestone northwest of Lecanto in west-central Florida, about 30 m above sea level. The cave, extending downward some 8–12 m, connects with the surface through two vertical shafts leading to a broad sink. Bird remains include three species of owls and various raptors; reptiles and a few amphibians have also been found, but no bats. The deer *Blastocerus* has southern affinities, whereas the gopher *Thomomys* is now strictly western. Most of the mammals are members of the wet lowland and swamp community. The fauna is tentatively regarded as Sangamonian. (Auffenberg, 1963; Holman, 1959; Kurtén, 1965; Leidy, 1889; Simpson, 1928; Webb, 1974a.) AMNH, FSM.

Seminole Field (46 spp.), Pinellas Co. (4–277)

This east coast marsh site near Seminole, west of St. Petersburg, has produced a large and varied fauna that includes a large representation of Pleistocene megafauna, as well as the domestic dog. The fossils come from a bone bed about 3 m above sea level and overlie the marine Pleistocene beds. Recent reworking of some of the deposits is suggested here, as at Melbourne and Vero. Forest, river, and plains-dwelling species are present, and 22 of the mammal species are extinct. The fauna is probably late Wisconsinan or early Holocene and is similar to that of Melbourne and Vero. (Auffenberg, 1963; Kurtén, 1965; Martin and Webb, 1974; Robertson, 1974; Simpson, 1929a, Webb, 1974a.) AMNH, FSM.

Vero (45 spp.), Indian River Co. (4–281)

The stratigraphic sequence at Vero closely resembles that of nearby Melbourne. The discovery of human fossils ("Vero man") associated with a Pleistocene fauna has lent additional interest to the site. Edentates, rodents, and carnivores are well represented, and 17 species are extinct. The age is late Wisconsinan and early Holocene. (Auffenberg, 1963; Hay, 1917c, 1923; Kurtén, 1965; Sellards, 1916b; Weigel, 1962.) USNM, FSM.

Waccasassa River (39 spp.), Levy Co. (3–130)

The Waccasassa River has yielded a rich Pleistocene fauna. Sites IIB and III, with upland *Terrapene*, are assigned to the Wisconsinan, and site VI, with lowland *Terrapene* and *Bison latifrons*, is regarded as Sangamonian in age. (Robertson, 1974; Webb, 1974a.) FSM.

Wakulla Springs (6 spp.), Wakulla Co. (4–270)

Wakulla Springs, located south of Tallahassee at about sea level, has yielded a Pleistocene megafauna. It is dated as Wisconsinan, primarily because of the presence of *Bison bison antiquus*. (Gunter, 1931; Robertson, 1974; Sellards, 1916a; Simpson, 1929b; Webb, 1974a.) FSM.

Williston IIIA (20 spp.), Levy Co. (3–131)

The fossils were found in a pipe in the Ocala Limestone, within the city limits of Williston at an elevation of 25 m. Amphibians, reptiles (including lowland *Terrapene*), birds, and mammals are represented. Rabbits (*Sylvilagus*) are the most abundant mammals. The habitat is thought to have been marshy pineland grading into well-drained pine forest with open sinks. A Sangamonian age is suggested. (Holman, 1959a,b; Webb, 1974a.) FSM.

Many other Rancholabrean sites in Florida are yielding fossils. Some of these are as follows: Chipola River (8 spp.), Jackson Co. (4–269); Joshua Creek (6 spp.), DeSoto Co. (4–279); Jupiter Inlet (11 spp.), Palm Beach Co. (4–280); North Havana Road (7 spp.), Sarasota Co. (4–278); Santa Fe River IIA (17 spp.), Gilchrist Co. (3–105); Sebastian Canal (12 spp.), Brevard Co. (4–283); and Withlacoochee River (22 spp.), Citrus Co. (3–129). Preliminary faunal lists for these localities are given in Webb (1974a).

Georgia

Ladds (48 spp.), Bartow Co. **(4–266)**

Fissure fillings in Quarry Mountain, about 3.7 km WSW of Cartersville in northern Georgia, have yielded a large (78 + spp. of vertebrates, 25 spp. of mollusks) late Pleistocene fauna. Quarry Mountain, an isolated dolomite ridge rising 152 m above the surrounding countryside, is pierced by caves formed through dissolution of the limestone by streams. Fossils occur in the materials deposited in solution cavities, and some are encrusted with flowstone. The bones were collected directly from the old land surface and from rubble resulting from commercial blasting, crevices among the boulders, and solution holes. Thus there is the possibility of faunal mixing and ecological incompatibility, but a more equable climate in the late Pleistocene (as shown by bog pollen studies in northern Georgia) would explain the anomalous elements (Ray, personal communication, 1975).

About 25 percent of the mammals are extinct (including *Tamias aristus, Peromyscus cumberlandensis, Dasypus bellus, Tremarctos floridanus,* and *Tapirus veroensis*). Several of the species (*Sorex cinereus, S. fumeus, Sylvilagus transitionalis, Synaptomys cooperi*) show northern affinities, whereas others (*Neofiber alleni, Conepatus leuconotus, Panthera onca*) now occur south of Ladds. The age of the fauna is probably late Wisconsinan. Work is continuing at the quarry and on the collections. (Lipps and Ray, 1967; Ray, 1967 and personal communication, 1975.) USNM.

Little Kettle Creek (6 spp.), Wilkes Co. **(4–267)**

This late Pleistocene local fauna came from Pleistocene alluvium along Little Kettle Creek 10 km SW of Washington, in the Piedmont of east-central Georgia. The fauna includes fish, the boreal rodents *Clethrionomys* (southernmost record) and *Synaptomys borealis, Mammut, Mammuthus, Odocoileus,* and *Bison*. The vertebrate and pollen evidence shows that a cool climate prevailed in the southern Appalachians, not only at high elevations but in adjacent lowlands as well. (Voorhies, 1974a.) UGV.

Watkin's Quarry (6 spp.), Glynn Co. **(4–268)**

This coastal Georgia site is located in a flat, swampy area near Brunswick, at an elevation of 3.0 to 4.5 m. The bone-producing layer is underlain by a clean quartz-sand layer and overlain by a layer of carbonaceous clay; the sediments are associated with the Princess Anne Pleistocene shoreline. The pollen includes grass and oak, which indicates a climate drier than that of today. The fauna includes fish, a crocodilian, and six species of mammals, four of which are extinct. A notable find was three individuals of the giant megathere, *Eremotherium*. (Voorhies, 1971, 1973.) UGV.

Idaho

American Falls (27+ spp.), Power, Bingham, and Bannock Cos. **(3–114)**

The American Falls fauna comes from the beach of the American Falls Reservoir, created in 1927 by the damming of the Snake River at the town of American Falls, about 16 km NW of Pocatello. The beach deposit represents the lower member of the American Falls Formation (Ridenour, 1969) and consists mostly of sands, silty sands, silts, and a small proportion of clayey silt beds. Above the beach deposit, and separated

4.5 In the Rancholabrean near American Falls, Idaho, *Bison latifrons, Megalonyx jeffersonii,* and *Mammuthus columbi* were members of the large-mammal fauna. Note the plantigrade hind feet of *Megalonyx*.

from it by a distinct change in texture and undulatory erosional surface, is the upper member, containing the Rainbow Beach fauna. The lower member represents a floodplain consisting mainly of point bar deposits. The molluscan fauna (Taylor in Carr and Trimble, 1963) suggests a cool, wet environment. The fauna has been traditionally dated as Illinoian, based on the presence of *Bison latifrons,* until recently not known from younger deposits but now positively identified in the Wisconsinan Rainbow Beach fauna (McDonald and Anderson, 1975). White (personal communication, 1976) believes that the American Falls fauna is more likely Sangamonian in age. Supporting this view is the presence of *Panthera leo atrox,* which makes its first appearance south of Alaska in the Sangamonian. Large mammals predominate, and *Bison latifrons* was probably the most abundant species. A representation of the fauna appears in figure 4.5. (Carr and Trimble, 1963; Hopkins, Bonnichsen, and Fortsch, 1969; McDonald and Anderson, 1975; Ridenour, 1969; White, personal communication, 1976.) ISUM, USNM.

Dam (Hop-Strawn Pit) (27 spp.), Power Co. **(4–187)**

A gravel pit on the southeast bank of American Falls Reservoir has yielded a late Pleistocene fauna. Most of the fossils were recovered when the Idaho Highway Department was removing fill for a new bridge near the dam. The fauna probably lived on the floodplain of the Snake River, where there was abundant grass, a riverine forest, and nearby water. Pleistocene megafauna, including ground sloths, mammoths, mastodonts, horses, camels, and bison, predominates. A radiocarbon date of 26,500 ± 3,500 years B.P. was obtained on bone. (Barton, 1975, and unpublished manuscript; White, personal communication.) ISUM.

Jaguar Cave (44 spp.), Lemhi Co. **(4–189)**

Situated at the western base of the Beaverhead Mountains in east-central Idaho, at an elevation of 2,257 m, Jaguar Cave was formed by frost-fracturing of Permocarboniferous limestone and concurrent gravitational removal of the resulting angular debris. Disturbance of the layers and periodic excavations of the depressions indicate occupation by man prior to 12,000 years B.P. and at intervals during the following 3,000 years. The oldest hearth is dated 11,580 ± 250 years B.P. The cave was self-sealed about 9,000 years ago. Six species are extinct. Included in the fauna are tundra (*Ochotona, Dicrostonyx, Rangifer*), forest (*Erethizon, Martes, Odocoileus*), and plains-dwelling (*Equus, Antilocapra, Bison*) forms, and one of the oldest records of *Canis familiaris* in North America comes from this site. (Dort, 1975; Guilday and Adam, 1967; Kurtén and Anderson, 1972; Lawrence, 1967; Sadek-Kooros, 1966.) IIU.

Middle Butte Cave (28 spp.), Bingham Co. (5–351)

Middle Butte Cave, a lava blister connected to a lava tube, is located 8 km E of Atomic City and 8 km SW of Moonshiner Cave. The entrance, measuring 1.0×2.0 m across, is 2 m above the floor. The opening slants through almost 2 m of smooth basalt that offers few toeholds for escape. The sediments were probably deposited by flowing water since cross-bedding is present. Like that of Moonshiner Cave, the early Holocene fauna consists predominantly of sagebrush and grassland species and a few forest dwellers; most are found in the region today. (McDonald, personal communication, 1973; McDonald et al., unpublished manuscript.) ISUM.

Moonshiner Cave (44 spp.), Bingham Co. (5–350)

Moonshiner Cave, named for the moonshiner who made illegal whiskey there during Prohibition, is located 50 km W of Idaho Falls and 1 km E of East Butte. The opening to this lava blister measures 1.5×2.0 m; 2.5 m below the entrance is a pile of rocks that resulted from roof fall. The cave is Y-shaped, and the arms can be traced on the surface. Almost no stratigraphy is present. In late spring, a luxuriant bed of ferns grows below the entrance, and in winter, a cone of snow rises to within 1.5 m of the cave opening. Both of these phenomena give the illusion of shallow depth and were (and are) major factors in the entrapment of animals. Grassland, sagebrush, meadow, and forest-living species are represented, and the deposit contains an abnormally high percentage of carnivore remains, including a minimum of 400 weasels and 25 wolverines. (McDonald et al., unpublished manuscript; White, personal communication.) ISUM.

Rainbow Beach (24 spp.), Power Co. (4–186)

The upper fossiliferous zone at American Falls Reservoir, called the Rainbow Beach local fauna, is late Wisconsinan in age. The strata are deltaic in character and were deposited in and along the shore of American Falls Lake by the spillover from glacial Lake Bonneville. Bone is scattered throughout the deposit, and the larger specimens are often broken and waterworn. Fish, amphibians, snakes, lizards, 8 species of birds, and 24 species of mammals have been identified, and aquatic, open-grassland, and forest-scrub habitats are represented. The last-known occurrence of *Bison latifrons* was reported from Rainbow Beach. Ten species are extinct. Carbon-14 dates obtained on collagen samples range between $21,500 \pm 700$ and $31,300 \pm 2,300$ years B.P.; the fauna is closely correlated with the Dam local fauna. (McDonald and Anderson, 1975.) ISUM.

Wasden (22 spp.), Bonneville Co. (4–188)

Located on the eastern Snake River Plain at an elevation of 1,584 m, the Wasden site consists of three caves formed by the collapse of a lava tube. Excavation has been carried out in Owl Cave, a stratified deposit. Major features uncovered to date include a deposition of Mazama Ash, a formation consisting of at least eight major ice wedge beds, a massive bison bone deposit (minimum of 75 individuals, dated about 8,000 years B.P.), and butchered and charred mammoth bone (dated 12,500 years B.P.). Microfaunal remains, mostly attributed to owl kills, show pronounced fluctuations in both total numbers and dominant species. (Guilday, 1969; S. Miller, personal communication, 1977.) ISUM.

Wilson Butte Cave (29+ spp., Strata E, D, and the lower and middle parts of Strata C), Jerome Co. **(4–185)**

A lava blister, Wilson Butte Cave is located on a broad basaltic ridge that was formed by successive outwellings of fluid lava flows. Five strata are recognized; the lower three are late Wisconsinan/early Holocene in age. *Thomomys talpoides* is the most abundant mammal in the lower levels. Several artifacts were found in the lowest layer, and Wilson Butte Cave is one of the earliest-dated archeological sites in North America. Changes in the deposits and faunas illustrate the past environments and climatic variations (from cool-moist to warm-dry) that have taken place during the last 12,000 years on the western part of the Snake River Plain. (Gruhn, 1961.) ISUM.

ILLINOIS

Alton (15 spp.), Madison Co. **(4–245)**

The fossils occur in concretions at the base of a loess deposit some 15 m thick and resting upon glacial till, on the Illinois–Missouri border about 25 km N of St. Louis. The fauna, mostly Pleistocene megafauna species, includes northern forms like caribou and *Symbos*. The only proboscidean is the mastodont; other, probably forest-living forms are *Marmota, Megalonyx, Cervalces,* and possibly the rare cervid *Sangamona*. A taiga environment is indicated. The fauna may date from a Wisconsinan interstadial. (Hay, 1920, 1923.) USNM.

Meyer Cave (42 spp.), Monroe Co. **(5–359)**

This vertical fissure, located in a Mississippi River bluff in southwestern Illinois, has probably been a natural death trap since early Holocene times. At least 115 species of vertebrates have been identified. Although the deposit is not stratified, the presence of northern or boreal species (*Microsorex hoyi, Mustela rixosa, Microtus xanthognathus, Clethrionomys gapperi, Erethizon dorsatum*) suggests a cool moist period in the early Holocene. (Parmalee, 1967.) ISM.

Polecat Creek (16 spp.), Coles Co. **(5–360)**

The fauna was collected in gravel pits just south of Ashmore, in east-central Illinois. Fossil wood (tamarack, elm, and hickory), mollusks, and vertebrates occur in all the postglacial deposits, which rest upon a glacial gravel bed of Wisconsinan age. The main fauna comes from an alluvial gravel bed and comprises five extinct mammals and numerous extant species, among which the domestic dog may be noted. The paleoenvironment was apparently a shallow, but not swampy, body of water surrounded by meadows, woodlands, and brushy thickets. A very late Wisconsinan to Holocene age is indicated. (Galbreath, 1938.) FMNH.

INDIANA

Evansville (6 spp.), Vanderburg Co. **(4–246)**

The fossils were first described by Leidy (1854), who stated that the bones were found sticking out of the bank of the Ohio River near the mouth of Pigeon Creek, a short distance below Evansville. This is a Rancholabrean megafauna with browsers (*Megalonyx, Odocoileus, Tapirus*), bison, horse, and dire wolf. (Hay, 1923.)

KANSAS

Butler Spring (18 spp.), Cragin Quarry (28 spp., not type locality), and
   Robert (15 spp.), Meade Co. **(3-102, 3-123, 4-230)**

The Butler Spring Basin, in Meade County, southwestern Kansas, has produced a series of four mammalian faunas: Adams, Butler Spring, Cragin Quarry, and Robert. (The Adams fauna is very late Irvingtonian; see description in chapter 3.) All the deposits are below the regional High Plains surface.

The sediments containing the Illinoian Adams fauna are overlain conformably by clayey sands, which have yielded a pollen flora as well as the Butler Spring fauna containing mollusks, ostracods, and vertebrates. The deposits were formed in quiet, shallow water off the main stream subsequent to a shift of the Cimarron River channel. Vertebrate remains include fishes, amphibians, reptiles, birds, and mammals. The latter include some forms present in the earlier Illinoian, as well as *Glossotherium, Cynomys ludovicianus,* and *Perognathus hispidus,* not present in the early Illinoian Adams fauna but found in later deposits in this area. Floral evidence suggests a somewhat warmer climate than in Adams times. The immediate habitat was marshy, with surrounding moist, low meadows having good grass cover; some distance away was upland prairie. The age is late Illinoian.

Renewed subsidence in the early Sangamonian led to the development of a basin in which a lake was formed. The silts containing the Cragin Quarry fauna were laid down in this lake and are overlain by a caliche which appears to have formed in more arid conditions at the climax of the interglacial. The type locality of this fauna is not, however, in this area but at Big Springs Ranch, about 24 km to the northeast, in a similar geological setting. The local habitat at Butler Spring in the time of the Cragin Quarry fauna probably was grassy upland, with some trees and shrubs, inhabited by Pleistocene herbivores, predators, and many species of rodents.

The Robert fauna (mollusks and small vertebrates) occurs in a 30-cm silt or soil zone below the surface, close to Butler Spring, and has a radiocarbon date of $11,100 \pm 300$ years B.P. All the species are extant, but a cool climate is indicated by presence of seven more-northern species which now do not range this far south. (Etheridge, 1958; Hibbard and Taylor, 1960; G. E. Schultz, 1965, 1967.) UMMP.

Duck Creek (17 spp.), Ellis Co. **(3-104)**

This fauna came from a gravel pit near the ancestral Smoky Hill River. Both the mollusks and the mammals indicate a period of time much cooler than the present. The small-mammal fauna includes the first record of *Clethrionomys gapperi* on the plains; in addition, *Synaptomys borealis* and *Sorex arcticus* are present. The fauna inhabited an area with a permanent stream and riparian forest. McMullen (1975) has suggested an equivalency with the Mt. Scott fauna (late Illinoian), but others (Kolb, Nelson, and Zakrzewski, 1975) feel that the climate indicates a somewhat earlier date, perhaps the same as Doby Springs or slightly younger. (Kolb, Nelson, and Zakrzewski, 1975; McMullen, 1975; Zakrzewski, personal communication, 1976; Zakrzewski and Maxfield, 1971.) FHSM.

Jinglebob (24 spp.), Meade Co. (3-124)

The sandy-silt terrace deposit containing this fauna is exposed along Shorts Creek at the Jinglebob pasture on Big Springs Ranch. The sediments were laid down by the ancestral Cimarron River, entrenched in the Rexroad Formation. The fauna comprises mollusks, fishes, amphibians, reptiles, and mammals. The presence of southern elements (*Terrapene, Oryzomys*), together with forms preferring moist habitats (some of them now more northerly in distribution), indicates warmer, more equable, and moister conditions than those of today. There is also a pollen flora. The environment appears to have been grassland with scattered pine and Osage orange trees. The age has been thought to be Sangamonian, though the fauna is probably younger than Cragin Quarry, but the equability model suggests an early phase of the Wisconsinan. Nine species are extinct. (Hibbard, 1955c; Hibbard and Taylor, 1960; Zakrzewski, 1975a.) UMMP.

Jones (14 spp.), Meade Co. (4-229)

The fauna occurs in a simple sinkhole fill, probably about 5 m in diameter. Mollusks, fishes, amphibians, reptiles, birds, and mammals are included. Apart from *Camelops* and *Platygonus*, all of the mammals are extant small species; three are northern or northwestern forms not now occurring in the area. The climate apparently was cooler and more humid than it is at present. Radiocarbon dates are 26,700 ± 1,500 and 29,000 ± 1,500 years B.P. (Downs, 1954; Goodrich, 1940; Hibbard, 1949a, 1970b.) UMMP.

Mt. Scott (26 spp.) and Cragin Quarry (32 spp., type locality), Meade Co. (3-103, 3-123)

Two superposed faunas occur in the Kingsdown Formation on the Big Springs Ranch. The formation is a channel fill of the ancestral Crooked Creek, entrenched in the Crooked Creek and Ballard formations.

The Mt. Scott fauna comes from three localities near the base of the Kingsdown Formation and comprises mollusks, ostracods, amphibians, reptiles, and mammals. Although, from its stratigraphic position, this fauna appears to be a close correlative of the Butler Spring fauna (late Illinoian), there are notable differences, which may be due to some difference in age as well as local environment; apparently, winters here were less severe than in the time of the Butler Spring and Doby Springs faunas. The paleohabitat was probably a river valley with marshes along the floor, tall grass, and stands of trees and shrubs. Several of the small mammals are extinct; four species are northern and do not occur in the area today. The sole large mammal present, a bison, appears to be *Bison latifrons*.

The Cragin Quarry fauna (mollusks, amphibians, reptiles, and mammals) is very similar to that occurring in a similar stratigraphic position (upper part of the Kingsdown Formation, beneath caliche) at Butler Spring but is somewhat richer in species; the total of mammalian species from the two areas is 41. The fossils are found in silts and fine sand deposited by a tributary of Crooked Creek and an artesian spring. Presence of southern forms like *Geochelone* and *Notiosorex* argues for a climate more equable and moister, and with much warmer winters, than that of the present day.

On top of the buried caliche occur sparse fossils that indicate a warm-humid climate of the same type as that under which the Jinglebob fauna lived. Thus it appears that the Cragin Quarry fauna represents an earlier phase of the Sangamonian than Jinglebob. (Hibbard, 1949a; Hibbard and Taylor, 1960; G. E. Schultz, 1965.) KUVP, UMMP.

Kentucky

Big Bone Lick (13 spp.), Boone Co. **(4–248)**

This was the first widely known fossil-vertebrate locality in North America; French soldiers picked up bones (still preserved in the Musée National d'Histoire Naturelle, Paris) here in 1739. Thomas Jefferson sent an expedition to Big Bone Lick in 1807 to collect fossils and study the natural history of the area. Some 300 bones were collected and shipped to Washington, where Jefferson and Caspar Wistar studied them. Today, this historic locality is largely included in Big Bone Lick State Park.

The site is located at the confluence of Big Bone Creek and Gum Branch, 33 km SW of Cincinnati, in a swampy area surrounding salt and sulfur springs. Although collections have been made through the years, it wasn't until the early 1960s that detailed studies were undertaken. At Site 1, three layers have been excavated: Zone A, 2.1–2.5 m below the surface, containing the bones of Recent animals and pieces of crockery; Zone B, 2.5–3.3 m, a dark humic silt containing the bones of wapiti, deer, modern bison, and reworked fossils; and Zone C, 3.0–4.5 m, a blue-gray sand silt containing the bones of a Wisconsinan megafauna. No microfauna has been found. Several type specimens, including *Bootherium bombifrons, Cervalces scotti, Bison antiquus,* and *Glossotherium harlani,* were described from Big Bone Lick. (C. B. Schultz, et al., 1963, 1967.) ANSP, MCZ, USNM.

Welsh Cave (27 spp.), Woodford Co. **(4–250)**

Situated on gently rolling farmland in the karst area just west of the Cumberland Plateau in central Kentucky, this small, featureless cave is a remnant of a collapsed cave system. Most of the material was laid down in relatively still water, and the skeletons, although in good condition, are disarticulated and scattered. The predominant member of the fauna is *Platygonus compressus* (31 individuals). The fauna indicates boreal-woodland conditions with a sympatric overlap today in the Transition–Canadian Life Zone ecotone of eastern Minnesota and western Wisconsin. A western faunal element is also present (*Taxidea taxus, Ursus arctos, Geomys, Spermophilus tridecemlineatus*), which indicates that tracts of open country existed nearby. A date of $12,950 \pm 550$ years B.P. has been obtained on bone. Excavation of the site was terminated by a change in ownership of the land. (Guilday, Hamilton, and McCrady, 1971.) CM.

Other Kentucky localities yielding Pleistocene faunas include Mammoth Cave (3 spp.), Edmonson Co. **(4–251)** (older than 38,000 years B.P.; Jegla and Hall, 1962); Savage Cave (31 spp.), Logan Co. **(4–252)**; Glass Cave (2 spp.), Franklin Co. **(4–249)**; and Woolper Creek Deposit (1 sp.), Boone Co. (type locality of *Bison latifrons;* C. B. Schultz, et al., 1963). (Guilday, personal communication.)

Louisiana

Avery Island (Petite Anse Island), (6 spp., Sangamonian; 9 spp., Wisconsinan), Iberia Parish **(4–243)**

Avery Island is one of a prominent string of southern Louisiana salt domes called Five Islands. It has been an important source of salt since prehistoric times, and the deposits continue to yield rock salt and petroleum today. The 3.2-km-wide, heavily forested hill arises abruptly from the salt marsh.

In the late Pleistocene, Avery Island was a much-frequented salt lick. Bones of extinct animals and human artifacts (matting found beneath proboscidean fossils) were brought to the attention of Joseph Leidy, who wrote several papers on the fauna beginning in 1866. Through the years, some 29 papers have been written about this rich assemblage. Two distinct faunas, one Sangamonian, the other Wisconsinan, are recognized. Large mammals, including *Mammut, Mammuthus, Glossotherium, Megalonyx, Equus,* and *Bison,* predominate. (Domning, 1969; Hay, 1924.) TU.

Maryland

Cavetown (16 spp.), Washington Co. **(4–292)**

A limestone quarry near Hagerstown, in north-central Maryland, has yielded a Wisconsinan-age fauna. Most of the bones are fractured. Five megafauna species are extinct, and woodland, plains, and marsh-dwelling species are represented. (Hay, 1920.) USNM.

Mexico (State of)

Tequixquiac (14 sp.), Mexico (State of) **(3–119)**

The Becerra Superior Formation, which overlies the Tacubaya Formation in the vicinity of Tequixquiac, in the northern part of the State of Mexico, has yielded a large-mammal fauna. The fauna, which may be heterochronous, has been considered to date from the late Sangamonian or the Wisconsinan. The presence of *Glyptotherium, Holmesina,* and *Euceratherium* is noted. A tallgrass and sedge paleohabitat is indicated. (Hibbard, 1955b.) IGM.

Mississippi

Natchez (13 spp.), Adams Co. **(4–244)**

Vertebrate fossils have been known from the Natchez area, in southwestern Mississippi, since before 1826, when the first note on them was published. Joseph Leidy's interest brought the Natchez fossils to the fore, and between 1847 and 1899, he published 14 papers dealing (in whole or in part) with this rich fauna and including the first description of *Felis atrox* (*Panthera leo atrox*).

The fossils were found in a blue clay deposit underlying a thick bed of loess in which Recent land snails have been discovered. The most controversial find was a human innominate bone ("Natchez man") found below *Megalonyx* material. Most authorities thought that the specimen was intrusive, perhaps washed down from an Indian grave near the surface, but fluorine tests (done in 1895) showed that it was more mineralized than the associated sloth material. However, this work was overlooked until 1951, when it was finally admitted that "Natchez man" and the extinct fauna were contemporaneous. No recent collections have been made at Natchez, and there has been no detailed study of the fauna. It is probably late Sangamonian or Wisconsinan in age and shows many similarities to the Avery Island fauna. (Domning, 1969; Hay, 1923; Leidy, 1853b.)

Missouri

Bat Cave (37 spp.), Pulaski Co. (4–237)

Pulaski County, in the Ozarks of south-central Missouri, contains more known caves (212) than any other county in the United States, and several of them have yielded late Pleistocene faunas. The main entrance of Bat Cave is located in the west-facing Gasconade River bluff and opens 10 m above the floodplain. The cave is developed in the Gasconade dolomite and contains about 1,100 m of passages at several levels. In "Bone Passage," bone has been recovered from the upper 30 cm of matrix, which consists of phreatic clay and residual material from the dolomite. Bones were deposited in the passage by moving water. Deposition probably occurred over a relatively short period of time since the assemblage is relatively homochronous. *Platygonus compressus* was the predominant animal in the fauna, and 27 percent of the mammals show northern or boreal affinities. (Hawksley, Reynolds, and Foley, 1973.) CMSU.

Boney Spring (23 spp.), Benton Co. (4–234)

Located along the Pomme de Terre River about 4.5 km NW of the Trollinger site, at an elevation of 212 m, this site contains a late Pleistocene fauna dominated by mastodont (31 individuals). Since it is an active spring, excavation was difficult. Besides mammals, the remains of fish, amphibians, and reptiles have been recovered. A representation of the fauna appears in figure 4.6. Pollen samples taken from the pulp cavities of mastodont tusks contain a high proportion of spruce pollen along with some of larch. The pollen spectrum is similar to that obtained recently from the western Great lakes region and northern Minnesota. Radiocarbon dates of 13,700 ± 600 and

4.6 Scene near Boney Spring, Missouri, in the late Rancholabrean. Shown are *Marmota monax, Mammut americanum,* and *Castoroides ohioensis.*

16,580±220 years B.P. have been obtained. Of the 23 species, 5 are extinct. (Lindsay, personal communication; Mehringer, King, and Lindsay, 1970; Saunders, 1977.) ISM.

Brynjulfson Caves (52 spp.), Boone Co. (4-232)

This site consists of two adjacent caves (horizontal shafts), located 4.5 m above Bonne Femme Creek, 10 km SSE of Columbia. The region around the cave is covered with glacial drift, mainly from the Kansan ice sheet, and supports a diversity of forest, prairie, and riverine habitats. Both caves were completely plugged with alluvial and loess-derived fill. For the most part, the faunas of the two caves are distinct. Cave No. 1 contains more extinct species (nine) than Cave No. 2 (three), and a radiocarbon date of 9,440 ± years B.P. was obtained for a bone sample from Cave No. 1. A rich microfauna was recovered by window-screen washing of fill from Cave No. 2. During the late Wisconsinian, a boreal forest covered much of the northern and central Ozark highland. In early Holocene times, a warm/moist, warm/dry period permitted shortgrass prairies and parklands to spread eastward, as reflected by the presence of *Taxidea, Geomys, Thomomys, Spermophilus franklini, Spermophilus tridecemlineatus,* and *Antilocapra americana* in Cave No. 2. (Parmalee and Oesch, 1972.) ISM.

Cherokee Cave (8 spp.), St. Louis Co. (4-238)

"Bones in the Brewery" was the title of Simpson's (1946) popular article describing the setting and fauna of Cherokee Cave, a site in southeast St. Louis that during the past 150 years has been used as a cool storeroom for beer, an underground beer parlor, and, during Prohibition, an illegal distillery. The cave was formed by solution localized along fissures in a limestone ridge, and its form suggests that it provided open underground runoff channels. Four distinct layers are present, with most of the bones concentrated in Bed No. 3. The bones probably accumulated in the fissures in which the animals were trapped and later washed into the cave. A discontinuity and dripstone above the fossil bed indicate considerable antiquity, and the deposit is late Wisconsinan/early Holocene in age. *Platygonus compressus* is the most abundant fossil species. (Simpson, 1946, 1949.) AMNH.

Crankshaft Cave (62 spp.), Jefferson Co. (4-240)

This site, named for the remains of at least two Model T Fords found at the bottom of the pit, is situated in a mixed-hardwood forest in an area of steep hills dissected by small streams. It is one of a series of caves developed in the Rockwoods anticlinal fold belt of east-central Missouri as a result of artesian circulation of water from a sandstone aquifer. The cave opening, measuring 0.6 × 1.4 m, resembles a shallow sink; the narrow shaft drops 18 m to the floor of the deposit. Bones were concentrated in the mud fill below the entrance. Groundwater has altered the chemical composition of the bones, so they cannot be dated, but the fauna suggests a late Wisconsinan/early Holocene age. Two distinct environments are represented. The earliest, a cool coniferous woodland, was replaced by a warm, dry period dominated by hardwoods. Of the 62 mammal species identified, 7 are extinct and 17 are no longer found in the state. (Parmalee, Oesch, and Guilday, 1969.) ISM.

Enon Sink (5 spp.), Cole Co. (4-233)

This sinkhole, in central Missouri, has yielded a predominantly large-herbivore fauna, only partially studied, as well as many turtles. It measures 12 m in depth and 30 m in diameter at the bottom, with a smaller surface opening. The fossils were found in blue clay intermixed with limestone blocks; above the fossiliferous zone, the hole is filled with cave earth. A late Wisconsinan age is likely. (Mehl, 1962; Saunders, 1977; Simpson, 1945b.)

Herculaneum (20 spp.), Jefferson Co. **(4–239)**

Located in east-central Missouri, near the town of Herculaneum, this limestone fissure has yielded a forest and plains-dwelling fauna. Seven species are extinct; most of the others are found in the vicinity of the cave today. *Odocoileus virginianus* is the most common large mammal. (Olson, 1940.)

Kimmswick (12 spp.), Jefferson Co. **(4–241)**

The Kimmswick bone bed, known for more than a century, is situated about 3 km W of the Mississippi River, at the base of a limestone bluff near a salt and sulfur spring. Most of the mammalian remains come from a locality near the confluence of Rock and Little Rock creeks and are overlain by a loess bed containing artifacts. The fauna consists of large mammals; only preliminary identifications are available. It is probably Wisconsinan in age. (Adams, 1953.)

Trollinger Spring (7 spp.), Hickory Co. **(4–235)**

Paleontological investigations of the spring deposits along the Pomme de Terre River have been going on for almost 200 years. A. K. Koch, among others, has made large collections of mastodont and other extinct mammals in the region. Trollinger Spring is located in a small field 16 m above and 1,800 m west of the Pomme de Terre River, at an elevation of 223 m, about 19.5 km SE of the town of Warsaw. It is a buried spring deposit with a feeder conduit. None of the bones is articulated, and some are waterworn. The pollen profile shows a significant change from the NAP-pine zone in the organic mud sediments (where the faunal remains are concentrated) to spruce dominance in the overlying blue-gray clay. Radiocarbon dates of 32,200 ± 1,900–1,600 and 25,650 ±700 years B.P. have been obtained from the pine zone and correspond to a mid-Wisconsinan interstadial. Mastodonts, horses, and muskoxen inhabited this parkland environment. (Mehringer, King, and Lindsay, 1970; Saunders, 1977.) ISM.

Zoo Cave (32 spp.), Taney Co. **(4–236)**

This previously undisturbed cave in central Taney County, about 1.6 km from Hilda at an elevation of 261.5 m, has yielded a late Pleistocene/early Holocene fauna. Two deposits, representing two time periods with different climates, habitats, and faunas, have been discovered. The fauna of the older one indicates a fairly warm, open prairie-like environment between 9,000 and 13,000 years ago; three extinct species (*Dasypus bellus, Canis dirus, Platygonus compressus*) have been found in this deposit. The younger deposit is stratified, and the fauna suggests a cooler climate with a prairie-parkland landscape, similar to that found in western Wisconsin and eastern Minnesota today. *Platygonus compressus* is the most abundant member of the fauna; as many as 81 (based on right upper canines) individuals may have been present. (Hood and Hawksley, 1975.) CMSU.

Many other Missouri caves, including Berone Moore Cave (1 sp.), Perry Co., (Oesch, 1969), Crevice Cave (7 spp.), Perry Co. (Oesch, 1969), Carroll Cave (2 spp.), Camden Co. (Hawksley Reynolds, and Foley, 1963), Perkins Cave (2 spp.), Camden Co., (Hawksley, 1965), and Powder Mill Creek Cave (1 sp.), Shannon Co. (dated at 13,170 ± 600 years B.P., Galbreath, 1964), have yielded late Pleistocene fossils.

Montana

Doeden (11 spp.), Custer Co. **(3–116)**

A Sangamonian-age local fauna has been recovered from a gravel pit located on a high terrace of the Yellowstone River just north of Miles City. Large grazing mammals predominate. The climate was probably warmer and drier than it is now. The fauna is under study. (M. Wilson, personal communication, 1977.)

Warm Springs 1 (10 spp.), Silver Bow Co. **(4–197)**

This fauna of small mammals occurs in a dissected alluvial fan sloping toward the Clark Fork River floodplain. Presence of an ash bed in the uppermost levels of a similar fan just north of the site suggests that both localities are of late Pleistocene, rather than Holocene, age. Several other sites in the same area have yielded small late Pleistocene/early Holocene faunas that include extinct species (*Camelops, Equus*). [See Rasmussen (1974) for preliminary details.]

Nebraska

Red Willow (24 sp.), Red Willow Co. **(4–231)**

Bones from wet gravel pits on the Upper Republican River in southwestern Nebraska have been collected for many years under the Paleontological Highway Salvage Program of the University of Nebraska State Museum. The fauna, sampled from several depositional cycles, is Rancholabrean, predominantly middle to late Wisconsinan in age, with a few Holocene fossils; associated artifacts indicate a maximum age of about 8,000 years B.P. for some of the Holocene fossils. Most fossils are Pleistocene megafauna, with *Bison* representing about 80 percent of the fauna. The presence of *Rangifer* and *Ovis* reflects cold conditions. Twelve species are extinct. (Corner, 1977; Corner and Myers, 1976.) UNSM.

Nevada

Glendale (14 spp.), Clark Co. **(4–178)**

A late Rancholabrean fauna has recently been recovered from a site about 70 km NE of Las Vegas, in extreme southern Nevada. A sequence of beaver dams, probably representing a former beaver meadow, and the presence of amphibians, snapping turtle (a range extension of 1,250 km for *Cheledra serpentina*), 11 species of water birds, beavers, and muskrats suggest a much wetter environment. The fauna is under study. (Tessman, personal communication, 1974.) UALP.

Gypsum Cave (14 spp.), Clark Co. **(4–177)**

The cave is located 30 km E of Las Vegas, in the foothills of Frenchman Mountain, at an elevation of 454 m. It consists of several chambers formed by differential solution in the Paleozoic limestone. At the top of the deposit of fragmented limestone is a gypsiferous layer, above which is a layer of sloth dung that reaches a maximum depth of 66 cm. *Nothrotheriops shastensis* was the most abundant member of the fauna and is represented by skeletal parts, hide, hair, and dung. Plants recovered from the dung now grow at higher elevations, which indicates that a somewhat more humid climate existed in this area in the late Pleistocene. A stone dart point of an atlatl, a worked stick, burnt dung, and ashes have been found in the same layer as a limb bone of *Nothrotheriops* and charred and broken limb bones of *Hemiauchenia*. Mehringer (1967) notes that dates of 8,527 ± 250 and 10,455 ± 340 years B.P. obtained on solid carbon may not be reliable; a more reliable date of 11,690 ± 250 years B.P. was obtained on a dung sample. (Laudermilk and Munz, 1934; Mehringer, 1967; Stock, 1931.) LACM.

Smith Creek Cave (22 spp.), White Pine Co. (4–175)

This huge limestone cavern, located near the Nevada–Utah border, 33 km N of Baker at an elevation of 1,878 m, contains a late Wisconsinan/early Holocene fauna. The cave is situated above the mouth of Smith Creek Canyon, 242 m above the valley floor. The avian fauna is extensive, with 50 species identified, 6 of which are extinct. *Oreamnos harringtoni* was described from this site. New excavations and faunal studies are in progress. (Howard, 1952; S. Miller, personal communication, 1975; Stock, 1936a.) LACM.

Tule Springs (20 spp.), Clark Co. (4–176)

The site is located in the Las Vegas valley, 16 km N of Las Vegas at an elevation of 699 m. Most of the bones were recovered during archeological investigations. Claims were made that human artifacts had been found in association with bones of extinct mammals dated at more than 23,000 years. Additional studies, however, showed that the bones and artifacts were not found in association, that the "burned bones" were actually discolored by mineral staining, and that the oldest artifacts were 13,000 years old. Pollen studies indicate that during pluvial periods (40,000+ to 34,000 years and 23,000 to 10,000 years B.P.) the area was cooler and wetter, with a juniper–piñon woodland and possibly a yellow pine parkland environment, which indicates a depression of vegetational zones of at least 1,200 m. A small but diverse avian fauna, including several species of ducks and the extinct giant condor, have been identified. Grazing animals predominate in the fauna, and *Mammuthus* and *Camelops* are the most abundant large mammals. (Mawby, 1967; Mehringer, 1965; Shutler, 1968.) UCMP.

NEW MEXICO

Blackwater Draw (Gray Sand Unit, 15 spp.; Brown Sand Wedge, 25 spp.), Lea Co. (4–205)

Late Pleistocene fossils have been recovered from two units at Blackwater Draw, a stratified early man site near Portales, in eastern New Mexico. The Gray Sand unit, the older of the two, appears to have been deposited in a spring-fed pond. Three skeletons of *Mammuthus* were semiarticulated and intimately associated with flint implements. Evidence indicates that the mammoths, representing at least two kills, were dispatched while drinking at the pond and were butchered there. Besides *Mammuthus,* the extinct fauna includes *Canis dirus, Smilodon, Platygonus, Camelops, Hemiauchenia, Bison,* and *Equus.* The age of the Gray Sand unit is perhaps more than 16,000 years.

The wedge shape of the overlying Brown Sand member suggests an outwash pediment or delta of a medium-sized feeder stream. The presence of *Dasypus bellus* indicates a climate with more moisture, cooler summers, and winters no more severe and perhaps milder than occur there today. Four species are extinct, and most of the others are not found in Blackwater Draw today. Carbonaceous material surrounding a mammoth skull has been dated at 11,170 ± 360 years B.P. (Lundelius, 1972b; Slaughter, 1964).

Burnet Cave (41 spp.), Eddy Co. (4-206)

Burnet Cave, a late Pleistocene deposit, is located on the east slope of the Guadalupe Mountains, about 80 km W of Carlsbad, at an elevation of 1,403 m. Several hearths and artifacts were found at the same level as the fossil fauna and indicate that the cave was used as a shelter or an occupational site. Rodents, rabbits, carnivores, and Pleistocene herbivores including *Equus, Camelops, Navahoceros, Tetrameryx, Euceratherium,* and *Bison* are represented. *Navahoceros fricki* was described from here. Above this level are the remains of a Basket Maker culture. (Schultz and Howard, 1935.) ANSP, UNSM.

Clovis (8 spp.), Curry and Roosevelt Cos. (4-204)

Artifacts associated with the bones of extinct mammals have been found in a shallow lake basin on a floodplain about 13 km S of Clovis, in east-central New Mexico. Several individual kills of mammoths are represented, and it is likely that the animals were killed while knee-deep in the pond drinking. A radiocarbon date of 11,170 ± 360 years B.P. has been obtained on silty clay surrounding a mammoth skull. Of the eight species identified, four are extinct. (Hester, 1967; Stock and Bode, 1936.)

Dry Cave (54 spp.), Eddy Co. (4-207)

A pluvial fauna was recovered from Dry Cave, an extensive maze cavern located about 24 km W of Carlsbad, in southeastern New Mexico, at an elevation of 1,281 m. Vegetation at the time of deposition probably included sagebrush grassland on the northern slopes, upper Sonoran grassland on the southern slopes, and heavy riparian growth along drainageways; groves of trees were limited to steep slopes. Cool summers, mild winters, and ample winter precipitation resulted in a complex mixture of northern and southern faunal elements. The fauna is especially rich in rodents and carnivores. Seven species are extinct. Most extralimital extant species are found today in the Transition Zone of central Wyoming; others inhabit the high mountains of the Southwest. A radiocarbon date of 14,470 ± 250 years B.P. has been obtained on woodrat dung. (Harris, 1970.) MALB.

Isleta Cave (42 spp.), Bernalillo Co. (4-203)

This late Pleistocene/early Holocene assemblage was found in two adjacent caves located on an old lava flow 13 km W of the Indian pueblo of Isleta, in central New Mexico, at an elevation of 1,717 m. Five species are extinct, and the extant fauna resembles that found in southeastern Wyoming and north-central Colorado today. This suggests that the climate was somewhat cooler, especially during the summer, than at present, with more precipitation, much of it falling during the spring. The environment around the cave was probably a sagebrush grassland with few trees. (Harris and Findley, 1964.) MSB.

NUEVO LEÓN, MEXICO

San Josecito Cave (45 spp.) (4-208)

The cave is in mountainous terrain, 2,300 m above sea level, near Aramberri in southern Nuevo León; it is now surrounded by pine and live oak forest. The rich ver-

4.7 Scene at San Josecito Cave, Nuevo León, Mexico in the late Rancholabrean. Shown is *Panthera onca* feeding on *Navahoceros fricki*. Other members of the fauna are *Oreamnos harringtoni*, *Stockoceros conklingi*, and *Desmodus stocki*.

tebrate fauna includes 2 species of iguanid reptiles and more than 43 species of birds. It may be heterochronous, dating mainly from the Wisconsinan, but with some Sangamonian elements, including *Pappogeomys* and *Thomomys*. The presence of these two taxa suggests a more arid interglacial environment, although a mesic boreal situation during Wisconsinan times may be indicated by the presence of *Sorex cinereus*, which is now found no nearer than 1,300 km to the north, in the mountains of New Mexico (Russell, 1960). The predominant large mammals are *Navahoceros fricki* and *Stockoceros conklingi*. the cave probably was a carnivore lair, the most common predators being *Canis dirus*, *Canis latrans*, and *Felis concolor*, but the bears *Ursus americanus* and *Tremarctos floridanus* are also well represented. Also present are the rare species *Cuon alpinus* (Nowak, personal communication) and *Oreamnos harringtoni*. A representation of the fauna appears in figure 4.7. (Jakway, 1958; L. H. Miller, 1943; Nowak, personal communication; Russell, 1960; Stock, 1950.) LACM.

OHIO

Carter (15 spp.), Darke Co. **(4–247)**

A peat bog on a farm near Ansonia, in west-central Ohio, has yielded a late Wisconsinan fauna that includes mollusks, turtles, birds, and mammals. Pollen analysis shows the presence of *Picea*, *Pinus*, *Quercus*, and *Populus*. The remains of *Ondatra*, *Castoroides*, and *Castor* indicate a marshy environment. The largest known specimen of *Megalonyx jeffersonii* was recovered here. Of the 15 species of mammals identified, 4 are extinct. (Mills, 1975 and personal communication, 1972; Mills and Guilday, 1972.) DOMNH.

Oklahoma

Afton (18 spp.), Ottawa Co. **(4–228)**

This site, in northeastern Oklahoma, is a spring in which remains of a Pleistocene fauna have been found together with artifacts. Most species are megafauna; aquatic, browsing, and grazing forms are present. The age is probably late Wisconsinan. (Hay, 1920.) USNM.

Doby Springs (23 spp.), Harper Co. **(3–101)**

The lake sediments which have produced this fauna were deposited in a collapse basin formed in early Illinoian times as a result of soaking out of the Permian bedrock. Harper County is located southeast of Meade County, Kansas, and early Rancholabrean faunas from the two areas show similarities. The fauna (vertebrates, mollusks, etc.) indicates a cooler environment than at present. Present-day sympatry of extant mammals is a small area straddling the eastern end of the North and South Dakota state line; sympatry of fishes extends from the same region east into western New York state. Communities represented in the fauna are lake and marsh border, lowland meadow, shrub and tree, and upland prairie. Extant mammals make up 58 percent of the fauna. An Illinoian date is indicated. (Stephens, 1960.) UMMP.

Domebo (8 spp.), Caddo Co. **(4–227)**

Domebo is a Paleo-Indian mammoth kill site located in a deep arroyo on the eastern margin of the Great Plains, in southwestern Oklahoma. Two Clovis-type projectile points have been discovered near the vertebrae and ribs, but no cut or hack marks were found on the mammoth bones. The associated fauna, including frogs, snakes, turtles, and rodents, suggests a more equable climate, with permanent water, tall grass, and nearby woods. The pollen profile is dominated by grasses and composites, and radiocarbon dates of about 11,200 years B.P. have been obtained on wood and bone. (Leonhardy, 1966.)

Ontario

Ottawa, Green's Creek area (3 spp.) **(4–297)**

The Champlain Sea was a prominent feature of the landscape in eastern North America during the late Wisconsinan/early Holocene. It was formed when the Laurentide ice sheet retreated north of the St. Lawrence Valley and the Atlantic Ocean flooded the depressed trough to create a large inland sea. At its maximum, about 11,500 years ago, it covered an area of about 53,100 km^2 in Ontario and Quebec, between Quebec City and Lake Ontario, the Lower Ottawa River and Lake Champlain valleys. At this time, the climate was probably similar to that of the Gulf of St. Lawrence today. Life was abundant and varied both in the sea and along its shores. Plant and mollusk fossils are very common, and remains of marine mammals, many adapted to breed on pack ice, have been found throughout the area; land vertebrates, though, are rare (see Harington, 1977).

At Green's Creek, near Ottawa, vertebrates including *Phoca groenlandica* and *Martes americana* have been preserved in clay nodules believed to be more than 10,000 years old. The marten specimen consists of the skull, neck, and forelimbs of an adult and suggests the former presence of boreal forests near the southwestern edge of this inland sea. During its later phases, the Champlain Sea became warmer and less saline, until it reached the freshwater stage, Lampsilis Lake, which persisted from about 10,000 to 8,500 years ago, at which time it drained and the landscape assumed its present form. (Harington, 1971c, 1977, 1978.) NMC.

Toronto, Don River Valley (7 spp., Sangamonian; 3 spp., Wisconsinan) **(3–137)**

Pleistocene fossils have been found in the Don Formation (Sangamonian) in the Toronto area. It is believed that temperatures were about 5°F warmer then than they are today. Although plant and invertebrate remains are common, only a few fish and mammal fossils have been collected.

Sand and gravel deposits, probably of Wisconsinan age, in the Toronto area have yielded mammoth, caribou, and muskox fossils. (Harington, 1971c, 1978.) NMC.

OREGON

Fossil Lake (22 spp.), Lake Co. **(4–174)**

This Wisconsinan fauna was recovered from a deflation basin in south-central Oregon, at an elevation of 1,300 m. Grassland, forest, and marsh-dwelling species are represented, and half of the species are extinct. The climate was probably more equable, with greater precipitation, than that of today. (Allison, 1966, Elftman, 1931.)

PENNSYLVANIA

Bootlegger Sink (40 spp.), York Co. **(4–295)**

Forty species of mammals have been recovered from a breccia deposit 8 cu m in extent in southeastern Pennsylvania. Although none of the species is extinct, 10 are no longer found in the area today. Included are such typical late Pleistocene Appalachian species as *Sorex arcticus, Synaptomys borealis, Microtus xanthognathus, Microtus chrotorrhinus,* and *Rangifer tarandus; Spermophilus tridecemlineatus,* a western species, is also present. The fauna is believed to be mixed, both chronologically and ecologically. (Guilday, Hamilton, and McCrady, 1966.) CM.

Durham Cave (13 spp.), Bucks Co. **(4–296)**

Although this east-central Pennsylvania cave has been known since the late 1890s, the fauna has never been written up. *Bison appalachicolus* Rhoads was described from Durham; it has now been tentatively referred to *Symbos cavifrons*. (Harington, personal communication; Ray, 1966a.)

Frankstown Cave (31 spp.), Blair Co. **(4–294)**

Limestone quarrying operations near Frankstown, in south-central Pennsylvania, revealed the presence of a large fissure approximately 12.1 m long, 1.8–2.4 m wide, and 3.0–3.6 m high, containing a Wisconsinan-age fauna. The bones had been disturbed and fractured by roof fall. Three partial skeletons of muskoxen were recovered, as well as other Pleistocene megafauna, shrews, bats, rodents, and carnivores. Forest forms predominate. Of the 31 species identified, 11 are extinct. (Guilday, 1961; Holland, 1908; Peterson, 1926.) CM.

Hosterman's Pit (11 spp.), Centre Co. **(5–364)**

The fauna of Hosterman's Pit, in central Pennsylvania, is modern and includes such temperate species as *Glaucomys volans, Synaptomys cooperi,* and *Microtus pinetorum.* Charcoal, presumably from forest fires, has been dated about 9,000 years B.P. Thus between about 11,300 years B.P. (New Paris No. 4) and 9,000 years B.P. (Hosterman's Pit), the regional fauna changed from boreal to temperate. (Guilday, 1967c.) CM.

New Paris No. 4 (43 spp.), Bedford Co. **(4–293)**

A sinkhole, New Paris No. 4 is located on the west flank of Chestnut Ridge, just east of the Appalachian Plateau, at an elevation of 465 m. Excavation of the 10-m column of surface-derived matrix has yielded a late Pleistocene biota consisting of 2,700 vertebrates, with an accompanying pollen profile and C-14 date of 11,300 ± 1,000 years B.P. The indicated environment is cool taiga parkland during the early stages of infilling (Hudsonian Zone fauna, *Dicrostonyx hudsonius, Synaptomys borealis, Microtus xanthognathus*), followed by a gradual transition to a boreal-temperate forest (Canadian Zone species, *Sorex arcticus, Clethrionomys gapperi, Phenacomys intermedius*). The rate of postglacial filling of the sinkhole is estimated to be from 0.5 to 1.0 or more m per 1,000 years. Besides mammals, the fauna includes arthropods, mollusks, amphibians, reptiles, and birds. Shrews, bats, and mice are numerous. Only *Mylohyus nasutus,* the largest species, is extinct. A representation of the fauna appears in figure 4.8. (Guilday, Martin and McCrady, 1964.) CM.

4.8 The late Rancholabrean Appalachian fauna near New Paris No. 4 included *Martes americana, Lepus americanus, Mylohyus nasutus, Rangifer tarandus, Tamiasciurus hudsonicus,* with *Pedioecetes phasianellus,* the sharp-tailed grouse, flying overhead.

Saskatchewan

Fort Qu'Appelle (10 spp.), (3–126)

Vertebrate fossils have been collected from the Bliss gravel pit, just southwest of Fort Qu'Appelle. The gravel is found between two till sheets, and drillholes in the vicinity indicate that the gravel is nearly 24 m thick. A steppe environment is envisioned, since grazing animals predominate. The age of the fauna is either Sangamonian or early Wisconsinan interstadial. (Kahn, 1970; M. Wilson, 1972.) SMNH, NMC.

Saskatoon area (7 spp.) (3–125)

Sites including Saskatoon, Sutherland gravel pit, and Riddell, on either side of the South Saskatchewan River downstream from the town of Saskatoon, have yielded a Sangamonian interglacial, or possibly Wisconsinian interstadial, fauna. Stratigraphic studies and carbon-14 dates suggest an age greater than 34,000 years B.P. for the Floral Formation. The fauna is similar to that of Fort Qu'Appelle, and most of the species are grassland forms. (Harington, 1978).

South Carolina

Ashley River (19 spp.), Dorchester Co. (3–106)

The fossiliferous beds along the Ashley River have been known and exploited for over a century. The principal locality is about 15 km above Charleston. The beds consist of a thin mud deposit (yielding most of the bones) overlain by clay-banded sands; they rest upon Miocene marls. The fauna may be heterochronous but is probably in part Illinoian or Sangamonian (*Bison latifrons, Arctodus pristinus*). Most species are megafauna, and included are *Eremotherium, Glossotherium, Mammut, Mammuthus, Equus,* and *Tapirus*. (Hay, 1923.) ANSP, CMSC.

South Dakota

Roosevelt Lake (5 spp.), Tripp Co. (4–298)

A small microfauna has been recovered from fine sands of a late Pleistocene stream channel that overlies Tertiary sediments in the south-central part of the state. It is the only Rancholabrean fauna thus far recorded from South Dakota. *Cynomys spispiza* Green (now referred to either *C. ludovicianus* or *C. niobrarius*) was described from Roosevelt Lake. (Green, 1963; R. A. Martin, 1973b.) SDSM.

Tennessee

Baker Bluff Cave (61 spp.), Sullivan Co. (4–264)

The cave is located high on the eastern-facing exposure of a ridge cut on the west bank of the South Fork Holston River, in extreme northeastern Tennessee, at an elevation of 439 m. It consists of a single large chamber measuring 3.6 × 9.0 m that was filled to within 1.2 m of the ceiling prior to excavation. There was no obvious stratification, and pollen analysis was negative. Radiocarbon dates run on charred bone fragments from below 1.8 m indicate a late Wisconsinan age. The extinct fauna includes *Dasypus bellus, Castoroides, Panthera onca augusta, Tapirus, Platygonus compressus,* and *Sangamona*. The extant fauna is composed of eastern temperate, midwestern grassland, and boreal forest species. Evidence of an early Archaic human occupation is present in the upper 90 cm of the deposit. (Guilday, personal communications; Guilday et al., 1978; Parmalee, Bogan, and Guilday, 1976.) CM.

Craighead Caverns (3 spp.), Monroe Co. **(4–257)**

Located 8.8 km SE of Sweetwater, in eastern Tennessee, Craighead Caverns is the largest cave in the region. It was formed by solution of the Paleozoic limestone and consists of a series of large rooms, the lowest of which contains a navigable underground lake. Bones and footprints of jaguar (also scratch marks on a wall, probably made by this cat) were found in one of the rooms. Today the cave is a tourist attraction. (Simpson, 1941b.) AMNH.

First American Bank Site (21 spp.), Davidson Co. **(4–299)**

During the excavation for the First American Bank of Nashville, a fissure containing late Pleistocene, late Prehistoric, and Recent bones was discovered. The site is located within the city of Nashville, on the south bank of the Cumberland River in north-central Tennessee, at an elevation of 133 m. Although the original configuration of the fissure was destroyed by the excavation, it appears that a vertical shaft extended from the surface downward through the Bigby–Cannon Limestone to the domed solution cavern. Animals probably tumbled down the shaft and could not escape. The deposit consisted of two sub-sites. The upper feature contained four Indian burials and remains of 16 species of Recent local vertebrates, dated between 2,400 and 1,600 years B.P. The lower feature contained bones of extinct animals, including *Smilodon, Equus, Mylohyus, Mammut,* and an ovibovid. Bone collagen and bone apatite from *Smilodon* were dated at 9,410 ± 155 and 10,035 ± 650 years B.P., respectively; these are the latest published dates for the extinction of the sabertooth. The first fossil occurrence of *Geomys* in Tennessee is also noted. (Guilday, 1977.) VU.

Kyle Quarry Cave (6 spp.), Monroe Co. **(4–258)**

This site is located 1.2 km south of the entrance of Craighead Caverns, and the two caves may be connected by underground passages. A late Pleistocene fauna, including *Tapirus* and an unidentified peccary, was recovered. (Simpson, 1941b.) AMNH.

Robinson Cave (52 spp.), Overton Co. **(4–253)**

This large, awesome cavern is highly decorated with cave formations. Bone has been recovered in two areas: the Sloth Pit, where two partial skeletons of *Megalonyx* were found, and the Armadillo Pit, where remains of at least 2,615 individual mammals, 14 birds, 10 species of reptiles and amphibians, and 461 individual gastropods were taken from approximately 4 cu m of matrix on the floor of the pit. Eighty-five percent of the mammalian fauna consists of species that habitually inhabit caves, and the remaining 15 percent includes animals that were trapped in the sinkhole or were the victims of predators. Six species are extinct, and 91 percent of the extant mammals of the Armadillo Pit inhabit the Minnesota–Wisconsin region today. The fauna of Robinson Cave is Wisconsinan in age. (Guilday, Hamilton, and McCrady, 1969.) CM.

Whitesburg (16 spp.), Hamblen Co. **(4–260)**

Collected in 1885, this poorly documented fauna was later studied by O. P. Hay. The matrix surrounding the bones indicates that the material came from a cave deposit. The genus *Sangamona* was first described from Whitesburg; in addition, *Equus, Tapirus, Mylohyus, Mammuthus,* and *Canis dirus* are present. The fauna is probably Wisconsinan in age. (Hay, 1920.) USNM.

Late Pleistocene/early Holocene faunas have been recovered from several other sites in Tennessee, including Big Bone Cave (9 spp.), Van Buren Co. **(4–255)**, (Guilday and McGinnis, 1972; Hay, 1923); Cave on Lookout Mountain (6 spp.), Monroe Co. **(4–259)**, (Hay, 1923); Saltpeter Cave (1 sp.), White Co. **(4–254)**, (McCrady, Kirby-Smith, and Templeton, 1951); Little Salt River Cave (1 sp.), Franklin Co. **(4–256)**, (McCrady, Kirby-Smith, and Templeton, 1951); Grassy Cove Saltpeter Cave (2 spp.), Cumberland Co. (Guilday and Irving, 1967); Guy Wilson Cave (27 spp.), Hamblen Co. **(4–263)**, (dated 19,700 ± 600 years B.P., Guilday, 1971b); Riverside Cave (24 spp.), Hamblen Co. **(4–262)**; Cave North of Whitesburg (17 sp.), Hamblen Co. **(4–261)**; and Carrier Quarry Cave (9 spp.), Sullivan Co. **(4–265)**; the last four sites are presently being studied. (Guilday, personal communication.)

Texas

Ben Franklin (22 spp.), Fannin and Delta Cos. **(4–226)**

The Sulphur River Formation in northeastern Texas consists of Pleistocene alluvium and has yielded a local fauna of mammals and mollusks. It has been exposed by downcutting of the riverbed subsequent to artificial channeling for reclamation purposes.

The fauna includes northern forms such as *Sorex cinereus* and *Microtus pennsylvanicus*. Cooler summer temperatures and greater rainfall than at present are suggested. Radiocarbon dates for mussel shells and a possibly man-made hearth are 11,135 ± 450 and 9,550 ± 375 years B.P., respectively. (Slaughter and Hoover, 1963.) SMUMP.

Cave Without A Name (29 spp.), Kendall Co. **(4–215)**

The fauna of this cave, on the margin of the Edwards Plateau in central Texas, includes four extinct species (*Dasypus bellus,* mastodont, an ovibovid, and a horse), as well as four mammals of northern affinities that are now extinct locally. For three of the latter (*Sorex cinereus, Microtus pennsylvanicus, Mustela erminea*) it is the last record in the area; the fauna is dated at 10,900 ± 190 years B.P. (Lundelius, 1967.) TMM.

Clamp Cave (13 spp.), San Saba Co. **(4–219)**

A late Rancholabrean fauna of mostly large mammals with horse, camelids, *Platygonus,* and several carnivores suggests that this cave was used as a carnivore den. (Kurtén, 1963a; Lundelius, 1967.) TMM.

Clear Creek (31 spp.), Denton Co. **(4–221)**

This varied fauna of mammals with associated mollusks was collected in two gravel pits in the second terrace above Clear Creek, in north-central Texas. Slaughter and Ritchie (1963) have referred it to the Sangamonian on the basis of the merging of the terrace with the T-2 terrace of the Trinity River and the presence of cf. *Bison latifrons*.

A radiocarbon date of 28,840 ± 4,740 years B.P., based on mussel shells, is too young if the age is Sangamonian; however, this date is now regarded as valueless (Lundelius, 1972a). The fauna suggests a warm, dry environment with grazers predominating. (Lundelius, 1972a; Slaughter and Ritchie, 1963.) SMUMP.

Easley Ranch (33 spp.), Foard Co. (3–121)

The Good Creek Formation, situated in a valley cut through the Irvingtonian-age Seymour Formation at the eastern base of the panhandle, consists of stream, marsh, and pond deposits and carries a fauna judged to be of Sangamonian age. The mollusk fauna includes several species now found only farther north; however, the mammals indicate a warmer climate than at present and compare closely with those from Jinglebob. *Bison latifrons* is present. (Dalquest, 1962.) MWU.

Friesenhahn Cave (32 spp.), Bexar Co. (4–214)

Situated 33 km N of San Antonio, this is the only richly fossiliferous cave known in the area. An inclined entrance leads to a chamber about 20 × 10 m in size, where a freshwater pond may have attracted the animals (mainly large carnivores) that used the cave as a den. *Homotherium* is the most abundant carnivore; cubs as well as adults are present. Hundreds of proboscidean milk molars (mostly mammoth, but also some mastodont) may represent their prey. Complete skeletons of *Homotherium* and *Mylohyus* are notable. Of edentates, only an undetermined sloth has been found. Rodents, all extant species, are common. The fauna, which also includes birds, reptiles, and amphibians, suggests a rocky, upland environment; a representation appears in figure 4.9. The age is Wisconsinan. A few flaked flints occur in the Pleistocene context, but their origin is controversial. (Evans, 1961; Hay, 1920; Lundelius, 1960; Meade, 1961; Milstead, 1956; Sellards, 1919.) TMM.

Hill-Shuler (47 spp.), Moore Pit (40 spp.), and Coppell (24 spp.), Dallas
    and Denton Cos. (4–222, 4–223, 3–122)

Several Pleistocene local faunas have been recovered from separate commercial sand and gravel pits in the second terrace (T-2) above the floodplain of the Trinity River in Dallas (see Slaughter *et al.*, 1962). The terrace fill consists of three members, the basal Hill gravels, the lower Shuler sand, and the upper Shuler sandy clay, a sequence that suggests gradually decreasing moisture and carrying power. Pollen and mollusks also indicate a transition from a heavily timbered to a more arid environment. A radiocarbon date from hearth charcoal at the Lewisville site 37,000+ years B.P., is contradicted by association with Llano culture remains.

Continued commercial excavations joined several of the pits, and continued scientific excavation at Moore Pit has produced every species known from the other localities, so they are brought together under a single assemblage called the Moore Pit local fauna. The sites occur in an old valley fill believed to have formed between 50,000 and 25,000 years ago. The climate was probably more equable than that now prevailing. The fauna includes mollusks, arthropods, fish, amphibians, reptiles, birds, and mammals including 19 extinct species.

Additional studies have revealed another assemblage, the Coppell local fauna, which is probably Sangamonian in age and includes both small and large mammals. It is under study. (Crook and Harris, 1957; Gillette, personal communication; Slaughter, 1966b; Slaughter et al., 1962.) SMUMP.

4.9 In south-central Texas near Friesenhahn Cave, *Didelphis virginiana, Homotherium serum, Mammuthus jeffersonii,* and *Mephitis mephitis* were faunal associates in the Rancholabrean. Baby mammoths were a favorite food of *Homotherium.*

Howard Ranch (40 spp.), Hardeman Co. (4–220)

This is a rich, predominantly small-mammal fauna from Groesbeck Creek, a tributary of the Red River at the base of the Texas Panhandle. The Groesbeck Formation, consisting of sands and gravels capped with gypsiferous clay, was deposited in a collapse basin in the underlying Permian rocks. The area was at that time covered by a lake of considerable size, with islands formed by limestone-capped hills. Crayfish, ostracods, mollusks, fishes, amphibians, and snakes are abundant as fossils, lizards, turtles, and birds less so. All the small mammals belong to extant species; however, many are not found in the area, and many have no range overlap today. Five species are northern, six eastern, and seven typical of the Great Plains. Seven species, all large, are extinct. Biota represented include lakeside mellow soil and woodland, as well as upland prairie. Rainfall was heavier, and summers were cooler, but winters were probably warmer than now. Shells have yielded the date $16,775 \pm 565$ years B.P. (Dalquest, 1965.) MWU.

Ingleside (31 spp.), San Patricio Co. (3–120)

The Ingleside fauna comes from freshwater pond deposits overlying a marine lagoonal clay, located 2 km E of the town of Ingleside on the southern Gulf Coast. The lagoon, bounded toward the sea by Live Oak Ridge, then a barrier island, was evidently in existence during the Sangamonian interglacial rise; thus the freshwater deposits are assumed to date from the late Sangamonian or early Wisconsinan. Freshwater fishes and mollusks, amphibians, and reptiles are also found, and presence of the tortoises *Gopherus* and *Geochelone* indicates mild winters. Among the mammals, large forms are in the majority. Apparent annual age grouping in the *Hemiauchenia* sample may suggest seasonality of climate. This llama, as well as the glyptodonts, pampatheres, and tapirs, has southern affinities and suggests forested conditions, as does the abundant presence of mastodonts. (Lundelius, 1972a.) TMM.

Kincaid Shelter (12 spp.), Uvalde Co. (4–213)

Kincaid Shelter, a late Rancholabrean site in central Texas, has yielded the remains of seven extinct species, including the only record of *Capromeryx* from this region. (Lundelius, 1967.) TMM.

Klein Cave (44 spp.), Kerr Co. (5–357)

This shallow limestone cave on the Edwards Plateau in central Texas has yielded a Holocene fauna about 8,000 years old. Compared to the semiarid conditions found on the plateau today, the environment was milder, wetter, and supported a richer vegetation. The cave served as a bat roost, and eight species of vespertilionids have been identified, five of which no longer occur in the area. Ten species from the fossil fauna are absent from the area today, and four of these (*Tamias striatus*, *Microtus pennsylvanicus*, *Synaptomys cooperi*, and *Mustela erminea*) no longer inhabit Texas. (Roth, 1972.) MWU.

Laubach Cave (13 spp.), Travis Co. (4–216)

This sealed cave was discovered during highway construction in Georgetown, about 50 km N of Austin. A sinkhole trap, the cave has mainly produced large mammals, with *Platygonus compressus* the dominant species. Two species of bats (*Myotis magnamolaris* and *M. rectidentis*) were described from Laubach. Radiocarbon dates (Valastro, Davis, and Varela, 1977) of $13{,}970 \pm 310$, $15{,}850 \pm 500$, and $23{,}230 \pm 450$ years B.P. indicate a late Wisconsinan age. (Choate and Hall, 1967; Lundelius, personal communication; Slaughter, 1966a; Valastro, Davis, and Varela, 1977.) SMUMP, TMM.

Levi Shelter (17 spp.), Travis Co. (4–217)

This Paleo-Indian campsite in central Texas has produced a fauna with four extinct species (*Canis dirus*, *Platygonus compressus*, *Tapirus veroensis*, and *Equus* sp.). Radiocarbon dates are available for two out of the four recognized stratigraphic zones (of which Zone 1 is the oldest): Zone II, $10{,}000 \pm 175$ years B.P.; and IV, $9{,}300 \pm 160$ and $6{,}750 \pm 150$ years B.P. The only extinct taxon ranging up to Zone II is *Platygonus*. (Alexander, 1963; Lundelius, 1967.) TMM.

Longhorn Cavern (29 spp.), Burnet Co. (4–218)

The basal "Longhorn Breccia" of this central Texas cave has a late Wisconsinan fauna, with *Equus*, *Camelops*, and *Hemiauchenia*; other mammals are extant. The overlying deposits contain a Holocene fauna. (Lundelius, 1967.) TMM.

Lubbock Lake (23 spp.), Lubbock Co. (4–210)

Lubbock Lake has long been famous as an early man site and was the first to provide a radiocarbon date ($9{,}883 \pm 350$ years B.P.) for a Folsom culture. Recent interdisciplinary studies have shown a sequence of changes in the climate, soil, topography, and inhabitants during the past 12,700 years. This complex series of sites is located in Yellowhouse Canyon, in the northwestern part of the city of Lubbock, on the Llano Estacado (southern High Plains). In Stratum 1 (the lowest level), bones of several extinct mammals, including *Mammuthus* and *Equus*, have been found along with Clovis points. A radiocarbon date of 12,650 ± 350 years B.P. has been obtained on clam shells from this level. Footbones of *Arctodus simus* showing distinct cut marks were also found in this stratum (E. Johnson, personal communication). Strata 2A and 2B contain the Folsom culture, along with bones of *Capromeryx* and *Bison bison antiquus;* the rest of the fauna from this level is essentially modern. Above it, in successive layers, the presence of Plainview, Archaic, and Ceramic cultures has been documented. At the time of the mammoths, camelids, peccaries, and horses, about 12,600 years ago, the climate in the area was cooler and more humid and supported a savanna grassland. (Black, 1974; E. Johnson, personal communication, 1978.) TTM.

Miller's Cave (24 spp.), Llano Co. (5–358)

Miller's Cave is located in the Riley Mountains, 21 km SE of the town of Llano, in central Texas, at an elevation of 412 m. The cave consists of two chambers; the south one is fossiliferous, and bone has been recovered from a lower travertine unit (dated 7,200 ± 300 years B.P.) and a brown clay unit (dated 3,008 ± 410 years B.P.). The only extinct species in the predominantly small-mammal fauna, *Dasypus bellus,* occurs in the lower unit. Several of the rodents are now found north or east of the Llano region. The age is early Holocene. (Patton, 1963.) TMM.

Quitaque (13 spp.), Motley Co. (4–211)

Broken and waterworn bones and freshwater mollusks have been found in the bed of a small arroyo tributary of Quitaque Creek, which is located in northeast Motley County, 216 km W of Wichita Falls. The fauna is believed to be early Wisconsinan in age, and four species are extinct. (Dalquest, 1964.) MWU.

Schulze Cave (62 spp.), Edwards Co. (4–212)

The extraordinarily rich mammalian fauna (5,878 catalogued specimens) of Schulze Cave, 45 km E of Rocksprings, comes from remnants of a Pleistocene deposit preserved on a travertine ledge within a vertical shaft. Most of the small mammals were probably accumulated by barn owls; other animals fell in or their bones were washed in or brought in by woodrats. The only extinct taxa are mammoth, horse, and possibly bison. The Pleistocene deposit is judged to have formed from 11,000 to 8,000 years B.P.; radiocarbon dates are 9,680 ± 700 and 9,310 ± 310 years B.P.

The Pleistocene deposit consists of two layers. Species now found in more northern or cooler regions are more common in the lower layer, whereas forms suggesting warmer and more arid conditions increase in numbers in the upper; human remains have also been found in the upper layers. Conditions in the late Wisconsinan are reconstructed as cooler, more uniform, and more humid than at the present, with extensive grasslands and open woodlands with broad-leaved trees.

In contrast to the Pleistocene deposit, an overlying deposit (dated 3,862 ± 208 years B.P.) contains only mammals still present on the Edwards Plateau. (Dalquest, Roth, and Judd, 1969.) MWU.

Sims Bayou (15 spp.), Houston Co. **(4-225)**

Flood-control channeling of Sims Bayou, within the city of Houston, led to erosion and the exposure of a gray fossiliferous, sandy clay deposit along the slope. Extinct taxa include horse, mammoth, *Capromeryx,* and pampathere. The small mammals point to a climate similar to that of the Gulf Coast today. Charcoal chips gave the date 23,000+ years B.P., which may be Wisconsinan interstadial. (Slaughter and McClure, 1965.) SMUMP.

Taylor Bayou (11 spp.), Harris Co. **(4-224)**

A lacustrine deposit, Taylor Bayou, near Houston, has yielded a late Pleistocene fauna dated between 35,000 and 50,000 years B.P. Fossils were recovered 6.1–8.5 m below the surface. The habitat was probably a fertile grassland with ample permanent water. Summers were probably milder than they are today. *Glyptotherium* is the only extinct species. (Kasper and McClure, 1976.)

Williams Cave (19 spp.), Culberson Co. **(4-209)**

The cave is in Carboniferous limestone in the southern Guadalupe Mountains, in western Texas, at an elevation of ca. 1,500 m (60 m above the canyon floor), and has an east-facing entrance. The deposit, also containing artifacts and Recent burials, has yielded remains of iguana, turkey, and a mammalian fauna with two extinct species (*Canis dirus, Equus*) and dung of the shasta ground sloth. Northern species not now found in the area include *Cynomys gunnisoni* and *Ursus arctos.* (Ayer, 1937.) ANSP.

UTAH

Silver Creek (25 spp.), Summit Co. **(3-113)**

This is the first terrestrial Pleistocene local fauna to be found in Utah. The site is located in a small basin just east of the crest of the Wasatch Range, 32 km E of Salt Lake City, at an elevation of 1,952 m. This is 366 m above the highest level of former Lake Bonneville, and the Silver Creek locality was 22.5 km E of the eastern shoreline of that lake. A large aquatic component in the fauna indicates that conditions were somewhat wetter than they are now, and the landscape probably featured a marsh within a grassland, with nearby wooded areas, that was inhabited by ground sloths, sabertooth cats, mammoths, camels, and giant bison. A radiocarbon date of 40,000+ years B.P. suggests a late Sangamonian/early Wisconsinan age. Of the 25 mammal species, 8 are extinct. (W. E. Miller, 1973, 1976, and personal communication.) BYU, UUVP.

VIRGINIA

Clark's Cave (53 spp.), Bath Co. **(4-289)**

This late Pleistocene owl roost has yielded 143 species of vertebrates. The cave is located 12 km SW of Williamsville, in north-central Virginia, at an elevation of 448 m. It is a large, complicated cave, with numerous entrances in the limestone cliff that rises about 30 m above the south bank of the Cowpasture River. Deposition took place during the late Wisconsinan and ceased prior to Recent times. The bulk of the deposit represents prey captured by diurnal and nocturnal raptors that roosted and nested in the cave and on the cliff face, and six raptor species, including three owls, have been identified. The mammalian fauna consists primarily of small rodents and bats, and *Myotis lucifugus, Myotis sodalis, Myotis keenii, Microtus pennsylvanicus,* and *Microtus xanthognathus* are the most abundant species. The ecological requirements of the fauna indicate a spruce–pine parkland with nearby bogs and meadows. The fauna of Clark's Cave is similar to that of New Paris No. 4, 240 km to the north. (Guilday, Parmalee, and Hamilton, 1977.) CM.

Natural Chimneys (55 spp.), Augusta Co. **(4–290)**

These chert-capped dolomite towers are erosion remnants of a former cave system located near the town of Mt. Solon, in the wide flat valley of the North River, west-central Virginia. The matrix grew in place by decomposition of the surrounding cave walls and filling from above. Accumulation of the small-mammal bones probably resulted from the activity of owls; wood rats probably dragged in the bones of the larger species. The fauna is similar to that of New Paris No. 4 and is early post-Wisconsinan in age. A boreal climate is indicated, and many of the species are found in the Canadian and Hudsonian life zones today. (Guilday, 1962b.) CM.

Saltville (9 spp.), Smyth Co. **(4–286)**

This primeval salt spring is located in the valley of the North Fork Holston River, in southwestern Virginia. Large-herbivore bones were recovered from the structureless muck as early as 1782, and the deposit continues to yield fossils. Analysis of a pollen sample taken from the endocranial cavity of an extinct muskox has shown *Picea* and *Pinus* to be the dominant species. The first records of *Bootherium* and *Symbos* in Virginia came from Saltville. The fauna is late Pleistocene in age. (Ray, Cooper, and Benninghoff, 1967.) USNM.

Other late Pleistocene/early Holocene faunas in Virginia include Meadowview Cave (11 spp.), Washington Co. **(4–285)**; Winding Stairs Cave (5 spp.), Scott Co. **(4–284)**; Gardner Cave (21 spp.), Wythe Co. **(4–287)**; and Early's Pit (15 spp.), Wythe Co. **(4–288)**, (Guilday, 1962a). The faunas are under study. (Guilday, personal communication.)

West Virginia

Eagle (Eagle Rock) Cave (30 spp.), Pendleton Co. **(5–363)**

The cave is located at an elevation of 757 m in a rugged area of the Appalachian Mountains, in north-central West Virginia. It is the smallest of several caves in the New Scotland–Coeymans Member of the Helderburg Limestone at the southern end of Cave Mountain. The cave floor had been disturbed by numerous visitors and was littered with debris from roof fall. There was no stratification. The cave has been used as an aerie by raptorial birds, and bone accumulation is probably due to their activities. All of the faunal remains are small enough to have been owl prey. A boreal element, represented by species now found in Canadian and Hudsonian life zones, is present, as well as a grasslands component. The age is early post-Wisconsinan. (Guilday and Hamilton, 1973.) CM.

Organ-Hedricks Cave (14 spp.), Greenbrier Co. (4–291)

Located in the Appalachian Mountains of southeastern West Virginia, at an elevation of 671 m, this limestone cave has yielded several extinct species and others that are no longer found in the area. The presence of *Dasypus bellus* and *Myotis grisescens* suggests relatively milder environmental conditions than those of the present day. The fauna is under study. (Guilday, personal communication; Guilday and McCrady, 1966; Handley, 1956.) CM.

WYOMING

Bell Cave (47 spp.), Albany Co. (4–200)

The cave, at an elevation of 2,379 m, is located directly across the canyon from Horned Owl Cave. The stratigraphy of the two caves is similar, but Bell Cave has a more extensive fauna, including 10 species of birds and 47 species of mammals. Numerous artifacts dating from the Late Middle Period to very late Prehistoric were recovered. The first fossil occurrence of *Falco rusticolus* Gmelin was reported from Bell Cave. The fauna shows many similarities to the faunas of Little Box Elder and Chimney Rock caves. Only a preliminary study of the fauna has been completed. (Walker, 1974; Zeimens and Walker, 1974.) UWYO.

Casper (11 spp.), Natrona Co. (5–354)

The predominant species at this Paleo-Indian kill site is *Bison bison antiquus;* in addition, a mollusk, a woodpecker, and a number of extant mammals have been found. Of the latter, only the red fox (*Vulpes vulpes*) is no longer found in the immediate vicinity. The remains come from a bone bed (Unit B) that rests upon Pleistocene sand and gravel and is overlain by pond and dune deposits. Unit B was deposited in a blowout trough in the underlying beds. Recently, postcranial material of *Camelops* showing definite butchering marks was found in the bone bed (Frison et al., 1978; Walker, personal communication, 1977). The bison and camel remains represent a death assemblage that resulted from human predation (Frison et al., 1978). Radiocarbon dates on charcoal and bone are $9,830 \pm 350$ and $10,080 \pm 170$ years B.P., respectively. (Frison, 1974; Frison et al., 1978; Walker, personal communication, 1977.) UWO.

Horned Owl Cave (25 spp.), Albany Co. (4–201)

Located on the south rim of Wall Rock Canyon, 32 km NE of Laramie, the cave contains an unstratified, loesslike fill. Animal remains are sparse and fragmentary; most of the bone fragments are probably the remains of predator kills. Of the 25 species, *Camelops* and *Equus* are extinct, *Oreamnos* is no longer found in Wyoming, and *Ochotona* and *Phenacomys* do not occur in the vicinity of the cave today. The age of the deposit is late Wisconsinan–early Holocene. (Guilday, Hamilton, and Adam, 1967.) CM.

Little Box Elder Cave (62 spp.), Converse Co. (4–199)

This is one of the best examples of a late Pleistocene–Holocene mammalian fauna of the High Plains. Little Box Elder Cave is located in the foothills of the Laramie Mountains, 30 km W of Douglas, at an elevation of approximately 1,677 m. The cave was formed either by solution or by the lateral cutting of the East Fork of Little Box Elder Creek. It is a simple sloping tunnel about 23 m long and 11–12 m wide. About two-thirds of the cave has been excavated, and more than 15,000 bones have been recovered. Of the 62 species, 7 are extinct, and 17 show boreal or tundra affinities. A mandible of *Equus* shows butchering marks (Robinson, personal communication). The fauna is especially rich in carnivores and contains the only known association of *Ursus arctos* and *Arctodus simus* south of Alaska. (Anderson, 1968; Kurtén and Anderson, 1974; Robinson, personal communication, 1974.) UCM.

Natural Trap Cave (23 spp.), Big Horn Co. **(4–198)**

Natural Trap Cave is a karst sinkhole situated on the west slope of the Big Horn Mountains, in north-central Wyoming, at an elevation of 1,510 m. The cave entrance, measuring 3.5–4.5 m in diameter, is hidden from view until the observer reaches the edge, at which point there is a sheer 20-m drop to the floor. For the victims, there was no way to escape; thus complete skeletons are common. There is no evidence of human or animal disturbance; instead, the actions of gravity and rainwater probably caused the disarticulation of the skeletons. Many of the species are highly cursorial forms such as *Acinonyx trumani*, a cheetahlike cat, and their presence indicates the prevalence of open country. A carbon-14 date of $12,770 \pm 900$ years B.P. has been obtained on *Equus* bones found at the deepest level. The fauna is under study. (L. D. Martin, Gilbert, and Adams, 1977.) KUVP.

Yukon Territory

Gold Run Creek (13 spp.) and Upper Hunker Creek (10 spp.), Dawson area **(4–156, 4–155)**

The unglaciated area on the maturely dissected Yukon Plateau just south of the Oglivie Mountains near Dawson formed the southeastern extremity of the Beringian refugium. Fossils are found in the frozen muck and on the surface of auriferous gravels exposed during placer mining operations. Several sites, including Gold Run Creek, Upper Hunker Creek, Dominion Creek (first Canadian record of *Panthera leo atrox*), and Miller Creek (viable lupine seeds found in the burrow of a Pleistocene lemming), are yielding interesting fossils. The faunas are similar to those found at other localities in the refugium. Bison, horse, and mammoth are the commonest animals recovered; this suggests a cool-grassland or open-parkland landscape. Bison and mammoth bones from Gold Run Creek have yielded radiocarbon dates of $22,000 \pm 1,400$ and $32,250 \pm 1,750$ years B.P., and a caribou antler from Upper Hunker Creek has been dated at $23,900 \pm 470$ years B.P. (Harington, 1969, 1970a, 1971c, 1978, and personal communication, 1977; Harington and Clulow, 1973.) NMC.

Old Crow River locs. 44 (20 spp.), 14N (20 spp.), and 11A (39+ spp.) **(3–110, 4–154)**

Pleistocene fossils have been recovered from a large number of sites along the meandering Old Crow River, which drains an intermontane basin in the northern Yukon. The basin is filled with unconsolidated fine sediments of unknown depth. High-level shorelines on the edge of the basin indicate the former presence of a large lake during the Pleistocene. Today, much of the area is covered with lakes and slow-moving streams, interspersed with patches of boggy tundra. There is no evidence that the area was ever glaciated, and the Old Crow area was part of the Beringian refugium. Fossils are usually derived from the oxidized sands and sandy gravels, and the bones are generally well preserved; fragile specimens were probably protected in blocks of frozen matrix or clumps of vegetation.

At Locality 44, a fauna of possible Sangamonian age has been recovered; radiocarbon dates ranging from more than 39,900 to more than 54,000 years B.P. have been obtained on bone and wood. A predominantly Wisconsinan fauna has been found at Locality 14N; radiocarbon dates on bone range from 22,000 to 36,000 years B.P. This fauna, consisting of both cold and warm-adapted components, inhabited a cool grassland environment with lakes, streams, and nearby spruce woodlands. Ten species are extinct, and several artifacts have been found in association with the extinct fauna. An extensive Pleistocene assemblage (probably ranging in age from early Pleistocene to early Holocene) has been collected at Locality 11A. Bison, horse, and mammoth are common, and flake tools have also been found. New, noteworthy finds include *Canis familiaris* and *Platygonus compressus* (northernmost record); a minimum age of 20,000 years is suggested for the specimens (B. Beebe, personal communication, 1978). The fauna is under study. In addition to the abundant mammal remains, plant, mollusk, insect, fish, and avian fossils have been identified from many sites along the Old Crow River. (Beebe, personal communication, 1978; Churcher, personal communication, 1976; Harington, 1971c, 1978, and personal communication, 1977; Irving and Harington, 1973; Matthews, 1975.) NMC.

CHAPTER FIVE

INTERCONTINENTAL CORRELATION AND MIGRATIONS

During the Pleistocene, the North American continent was intermittently or continuously in contact with other areas by water, ice, and land routes. Many North American mammals made their way into other continents, and the North American fauna was also enriched by the arrival of immigrant forms. By means of migration data and radiometric or magnetic chronologies, the faunal sequence in North America may be correlated with that of other continents. This topic is briefly reviewed in the present chapter.

SOUTH AMERICA

For most of the Tertiary, South America was isolated from other continents by sea barriers. Its mammalian faunas were made up of descendants of the mammals which had populated South America in Cretaceous or early Paleocene times or which succeeded in reaching the continent from across the sea. Early South American groups include marsupials, edentates, and the hoofed mammals that gave rise to the now extinct "South American ungulates"—a varied group which, in its heyday, comprised several orders. Mid-Tertiary immigrants from North America were caviomorph rodents and New World monkeys, both of which probably spread into South America by waif dispersal across the sea. These animals were followed in the late Tertiary by procyonid carnivores such as the *Cyonasua* group, presumably in the same manner. When a land connection to the north was formed, there was a massive immigration of mammals from North America. In a corresponding manner, South American mammals made their way into North America, at first by waif dispersal, and later along a land route. An integrated inter-American fauna was thus formed.

The fossil record of mammalian life in South America is excellent, and a continental sequence of Land Mammal Ages has been established (Pascual, et al., 1965; Patterson and Pascual, 1972). Unfortunately, few absolute dates are available, and correlations with other continents are still tentative. A correlation of the late Cenozoic Land Mammal Ages in North and South America, essentially following Webb (1976), is given in table 5.1.

TABLE 5.1
CORRELATION OF LAND MAMMAL AGES IN NORTH
AND SOUTH AMERICA

Million years B.P.	North America	South America
	Rancholabrean	Lujanian
0.4		
	Irvingtonian	Ensenadan
1.9		Uquian
	Blancan	
3.5		Chapadmalalan
	Hemphillian	
		Huayquerian

The Huayquerian faunas are still completely dominated by old endemics and old immigrants. The only new immigrant of North American origin is *Cyonasua*, which probably reached South America by waif dispersal, as noted above.

The beginning of the overland faunal exchange with North America apparently came in Chapadmalalan times. In these faunas appear a number of migrants, such as the mustelid *Conepatus*, the peccary *Argyrohyus*, and the cricetid rodent *Proreithrodon*. However, the endemic groups—marsupials, edentates, native ungulates, and caviomorph rodents—remained predominant. Although the Chapadmalalan overlaps broadly with the Hemphillian, late Chapadmalalan faunas were probably contemporary with those of the earlier Blancan.

In contrast, Uquian faunas have a very strong northern element. We may note the presence of the canids *Dusicyon* and *Protocyon* (the latter also occurring in the early Irvingtonian), the ursid *Arctotherium* (closely related to *Arctodus*, known from the early Irvingtonian on), the mustelid *Galictis* (probably descended from the Blancan *Trigonictis*), the felid *Megantereon* ('*Smilodontidion*') and, possibly, a primitive *Smilodon* (affinities Blancan to early Irvingtonian), *Tapirus*, the equids *Hippidion* and *Onohippidium* (one-toed horses related to *Equus*, of Blancan origin), the camelid *Hemiauchenia* (Hemphillian and later), the peccary *Platygonus* (Blancan and later), cervids related to *Odocoileus* (Blancan and later), and a stegomastodont. Clearly, by Uquian times a good land connection was in existence. Among the endemic forms, the edentates and caviomorph rodents remained numerous and varied, but the native ungulates and marsupials decreased in numbers. We agree with Webb (1976) that the Uquian probably corresponds to the late Blancan and early Irvingtonian.

In the Ensenadan faunas, the northern elements show further diversification and are reinforced by additional migrations. New arrivals include the mustelid *Lutra*, the felids *Felis* and *Panthera*, and the true horses. Native ungulates are reduced to three genera, but the edentates and caviomorph rodents enjoy a continued success. The Ensenadan is probably equivalent to the later Irvingtonian.

The Lujanian, or late Pleistocene, faunas still include a great and varied assemblage of edentates, but only three genera of native ungulates persist. Llamas and cervids are varied. The Lujanian may be correlated with the Rancholabrean.

The event which triggered the massive faunal interchange was evidently the drying up of the narrow waterway formed by the Bolívar geosyncline, between Panama and the northwestern corner of South America. Tertiary faunas in Central America are northern in character and show that the area was part of the North American continent. The fossil record in tropical North America is as yet very scanty, however, and the history of its small mammals, in particular, is almost unknown. It is probable that this area was an important evolutionary center for the various mammalian groups that later made their entry into temperate North America, as well as into South America (Patterson and Pascual, 1972).

Among the southern immigrants to North America, several groups extended their range into the tropical area only. These are the New World monkeys, the tree sloths, the anteaters, three families of caviomorph rodents, and the toxodonts (which belong to the South American ungulate order Notoungulata).

The area north of tropical North America was reached by the following families (with large mammals in the great majority): the opossums (Didelphidae), the armadillos (Dasypodidae), the glyptodonts (Glyptodontidae), the various ground sloths (Megalonychidae, Megatheriidae, Mylodontidae), the capybaras (Hydrochoeridae), and the porcupines (Erethizontidae). In most or all cases, these animals would have had first to traverse the area here referred to as tropical North America, although an Antillean route cannot be definitively ruled out for those groups with a southeastern distribution (e.g., *Eremotherium,* Hydrochoeridae). A more probably alternative route for the latter, however, may be along the Gulf Coast. In any case, the history in intervening areas is largely unknown, and dates of first appearance in the record may postdate the actual entrances in North America by unknown lengths of time.

The first appearances in North America of taxa of South American origin are temporally distributed into four rather well-defined groups:

1. Pre-Blancan, ca. 8 m.y. B.P. Appearance of Mylodontidae and Megalonychidae. This is regarded as the result of waif dispersal in Bolívar Trench times, correlated with the entrance of *Cyonasua* in South America.

2. Transition between early and late Blancan, ca. 3.0 m.y. B.P. Appearance of *Didelphis, Glyptotherium, Kraglievichia, Dasypus, Glossotherium,* and, possibly, *Neochoerus* or a related form. Generic identity with South American forms suggests that spread was rapid and occurred soon after entrance.

3. Transitional Blancan–Irvingtonian. Appearances of *Hydrochoerus* and *Coendou* (ca. 2.0 m.y. B.P.) and *Holmesina, Nothrotheriops,* and *Eremotherium* (ca. 1.8 m.y. B.P.). These taxa may be later immigrants than group 2 or else forms that entered tropical North America at the same time but extended their range only later. *Erethizon,* a probable descendant of *Coendou,* appears at about 0.7 m.y. B.P.

4. Rancholabrean, 0.4 m.y. B.P. and later. Appearance of *Neochoerus, Blastocerus,* and *Palaeolama,* all having an eastern distribution. The group may include *Conepatus,* which appears in North America late in the Irvingtonian; the genus was present in South America from the Chapadmalalan.

The large group of complex-penis Cricetinae, thought to have deployed in South America and reimmigrated to North America (Hershkovitz, 1966), most probably had their evolutionary center in tropical North America in pre–land bridge times (Pascual and Patterson, 1972).

For further discussions of the faunal interchange between North and South America, see Webb (1976).

Eurasia

Faunal interchange between North America and Eurasia in the late Cenozoic appears to have taken place only by way of Beringia, as far as land mammals are concerned. Even a moderate eustatic lowering of the sea level would have exposed a broad land area connecting Alaska with eastern Siberia and crustal movements may have occurred in addition. In the Pleistocene, however, glacial-controlled eustatic fluctuation was the main factor in the opening and closing of this passage (see Hopkins, 1967; Kontrimavichus, 1976). The rigorous climate of the land bridge denied passage to mammals lacking appropriate adaptation; Beringia thus acted as a "filter" bridge (Simpson, 1947), an effect which may be observed well back in the Tertiary and which became highly selective in the Pleistocene. As a result, the resemblance between Palearctic and Nearctic faunas is particularly great with regards to cold-adapted mammals.

Glacial Beringia, however, should not be thought of only as a migration bridge. A subcontinent in itself, Beringia probably played an important role as an evolutionary center for arctic mammals, commensurate with that of tropical North America for the southern fauna.

Because the theory of glacial–eustatic control leads to the conclusion that the land bridge was exposed only in the cold intervals, it might be thought that inland ice would be an effective barrier to migration. However, a large area of Beringia—most of Alaska and part of the Yukon—was in fact icefree and populated by mammals of Eurasian as well as North American origin, even when the North American land ice formed a barrier all the way from the Pacific to the Atlantic. In this way, species originating in one continent could obtain a foothold on the other and continue their spread as conditions improved. Evidence for relay-type migrations is good, especially for the Wisconsinan. The fragmentation and isolation of marginal populations may also have played an important role in the mode of evolution outlined in the "punctuated equilibrium" model (Eldredge and Gould, 1972).

In western Europe, a sequence of provincial Land Mammal Ages has been set up (Fahlbusch, 1976). The faunal record is good, and a number of radiometric and paleomagnetic dates are at hand, making a correlation with North America possible (table 5.2). Correlation of the classical Alpine and Scandinavian glacial-interglacial stages is still controversial (see table 1.1).

The Ruscinian, or early Pliocene, is subdivided by some authors (Kretzoi, 1962; Tobien, 1970) into an early (the Ruscinian *sensu stricto*) and a late age, or subage (the Csarnotan). The transition from the Ruscinian *sensu lato* to the Villafranchian occurred at about 3.5–4.0 m.y. B.P. and so may antedate the Hemphillian–Blancan transition to

TABLE 5.2
CORRELATION OF PLIO-PLEISTOCENE LAND MAMMAL FAUNAS IN
EUROPE AND NORTH AMERICA

European Land Mammal Ages	Million Years B.P.	European Local Faunas	GL/IGL[a]		North American Land Mammal Ages
			Europe	No. Amer.	
		Cave & Loess Faunas	Weichsel	Wisconsin 2	
	0.08	Travertines, Caves	Eem	Sangamon ?	Rancholabrean
	0.2	Cave & Loess Faunas	Saale	Wisconsin 1	
	0.3	Swanscombe, Lunel-Viel	Holstein	Sangamon ?	
	0.4		Elster	Illinoian	
		Mosbach			
Cromerian	0.7	Cromer, Mauer	Cromer	Yarmouth ?	
	0.8	Süssenborn	Günz	Kansan	
	1.0		Waal	Aftonian	Irvingtonian
Late Villa-franchian	1.3	Valdarno superiore	Donau	Nebraskan	
	1.5	Tegelen	Tegelen		
	1.8	Leffe, Senèze 1.6			
	1.9	—Coupet 1.9			
Middle Villa-franchian		St. Vallier, Chagny Puebla de Valverde Pardines	Pre-glacial	Sierran	Blancan
	2.5	—Roccaneyra 2.5			
Early Villa-franchian		Etouaires 3.3			
	3.5	Villafranca d'Asti Vialette 3.8			
	4.0				Hemphillian
Ruscinian					

[a]GL, glacial; IGL, interglacial

TABLE 5.3.
CORRELATION OF PLIO-PLEISTOCENE FAUNAS IN THE USSR AND CHINA

Provincial Ages		Million years B.P.	Local Faunas				North American Ages
Western Europe	Moldavia, Ukraine		Transbaikalia	Western Siberia	Kolyma Lowland	Northern China	
Late Cromerian	Tiraspolian	0.4	Tologoi	Viatkino	Olior	Choukoutien	
Early Cromerian	Tamanian	0.7		Itantsa Razdolie	Krestovka		Irvingtonian
Late Villafranchian	Odessan	0.9		Kizikha		Nihowan	
Middle Villafranchian	Khaprovian	1.9	Beregovaia	Lebiazhie			
Early Villafranchian	Moldavian	2.5	Chikoi	Betekey		Chinglo	Blancan
		3.5					

some extent. The Villafranchian as a whole represents a long span of time, transgressing the Pliocene–Pleistocene boundary and overlapping not only with all of the Blancan but also with part of the Irvingtonian.

Villafranchian faunas are characterized by the appearance of such genera as *Mammuthus, Elephas, Leptobos, Equus,* and *Hypolagus,* all of which are immigrants, and the great evolution of the rooted-tooth voles of the *Mimomys* group. Based on the evolution of various lineages (e.g., carnivores, microtines, cervids), the Villafranchian may be divided into three subages (Tobien, 1970). The transition from the middle to the late Villafranchian is dated at about 1.9 m.y. B.P. and thus coincides with the Pliocene–Pleistocene and Blancan–Irvingtonian transitions.

The early Middle Pleistocene in Europe, comprising the Cromerian interglacial and the preceding and succeeding glaciations, is referred to as the Biharian Age (the Cromerian or Mosbachian of some authors). It is characterized by decline of the *Mimomys* group and appearance of modern voles and lemmings like *Arvicola, Microtus, Lagurus, Lemmus,* and *Dicrostonyx* and the first appearance of cold-adapted forms like reindeer, muskoxen, and advanced mammoths and early forms of the typical carnivores, ungulates, and proboscideans which were to dominate the European scene in the later Pleistocene.

The late Middle Pleistocene and late Pleistocene faunas, with a strongly marked alternation of "cold" and "temperate" forms, have not yet been grouped into generally recognized provincial ages. This role is here played by the glacial–interglacial units.

"Villafranchian," "middle Pleistocene" (or "Cromerian"), and "late Pleistocene" faunas, under distinct provincial age names, have also been recognized in the USSR, China, and other areas. Table 5.3 indicates the classification of provincial ages, or "faunal complexes," in Moldavia and the Ukraine (European USSR), western Siberia, Transbaikalia, and the Kolyma lowland (Asiatic USSR), and northern China. The correlations are according to Alekseev et al. (1973); see also Gabunia (1972); Kahlke (1972); and Vangengeim and Zazhigin (1972).

There seems to have been an episode of intermigration between the Palearctic and Nearctic around 3.5 m.y. B.P., at or near the beginning of the Blancan and Villafranchian. Migrants from the Old World to the New include genera such as *Trigonictis, Parailurus, Ursus, Megantereon,* and *Chasmaporthetes;* in addition, *Homotherium* and *Dinofelis* may have traversed the Bering Bridge at this time. Migration of *Lynx* during this period is also possible, but whether lynxes originated in the New or the Old World is still uncertain. Camelidae, *Equus,* and cheetahs migrated to Eurasia during this time also. Evidently, the filtering effect of the Bering Bridge was somewhat less pronounced at this time than in the Pleistocene. After this early pulse of migration, interchange appears to have abated. Coyotes, however, which are evidently of American origin, do not appear to have reached Europe until late Villafranchian times. A late Blancan immigration of *Synaptomys* from Eurasia also seems indicated.

A second phase of intense intermigration inaugurated the Irvingtonian. There is an influx in North America of Eurasian forms such as jaguars, *Euceratherium, Soergelia,* and *Mammuthus.* Other migrating forms, of uncertain provenance, are *Lepus* and *Gulo.* This episode started about 1.8 m.y. B.P. Other Irvingtonian migrants, probably arriving at a somewhat later date, include the following: *Rangifer, Lemmus,* and the *Dicrostonyx*-like lemmings, arctic forms which may well have originated in the Beringian region; *Clethrionomys, Canis lupus, Mustela erminea,* and *Lutra,* which probably migrated from the Palearctic to the Nearctic; and *Microtus,* source area debated. These mammals are arctic or boreal, and their presence suggests that the climate in Beringia was more rigorous than in pre-Pleistocene times.

At the beginning of the Rancholabrean, *Ovibos* and *Alces* reached Alaska, and *Bison* and *Mustela nigripes* appeared south of the ice; all are of Old World origin. At a somewhat later date, though probably before the Wisconsinan glaciation, *Vulpes vulpes* and *Panthera leo* immigrated to North America. Wisconsinan immigrants include *Mammuthus primigenius, Cuon alpinus, Martes americana, Mustela rixosa, Ursus arctos, Alces alces,* and *Homo sapiens*. The three last-mentioned species may have arrived only after the bipartition of the ice sheet, although the evidence is conflicting with regard to man. It may be noted that few of the Rancholabrean immigrants become specifically differentiated from their Old World counterparts; the supposedly distinct *Martes americana* and *Mustela nigripes,* for example, are very similar to *Martes zibellina* and *Mustela eversmanni,* respectively.

Each of the three North American Land Mammal Ages, the Blancan, the Irvingtonian, and the Rancholabrean, can thus be seen as inaugurated by an intense phase of migrations across the Bering Bridge.

PART TWO

MAMMALIAN SPECIES OF THE BLANCAN AND PLEISTOCENE

CHAPTER SIX

Introductory Notes

In this section, the Blancan and Pleistocene mammalian species of the area under study (the United States, Canada, and northern Mexico) are described. The numerous living North American species that are unknown in the fossil state are not included. Our aim has been to provide an annotated list of those species that we consider certainly or probably valid. For the sake of completeness, however, it has been necessary to include a few species of doubtful status. In this endeavor, we have followed the principle of saddling heterogeneity, rather than homogeneity, with the burden of proof. Consequently, many species have been reduced to synonymy.

As noted in the Introduction, our first-hand research has focused upon selected taxonomic groups, and our treatment of these taxa is in many cases based upon unpublished data. Treatment of other taxa follows published sources as well as being drawn from information and advice kindly furnished by our colleagues. For some taxa, no comprehensive revisions are available, and the difficulties that such cases present should be evident. It must be conceded that a fully balanced systematic account is not possible at present. We hope that the shortcomings of the book in this respect will stimulate further research.

Of the supraspecific taxa, orders and families are treated in brief presentations, which include main characters, nature of the fossil record, and major features of infraordinal and infrafamilial taxonomy. Subfamilies are characterized summarily in the family presentations. More-detailed information on these topics is available in standard textbooks (e.g., Gromova, 1962; Romer, 1966; Vaughan, 1972).

The species descriptions follow a common format:

1. Name of the species († indicates extinct species). Included are the (a) vernacular name, (b) scientific name, with the author and date, and (c) synonymy (in brackets, with the original designation of the species, if different from the present form, given first). For living species, only synonyms based on fossil material are included, with certain exceptions for very commonly used names. References to descriptions of living species made from Recent material are not included in the bibliography; such references may be found in Hall and Kelson (1959) and other zoological handbooks.

2. Brief discussion of the genus and subgenus (appears with the first species in each genus or subgenus).

3. Case history of the species. The following information is included: geographic and stratigraphic range (including absolute dates pertaining to migration or extinction), distinguishing characters, probable ancestry and source of migration, trends of intraspecific evolution, important subspecies, daughter species, mode of life, functional significance of evolutionary changes, factors relating to fossil preservation, and factors possibly involved in extinction or survival. Illustrations show osteological and odontological characters and geographic distribution of selected species; a number of life restorations, based on complete or composite skeletons, are also included. More-detailed references are given for localities not treated in the site descriptions. Some species occur at a very large number of sites, especially in the Rancholabrean. For these species, only the states and provinces in which they have been found are mentioned.

Naturally, it is only in exceptional cases that all this information is available. Many species are known from fragmentary remains, and in this case, relationships may be obscure. In addition, some specimens come from deposits of uncertain age.

4. References. They relate to the information given and are intended to guide the reader to the pertinent literature but do not constitute a complete bibliography of the species.

We have endeavored to group the species by their affinities as well as chronologically and according to evolutionary lineages, where such are recognized. Stratigraphic ranges are given in Appendix 2.

CHAPTER SEVEN

ORDER MARSUPIALIA

The earliest known marsupials, the didelphoids, have been found in early Cretaceous deposits in North America. From this center of origin, they dispersed to South America in the late Cretaceous or early Paleocene, before placental mammals spread into the area, and reached Australia at about the same time. In both these areas, marsupials underwent spectacular adaptive radiation, filling almost all available niches that on other continents were occupied by orders of placental mammals. Today, these two geographic regions remain the stronghold of marsupials, although one genus, *Didelphis*, continues to inhabit North America.

Marsupials differ from placentals in having an inflected angle on the dentary, auditory bullae (when present) formed primarily by the alisphenoid instead of the tympanic, and epipubic bones extending forward from the pubis, as well as in their reproductive system. Female marsupials have, on the abdomen, a marsupium or marsupium folds in which the nipples are located. The young are born in an immature condition, make their way to the pouch, become firmly attached to the teats, and complete development there. Marsupials have a small, narrow braincase and a long facial region; the cerebral hemispheres are relatively smooth. Body temperature and metabolic rate are lower than in placentals. The teeth and feet are often highly modified for specialized modes of life.

Although marsupials are usually placed in a single order, recent studies have shown the differences between the various families to be of ordinal rank. Many workers now recognize the orders Marsupicarnivora, the opossums and marsupial carnivores (North and South America and Australia), Paucituberculata, the opossum-rats (South America), Peramelina, the bandicoots (Australia), and Diprotodonta, the phalangers and kangaroos (Australia). Many of the Australian forms are close to extinction.

FAMILY DIDELPHIDAE—OPOSSUMS

With a fossil record extending back to the early Cretaceous, the Didelphidae are structurally the most primitive and generalized marsupials. Although primarily tropical and subtropical in distribution, opossums today range from southeastern Canada to southern Argentina.

Opossums have more teeth (50) than any other American mammal, and the dental series includes small incisors, large canines, and sharp-cusped cheek teeth. They are omnivores. Skull characteristics include a long rostrum and a narrow braincase with a prominent sagittal crest. Adapted, but not restricted, to an arboreal way of life, opossums have an opposable, clawless hallux and a long, usually prehensile tail. A high reproduction rate is a contributing factor in the continuing success of these primitive but well-adapted mammals.

Opossum, *Didelphis virginiana* Kerr, 1792

The only marsupial in North America during the Pleistocene was the opossum. Until the Wisconsinan, their fossil record is spotty. Early records include Santa Fe River IB

(Blancan) and Coleman IIA (Irvingtonian). In the late Pleistocene, the opossum appears in some 30 faunas, in Arkansas, Florida, Georgia, Kentucky, Louisiana, Missouri, New Mexico, South Carolina, Tennessee, Texas, and Virginia. Holocene occurrences are numerous, and presence in archeological sites indicates that opossums were hunted by Indians. *Didelphis virginiana* has extended its range northward in historic times and, although cold temperatures are a limiting factor, is now found as far north as southern Ontario. It inhabits woodlands and cultivated areas near water.

About the size of a house cat, this animal shows considerable individual variation in size. The skull and certain postcranial skeletal elements increase in size throughout the life of the animal. Opossums are opportunistic omnivores and are able to utilize all kinds and sources of food, but the availability and quality of the food ultimately affects their size and longevity.

The earliest record of the genus is from the Pliocene of South America, and the taxon spread northward after the establishment of the Panamanian land bridge. Gardner (1973) believes that *Didelphis virginiana* arose from a Pleistocene isolate of *Didelphis marsupialis,* probably in western Mexico. An ability to tolerate temperate climatic conditions permitted the species to spread northward. (Gardner, 1973; Guilday, 1958.)

CHAPTER EIGHT

Order Insectivora

Insectivores first appear in the fossil record in the late Cretaceous but are not well known until the mid-Tertiary. McKenna (1975), in his phylogenetic classification of mammals, recently, called the Insectivora a "superorder"; he considers the Erinaceomorpha (hedgehogs) and Soriciomorpha (shrews, moles, tenrecs, golden moles, solenodons, and several extinct families) to be distinct orders. All are characterized by a reduced jugal bone, an expansion of the maxillary bone in the orbital wall, a reduction of the pubic symphysis, loss of the cecum in the gut, and loss of the medial branch of the carotid artery. Living insectivores are found throughout the world, with the exception of Australia, most of South America, and the polar regions.

Family Soricidae—Shrews

Shrews, the most abundant and widespread of the insectivores, are among the smallest mammals. Rare as fossils, shrews have an obscure early history, but common ancestry with the Talpidae and the Chiroptera, in a lineage perhaps extending back to the Mesozoic, is indicated.

The stratigraphic range of the Holarctic subfamily Soricinae extends from the late Oligocene to the Recent in the Old World and from the late Miocene to the Recent in the New World. The skull is narrow and elongate and lacks zygomatic arches and tympanic bullae. Dental chracteristics include a large, hooked first upper incisor, an enlarged, procumbent lower incisor, and W-shaped ectolophs on the upper molars. Tooth pigmentation is usually present.

On the basis of masticatory modifications, Repenning (1967b) divided the Soricinae into three tribes—the Soricini, Blarinini, and Neomyini—all of which are represented in the North American Blancan and Pleistocene. In the Soricini (*Sorex* and *Microsorex*), the articular facets of the mandibular condyle are continuous or only slightly separated, M^2 is rectangular in occlusal view, and M_3 is unreduced or only the talonid is reduced. This is the most primitive group and ranges from the early Miocene to the Recent. In the tribe Blarinini (*Paracryptotis, Planisorex, Blarina, Cryptotis*), the articular facets of the condyle are widely separated with a broad interarticular area, M^2 is trapezoidal in shape, M_3 is reduced, with the talonid forming a crescentic crest, and the teeth are stout and heavily pigmented. The range is late Miocene to Recent. In the tribe Neomyini (*Notiosorex*), the lingual condylar emargination is open with a very narrow interarticular area, M^2 is rectangular, and the teeth are lightly pigmented. The range is early Pliocene to Recent. (Repenning, 1967b.)

†Rexroad Shrew, *Sorex rexroadensis* Hibbard, 1941

Sorex is the most generalized and primitive of the living soricine shrews, and its stratigraphic range extends from the late Pliocene to the Recent. All of the species have 32 pigmented teeth.

The smallest known Pliocene shrew is *Sorex rexroadensis,* a rare species found in the early Blancan Fox Canyon and Hagerman faunas. On M_1 and M_2, the entoconid is more distinctly separated from the metaconid and the talonid is not as wide in comparison to the trigonid than is the case in Recent species. At Fox Canyon, the species probably inhabited upland areas, in contrast to *Sorex taylori*, which was found in moist lowlands. (Hibbard, 1953a.)

†Taylor's Shrew, *Sorex taylori* Hibbard, 1937

A Blancan species (Rexroad, Borchers, Blanco), *Sorex taylori* has a P_4 that is relatively wider than that of *Sorex sandersi*, and the talonids and trigonids of the molars are of equal width. (Hibbard, 1953a.)

†Hagerman Shrew, *Sorex hagermanensis* Hibbard and Bjork, 1971

Found only at Hagerman (early Blancan), *Sorex hagermanensis* was about the size of *Sorex obscurus*. The anterior mandibular foramen is more anteriorly placed than that of *Sorex arcticus* or *Sorex obscurus*, and a small posterior mandibular foramen is present. The species is distinguished from *Sorex powersi* by a heavier and larger jaw, a broader talonid on M_3 with a distinct entoconid, and the position of the mandibular foramen. (Hibbard and Bjork, 1971.)

†Powers's Shrew, *Sorex powersi* Hibbard and Bjork, 1971

In this species, the posterior mandibular foramen is larger than the anterior one, and both are situated in a deep depression. The talonids of M_1 and M_2 are approximately the same width as the trigonids. It is known only from the early Blancan Hagerman fauna. (Hibbard and Bjork, 1971.)

†Melton's Shrew, *Sorex meltoni* Hibbard and Bjork, 1971

This shrew was approximately the same size as *Sorex cinereus*. The mandible is not as deep or wide as that of *Sorex hagermanensis*. The anterior and posterior mandibular foramina are distinct and situated in a depression. It has been found only at Hagerman (early Blancan). (Hibbard and Bjork, 1971.)

†Leahy Shrew, *Sorex leahyi* Hibbard, 1956

Another Blancan shrew is *Sorex leahyi*, known only from the Dixon local fauna. It is characterized by having the mandibular condyle more elongate lingually than that of most other species of *Sorex*. (Hibbard, 1956b.)

†Sanders's Shrew, *Sorex sandersi* Hibbard, 1956

This late Blancan shrew has been found at Sanders, Sand Draw, and White Rock. Only the anterior mandibular foramen is present, and the talonids of the lower molars are not as wide as the trigonids. The jaw is heavier than that of *Sorex taylori*. *Sorex sandersi* is probably derived from *Sorex taylori*. (Eshelman, 1975; Skinner and Hibbard, 1972.)

†Cudahy Shrew, *Sorex cudahyensis* Hibbard, 1944

This species, larger than *Sorex taylori* or *Sorex cinereus*, was about the size of *Sorex merriami*. The robust P_4 lacks both the depression on the lingual side and the small anterior cusp in front of the principle cusp and has a broad heavy cingulum along the labial border. The mandibular foramen is below the posterior part of the trigonid of M_1. It is known only from the late Irvingtonian Cudahy fauna. (Hibbard, 1944.)

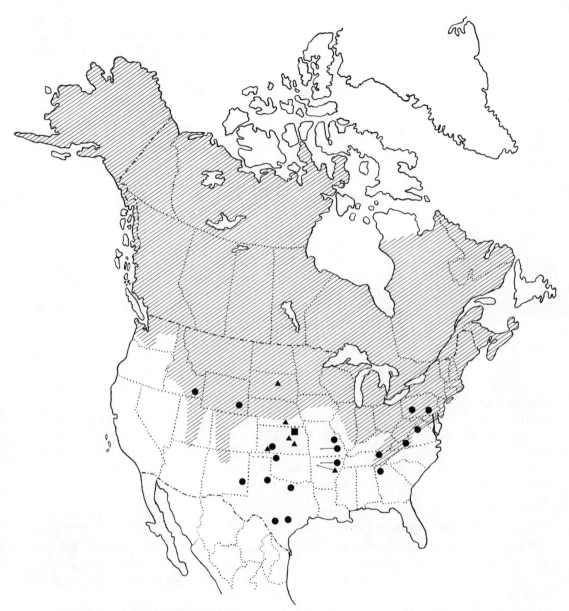

8.1 Pleistocene and Recent distribution of *Sorex cinereus*. ■ = Blancan; ▲ = Irvingtonian; ● = Rancholabrean; *hatching* = Recent.

Masked Shrew, *Sorex cinereus* Kerr, 1792 (*S. franktownensis* Peterson, 1926; *S. personatus* Brown, 1908)

The stratigraphic range of the living *Sorex cinereus* extends back to the late Kansan (Cudahy) and perhaps to the very late Blancan (White Rock). The species has been identified in more than 30 faunas (see fig. 8.1). Recently, Jammot (1972) described the specimens from Cudahy as an extinct subspecies, *Sorex cinereus meadensis,* and Eshelman (1975) believes that the specimens from White Rock are closely related and perhaps ancestral to the extinct subspecies.

Although *Sorex cinereus* has the widest range of any extant North American soricid, its Pleistocene range was even larger and included Kansas, Missouri, Arkansas, Texas, New Mexico, and northern Mexico (fig. 8.1). Masked shrews live in moist habitats with abundant ground cover in northern North America, with southward extensions into the Rocky Mountains and the Appalachians. There is a positive correlation between size and latitude (Bergmann's response) in this species, except in tundra areas of the western Nearctic.

Sorex cinereus is one of the smaller shrews, with a head and body length of 50–65 mm and a tail 30–50 mm long. It is distinguished from *Microsorex hoyi* by the unreduced condition of the third and fifth upper antemolars (unicuspids). Like other shrews, it has a high rate of metabolism and feeds voraciously on insects and other small animals (Eshelman, 1975; Guilday, Martin, and McCrady, 1964; Jammot, 1972; G. E. Schultz, 1969.)

†Scott Shrew, *Sorex scottensis* Jammot, 1972

Originally described from Mt. Scott, this species has also been identified in the Duck Creek local fauna. Both deposits are late Illinoian in age. *Sorex scottensis* is distinguished from *Sorex cinereus* by the greater height and width of the horizontal and ascending rami and the position of the mental foramen, located under the posterior portion of the protoconid of M_1. (Jammot, 1972; McMullen, 1975.)

†Kansas Shrew, *Sorex kansasensis* McMullen, 1975

An Illinoian form, *Sorex kansasensis* was described from the Williams local fauna, Rice County, Kansas, and is also known from Duck Creek. The species was intermediate in size between *Sorex arcticus* and *Sorex palustris*. Differences from these species include a more robust P_4 lacking the prominent ridge extending from the paraconid to the anterior cingulum and a broader talonid on M_3. The species probably fed on mollusks and hard-bodied insects. At Duck Creek, it was found in association with *Sorex scottensis, Sorex arcticus,* and *Sorex palustris;* presence of the latter two species indicates a stream/marsh environment with marshy meadows and deciduous forests. (McMullen, 1975.)

Southeastern Shrew, *Sorex longirostris* Bachman, 1837

Found in the fossil state only at Haile XIB (Sangamonian), the southeastern shrew is an inhabitant of moist, open fields and woodlots from central Florida to northern Virginia and west to northern Arkansas. It is a small shrew (50–63 mm) with a fairly long tail (25.4–38 mm).

Vagrant Shrew, *Sorex vagrans* Baird, 1858

Pleistocene occurrences of the vagrant shrew include Slaton, Schulze, Conard Fissure, and Dry caves. Of these sites, only Dry Cave is within the present range of the species, an area that includes western United States, except desert regions. Like *Sorex palustris,* it possesses a third upper antemolar that is smaller than the fourth.

Ornate Shrew, *Sorex ornatus* C. H. Merriam, 1895

An inhabitant of stream banks and wet meadows, the ornate shrew is found from central California south into Baja California. It has been identified at Rancho La Brea, Carpinteria, McKittrick, and Newport Bay Mesa. It is about the same size as *Sorex vagrans.* (Compton, 1937.)

†Lake Shrew, *Sorex lacustris* (Hibbard), 1944 *(Neosorex lacustris)*

Known only from the late Kansan Cudahy fauna, *Sorex lacustris* had a large mandibular foramen under the anterior root of M_1, and the talonid of M_3 was well developed. The characters of the teeth and the ramus are those of the subgenus *Neosorex* (water shrews). (Hibbard, 1944.)

†Giant Water Shrew, *Sorex megapalustris* Paulson, 1961

Another shrew known only from Cudahy is *Sorex megapalustris*. It is similar to *Sorex palustris* but more robust. (Paulson, 1961.)

Water Shrew, *Sorex palustris* Richardson, 1828

The stratigraphic range of *Sorex palustris* extends from the Illinoian (Doby Springs) to the Recent. It has been reported from Little Box Elder, Bell, Silver Creek, Natural Chimneys, New Paris No. 4, Welsh, Robinson, Mt. Scott, Doby Springs, Robert, Crankshaft Pit, Peccary, and Howard Ranch. The last six sites are outside the present range of the species, an area that includes Canada, the mountains of the West, and the Appalachians. Semiaquatic in habits, the water shrew lives along cold streams where there is adequate cover.

A large shrew, it has a head and body length of about 95 mm and a tail length of 65–75 mm. The large skull, small third upper antemolar, and medially directed pigmented ridges on the antemolars are distinctive. A fringe of stiff hairs on the hind feet aids the animal in swimming and diving. *Sorex lacustris* may have been ancestral. (Guilday, Hamilton, and McCrady, 1969, 1971; Guilday, Martin, and McCrady, 1964; G. E. Schultz, 1969; Walker, 1975.)

Smoky Shrew, *Sorex fumeus* Miller, 1895

The smoky shrew, intermediate in size between *Sorex cinereus* and *Sorex arcticus*, inhabits both coniferous and deciduous forests of the Appalachian Mountain chain and its flanking plateau areas, from the Great Smoky Mountains north to the Gaspé Peninsula and west to the north shore of Lake Superior. It is a common animal over much of its range. The stratigraphic range of *Sorex fumeus* extends back to the late Irvingtonian (Conrad Fissure), and the species has been found at several Rancholabrean localities, including New Paris No. 4, Eagle, Natural Chimneys, Robinson, Baker Bluff, Ladds and Clark's Cave. (Guilday, Martin, and McCrady, 1964; Guilday, Hamilton, and McCrady, 1969.)

Arctic Shrew, *Sorex arcticus* Kerr, 1792

The Arctic shrew has been identified at Natural Chimneys, New Paris No. 4, Clark's, Robinson, Baker Bluff, Trout (earliest occurrence—late Irvingtonian), Bootlegger Sink, Eagle, Crankshaft Pit, Peccary, Mt. Scott, and Doby Springs—all of which are south of the present range of the species (see fig. 8.2), an area extending across Alaska, Canada, North Dakota, Wisconsin, and Minnesota. It inhabits tamarack and spruce swamps.

Sorex arcticus is sharply tricolored with a wide blackish dorsal stripe, brownish sides, and grayish underparts. The pattern brightens in winter. A medium-sized shrew, it has a head and body length of 70–76 mm and a tail length of 31–42 mm.

The Arctic shrew is an indicator of boreal conditions, and at a stratified site such as New Paris No. 4, it is found at the deeper levels, replacing the more southerly forms

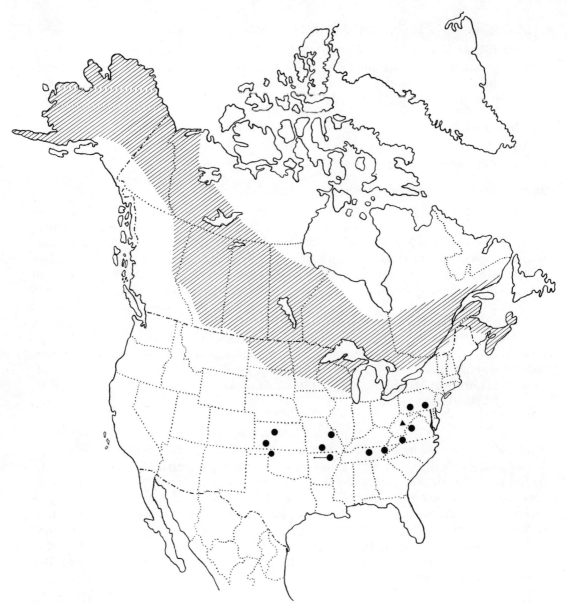

8.2 Pleistocene and Recent distribution of *Sorex arcticus*. ▲ = Irvingtonian; ● = Rancholabrean; *hatching* = Recent.

such as *Sorex fumeus, Sorex dispar,* and *Blarina brevicaua*. (Guilday, Martin, and McCrady, 1964; Guilday, Hamilton, and McCrady, 1969.)

Longtail Shrew, *Sorex dispar* Batchelder, 1896

The longtail shrew is endemic to the Appalachian Mountains, where it is found in cool, moist, and rocky forests. *Sorex dispar* has been identified at New Paris No. 4, Clark's, Baker Bluff, and Robinson caves, all late Pleistocene in age. The tail is nearly as long as the combined head and body length. The unique skull is flattened, with the rostrum and mandibles long and delicate, almost forcepslike. (Guilday, Martin, and McCrady, 1964; Guilday, Hamilton and McCrady, 1969.)

Trowbridge Shrew, *Sorex trowbridgii* Baird, 1858

A fairly large species, the Trowbridge shrew lives in coniferous forests from southern British Columbia to central California. It has been recognized in fossil form only at Carpinteria, a Wisconsinan deposit.

Merriam's Shrew, *Sorex merriami* Dobson, 1890

Reported in the fossil state only from the Wisconsinan Dry Cave, this shrew inhabits arid parts of the Great Basin and Rocky Mountain regions. Head and body length is 57–64 mm, and tail length is about 38 mm. The third upper antemolar is larger than the fourth (Armstrong and Jones, 1971.)

Saussure's Shrew, *Sorex saussurei* Merriam, 1892

The only Pleistocene occurrence of Saussure's shrew is San Josecito Cave. *Sorex saussurei* is a widespread Mexican species and is found in the vicinity of the cave today. Morphologically, it is similar to *Sorex trowbridgii*.

†Meadow Shrew, *Microsorex pratensis* Hibbard, 1944

Microsorex is an unspecialized lineage of the tribe Soricini; the genus has minute third and fifth upper antemolars. Three species are known from the North American Pleistocene.

Larger than *Microsorex hoyi*, the meadow shrew has a heavier ramus and larger teeth. A late Irvingtonian species, *Microsorex pratensis* is known only from Cudahy. (Hibbard, 1944.)

†Minute Shrew, *Microsorex minutus* Brown, 1908

Known only from the late Irvingtonian Conard Fissure, *Microsorex minutus* was more robust than *Microsorex hoyi* and had a longer tooth row and larger teeth. (Brown, 1908.)

Pygmy Shrew, *Microsorex hoyi* (Baird), 1858 (*Sorex hoyi*)

The pygmy shrew ranged much further south in the late Pleistocene and has been identified at Natural Chimneys, New Paris No. 4, Clark's, Bootlegger Sink, Robinson, Baker Bluff, Welsh, Peccary, Crankshaft Pit, and Little Box Elder; the last three sites are outside the present range of the species. Today, the pygmy shrew is found in wooded and open areas of the Canadian and Hudsonian life zones of Alaska, Canada, and parts of northern and eastern United States.

By weight, *Microsorex hoyi* is the smallest American mammal; it tips the scale at between 2.2 and 3.5 g. The skull resembles that of *Sorex* but is relatively narrower and flatter. There are 32 teeth; the upper third antemolar is greatly reduced and not visible except under magnification, and the talonid of M_3 is more reduced than that of *Sorex*. (See fig. 8.3.)

Considered rare by mammalogists, *Microsorex hoyi* was apparently more common in the late Pleistocene; its remains make up 10 percent of the shrew population at New Paris No. 4. (Guilday, Hamilton, and McCrady, 1969; Repenning, 1967b; Walker, 1975.)

†Dixon Shrew, *Planisorex dixonensis* (Hibbard), 1956 (*Sorex dixonensis*)

This late Blancan shrew is known from Dixon and Sand Draw. M^1 and M^2 possess distinct hypocones, a character not found in any other genus of shrew. M_1 and M_2 are rectangular, with heavy labial and lingual cingula, and the talonid of M_3 is broad-

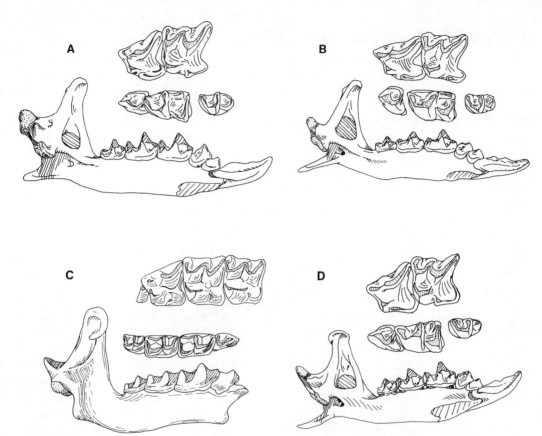

8.3 Soricidae. *A, Notiosorex crawfordi*, Recent, left P^4–M^1, P_4–M_1, M_3, occlusal views, and left mandible, internal view. *B, Microsorex hoyi*, Recent, same views. *C, Planisorex dixonensis*, Blancan, Sand Draw, left P^4–M^2, right P_4–M_3, occlusal views, right mandible, external view. *D, Cryptotis parva*, Recent, same views as *A*. *A, B, D*, after Repenning (1967); *C*, after Skinner and Hibbard (1972). Teeth ca. 10×, jaws ca. 6×.

basined. Only the anterior mandibular foramen is present. The lingual emargination of the mandibular condyle is intermediate between that of *Blarina* and *Sorex*. (Skinner and Hibbard, 1972.)

†King Shrew, *Paracryptotis rex* Hibbard, 1950

Paracryptotis shows similarities to *Blarina* in jaw articulation, size of M_3, and trapezoidal outline of M^2, but the dental formula and the structure of the talonid of M_3 show the genus to be more closely related to *Cryptotis*. The teeth, 30 in number, show slightly less pigmentation than those of *Blarina*. *Paracryptotis* was probably too specialized to have been ancestral to either *Blarina* or *Cryptotis*. Two species are recognized.

Paracryptotis rex is known from Hemphillian-age faunas in Kansas and Oregon and from the early Blancan Fox Canyon local fauna. Closely related to *Paracryptotis gidleyi*, it differs from the latter in its larger size and the presence of a larger talonid on M_3. Both species show a reduction of the length of M_2 and M_3 relative to M_1. (Hibbard, 1950; Hibbard and Bjork, 1971; Repenning, 1967b.)

†Gidley's Shrew, *Paracryptotis gidleyi* (Gazin), 1933 (*Blarina gidleyi*)

Paracryptotis gidleyi is known only from the early Blancan Hagerman fauna, where more than 120 individuals have been found. Until the recovery of upper dentitions a few years ago, this species was thought to belong to the genus *Blarina*. The teeth are not as robust as those of *Paracryptotis rex:* P^4 is narrow anteriorly with moderate posterior emargination, the talonids of M_1 and M_2 are short, and M_3 has a shorter talonid but is relatively longer than that of *P. rex*. (Hibbard and Bjork, 1971; Repenning, 1967b.)

†Meade Shrew, *Cryptotis? meadensis* Hibbard, 1953

Cryptotis is a primitive member of the tribe Blarinini, and Repenning (1967b) considered it to be less specialized than other members of the tribe (except for the late Miocene *Adeloblarina*) because of its mandibular structure and the position of the zygomatic processes. Characteristic of the genus is the extreme reduction of the talonid of M_3. *Cryptotis* inhabits forests and open areas of the eastern United States, Mexico, Central America, and northern South America. It is the only genus of insectivore in South America. The stratigraphic range of *Cryptotis* in North America extends from the Middle Pliocene to the Recent.

Known only from the early Blancan Fox Canyon fauna, this species is intermediate in size between *Sorex taylori* and *Sorex rexroadensis*. Some of its features appear to be less advanced than but similar to those of the living *Cryptotis*. The entoconid on M_1 is not distally isolated, however, and the P_4 resembles that found in the extinct subfamily Limnoecinae. (Hibbard, 1953a; Repenning, 1967b.)

†Adam's Shrew, *Cryptotis adamsi* (Hibbard), 1953 (*Blarina adamsi*)

Originally described from Fox Canyon, this species has also been recognized in a Hemphillian fauna in Lake County, Oregon. It was about the size of *Notiosorex jacksoni*. Except for the retention of a minute P^2, its dentition and mandible are nearly identical with those of the living *Cryptotis*. The species differs from *Blarina* in the structure of M_1 and M_3. (Hibbard, 1953a; Repenning, 1967b.)

Least Shrew, *Cryptotis parva* (Say), 1823 (*Sorex parvus*)

The stratigraphic range of *Cryptotis parva* extends from the late Blancan (Haile XVA) to the Recent, and the species has been identified from more than 30 Pleistocene sites in Florida, Kansas, Missouri, Pennsylvania, Tennessee, Texas, Virginia, and West Virginia. The living animal is found in open, grassy areas and marshes in the eastern half of the country south of the Great Lakes.

Characterized by an extremely short tail and chunky body, the least shrew is America's shortest mammal (head and body length, 56–64 mm; total length 64–83 mm, weight 4–7g). The skull is small but broader and higher than that of *Sorex*. It has 30 teeth, with slightly pigmented cusps. The talonid of M_3 is greatly reduced. Feeding on insects and other small animals, the least shrew may eat more than its own weight each day. (Burt and Grossenheider, 1976; Guilday, Hamilton, and McCrady, 1969; Repenning, 1967b; Whitaker, 1974.)

Mexican Shrew, *Cryptotis mexicana* (Coues), 1877 [*Blarina (Sorisicus) mexicana*]

The only Pleistocene record of *Cryptotis* in Mexico is from San Josecito Cave, where *Cryptotis mexicana* was identified. The extant species has a wide distribution in Mexico and elsewhere. (Findley, 1953.)

†Ozark Short-tailed Shrew, *Blarina ozarkensis* Brown, 1908

The stratigraphic range of the genus *Blarina* extends from the early Irvingtonian to the Recent. The genus shows more intraspecific variation, especially in the structure of the talonid of M_3, the location of the mental foramen, and the form of the cingulum surrounding the lower cheek teeth, than any other extant shrew. *Blarina* is derived from the late Miocene form *Adeloblarina*.

This Irvingtonian shrew is known only from Conard Fissure but may be present in other Kansan-age faunas. First described as a subspecies of *Blarina brevicauda*, it is distinguished from the latter by the absence of the accessory cone (which, in the living species, is situated lingually and posterior to the protoconid of the first antemolar), the reduced condition of fifth upper antemolar and the talonid of M_3, and the larger and more bulbous cingula surrounding the lower teeth. In size, *Blarina ozarkensis* is intermediate between *Blarina b. brevicauda* and *Blarina b. kirtlandi*. (Graham and Semken, 1976.)

Short-tailed Shrew, *Blarina brevicauda* (Say), 1823 (*Sorex brevicauda;* *Blarina simplicidens* Cope, 1899; *B. fossilis* Hibbard, 1943)

The short-tailed shrew has been found at more than 50 sites, the earliest of which are early Irvingtonian (Inglis IA and Port Kennedy). Several localities (Jinglebob, Robert, Mt. Scott, Berends, Doby Springs, Schulze, Howard Ranch) are outside the present range of the species, an area that includes a wide variety of habitats in the eastern half of the United States. The species is common throughout much of its range.

Blarina brevicauda is a large, stocky shrew, measuring 95 to 130 mm and weighing 15 to 30 g. It has 32 teeth. The dentition is specialized except for the retention of six upper antemolars; the talonids of M_1 and M_2 are shortened, and M_3 shows the least reduction of that of any of the Blarinini. The saliva from the submaxillary glands is poisonous and acts on the nerves of mice, insects, and other small invertebrates on which *Blarina* feeds.

Several authors have noted that *Blarina* remains from Pleistocene faunas are larger than the extant members of the population living in the same area (Dalquest, 1965; Guilday, Hamilton, and McCrady, 1969; Hibbard, 1943a, 1963a; Paulson, 1961; Stephens, 1960). Generally, this has been attributed to the *Blarina* topocline, which exhibits a positive Bergmann's response. Specimens from Robinson Cave, for example, are as large as those found in Wisconsin today. Moreover, Natural Chimneys, Crankshaft Pit, Meyer, and Thurman local faunas contain two *Blarina* populations which have been compared to different portions of the cline and taxonomically designated as some combination of *Blarina b. brevicauda,* *Blarina b. kirtlandi,* and *Blarina b. carolinenis.* The modern distributions of these three subspecies correspond, respectively, to the subhumid microthermal, humid microthermal, and humid mesothermal climatic regions. Graham and Semken (1976) have identified all three ecotypes at Cumberland, New Paris No. 4, and Peccary caves and two ecotypes at Natural Chimneys, Crankshaft, Baker Bluff, Meyer, and Thurman caves. Coexistence of two or three ecotypes in the same deposit without intermediates or apparent interbreeding suggests to Graham and Semken that the populations are not the result of clinal migrations in response to glacial fluctuations but rather that the subspecies were behaving as species at that time. This assumes, however, that there were no ecological barriers (Guilday et al., 1978). Recently, Jones, Carter, and Genoways (1973) recognized *Blarina brevicauda* and *Blarina carolinensis* as distinct species because they have been taken in the same trapline in Nebraska. The presence of two size forms in Pleistocene local faunas supports this view. Specific rather than subspecific relationships alleviate the problem of extreme intraspecific variation previously attributed to *Blarina* (Semken, personal communication, 1975). (Dalquest, 1965; Graham and Semken, 1976; Guilday, Hamilton, and McCrady, 1969; Guilday, Martin, and McCrady, 1964; Guilday et al., 1978; Hibbard, 1943a,

1963a; Jones, Carter, and Genoways, 1973; Paulson, 1961; Repenning, 1967b; Semken, personal communication, 1975.)

†Jackson's Shrew, *Notiosorex jacksoni* Hibbard, 1950

Notiosorex is characterized by having a nonbifid upper incisor, a reduced heel on M_3, and lightly pigmented teeth. Two species are recognized, with a stratigraphic range from the late Pliocene to the Recent.

A Blancan/early Irvingtonian species, *Notiosorex jacksoni* was a member of the Rexroad fauna and has also been identified at Benson and Vallecito Creek. It was larger and had more generalized teeth than those of *Notiosorex crawfordi*. (Hibbard, 1950.)

Desert Shrew, *Notiosorex crawfordi* (Coues), 1877 [*Sorex (Notiosorex) crawfordi*]

The stratigraphic range of the desert shrew extends back to the Sangamonian, and the species has been identified at Rancho La Brea, Newport Bay Mesa, Costeau Pit, Dry, Friesenhahn, Longhorn Cavern, Clear Creek, Schulze, Klein, and Cragin Quarry. *Notiosorex crawfordi* is an inhabitant of the semidesert country of the southwestern and south-central United States and northern and central Mexico. Weighing only 3.0–3.5 g, this small shrew feeds on insects and small invertebrates. It has 28 teeth (with 3 upper antemolars instead of 5 as in *Sorex*). Its ancestry can be traced back to *Notiosorex jacksoni*. (Armstrong and Jones, 1972; Compton, 1937; Repenning, 1967b.)

FAMILY TALPIDAE—MOLES

The talpid family comprises insectivores with a wide range of adaptations; it includes aquatic and nonburrowing species, as well as burrowing forms at various stages of specialization. All of the Blancan-to-Recent North American talpids are classed within the subfamily Talpinae, which may be further subdivided into a number of tribes. Of these, the Urotrichini is represented by the shrew mole genus *Neurotrichus*, which has no fossil history (except for an uncertain Hemphillian record). The Condylurini, not known with certainty prior to the Pleistocene, comprises the single genus *Condylura*, an aquatic mole. Finally, the genera *Parascalops, Scalopus,* and *Scapanus,* together with a number of pre-Blancan genera, make up the Scalopini, to which the majority of the Blancan-to-Recent talpids belong; these animals show the greatest fossorial specialization, with forefeet developed as extremely powerful digging organs. Fossil skulls are rare, but the short, stout arm bones preserve well and show important taxonomic characters (fig. 8.4). Many talpid fossils have been found in caves, to which they had probably been brought by raptorial birds. But on the whole, the talpids have a relatively poor fossil record. Their history begins in North America in the late Eocene. (Hutchison, 1968; Van Valen, 1967.)

Star-nosed Mole, *Condylura cristata* (Linnaeus), 1758 (*Sorex cristatus*)

Condylura is distinguished externally by the ring of 22 sensitive tentacles around its nose and the long tail, which is typical of aquatic moles. The arm bones are slender, and the skull has a long, tapering rostrum with small, widely spaced teeth.

Condylura cristata is the sole known species, and it is not particularly common in the fossil state. Pleistocene records come from Arkansas (Peccary Cave), Missouri (Crankshaft Cave), Pennsylvania (New Paris No. 4, Bootlegger Sink), Tennessee (Baker Bluff

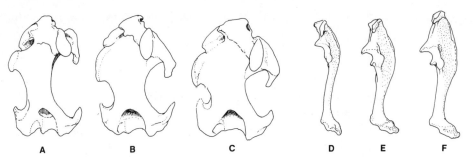

8.4 Talpidae. *A–C*, humeri; *D–F*, ulnae. *A, D, Condylura cristata. B, E, Parascalops breweri. C, F, Scapanus latimanus.* All Recent. After Hutchison (1968). Not to scale.

Cave), Virginia (Natural Chimneys), and West Virginia (Trout Cave). The last-mentioned site dates from the late Irvingtonian; the others are Wisconsinan. Although most of the records are within the modern range of the species, those from Peccary Cave and Crankshaft Cave are far outside it and indicate a very considerable range extension to the southwest in late Wisconsinan times.

Condylura cristata is an eastern form, adapted to very damp soil and spending much of its time in the water. Its presence in Missouri and northern Arkansas may indicate wetter conditions than those prevailing today.

The origin of this species is unknown; it is the sole known member of the Condylurini. (Hutchison, 1968; Parmalee, Oesch, and Guilday, 1969.)

Hairy-tailed Mole, *Parascalops breweri* (Bachman), 1842 (*Scalopus breweri*)

The tribe Scalopini is divided by Hutchison into two subtribes, the Parascalopina and Scalopina, which have been separate since well back in the Miocene. *Parascalops* is the only Pleistocene and Recent North American member of the Parascalopina. Its fossorial adaptation is somewhat less extreme than that of the subtribe Scalopina, and this is reflected in the structure of its humerus, which is somewhat narrower. Odontologically, the genus is characterized by the basal shelves of M^1 and M^2 being trilobed, with that of M^3 bilobed. The ancestry of the genus is unknown.

Parascalops breweri, the sole known species of its genus, has been found at a number of localities in Missouri, Pennsylvania, Tennessee, Virginia, and West Virginia, ranging in age from the late Irvingtonian (Trout Cave) to the late Wisconsin. As in the case of *Condylura*, the Missouri record (Crankshaft Cave) is far west of the present-day range. It has been suggested that the species spread southward from Lake Superior along the Mississippi River, perhaps in late Wisconsinan times. The species inhabits light, sandy soils. (Hutchison, 1968; Parmalee, Oesch, and Guilday, 1969.)

Broad-footed Mole, *Scapanus latimanus* (Bachman), 1842 (*Scalopus latimanus*)

A member of the subtribe Scalopina, the genus *Scapanus* usually has the full complement of teeth; the humerus is not as broad as in *Scalopus*. The genus dates back to the late Miocene (Clarendonian), and there are records of *Scapanus* sp. from the Blancan (Hagerman, Grand View, White Bluffs).

Material referred to *Scapanus latimanus* has been recorded from the Los Angeles area and from Potter Creek and Hawver caves; the earliest finds are from Newport Bay Mesa and Sacramento, dated at 103,000 B.P.

In this species, the rostrum is moderately long, with crowded teeth and well-developed I^1. The present-day species of *Scapanus* are distributed along the Pacific coast; the range of *Scapanus latimanus* extends from central Oregon to northern Baja California. Its preferred habitat is porous soil in valleys and mountain meadows. (Hutchison, personal communication, 1975; W. E. Miller, 1971.)

†Rexroad Mole, *Scalopus rexroadi* (Hibbard), 1941 (*Hesperoscalops rexroadi*)

Fossorial adaptation within the Scalopini attains its acme in *Scalopus*, as reflected in the very stout arm bones. The cheek teeth are hypsodont, and I¹/I₂ are large. There are only two incisors in each jaw half, as compared to three in *Scapanus*. The genus appears in the Hemphillian but is thought to have a fairly long independent history before that.

The genus *Hesperoscalops* was established because of the presence of large basal accessory cusps in the lower molars. Similar but smaller cusps occur in some *Scalopus*, however, and *Hesperoscalops* is now regarded as a subgenus. The species *Scalopus rexroadi* occurs in the early Blancan Rexroad fauna. A somewhat larger species, *Scalopus sewardensis* (Reed), is known from the Hemphillian of Saw Rock Canyon and also from a site in Nevada but is not thought to be directly ancestral. *Scalopus rexroadi* may have given rise to *Scalopus aquaticus*. (Hibbard, 1941b; Hutchison, 1968; K. M. Reed, 1962.)

†Mt. Blanco Mole, *Scalopus blancoensis* (Dalquest), 1975
(*Hesperoscalops blancoensis*)

Known only from the late Blancan fauna of Blanco, this species resembles *Scalopus rexroadi* but has larger, broader teeth and more strongly developed basal accessory cusps, especially on M_3; M_3 is more reduced than M_2. (Dalquest, 1975.)

Eastern Mole, *Scalopus aquaticus* (Linnaeus), 1758 (*Sorex aquaticus*)

This species, which today is widely distributed from the eastern and central United States west approximately to the Nebraska–Colorado line, is common in the fossil state. Records range from the late Blancan and Irvingtonian (Sand Draw, Haile XVA, White Rock, Inglis IA, Mullen, Conard Fissure) to the Rancholabrean, with more than 35 sites in Arkansas, Florida, Georgia, Kansas, Kentucky, Missouri, Nebraska, Pennsylvania, Tennessee, Texas, and Virginia. The finds are generally within the present-day range of the species. However, Holocene records of *Scalopus aquaticus* are available from the Upper Ohio Valley in western Pennsylvania, where the species is not found today.

The species may be derived from *Scalopus rexroadi*, from which it differs by weaker development of the accessory cusps in the lower molars. It prefers moist, sandy loam. Although the eastern mole is a perfectly able swimmer, the name *aquaticus* is due to a misunderstanding and should not be taken literally. The species is not found in very dry soil, however. (Guilday, 1961; Hibbard, 1941b; Hutchison, 1968.)

CHAPTER NINE

ORDER CHIROPTERA

Bats, the only mammals having true flight, are a highly successful group and comprise the second largest order of mammals today. Their most distinctive feature, the membranous wings, is formed by webs of skin stretched between four elongated, clawless fingers; the free clawed thumb is used in clinging. Fruit bats have a more primitive wing with two clawed digits. Bats are probably derived from an arboreal insectivore group and developed their flying adaptations in the Paleocene or very early Eocene, for middle Eocene deposits in Europe and North America have yielded microchiropterans with well-developed wings. Their fossil history, however, is incomplete owing to their small size, fragile bones, nonterrestrial habits, and mainly tropical distribution. Bats are common as fossils only in Pleistocene cave deposits.

The order is divided into two suborders: Megachiroptera, the fruit bats, or flying foxes, of the Old World Tropics, and Microchiroptera, the nearly cosmopolitan, typically small, insect-eating bats. The microchiropterans, many of which hibernate, use echolocation (emission of ultrasonic pulses, which are reflected back from objects, enabling the bat to "see") as their primary means of orientation and can fly and capture their prey in total darkness. Fruit bats, on the other hand, are not known to hibernate, use vision as their primary means of orientation, and cannot fly in the dark.

FAMILY MORMOOPIDAE—LEAF-CHINNED BATS

Formerly considered to be members of the Family Phyllostomatidae, the two genera and eight species of leaf-chinned bats differ markedly from that family in external, anatomical, and behavioral specializations. Adaptations associated with their swift, highly maneuverable flight and ability to remain airborne for long periods of time include reduction in the weight of the limbs by simplification and loss of elasticity in muscles. The greater tuberosity of the humerus does not form a locking device with the scapula.

These small, very abundant bats are found in the West Indies and southwestern United States south to Brazil. Vaughan and Bateman (1970) believe that intense competition between the already established Mormoopidae and the later-arriving Vespertilionidae explains the limited success of the vespertilionids in the Neotropics. Their fossil history is little known.

Ghost-faced Bat, *Mormoops megaphylla* (Peters), 1864 (*Mormops megaphylla*)

Characteristic of the genus *Mormoops* is a skull so greatly shortened that both the braincase and rostrum are wider than long, and the floor of the deepened braincase is so elevated that the lower rim of the foramen magnum is above the level of the rostrum. Two extant species are recognized.

The only Pleistocene occurrence of the ghost-faced bat is from Rock Springs, a Sangamonian deposit in Florida. Today, the species ranges from southern Arizona and southern Texas to northern South America. It is colonial and roosts singly (not in clusters) in caves and tunnels, often in association with other species of bats.

A large, puglike face with well-developed folds of skin on the lower lip and chin are characteristic of *Mormoops megaphylla*. The coronoid process of the mandible is reduced, allowing the jaws to gape widely. There are 34 insectivorous-type teeth. A strong, swift flier, the ghost-faced bat forages over wide areas. (Vaughan, 1972; Vaughan and Bateman, 1970.)

Family Phyllostomatidae—Leaf-nosed Bats

The Neotropical leaf-nosed bats are the most diverse family of chiropterans with respect to structural variations associated with feeding adaptations. Some of these bats are insectivorous or carnivorous, whereas others feed on fruit, nectar, or pollen, and the Desmodontinae feed on blood. Unlike the mormoopids, the phyllostomatids remain on the wing only for short periods during foraging. The shoulder joint has a moderately well-developed locking device formed between the greater tuberosity of the humerus and the scapula. The forelimb is generalized and is used in food-handling and climbing.

The stratigraphic record of the Phyllostomatidae extends back to the Miocene in Colombia. The ancestral type had tuberculosectorial teeth adapted for a diet of insects.

Subfamily Phyllostomatinae

Mexican Long-nosed Bat, *Leptonycteris nivalis* (Saussure), 1860
   (*Ischnoglossa nivalis*)

Leptonycteris is characterized by the absence of the third molar and the presence of lower incisors. Bats of this genus feed on a diet of nectar, pollen, fruit, and insects. Three extant species are recognized.

The only Pleistocene occurrence of *Leptonycteris nivalis* is at San Josecito, a Wisconsinan deposit. Today, the rare Mexican long-nosed bat occurs in Big Bend National Park, Texas, and northern Mexico. Colonial and cave-dwelling, it inhabits the arid pine–oak country above 1,500 m.

A rather large species, *Leptonycteris nivalis* has a triangular, leaflike flap on the nose and a minute tail, not visible externally. It has 30 teeth; the lower incisors are usually present, but the third molar is absent. *Leptonycteris* feeds on nectar and pollen. Its flight is strong and highly maneuverable. Little is known about its habits. (Barbour and Davis, 1969; Vaughan, 1972.)

Subfamily Desmodontinae

†Stock's Vampire Bat, *Desmodus stocki* Jones, 1958 (*D. magnus* Gut, 1959)

The highly specialized vampire bats are adapted to feed on the blood of birds and mammals. They are of great economic importance in many Neotropical areas because they are vectors of rabies and other diseases. Three living genera, each containing a single species, inhabit subtropical and tropical regions from northern Mexico south to northern Argentina, Uruguay, and central Chile. Their fossil history is incomplete, but Pleistocene vampire bats are known from deposits in the United States, Cuba, Mexico, and South America.

The stratigraphic range of *Desmodus stocki* extends from the Sangamonian to the Wisconsinan and its remains have been identified at Arredondo, Big Bend (Texas), Haile XIB, Potter Creek, Reddick, and San Josecito. The Pleistocene species is characterized by a larger, more massive skull, wider teeth, and longer limb bones than those of the living species, *Desmodus rotundus*.

Desmodus is the most specialized of the vampire bats. It has a short rostrum and a highly arched skull. The reduced dentition is characterized by enlarged, bladelike incisors and canines and small, apparently nonfunctional cheek teeth. *Desmodus* feeds solely on fresh mammal blood. A long, sturdy thumb and large, robust hind limbs aid the animals in running rapidly and making short jumps. Their flight is strong and direct but not highly maneuverable. Vampire bats roost in caves, hollow trees, and fissures.

The living species cannot tolerate low temperatures and is not found north of the 10°C winter isotherm. Whether *Desmodus stocki* had similar environmental tolerances is, of course, unknown. Olsen (1960) has suggested that the transmission of rabies by vampire bats may have been a contributing factor in the extinction of some of Florida's large Pleistocene mammals. (Gut, 1959; Hutchison, 1967; J. K. Jones, 1958; Koopman, personal communication, 1975; Olsen, 1960; Vaughan, 1972; Walker, 1975.)

Family Vespertilionidae—Vespertilionid Bats

The Vespertilionidae is the largest and most widely distributed family of bats. Its fossil history can be traced back to the middle Eocene in both North America and Europe, to the Pliocene in Asia, and to the Pleistocene in South America and Africa. Today, these rather small, plain-looking, broad-winged bats occupy a wide variety of habitats, ranging from tropical jungles to sandy deserts and boreal coniferous forests, on all continents except Antarctica and isolated oceanic islands.

Vespertilionid bats have tuberculosectorial teeth with well-developed W-shaped ectolophs on the upper molars. The shoulder joint is of an advanced type, with the large greater tuberosity of the humerus locking against the scapula. Their flight is highly maneuverable, and the animals pursue and capture flying insects on the wing. Echolocation is highly developed. These bats often congregate in huge colonies, and many species inhabit caves. Vespertilionids are well represented in Quaternary deposits; 1 extinct genus and 8 extinct species are recognized, and the fossil history of 21 extant species extends back into the Pleistocene.

Subfamily Vespertilionae

Fringed Myotis, *Myotis thysanodes* Miller, 1897

Bats of the genus *Myotis* have the widest distribution (being absent only in arctic, subarctic, and antarctic regions and most oceanic islands) and the greatest number of species of any genus of chiropterans. They are relatively small bats and have a simple snout, a long pointed tragus, 38 teeth, and a tail extending to the edge of the interfemoral membrane. Most species are cave-dwellers, and their remains are quite common in Pleistocene deposits.

Ranging from British Columbia south to Vera Cruz and Chiapas, with an isolated population in the Black Hills, this southwestern species inhabits oak, piñon, and juniper forests and desert scrub areas, usually at elevations between 1,200 and 2,100 m. *Myotis thysanodes* has been identified in the Wisconsinan faunas of Isleta, Papago Springs, and Little Box Elder; the latter site is outside its present range, being within the hiatus between the main and Black Hills populations. A conspicuous fringe of hairs along the edge of the interfemoral membrane gives this bat its name. It is a highly colonial species, but little is known about its habits. (Barbour and Davis, 1969.)

Long-eared Myotis, *Myotis evotis* (H. Allen), 1864 *(Vespertilio evotis)*

A western species, the well-named long-eared myotis is found in coniferous and coastal forests from central British Columbia to Baja California and east to the Dakotas and central New Mexico. It has been found at Little Box Elder, Papago Springs, Schulze, and Klein; the latter two sites are east of the animal's present range. It seldom resides in caves but may use them at night. Although widespread, it is nowhere abundant. (Barbour and Davis, 1969; Hall and Kelson, 1959.)

Keen's Myotis, *Myotis keenii* (Merriam), 1895 *(Vespertilio subulatus keenii)*

This extant species has been identified at New Paris No. 4, Robinson, Natural Chimneys, Clark's, Organ-Hedricks, and Bootlegger Sink, all Wisconsinan in age. Two populations of *Myotis keenii* are known: one is found from southern Alaska to Puget Sound, Washington, and the other is widely distributed across eastern North America, from Saskatchewan to northern Florida. The species is found in small, scattered colonies and is nowhere abundant. At Natural Chimneys and New Paris No. 4, it is the commonest bat; today, *Myotis lucifugus* is more common in the Appalachian region. Perhaps *Myotis keenii* was better adapted to the boreal conditions of the late Pleistocene in the Appalachians. (Barbour and Davis, 1969; Guilday, Martin, and McCrady, 1964.)

Small-footed Myotis, *Myotis leibi* (Audubon and Bachman), 1842
(Vespertilio subulatus leibi)

This species is common and widespread in the western half of the continent from southern Canada to Michoacán, Mexico; in the East, it is found from New England to Georgia. Its Pleistocene record extends back to the Irvingtonian (Cumberland and Conard Fissure), and the species has also been identified at Little Box Elder and Clark's caves. It is the smallest myotis in the East, and in the West only *Myotis californicus* is smaller. It has a flattened skull. A hardy species, *Myotis leibi* moves into caves only in the late fall and emerges by late March. (Barbour and Davis, 1969.)

California Myotis, *Myotis californicus* (Audubon and Bachman), 1842
(Vespertilio californicus)

A western species, the California myotis inhabits the Sonoran and Transition life zones from sea level to about 1,800 m, from southern Alaska to Chiapas and east to Colorado. Its only reported Holocene occurrence is from Klein Cave. Small in size, the species has a forearm length of 29–36 mm and a wingspread of 230 mm. A slow, erratic flier, it commonly feeds 1.5–3.0 m above the ground. This bat roosts singly or in small colonies and frequently uses man-made structures for night roosts. (Barbour and Davis, 1969.)

Indiana Myotis, *Myotis sodalis* Miller and G. M. Allen, 1928

This eastern species ranges from Oklahoma to Vermont and from Wisconsin to northern Florida. It is a member of the Wisconsinan Crankshaft and Bat caves faunas.

Myotis sodalis hibernates in caves in densely packed clusters which are characteristic of the species. Little is known about its habits, but a drastic decline in numbers has been noticed in recent years. (Barbour and Davis, 1969.)

Gray Myotis, *Myotis grisecens* A. H. Howell, 1909

The stratigraphic record of this species extends back to the late Irvingtonian (Cumberland), and during the Pleistocene–Holocene, the species ranged farther south (Devil's Den) and east (Patton, Windy Mouth, and Organ-Hendricks caves, all in West Virginia; Clark's Cave, Virginia; and Cumberland Cave, Maryland) than at present. Other Pleistocene occurrences include Ladds, Robinson, Crankshaft, Brynjulfson, and Bat. Today, the gray myotis inhabits cave regions from eastern Oklahoma to western Virginia and south to northern Florida. Unique among North American *Myotis* is the attachment of the wing membrane at the ankle. The skull shows a distinct sagittal crest. Apparently, almost the whole population winters in five or six caves, and recent human disturbance is threatening the entire species. (Barbour and Davis, 1969; Guilday, Hamilton, and McCrady, 1969; Handley, 1956.)

† Straight-toothed Myotis, *Myotis rectidentis* Choate and Hall, 1967

Found only at Laubach Cave (Wisconsinan), *Myotis rectidentis* is distinguished by a nearly straight lower canine and a relatively long, thick, massive mandible. It differs from *Myotis magnamolaris* in its smaller size and in the form of the lower canine. The dentition, except for smaller size, resembles that of the living *Myotis velifer*, and the mandible is similar to that of *Myotis evotis* in size and shape. (Choate and Hall, 1967.)

† Large Myotis, *Myotis magnamolaris* Choate and Hall, 1967

Another extinct species known only from Laubach Cave is *Myotis magnamolaris*, the largest American species of *Myotis*. It most closely resembles *Myotis velifer* and may be its direct ancestor but differs from the latter in body size and in the relative size of the lower canine, which is longer and more massive. (Choate and Hall, 1967.)

Cave Myotis, *Myotis velifer* (J. A. Allen), 1890 (*Vespertilio velifer*)

An inhabitant of the Sonoran and Transition life zones of the arid Southwest, from Nevada to Kansas and south to Honduras, *Myotis velifer* is tolerant of high temperatures and low humidities. It is a year-round resident of caves. Pleistocene occurrences include Papago Springs, Cave Without A Name, and Schulze (all late Wisconsinan) and Klein and Miller's (early Holocene). A large species, the cave myotis has a forearm length of 37–47 mm and a wingspread of 280–315 mm. It has robust teeth and a pronounced sagittal crest. Colonies of *Myotis velifer* and *Tadarida brasiliensis* are often found clustered together. (Barbour and Davis, 1969.)

Little Brown Myotis, *Myotis lucifugus* (Le Conte), 1831 (*Vespertilio lucifugus*)

A widely distributed northern species, *Myotis lucifugus* is found from central Alaska to Labrador and south to southern California, northern Chihuahua, and southern Georgia. It is probably the most numerous bat in the United States. Pleistocene occurrences include New Paris No. 4, Robinson, Natural Chimneys, Ladds, Crankshaft, Bat, Klein, and Bell. Medium-sized, the little brown myotis has a forearm length of 34–41 mm and a wingspread of 229–269 mm. The skull lacks a sagittal crest, and there is an evolutionary trend toward loss of the tiny upper premolars. Using echolocation, this species takes insects on the wing. In the fall, these bats move by the thousands into caves and mine tunnels for winter hibernation. (Barbour and Davis, 1969; Guilday, Martin, and McCrady, 1964.)

Southeastern Myotis, *Myotis austroriparius* (Rhoads), 1897
(*Vespertilio lucifugus austroriparius*)

A southeastern species, *Myotis austroriparius* ranges from coastal North Carolina west to eastern Texas and north to southern Indiana. Pleistocene occurrences are all from Florida—Coleman IIA (late Irvingtonian), Reddick, Vero, Kendrick, Arredondo, and Devil's Den—and the species is very abundant in the state today. A medium-sized species, it has a globose braincase with a slight sagittal crest. Colonies numbering from 2,000 to 90,000 inhabit caves where there is standing water. (Barbour and Davis, 1969.)

Long-legged Myotis, *Myotis volans* (H. Allen), 1866 (*Vespertilio volans*)

This is the common myotis of the transition and Canadian life zones of the West, from southern Alaska to Verz Cruz and east to Nebraska and the Dakotas. The only Pleistocene occurrence is from Little Box Elder Cave (Wisconsinan). It has a short rostrum and a globose braincase. The long-legged myotis forms large nursery colonies in trees, rock crevices, and buildings and frequently enters caves at night. (Barbour and Davis, 1969.)

Silver-haired Bat, *Lasionycteris noctivagans* (Le Conte), 1831
(*Vespertilio noctivagens*)

The single species, *Lasionycteris noctivagans* is a widespread, primarily northern, forest species. It is found from southern Alaska and central Canada to the southern United States, though it is most abundant in the northern Rocky Mountains, New York, and parts of New England. These migratory tree bats roost singly or in small groups. The silver-haired bat has been identified at Little Box Elder and Bell caves, both Wisconsinan in age. It has 36 teeth and feeds on soft-bodied insects. (Barbour and Davis, 1969.)

Eastern Pipistrelle, *Pipistrellus subflavus* (F. Cuvier), 1832 (*Vespertilio subflavus*)

Smallest of the North American bats are the pipistrelles, genus *Pipistrellus*, which are sometimes mistaken for large moths. Our two species differ from each other in the shape of the baculum and in the number of chromosomes but are more similar than many Old World species of *Pipistrellus*. Both species have 34 teeth, including a small anterior upper premolar behind the canine. This premolar is readily visible in the tooth row of *Pipistrellus subflavus* but is greatly reduced and displaced inward in *Pipistrellus hesperus*.

The eastern pipistrelle, *Pipistrellus subflavus*, is found in eastern North America, from Nova Scotia and Minnesota south to Florida and Honduras. It is abundant over much of its range and probably inhabits more caves in eastern North America than any other species of bat. *Pipistrellus subflavus* has been identified from some 15 Pleistocene localities in Florida, Kentucky, Missouri, Pennsylvania, Tennessee, Texas, and Virginia. The earliest occurrences are late Irvingtonian (Coleman IIA, Trout). The species is slightly larger than *Pipistrellus hesperus*. (Barbour and Davis, 1969; Guilday, Martin, and McCrady, 1964.)

Western Pipistrelle, *Pipistrellus hesperus* (H. Allen), 1864 *(Scotophilus hersperus)*

With a forearm measurement of 27–33 mm and a wingspread of 190–215 mm, *Pipistrellus hesperus* is the smallest bat in the United States. An inhabitant of desert lowlands and rocky canyons, it ranges from central Mexico north to Washington and east to southern Colorado and western Oklahoma. The only fossil occurrence is from Klein Cave, an early Holocene deposit. Both species of *Pipistrellus* have been identified at that site. (Barbour and Davis, 1969.)

Big Brown Bat, *Eptesicus fuscus* (Palisot de Beauvois), 1976 *(Vespertilio fuscus; Vespertilio fuscus grandis* Brown, 1908)

Eptesicus, a nearly cosmopolitan genus of about 30 extant species, is characterized by large size and a rather slow, fluttering flight. These bats feed on a variety of insects.

This widespread species is abundant over much of its range, an area that encompasses southern Canada, all of the United States except southern Florida and central Texas, the Greater Antilles, Middle America, and northern South America. There is even a record from Alaska. Its Pleistocene record extends back to the Irvingtonian (Conard Fissure, Cumberland, Trout), and the species has been identified at some 25 Rancholabrean sites. An extinct form, *Eptesicus grandis* (Brown) was described from Conard Fissure, and specimens from several other sites have been referred to it; however, Guilday (1967b) has shown that, except for slightly larger size, *Eptesicus grandis* cannot be adequately distinguished from its modern counterparts at either the specific or subspecific level.

Eptesicus fuscus is a large bat with a forearm length of 42–51 mm and a wingspread of 325–350 mm; females are 5 percent larger than males. It has 32 teeth and feeds chiefly on beetles. A hardy species, it hibernates as far north as the north shore of Lake Superior. (Barbour and Davis, 1969; Guilday, 1967b; Guilday, Martin, and McCrady, 1964.)

† Stock's Snub-nosed Bat, *Histiotus stocki* (Stirton), 1931 *(Simonycteris stocki)*

This extinct species is known only from the early Irvingonian Curtis Ranch fauna. Stirton (1931) described it as a new genus, *Simonycteris,* distinct from *Eptesicus, Chalinobus,* and *Lasiurus.* Unfortunately, he did not compare it with the living South American genus *Histiotus;* Koopman (personal communication, 1975) could find no differences between *Simonycteris* and *Histiotus* and referred the fossil to the living genus. *Histiotus* closely resembles *Eptesicus* in skull and dental characters but has much larger ears. *Histiotus stocki* has a short rostrum and an abrupt facial angle. (Koopman, personal communication, 1975; Stirton, 1931.)

Red Bat, *Lasiurus borealis* (Muller), 1776 *(Vespertilio borealis)*

The New World genus *Lasiurus* is found over most of North America, the West Indies and other islands, and Central South America. These bats are strong fliers, and at least the northern mainland species migrate. *Lasiurus* is the only bat that regularly has more than 2 young at birth.

Ranging from Canada over much of the United States, Mexico, the Greater Antilles, and Central America and throughout South America, the red bat is most abundant in the northern part of its range. Pleistocene records include Vero, Reddick (Sangamonian), Bat, Organ-Hedricks, and Natural Chimneys. A solitary species, it spends the day hidden in the foliage of trees and comes forth in the evening to feed. Red bats are

ORDER CHIROPTERA

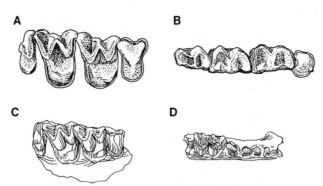

9.1 Vespertilionidae. *A, B, Lasiurus cinereus,* Recent. *A*, right P^4-M^3; *B*, left P_4-M_3. *C, D, Anzanycteris anzensis,* Blancan, Arroyo Seco. *C*, P^4-M^3; *D*, right mandible fragment with M_2. *A, B*, after Vaughan (1972); *C, D*, after White (1969). *A, B*, 7×; *C, D*, ca. 10×.

heavily furred and can withstand drastic temperature changes. (Barbour and Davis, 1969; Guilday, 1962.)

† Extinct Hoary Bat, *Lasiurus fossilis* Hibbard, 1950

Known only from the type locality, Fox Canyon (very early Blancan), this species is larger than *Lasiurus borealis* and smaller than *Lasiurus cinereus*. The transverse width of M_1 compared to the length of the tooth is less in the extinct species than in living forms. Characters of the jaw and dentition of *Lasiurus fossilis* indicate that it was ancestral to *Lasiurus cinereus*. *Lasiurus* sp. has been reported from Vero and Slaton. (Hibbard, 1950.)

Hoary Bat, *Lasiurus cinereus* (Palisot de Beauvois), 1796 (*Vespertilio cinereus*)

Ranging from Canada to Argentina and Chile (though absent from most of Central America) and from Hawaii to Bermuda, *Lasiurus cinereus* has one of the largest distributions of any American bat. Its stratigraphic range extends back to the late Illinoian (Mt. Scott), and the species has also been found at Cragin Quarry, Dry, Schulze, and San Josecito. A large bat (fig. 9.1), it has a forearm length of 46–58 mm and a wingspread of 380–410 mm. Generally solitary, the hoary bat hangs in the foliage of trees during the day and seldom enters caves. It is a strong, swift flier and is migratory. Its ancestry can be traced back to the early Blancan species, *Lasiurus fossilis*. (Barbour and Davis, 1969; Hibbard, 1963a.)

† Golliher's Bat, *Lasiurus golliheri* (Hibbard), 1960 (*Dasypterus golliheri*)

A member of the Sangamonian Cragin Quarry fauna, *Lasiurus golliheri* was smaller than *Lasiurus borealis* and just barely smaller than *Lasiurus ega*, the smallest living member of the genus. The species lacks P^3, and the position and development of the infraorbital foramen is similar to the condition found in extant yellow bats. These bats are tree dwellers and apparently prefer hot, dry country. (Hibbard and Taylor, 1960.)

Northern Yellow Bat, *Lasiurus intermedius* H. A. Allen, 1862 (*Dasypterus floridanus* Miller, 1902)

This southeastern species is found from South Carolina to Texas and south to the lowlands of Mexico and Honduras and also Cuba. There are also single records from Virginia and New Jersey. In the Southeast, it is closely associated with Spanish moss

(*Tillandsia usneoides*), in which it roosts and bears its young. Pleistocene occurrences include Reddick (Sangamonian), Arredondo, Haile XIB, and Devil's Den. It has large, long wings and 30 teeth. The northern yellow bat forages in open areas and along forest edges. (Barbour and Davis, 1969; Gut and Ray, 1963.)

Evening Bat, *Nycticeius humeralis* (Rafinesque), 1818 (*Vespertilio humeralis*)

Evening bats, genus *Nycticeius*, are found in the eastern United States, northeastern Mexico, Cuba, Africa, India, Australia, and Papua. These bats resemble *Eptesicus*, and their ears and tragus are more rounded than in *Myotis*.

This small, rather nondescript brown bat has been identified in the fossil states only at Baker Bluff (late Wisconsinan). Common in the southern coastal states, the evening bat ranges from Verz Cruz to southern Ontario and from the Atlantic Coast west to eastern Kansas. Characteristic is the short, blunt tragus and the presence of only two upper incisors (all other brown bats have four). The skull is low, broad, and short. Evening bats are slow, steady fliers and roost in trees, crevices, and houses, seldom caves. (Barbour and Davis, 1969; Guilday, et al., 1978; Hall and Kelson, 1959.)

† Four-lophed Big-eared Bat, *Plecotus tetralophodon* (Handley), 1955
(*Corynorhinus tetralophodon*)

Bats of the genus *Plecotus* have large ears that are joined across the forehead, 36 teeth, including a bicuspid upper first incisor, and are medium-sized, with males smaller than females. They are colonial and frequently inhabit caves. Four species have been identified in Pleistocene deposits.

Known only from the late Pleistocene of San Josecito, *Plecotus tetralophodon* resembles living species of *Plecotus* in most cranial characters, but the retention of the fourth commissure (the ridge extending posteroexternally from the metacone) on the last upper molar distinguishes it from other plecotine bats. (Handley, 1955; J. K. Jones, 1958.)

† Allegany Big-eared Bat, *Plecotus alleganiensis* (Gidley and Gazin), 1933
(*Corynorhinus alleganiensis*)

Known only from the late Irvingonian Cumberland Cave fauna, *Plecotus alleganiensis* closely resembles the extant *Plecotus rafinesquii*, but the braincase of the extinct species is less inflated, and the temporal ridges do not unite posteriorly to form a sagittal crest. The dentition is similar, and the mandible is indistinguishable from that of the living species. *Plecotus* sp. has been reported from Trout, Robinson, Frankstown, Crankshaft Pit, and Bootlegger Sink. (Gidley and Gazin, 1933, 1938.)

Rafinesque's Big-eared Bat, *Plecotus rafinesquii* (Lesson), 1818
(*Vespertilio rafinesquii*)

The only Pleistocene occurrence of this extant species is from Coleman IIA (late Irvingonian); the living animal is found in the forested regions of the southeastern quarter of the United States. Its habits are little known. (Barbour and Davis, 1969.)

Townsend's Big-eared Bat, *Plecotus townsendii* Cooper, 1837

A chiefly western species, *Plecotus townsendii* is found in a variety of habitats, from arid desert scrub and pine forests in the West and oak–hickory forests in the Midwest. It ranges from British Columbia to Oaxaca, with isolated populations in the Ozarks and the central Appalachians. It has been identified at Papago Springs, Dry and Clark's caves, all Wisconsinan in age. This species is a versatile flier and feeds primarily on moths. The isolated populations found in the gypsum-cave regions in the Midwest, the

Ozarks, and the Appalachians are decreasing in numbers because of disturbance by man. (Barbour and Davis, 1969.)

Pallid Bat, *Antrozous pallidus* (Le Conte), 1856 (*Vespertilio pallidus*)

Pallid bats, genus *Antrozous*, are found in western North America from southwestern Canada to central Mexico, and in Cuba. Characteristic of the genus are the large, separate ears and the presence of a small, horseshoe-shaped ridge on the truncate muzzle.

This big-eared bat is found in western and southwestern United States, from southern British Columbia south to the central plateau of Mexico; disjunct populations occur in the gypsum-cave regions of southern Kansas and Oklahoma. The species has been found at Newport Bay Mesa (Sangamonian), Potter Creek, McKittrick, Isleta, and Papago Springs. It has 28 teeth and feeds near the ground, often landing to pick up beetles, crickets, and other large insects. Apparently nonmigratory, the pallid bat hibernates throughout much of its range. (Barbour and Davis, 1969.)

Subfamily Nyctophylinae

†Anza-Borrego Bat, *Anzanycteris anzensis* White, 1969

The monotypic genus *Anzanycteris* is known only from the type locality, Arroyo Seco (late Blancan). *Anzanycteris anzensis* is most closely allied to *Antrozous* and *Bauerus* (usually considered a subgenus of *Antrozous*), and they are all placed in the subfamily Nyctophylinae. It has two lower incisors; I_2 is greatly reduced with a simple crown and is appressed into the identation near the base of the canine. The other lower teeth are weakly developed compared to those of the latter two genera, and the upper tooth row was probably slightly upturned as in *Antrozous* (see fig. 9.1). Judging from the associated faunal assemblage, a subtropical or tropical savanna was the probable habitat of *Anzanycteris anzensis*. (White, 1969.)

Family Molossidae—Free-tailed Bats

Free-tailed bats inhabit most of the subtropical and tropical regions of the world; in the New World, their range extends from the southern and southwestern United States to the West Indies and South America, with only southern Chile and southern Argentina lying outside the range. The Molossidae show extreme modifications for flight: the wings are long and narrow, and the leathery membrane is reinforced by numerous bundles of elastic fibers. The greater tuberosity of the humerus is large, with a highly developed locking device between it and the scapula. Molossids are high, fast fliers, capable of long flights, but since they live in warm areas, long migrations are seldom undertaken, some populations of *Tadarida brasiliensis mexicana* being an exception.

Free-tailed bats are known from Miocene deposits in Europe and Pleistocene deposits in the New World. Four species have been identified in the North American Pleistocene.

†Constantine's Free-tailed Bat, *Tadarida constantinei* Lawrence, 1960

About 35 extant species of *Tadarida* are found in the tropical and subtropical regions of the world. Most species roost in caves in enormous colonies. Bat guano deposits in some of these caves have been mined commercially.

The extinct Constantine's free-tailed bat is known only from New Cave, Carlsbad Caverns National Park, Eddy Co., New Mexico. It was about 10 percent larger than its living relatives. The braincase is narrower and lower in proportion to the length of the skull and the rostrum relatively deeper than those of *Tadarida brasiliensis,* its closest relative. Though not compared with *Tadarida femorosacca,* which occurs (along with *Tadarida brasiliensis*) in the Carlsbad area today, *Tadarida constantinei* would seem to be distinguishable from it by its flatter skull and broader rostrum. (Koopman, personal communication, 1975; Lawrence, 1960.)

Brazilian Free-tailed Bat, *Tadarida brasiliensis* (I. Geoffry St. Hilaire), 1824
   (*Nyctinomus brasiliensis*)

This small free-tailed bat has been identified at Reddick (Sangamonian), Nichol's Hammock, Papago Springs, and Mammoth Cave, Kentucky. The latter site is outside the present range of the species, and the specimen has been dated at 38,000 years B.P. (Jegla and Hall, 1962). A bat resembling *Tadarida brasiliensis* has been found at Blanco. Today, *Tadarida brasiliensis* ranges from northern South America north to the southern and southwestern United States. Highly gregarious, the Brazilian free-tailed bat is often found in colonies that number in the millions in some southwestern caves.

Tadarida brasiliensis has long, narrow wings and long stiff hairs on the toes, and half of the tail extends beyond the uropatagium. It has 32 tuberculosectorial teeth and feeds almost exclusively on small moths captured on the wing. Sexual dimorphism has been noted; males are slightly larger and have larger canines than females. Their flight is swift, high, and direct, and the animals may fly 80 km each night to reach their feeding areas. These bats make seasonal migrations, and populations living in the Southwest winter in Mexico. (Barbour and Davis, 1969; Dalquest, 1975; Jegla and Hall, 1962; Vaughan, 1972.)

Western Mastiff Bat, *Eumops perotis* (Schinz), 1821 (*Molossus perotis*)

The genus *Eumops* is found in the southwestern United States, Cuba, and Central and South America. Mastiff bats have large ears that are usually connected at the base, and long, narrow wings.

The only Pleistocene occurrence of *Eumops perotis* is from Centipede Cave, Val Verde Co., Texas, an early Holocene deposit. Three widely separated populations of this nonmigratory species are known, one in the southwestern United States and northern Mexico, another in Cuba, and the third over most of tropical and subtropical South America.

Our largest native bat, *Eumops perotis* has long narrow wings and large ears that are joined at the midline and extend forward over the face, and the distal half of the long tail is free from the uropatagium. The species has 30 teeth and feeds primarily on hymenopterous insects. The mastiff bat roosts in crevices high above the ground. Its fossil history is largely unknown. (Barbour and Davis, 1969; Vaughan, 1972.)

Florida Mastiff Bat, *Eumops glaucinus* (Wagner), 1843 (*Dysopes glaucinus;*
   Molossides floridanus Allen, 1932)

This living species is known in the fossil state only from the late Pleistocene of Melbourne. Allen (1932) described the specimen as a new species, *Molossides floridanus,* because the mandible had but a single lower incisor, but Ray, Olsen, and Gut (1963) have determined that the alveolus of I_2 had been broken off and found the specimen's other characteristics to be those of *Eumops.* Koopman (1971) has shown that the specimen is indistinguishable from the living *Eumops* of Florida; he regards it as a well-marked subspecies of *Eumops glaucinus*. No explanation has been found for the restriction of the living animal to extreme southern Florida since the Pleistocene. (G. M. Allen, 1932; Koopman, 1971; Ray, Olsen, and Gut, 1963.)

CHAPTER TEN

ORDER EDENTATA

The living armadillos, anteaters, and tree sloths and the extinct glyptodonts, pampatheres, and ground sloths, all inhabitants of the New World, belong to the order Edentata. Their known fossil record extends back to the late Paleocene in South America, and the edentates evolved in isolation on that continent throughout most of the Tertiary. At the end of the Tertiary, ground sloths, glyptodonts, and armadillos expanded their range north into the United States, and one genus, *Megalonyx*, even reached Alaska. Extinction took a heavy toll of edentates at the end of the Wisconsinan, and today only a small armadillo, *Dasypus novemcinctus*, survives in the United States. Edentates are divided into two main groups, the armored cingulates (glyptodonts and armadillos) and the hirsute pilosans (ground sloths, tree sloths, and anteaters).

Edentates retain many archaic features, and McKenna (1975) believes they probably separated from other eutherians well back in the Cretaceous. Characteristic of edentates are xenarthrous vertebrae (extra articulations between successive arches of the lumbar vertebrae), a varying number of cervical vertebrae (six to nine instead of the usual seven), ossified ribs that reach the sternum, an elongated sacrum, loss of enamel on the teeth, a small brain with a low level of organization, and low body temperature with poor thermoregulation. The limb bones are short and massive, and the claws are often excessively developed.

These clumsy, slow-moving herbivores are well represented in the North American Blancan and Pleistocene (5 families, 14 species). In South America, the fossil record is good, but much comparative work remains to be done; the extant tree sloths, anteaters, and armadillos (31 spp.), though not of great importance, are fascinating members of the Neotropical fauna.

FAMILY DASYPODIDAE—ARMADILLOS

The armadillo family represents the main stem of the South American edentates and probably gave rise to the other edentate families. The earliest representatives are found in late Paleocene and Eocene beds in Patagonia. The Dasypodidae is the most successful edentate family, with 21 living species and a geographic range extending from Argentina to Kansas. The family can be split into the dasypodines and the extinct pampatheriines; each had a long, independent history in South America before reaching North America. Characteristic of the family is a protective coat of jointed armor that covers the head, body, limbs, and tail (fig. 10.1). Blancan and Pleistocene representatives in North America include *Kraglievichia floridanus, Holmesina septentrionalis* and *Dasypus bellus;* today, *Dasypus novemcinctus* inhabits the same area.

SUBFAMILY PAMPATHERIINAE

†Florida Pampathere, *Kraglievichia floridanus* Robertson, 1976

Pampatheres are the largest and most specialized of the dasypodids. Although predominantly South American in distribution, two genera are known from North America, *Kraglievichia* (late Blancan) and *Holmesina* (Irvingtonian and Rancholabrean).

ORDER EDENTATA

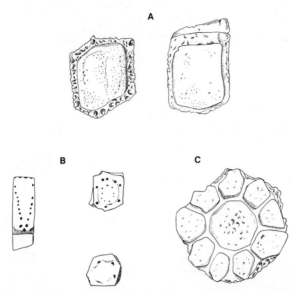

10.1 Dasypodidae and Glyptodontidae. Isolated scutes of armored edentates. *A, Holmesina septentrionalis. B, Dasypus bellus. C, Glyptotherium floridanum.* All from the late Rancholabrean of Florida. 0.5×.

Kraglievichia, previously known from Chapadmalalan deposits (early Pleistocene) in Argentina, has recently been recognized in late Blancan (Haile XVA, Santa Fe River IB) and Irvingtonian (Inglis IA, Punta Gorda) faunas in Florida. Robertson (1976) described the material as a new species, *Kraglievichia floridanus,* differing from the South American species *Kraglievichia paranense* (Kraglievich) in having a reniform-shaped upper fourth tooth instead of a peglike one and more parallel tooth rows. It is the earliest known North American pampathere.

A trend in pampathere evolution was a consistant increase in size from the small *Vassalia* (Miocene and early Pliocene, South America) to the medium-sized *Kraglievichia* (Pliocene, early Pleistocene, South America; late Blancan, early Pleistocene, Florida) to the large *Pampatherium* (middle and late Pleistocene, South America) and *Holmesina* (Pleistocene, North America). *Kraglievichia floridanus* was about the size of a Rancholabrean *Dasypus bellus* but, since the teeth and jaws are quite different, probably had different habits and diet. Apparently, the Florida pampathere was restricted to the Gulf Coastal Plain, where warm temperatures and high rainfall predominated. *Kraglievichia* was probably ancestral to *Pampatherium* and *Holmesina.* (Edmund, personal communication, 1975; Robertson, 1976; Webb, personal communications, 1974.)

†Northern Pampathere, *Holmesina septentrionalis* (Leidy), 1889 (*Glyptodon septentrionalis; Chlamytherium septentrionalis* Sellards, 1915; *Holmesina septentrionalis* Simpson, 1930)

North American pampatheres are now referred to *Holmesina,* a monotypic genus. *Holmesina septentrionalis* inhabited the southeastern United States from the Irvingtonian (Gilliland) until the end of the Pleistocene. It has been taken at some 65 localities in Florida, Texas, Oklahoma, and Kansas (the latter, from Kanopolis, is the northernmost record).

The northern pampathere has been known under several generic names, including *Chlamytherium* (invalid), *Holmesina* (the currently accepted name), and *Pampatherium* (now considered to be a strictly South American genus). Needless to say, pampathere

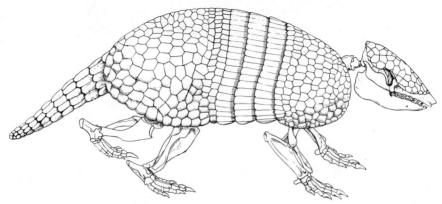

10.2 Dasypodidae. *Holmesina septentrionalis,* nearly complete specimen found in a late Rancholabrean deposit in Houston, Texas. After James (1957). Not to scale.

nomenclature is in need of revision (see Paula Couto, 1954, for a summary of name changes). *Holmesina* is closely related to *Kraglievichia;* in fact, the four terminal populations of pampatheres (including *Holmesina*) differ only slightly from each other and differ from *Kraglievichia* only in size and bilobations of the teeth (Edmund, personal communication, 1975).

Most of the records of *Holmesina* consist of scutes, teeth, and limb bones. In 1955, however, a nearly complete skeleton was found in Houston, Texas (see James, 1957). From it, we know that *Holmesina septentrionalis* stood about 1 m high and was 2 m long. The carapace is divided into anterior and posterior bucklers by three bands of imbricating dermal scutes (similar to the nine bands of *Dasypus novemcinctus*). A shield covers the head, and the tail is completely encased in a rigid tube. The skull is rather long and narrow and resembles the skull of an armadillo. There is no postorbital bar, a structure present in glyptodonts but absent in armadillos. Nine teeth are present in each half of the upper and lower jaws. The cheek teeth are bilobate, open-rooted, and, like other edentate teeth, lack enamel. The skeleton is similar to an armadillo's, only much larger (fig. 10.2).

Little is known about the habits of pampatheres. They probably fed upon insects and other invertebrates; this diet would restrict them to mild climates where food was available throughout the year. Considering their size, pampatheres were probably not as fossorial as armadillos. A terminal date for the northern pampathere is 9,880 years B.P., at Hornsby Springs. (Bader, 1957; Cahn, 1922; Edmund, personal communication, 1975; Hibbard et al., 1978; James, 1957; Paula Couto, 1954; Sellards, 1915; Simpson, 1930b.)

Subfamily Dasypodinae

†Beautiful Armadillo, *Dasypus bellus* (Simpson), 1930 *(Tatu bellus)*

The genus *Dasypus* is known from Pleistocene faunas in both North and South America and today is found from southern Kansas to northern South America.

Apparently, Florida was the Pleistocene stronghold of *Dasypus bellus,* for it has been found at the Blancan sites of Haile XVA and Santa Fe River IB, in the Irvingtonian faunas of Coleman IIA and Inglis IA, and at many Rancholabrean localities. The earliest record outside of Florida is probably Slaton Quarry, which is Illinoian in age; the species has also been reported from the Sangamonian (Clear Creek, Coppell). In the Wisconsinan, the species ranged as far north as Missouri (Zoo, Cherokee, Crankshaft, Brynjulfson) and has also been reported from Tennessee (Robinson, Baker Bluff), West Virginia (Organ-Hedrick), Arkansas (Peccary), Georgia (Ladds), Texas (Miller's Cave, Ben Franklin, Kincaid Shelter, Cave Without A Name, Hill-Shuler), and New Mexico (Blackwater Draw).

A trend in the evolution of *Dasypus bellus* is a striking increase in size (possibly a doubling in volume) from the Blancan to the Rancholabrean. Late Pleistocene specimens of *Dasypus bellus* were twice as large as, but otherwise identical with, the extant nine-banded armadillo, *Dasypus novemcinctus*. From a nearly complete specimen found in Medford Cave near Reddick, Florida, we know that the animal measured about 1.2 m in length. This specimen was a mature female, and unborn young were found within the carapace.

The availability of insect food throughout the year was undoubtedly the limiting factor in the distribution of *Dasypus bellus*. Its presence in a fauna indicates winters no more severe than occur in north-central Texas today (Slaughter, 1961b). Although its ancestry is unknown, *Dasypus bellus* resembles Pleistocene species found in Brazil and Argentina. Deteriorating climate at the end of the Pleistocene was probably a factor in its extinction. Its niche was later filled by *Dasypus novemcinctus*. (Auffenburg, 1957; Edmund, personal communication, 1975; Simpson, 1930b; Slaughter, 1961b; Webb, personal communication, 1974.)

Nine-banded Armadillo, *Dasypus novemcinctus* Linnaeus, 1758

The record of *Dasypus novemcinctus* in the Pleistocene is tenuous. On the basis of a single isolated scute, the species was reported from Slaton (Illinoian) (see Dalquest, 1967). There are no other Pleistocene records. Remains of the animal first appear in the archeological record in Texas about 3,000 years ago (Miller's Cave). The scute from Slaton may have belonged to a small *Dasypus bellus*.

The nine-banded armadillo has been expanding its range northward since the 1850s, when it was first observed in extreme southern Texas. By the early 1960s, it ranged as far north as southern Nebraska, but during the last 10 years, armadillos have retreated to southern Kansas because of colder winters on the Great Plains. Armadillos have greatly expanded their range in Florida, where they were introduced. Limiting factors to their distribution are scarcity of insects and other invertebrates in winter, lack of moist soil for digging (range expansion has ceased in areas receiving less than 45–50 cm of precipitation annually), and cold temperatures (populations have suffered 80 percent, or higher, mortality during prolonged cold spells).

This animal, about the size of a house cat, is covered by a flexible carapace made up of usually nine moveable bands on the back with flexible bucklers over the shoulders and pelvis; a separate shield covers the head, and the tail is encased in a tube. The armor consists of bony scutes covered by a horny epidermis. Sparse hair grows from the flexible skin between the scutes and on the limbs and ventral surface of the body. Although there is no bony contact between the carapace and the skeleton, the shell is partially supported by prominent dorsolateral processes arising from the ilium and ischium and by modified tips of neural spines on the thoracic and lumbar vertebrae. Some of the cervical vertebrae are fused. The broad ribs, ossified costal cartilage, and heavy intercostal muscles provide a rigid ribcage. These fossorial mammals have powerfully built limbs, and their plantigrade feet bear large, stout claws.

Even though definite records of *Dasypus novemcinctus* do not appear until Holocene times, there is no doubt that it is the ecological equivalent of *Dasypus bellus,* the large Pleistocene armadillo. (Dalquest, 1967; Patterson, personal communications, 1972; Patton, 1963; C. B. Schultz, 1972; Vaughan, 1972; Webb, personal communication.)

Family Glyptodontidae

The Glyptodontidae were one of the most aberrant families of the Edentata. Clumsy and slow-moving, these large turtlelike mammals were covered by a nearly immobile carapace made up of distinctive polygonal scutes with rosette sculpturing (see fig. 10.3). Internally, there was extensive fusion of the vertebrae, and the pelvic girdle was fused to the carapace. The pillarlike limbs were adapted to support the great weight of the carapace and internal organs.

Glyptodonts, a predominantly South American group, invaded North America at the end of the Pliocene or in the very early Pleistocene and survived until the end of the Pleistocene (see fig. 10.4). The North American species are all closely related and seemingly represent radiation and evolution from a single early Pleistocene population. Semiaquatic in habits, they inhabited tropical and subtropical regions where there was quiet water and lush vegetation.

The taxonomy of glyptodonts has been based primarily on characteristics of the pelvis (an element well represented in collections) and the carapace or isolated scutes (also numerous, since there were over 1800 scutes per carapace). Until Gillette's recent study (1973), skulls and dentitions were ignored. Gillette recognizes one genus and three species in the United States.

†Osborn's Glyptodont, *Glyptotherium texanum* Osborn, 1903

Glyptodonts (genus *Glyptotherium*) had heavy, massive limbs and are considered to have been the most graviportal of all mammals. The bones of the manus and pes are short, broad, and stout, and the terminal phalanges were encased in an ungual sheath, a structure intermediate between a claw and a nail. Locomotion was undoubtedly slow and clumsy. The combination of limited axial mobility and a rigid carapace made glyptodonts the mammalian analog to turtles. Glyptodonts had a heavy but flexible tail which, when swung from side to side, probably served as a counterbalance during locomotion. Unlike some of the South American glyptodonts, none of the North Amer-

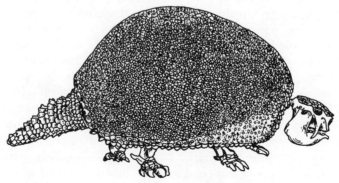

10.3 Glyptodontidae. *Glyptotherium* sp., Pleistocene, carapace. After Romer (1966). 0.03×.

ORDER EDENTATA

10.4 Glyptodontidae. Blancan and Pleistocene distribution of *Glyptotherium:* □ = *G. texanum;* △ = *G. arizonae;* ○ = *G. floridanus.*

ican species had a terminal "mace"; instead the tail ended in a tube formed by a single terminal ring or by fusion of the last two caudal vertebrae.

Glyptotherium had 32 flat-crowned hypsodont cheek teeth that formed an effective grinding mill similar to that of microtine rodents. Incisors and canines were absent. The maxilla and mandible were massive and deep-bodied to accomodate the long, hypsodont teeth. Chewing was accomplished by a simple forward grinding movement of the mandible, with the upper and lower cheek teeth maintaining surface-to-surface contact. The enormous descending process of the zygomatic arch provided a large area of attachment for the massive masseter muscle complex which probably made up 90 percent of the jaw musculature. Glyptodonts presumably were browsers and fed on the lush vegetation growing along waterways and in shallow water. Gillette (1973) believes that they had a short proboscis.

Adult glyptodonts probably had few enemies. Their dermal armor was their main line of defense, since they could neither run nor fight. However, it was not always effective: a skull of *Glyptotherium texanum* (F:AMNH 95737) shows two well-positioned holes in the skull roof that probably were made by the canines of an attacking cat.

A frequent faunal associate of glyptodonts was the capybara (*Neochoerus* and *Hydrochoerus*). Both capybaras and glyptodonts had a Neotropical origin and center of dispersal. Capybaras are semiaquatic, and it has been suggested that glyptodonts were too. Their habitat was probably semitropical swamps with a warm, humid climate, luxurious vegetation, and permanent watercourses.

The earliest North American glyptodont, *Glyptotherium texanum* has been reported from late Blancan deposits in Texas (Blanco, Cita Canyon, Hudspeth, Red Light) and Arizona (Tusker). It was ancestral to *Glyptotherium arizonae*.

Glyptotherium texanum was the smallest North American glyptodont, with a carapace length of approximately 1.3 m, a total length of about 2.2 m, and a standing height of about 1 m. It probably weighed between 450 and 550 kg (Gillette, personal communication, 1974). It differed from *G. arizonae* in its small size, slightly arched carapace, and small scutes. Details of its anatomy are better known now that the excellent collection from Tusker, including three carapaces and a complete skeleton, has been studied (Gillette, 1973).

Glyptotherium texanum reached North America in the early Pleistocene from South America and gave rise to the descendant species, *Glyptotherium arizonae*. (Gillette, 1973, and personal communication, 1974; Osborn, 1903.)

†Gidley's Glyptodont, *Glyptotherium arizonae* Gidley, 1926 (*Xenoglyptodon fredericensis* Meade, 1953)

During the Irvingtonian, a huge glyptodont inhabited Arizona (Curtis Ranch), Texas (Gilliland, Rock Creek), Oklahoma (Holloman), and Florida (Inglis IA). The species has also been identified tentatively at Santa Fe River IB; this may be the earliest record of *Glyptotherium arizonae*, or the specimen may belong to *Glyptotherium texanum*, the Blancan species. Although the two species occupied almost the same geographical area, they did not exist contemporaneously in any known local fauna.

Gyptotherium arizonae was the largest of the North American glyptodonts, with a carapace length of about 1.8 m, a total length of 3 m, and a standing height of about 1.5 m; it probably weighed at least a ton (Gillette, personal communication, 1974). It differed from *Glyptotherium texanum* primarily in its larger size and the resulting exaggeration of characters. For example, the carapace was highly arched, the skull was more massive, and the limbs and feet showed extreme graviportal tendencies. The species is well represented in collections from Curtis Ranch and Gilliland.

Sometime in the Irvingtonian, *Glyptotherium arizonae* retreated south into Mexico, probably as a result of deteriorating climatic conditions, and did not reappear in the United States until late Rancholabrean times. Gillette (1973) believes that *Glyptotherium arizonae* was ancestral to *Glyptotherium floridanum*. (Gidley, 1926; Gillette, 1973, and personal communication, 1974; Meade, 1953; Melton, 1964.)

†Simpson's Glyptodont, *Glyptotherium floridanum* (Simpson), 1929
  [*Boreostracon floridanus; Glyptodon petaliferus* Cope, 1888 (*nomen nudum*)]

Glyptotherium floridanum inhabited the Gulf Coastal Plain and the southern Atlantic Coast during the late Pleistocene, and although it is known from many localities in Florida, South Carolina (northernmost occurrence), and Texas, the material is fragmentary, and no complete skeletons have been found. All of the sites are Wisconsinan in age. The apparent geographical separation between the localities is probably due to incomplete sampling since the Pleistocene of the intervening states is poorly known. These glyptodonts have been found almost exclusively in warm, humid, coastal environments; the only exceptions are the two inland sites of Laubach Cave in Williamson County and Wolfe City in Hunt County, Texas. This may indicate the presence of riparian corridors extending inland from the coast.

The type material is from Seminole Field and consists entirely of scutes, which Simpson (1929) believed belonged to a new genus, *Boreostracon,* because of their small size. It is now thought that *Glyptotherium floridanum,* unlike the other species, showed sexual dimorphism, with females smaller than males, so that Simpson's material apparently belonged to a female. *Glyptotherium floridanum* was intermediate in size between the Blancan and Irvingtonian species, though the males may have approached *Glyptotherium arizonae* in bulk.

At the end of the Wisconsinan, glyptodonts became extinct. The most probable cause of their extinction was climatic deterioration, for these animals were so specialized and lived in such a restricted environment that local populations could have been easily wiped out by climatic changes. (Gillette, 1973; Simpson, 1929a.)

†Mexican Glyptodont, *Glyptotherium mexicanum* (Cuatáparo and Ramirez),
  1875 (*Glyptodon mexicanus*)

Two species of glyptodonts have been described from Mexico, but the validity of *Glyptotherium mexicanum* is in doubt since the type specimen has been lost, and there are pronounced discrepancies in the measurements and figures. It was originally described from the late Pleistocene of Tequixquiac, State of Mexico, and specimens from Vera Cruz and Cedazo have been referred to it. Brown (1912) transferred it to his new genus *Brachyostracon* (= *Glyptotherium*). The carapace closely resembles that of *Glyptotherium cylindricum* (Brown), a large early Rancholabrean species known only from Ameca, State of Jalisco, but the skull from Tequixquiac is like no other glyptodont skull, and it is now believed that owing to an admixture of parts, a pampathere skull was illustrated with the glyptodont carapace. (Brown, 1912; Gillette, 1973.)

FAMILY MEGALONYCHIDAE—MEGALONYCHID GROUND SLOTHS

This family of small to medium-sized ground sloths first appeared in the early Oligocene (Deseadan) of Patagonia, and several genera are recognized in the middle Miocene (Santacrucian). The post-Miocene history of the family in South America is obscure. In the mid-Tertiary, one or more megalonychids reached the Antilles as waif immigrants and underwent insular radiation there; as many as nine genera have been identified from Pleistocene deposits in the West Indies (see Paula Couto, 1967, for a discussion of these forms). The mid-Tertiary appearance of the family in North America antedates the establishment of a land connection between the continents, which suggests that these immigrants probably rafted to North America.

Pliometanastes protistus Hirschfeld and Webb, the earliest known North American megalonychid, was described from early Hemphillian deposits in Florida; it shows no relationship to *Megalonyx* or to any of the West Indian ground sloths. *Megalonyx*, a widespread North American genus, ranged stratigraphically from the late Hemphillian (*Megalonyx curvidens* Matthew and *Megalonyx mathisi* Hirschfeld and Webb) to the late Wisconsinan.

An evolutionary trend in the Megalonychidae was increase in size; with this size increase, many skeletal proportions changed allometrically, which resulted in the presence of some intermediate characteristics in the Irvingtonian forms. The first megalonychids were small and may have been partially arboreal, and the Pliocene species were about half the size of the late Pleistocene species, *Megalonyx jeffersonii* (Desmarest); however, the West Indian species were cat- to bear-sized, their dwarfed condition reflecting their insular environment. Diagnostic of the Megalonychidae is the presence of a pair of secant, self-sharpening caniniform teeth in front of which is a much-reduced predentary spout.

A number of species of *Megalonyx* have been named; in fact, as Hirschfeld and Webb (1968, p. 216) noted, "nearly every good specimen has been described as a different species." Isolated teeth (an extremely variable element in sloths) were designated as the types of several species. Unfortunately, ontogenetic, sexual, and geographical variation was not considered; later studies have reduced most of these "species" to synonymy. In his recent revision of the genus, McDonald (1977) recognizes three species, *Megalonyx leptostomus*, *Megalonyx wheatleyi*, and *Megalonyx jeffersonii*, in the Blancan and Pleistocene of North America. (Hirschfeld and Webb, 1968; McDonald, 1977, and personal communication, 1975; Patterson and Pascual, 1972; Paula Couto, 1967.)

† Narrow-mouthed Ground Sloth, *Megalonyx leptostomus* Cope, 1893
   (*Morotherium leptonyx* Marsh, 1874)

The genus *Megalonyx* was widely distributed in North America by the middle Pliocene. The Blancan–Nebraskan species *Megalonyx leptostomus* has been reported from Blanco (type locality), Cita Canyon, Hudspeth, El Casco (Calif.), Hagerman, Inglis IA, Mabel (Fla.), Rexroad, White Bluffs, and White Rock. *Megalonyx* material from other faunas of the same age (Broadwater, Grand View, Keefe Canyon, Lisco, Proctor Pits, Red Light, Santa Fe River IB) is probably referrable to this species. Hirschfeld and Webb (1968) redescribed the species from new material found at Cita Canyon, and McDonald (1977) has recently reviewed its distribution.

Megalonyx leptostomus was intermediate in size between the larger *Megalonyx wheatleyi* and the smaller *Megalonyx curvidens*, except for the small-sized endemic population at Inglis IA. The facial region of *Megalonyx leptostomus* is not as deep as that of later species. The caniniform teeth are broad and blunt and have a prominent median bulge. M^2 and M^3 have a greater mediolateral diameter than the teeth of the other species. Phalanges I and II of the third digit of the pes are separate and distinct; in *Megalonyx jeffersonii*, these phalanges are completely coossified.

Megalonyx leptostomus was probably ancestral to *Megalonyx wheatleyi*. (Hirschfeld and Webb, 1968; McDonald, 1977 and personal communication, 1976.)

† Wheatley's Ground Sloth, *Megalonyx wheatleyi* Cope, 1871 (*Megalonyx loxodon* Cope, 1871; *M. sphenodon* Cope, 1871; *M. tortulus* Cope, 1871; *M. scalper* Cope, 1899)

Cope (1871, 1899) named five species of *Megalonyx* from Port Kennedy, four of them from isolated teeth; of these, only *Megalonyx wheatleyi* is now recognized as valid. Besides Port Kennedy, it has also been identified at Cudahy (Kansas), Haile XVI and McLeod (Fla.); and Vallecito Creek. Its geologic range is early Irvingtonian to Yarmouthian.

Cope (1871) distinguished *Megalonyx wheatleyi* from *Megalonyx jeffersonii* by its smaller size and the absence of a groove anterior to the lingual bulge on the upper caniniform. McDonald (1977) noted that the consistent small size of *Megalonyx wheatleyi* is due to its earlier age and ancestral position. Some specimens formerly assigned to *Megalonyx wheatleyi* are immature individuals, and some Florida specimens have been referred to *Megalonyx wheatleyi* because of their small size. Later studies have shown, however, that individuals from Florida populations were smaller than those farther north. *Megalonyx wheatleyi* is intermediate in size and diagnostic characters between the smaller *Megalonyx leptostomus* and the larger *Megalonyx jeffersonii*.

Megalonyx wheatleyi probably gave rise to the Rancholabrean species, *Megalonyx jeffersonii*. (Cope, 1871, 1899; McDonald, 1977 and personal communication, 1976.)

† Jefferson's Ground Sloth, *Megalonyx jeffersonii* (Desmarest), 1822
  (*Megatherium jeffersonii*; *Megalonyx laqueatus* Harlan, 1843; *Megalonyx dissimilis* Leidy, 1855; *M. leidyi* Lindahl, 1893; *M. sierrensis* Sinclair, 1905; *M. californicus* Stock, 1913; *M. milleri* Lyon, 1938; *M. hogani* Stovall, 1940; *M. brachycephalus* Stovall and McAnulty, 1950)

"An animal of the clawed kind" was Thomas Jefferson's description of the animal whose fragmentary bones he possessed. Jefferson called it *Megalonyx* ("great claw") and at first thought (and hoped) that the bones were those of a giant carnivore three times the size of a lion. When he came across a drawing of a recently discovered South American megathere, however, he realized that *Megalonyx* was related to the "bradypus, dasypus, and pangolin" and was not a carnivore, possessing a "phosphorus eye" and "leonine roar" (Jefferson, 1799). Jefferson's talk to the American Philosophical Society in 1797 about *Megalonyx* marked the beginning of vertebrate paleontology in North America, and it is fitting that *Megalonyx jeffersonii* was named for him.

The stratigraphic range of *Megalonyx jeffersonii* extends from the Illinoian to the late Wisconsinan, and it has been identified at more than 75 localities. Geographically, it was found in the eastern two-thirds of the United States, along the West Coast, from Alaska southward, and inland to Idaho; apparently, it was absent from the Great Basin, the desert Southwest, and the Rocky Mountains. *Megalonyx jeffersonii* was the only ground sloth that roamed north into what is now Canada and Alaska, and it has been identified at Fairbanks, Old Crow River, Lower Carp Lake, Northwest Territories, Quesnel Forks, British Columbia, and Medicine Hat (Sangamonian). At Rancho La Brea, it was less abundant than either *Glossotherium* or *Nothrotheriops*.

Although most workers now regard *Megalonyx jeffersonii* as the only Rancholabrean species, there remains the possibility that the eastern and western populations were geographically isolated throughout the Pleistocene and were specifically distinct (as they were in the Hemphillian), but this has not been documented yet (McDonald, personal communication, 1975). If the western population is shown to be distinct, the valid name would be *Megalonyx sierrensis* Sinclair. Specimens of *Megalonyx jeffersonii* from the northern parts of its range are usually larger than specimens of the same age found farther south (the largest known specimen is from the Carter site, Darke Co., Ohio). Material from several Rancholabrean localities in Florida that had been referred to *Megalonyx wheatleyi* because of their smaller size are now considered to belong to *Megalonyx jeffersonii*.

Jefferson's ground sloth, the largest known species of *Megalonyx*, was larger and more robust than *Nothrotheriops* and nearly as large as *Glossotherium;* when fully grown, it was about the size of an ox. The skull is short, broad, and deep, with a sharply truncated anterior end (see fig. 10.5). A Y-shaped bifurcation at the anterior end of the palate is characteristic. The temporalis and masseter muscles (both used in chewing) were well developed, and the pterygoid muscles were weak, whereas the opposite condition, enlarged pterygoids and small temporalis and masseters, prevailed in *Glossotherium* and *Nothrotheriops.* The deep, heavy mandible lacks the spoutlike predental region characteristic of the megatheriids. Dentition is $^5/_4$; a diastema separates the blunt, broad, ovate caniniforms from the cheek teeth, which are meniscoid-shaped with a prominent inner bulge; the last upper cheek tooth is subtriangular in cross section. The teeth of *Megalonyx,* and of sloths in general, taper to the occlusal surface from a wider base in young animals and are parallel-sided in adults. No deciduous teeth are known (true also for the living tree sloths).

The skeleton of *Megalonyx jeffersonii* is intermediate in robustness between *Glossotherium* and *Nothrotheriops,* and the radius, ulna, tibia, and fibula are longer than the same bones in *Glossotherium.* Diagnostic characters of the skeleton include broad, winglike processes on the calcaneum, a short third metatarsal, a relatively slender, V-shaped fifth metatarsal, and a laterally bowed fibula. Like other ground sloths, *Megalonyx* had well-developed clavicles and sternal ribs. The presence of the entepicondylar foramen on the humerus is variable—in a nearly complete skeleton from American Falls, it is present on one humerus but absent on the other. A distinct third trochanter is present on the femur (absent in *Glossotherium* and present, but not as a separate entity, in *Nothrotheriops*).

Unlike the other ground sloths, *Megalonyx* apparently had a plantigrade hind foot (see the restoration in fig. 4.5) characterized by a relatively slender fifth metatarsal (broad and massive in *Glossotherium* to support body weight), a large, supportive third metatarsal (smaller in *Glossotherium*), well-developed claws on digits II, III, and IV which functionally touched the ground (claws smaller and elevated in *Glossotherium* and *Nothrotheriops*), absence of the tibial knob on the astragalus (present in the other ground sloths), and a shorter fibula (extended distally in *Glossotherium* and *Nothrotheriops*) (McDonald, 1977 and personal communication, 1975).

Dwelling primarily in woodlands and forests, where it browsed on leaves, twigs, and perhaps nuts, Jefferson's ground sloth survived until the late Wisconsinan; a somewhat questionable terminal date is $9,400 \pm 250$ years B.P. at Evansville, Indiana. (Harington, 1970a; Hirschfeld and Webb, 1968; Jefferson, 1799; Lillegraven, 1967; Lyon, 1938; McDonald, 1977, and personal communications; W. Miller, 1971; Paula Couto, 1967; Stock, 1925, 1942; Stovall, 1940; Stovall and McAnulty, 1950.)

Family Megatheriidae—Megathere Ground Sloths

First appearing in late Oligocene beds in South America, the megatheriids underwent their major development on the southern continent. Although a common ancestry with the Megalonychoidea is indicated, the megatheres and nothrotheres, despite their disparate size, are more similar to each other especially in cranial structure, than either is to the ancestral megalonychids. There are two major groups: the megatheres (represented by *Megatherium* and *Eremotherium*), heavily-built animals that reached a length of 6 m, and the nothrotheres (represented by the *Hapalops–Nothrotheriops* line), slightly-built ground sloths that reached a length of about 1.2 m. In North America,

10.5 Megalonychidae. *Megalonyx jeffersonii*, Rancholabrean, American Falls, skull. ca. 0.33×.

Eremotherium is known from early Irvingtonian to late Rancholabrean deposits; *Nothrotheriops* appears first in the Irvingtonian and survived until about 10,000 years ago. (Patterson and Pascual, 1972.)

†Rusconi's Ground Sloth, *Eremotherium rusconii* (Schaub), 1935 (*Megatherium rusconii; Megatherium mirabile,* Leidy, 1855; *Eremotherium carolinense* Spillman, 1948)

The genera *Eremotherium* and *Megatherium* have often been confused. The main difference between the two genera is in the structure of the forefoot. *Eremotherium* has only three fully developed digits, with the third and fourth bearing claws. *Megatherium* has four digits, with the second, third, and fourth bearing well-developed claws. In *Eremotherium,* the premaxillae are weakly developed, forming a dorsoventrally compressed triangle loosely sutured to the maxilla (and thus usually lost in fossils); in *Megatherium,* the premaxillae form a narrow but thick bar protruding some distance anteriorly and are solidly fused to the maxillae. Other skull differences are the shorter predental region, lower-placed orbits, and shallower jaw with a less convex lower margin in *Eremotherium. Megatherium* never got very far north in South America, and all of the North American records are definitely *Eremotherium.*

Most of the *Eremotherium* material consists of isolated teeth, manus and pes elements, and parts of long bones; skulls are rare. The best preserved and most complete material known was found at El Hatillo, Panama, where the remains of at least eight of the giant megatheriids have been recovered (Gazin, 1956).

Eremotherium probably reached North America in the late Blancan but didn't flourish until Rancholabrean times. Of course, this may be a reflection of sampling and preservation. Edmund (personal communication, 1975) believes that *Eremotherium* was in many ways a more primitive animal than *Megatherium,* and this seems to be borne out by the close resemblance of *Eremotherium* to Pliocene forms.

Remains of *Eremotherium rusconii* have been found at some 14 sites in Georgia (Skidaway Island, Watkins's Quarry), South Carolina (Ashley River), Florida [Inglis IA (earliest occurrence), Waccasassa River, Crystal River Power Plant, Haile XVI, and Polk, Sarasota, St. Lucia, Volusia, and Hendry counties], and Texas (Sinton, Galveston). The genus is also found in Central and South America. Two other Rancholabrean-age species have been described from South America, but only the larger one, *Eremotherium carolinense* Spillman (probably conspecific with *E. rusconii*), is found north of Ecuador. *Eremotherium elense* Hoffstetter is much smaller.

Eremotherium rusconii reached a length of about 6 m and weighed more than 3 tons. Pronounced sexual dimorphism is indicated, with males weighing perhaps 50 percent more than females (Voorhies, 1971). The skeletal anatomy suggests that the animal frequently assumed an upright or sitting position, the massive tail serving as a prop. The pes is huge, with a tremendous astragalus and a great tuber calcanii reaching almost as far to the rear as the unguals reach forward, which aided in maintaining balance in the upright position. *Eremotherium* probably walked less on the side of its hind foot than *Megatherium,* which has a well-developed falciform bone.

In *Eremotherium rusconii,* the jaw is loosely slung but provided with huge muscles attached both to the cranium and the massive zygoma. Mastication consisted of some back-and-forth component added to vertical shear, but there was no true grinding motion. A browsing habit is confirmed by the discovery of large quantities of cleanly chopped twigs associated with *Eremotherium* remains in Ecuadorian and Peruvian tar pits. The length of the twigs is exactly the same as the length between the chisel-shaped ridges on the teeth. The vegetation consumed indicates a savanna habitat with numerous thornbushes. (Edmund, personal communication, 1975; Gazin, 1956; Voorhies, 1971.)

†Shasta Ground Sloth, *Nothrotheriops shastensis* (Sinclair), 1905 (*Nothrotherium shastense; N. graciliceps* Stock, 1913; *N. texanum* Hay, 1916)

Nothrotheriops (a monotypic genus) was a medium-sized North American nothrothere with a range that extended from northern Mexico to southern Alberta. Recently, Paula Couto (1971) showed the South American *Nothrotherium* to be generically distinct from the North American animal and transferred the latter to the genus *Nothrotheriops* Hoffstetter. According to this author, *Nothrotherium* differs from *Nothrotheriops* in its smaller size and in characters of the skull and hind limb bones; it was confined to the Pleistocene of tropical South America and may have been semiarboreal.

The ancestry of *Nothrotheriops* and *Nothrotherium* can probably be traced back to *Hapalops,* a South American Miocene nothrothere. *Hapalops* was about 1.2 m in length, including an elongated tail. Diagnostic is the ⁵/₄ dentition, including well-developed caniniforms, a long, slender muzzle, and a small, spoutlike predentary region. This lineage of nothrotheres is characterized by small size (for ground sloths), reduced dentition, including reduction and, finally, loss of the caniniform teeth, enlargement of the predentary scoop, and fusion of phalanges I and II of the second digit. *Nothrotheriops* and *Nothrotherium* represent two divergent phylogenetic lines that underwent parallel evolution but were adapted to different ecological conditions.

The earliest appearance of *Nothrotheriops shastensis* is in Irvingtonian faunas (Gilliland, Vallecito Creek, Medicine Hat). The species was described from Potter Creek Cave and has been found in Alberta, Arizona, California, Mexico (Aguascalientes, Jalisco, Nuevo León), Nevada, New Mexico, and Texas. At San Josecito Cave, 35 individuals have been recovered.

More is known about the external appearance of *Nothrotheriops shastensis* than any other ground sloth because several well-preserved specimens have been found in dry caves in the Southwest. At Aden Crater (Dona Ana Co., N. Mex.), a completely articulated skeleton still held together by tendons and sinews and including patches of skin and hair was recovered from the bottom of a 30-m fumarole. The animal was immature and measured 2.45 m long and stood 1 m high at the withers. A full-grown animal probably weighed between 135 and 180 kg. Thus although *Nothrotheriops shastensis* was the smallest North American ground sloth, it was about twice as large as the South American *Nothrotherium.* The pale yellowish hair is long and coarse, and close examination reveals the presence of ovate bodies similar to the algal cells found on the hairs of living tree sloths. These cells impart a greenish tinge to the coat, enabling the sloth to blend into its environment. The shasta ground sloth had a relatively small head, prehensile lips, and a long flexible neck. The forelimbs were long and slender, and the hindquarters were not as massive as those of *Megalonyx.* Like *Myrmecophaga,* the gaint anteater, *Nothrotheriops* walked on its knuckles with its toes partly flexed.

The dentition was reduced to ⁴/₃, and there was no caniniform tooth. The crowns of the cheek teeth usually show a wear pattern of anterior and posterior ridges separated by intervening valleys. The predentary portion of the mandible formed a long, spoutlike beak containing a horny cropping plate; another horny pad opposed it in the upper jaw. These features, together with the long tongue and prehensile lips, aided the sloth in procuring food (see fig. 10.6).

Large numbers of perfectly preserved coprolites of *Nothrotheriops shastensis* have been

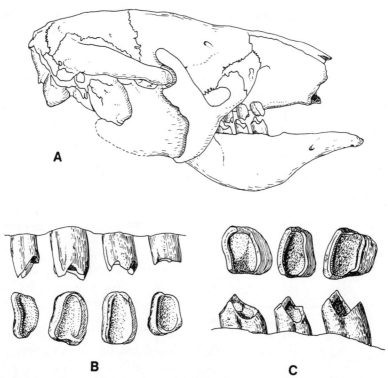

10.6 Megatheriidae. *Nothrotheriops shastensis*, Rancholabrean, Rancho La Brea. *A*, skull and mandible; *B*, right upper, and *C*, right lower cheek teeth, external and occlusal views. After Stock (1925). *A*, 0.25×. *B*, *C*, 0.75×.

found in the arid caves of the Southwest, and analysis of them has shown us what these browsing ground sloths ate. The dung contains fragments of roots, stems, seeds, flowers, and fruits of desert plants, including *Ephedra, Atriplex, Prosopsis, Sphaeralcea, Agave, Yucca,* and *Opuntia,* plants still found in the area today. At Rampart Cave, successive layers of dung have been dated; the oldest is more than 35,000 years old, and the sloths flourished in the cave between 13,000 and 11,000 years ago, judging from the amount of dung they left behind. The dung samples are rich in pollen, and analysis shows not only a seasonal change in available plant food but a long-term shift in climate from warm-dry to cool-moist and back to warm-dry. Martin (1975) believes that the sloths inhabited the desert caves in winter and early spring, then summered at higher elevations.

The shasta ground cloth died out about 11,000 years ago; the youngest dung sample at Rampart Cave is dated 10,780 years B.P. No archeological material has been found in direct association with any bones or dung of the species. (Eames, 1930; Edmund, personal communication, 1975; Hausman, 1929; Laudermilk, 1938; Long and Martin, 1974; Lull, 1929, 1930; P. S. Martin, 1975; Martin, Sabels, and Shutler, 1961; Paula Couto, 1971; Romer, 1966; Stock, 1917b, 1925.)

Family Mylodontidae—Mylodont Ground Sloths

Mylodont ground sloths apparently have had an independent history since the Oligocene. Their heyday was in the Plio–Pleistocene, when they diversified in form, increased in number, and expanded their range to become the dominant ground sloths in South America and a widespread group in North America. A trend in their evolution is an increase in size and stoutness of build. Characteristic of the family is the presence of dermal ossicles, small pebblelike bones deeply embedded in the skin, which probably afforded protection against predators. Mylodont sloths have a single astragular facet on the calcaneum.

Several genera of mylodont sloths have been described from the Pleistocene of South America. Of these, at least *Mylodon* and man coexisted—pieces of hide covered with long, reddish hair and given the name *Mylodon listai* have been found in direct association with human remains in a cave in Patagonia. Two species of mylodonts have been identified in the Blancan and Pleistocene of North America.

†Chapadmalalan Ground Sloth, *Glossotherium chapadmalense* Kraglievich, 1925

A partial skeleton of the small South American mylodont, *Glossotherium chapadmalense*, has recently been identified in Florida at Haile XVA, and material from Santa Fe River IB, Inglis IA, and Blanco has been referred to this species. The stratigraphic range is late Blancan to early Irvingtonian. Robertson (1976) said that the specimen from Haile XVA is identical to *Glossotherium* from the Chapadmalalan (early Pleistocene) of Argentina.

A striking difference between *Glossotherium chapadmalense* and *Glossotherium harlani* is the presence of well-developed caniniform teeth in the former species. The stoutness of the tooth is reflected in the transverse expansion of the anterior portion of the maxilla. The large lower caniniform has a chisellike tip owing to the double occlusion with the upper caniniform and the first upper molariform tooth. In *Glossotherium harlani*, the caniniform is often reduced or absent.

Glossotherium chapadmalense, as it existed in North or South America, may have been ancestral to both *Glossotherium harlani* and *Glossotherium robustum* (Owen).

Glossotherium sp. has been reported from several other Blancan localities (Cita Canyon, Red Light, Donnelly Ranch, Broadwater, White Bluffs); whether this material should be referred to the smaller *Glossotherium chapadmalense* or the larger *Glossotherium harlani* is at present uncertain. (Robertson, 1976.)

†Harlan's Ground Sloth, *Glossotherium harlani* (Owen), 1840 (*Mylodon harlani; Paramylodon nebrascensis* Brown, 1903; *Mylodon garmani* Allen, 1913)

Best known of the mylodont sloths is *Glossotherium harlani*, a grasslands species that ranged from coast to coast and from Florida to Washington during Irvingtonian and Rancholabrean times. It has been identified at some 40 sites. Apparently, the species was fairly common over much of its range: at Rancho La Brea, a minimum of 76 individuals have been recovered, and its remains are more abundant than those of *Nothrotheriops* or *Megalonyx*.

Harlan's ground sloth was first described as a species of *Mylodon* from material found at Big Bone Lick. Brown (1903) called the sloth material from Hay Springs *Paramylodon* and distinguished it from *Mylodon* by the presence of four upper teeth instead of five. Stock (1914a, 1917a) showed that the presence or absence of the first upper tooth is variable (usually absent in the Rancho La Brea specimens and present in the Florida specimens) and relegated the name *Paramylodon* to synonymy. Simpson (1945a) argued that if the North and South American genera of mylodont ground sloths are not distinct, they all belong to the genus *Glossotherium* Owen. Hoffstetter (1952) recognized

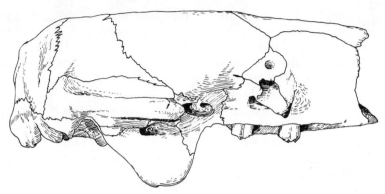

10.7 Mylodontidae. *Glossotherium harlani*, Rancholabrean, Rancho La Brea, skull. After Stock (1925). 0.25×.

Paramylodon as a subgenus of *Glossotherium*. *Glossotherium harlani* is closely related to the South American species *Glossotherium robustum*, and both of them may be descended from *Glossotherium chapadmalense*.

Glossotherium harlani is distinguished from *Nothrotheriops* and *Megalonyx* by larger size, lobate teeth, and the presence of dermal ossicles. The animal was covered with coarse, shaggy hair. Having a height of about 130 cm over the back, *Glossotherium harlani* was a powerfully built animal, with a massive chest, enormous forelimbs, stout hind limbs, strong claws, and a robust tail. Clumsy and slow-moving, Harlan's ground sloth walked on the outside of its feet, and much of the body weight was supported by the large calcaneum and the outer side of the fifth metatarsal. Tracks of *Glossotherium harlani* have been preserved in the silts and sandstones of what is now the yard of the Nevada State Prison near Carson City; the tracks measure 204 mm in width and between 457 and 483 mm in length.

Glossotherium harlani had an elongated skull with a blunt nose (see fig. 10.7). The largest known skull was found near Walsenburg, Colorado, and measured 540 mm in length (Cockerell, 1909); skull measurements of the sample from Rancho La Brea are somewhat smaller. The well-developed turbinals indicate that the animal had a keen sense of smell. Surrounding the small brain cavity are a large number of sinuses. The open-rooted, lobate teeth consist of an outer layer of cement surrounding a dense, resistant layer of dentine that encloses a vascular dentinal core. Differential wear of the three layers resulted in an efficient grinding surface. The dental formula is $^{4-5}/_5$, and there is no diastema between the first tooth and the rest of the dental series. The animal probably fed on grass and small shrubs and possibly used its claws to dig up roots.

The skeleton is characterized by massiveness. The lumbar and sacral vertebrae are fused, and the neural spines, usually nine in number, form a continuous ridge over the series. Ten sternal ribs that articulate with each other at points along their length enclose the thoracic cavity ventrally and, with the sternum, form a solid framework around the chest. The sternebrae articulate with each other and, by complex joints, with the sternal ribs. The limb bones are heavy—the humerus is broadened distally, the femur lacks a third trochanter (present in *Nothrotheriops* and *Megalonyx*), and the tibia is about one-half the length of the femur. Of the five metacarpals, II, III, and IV bear stout claws. The pes is characterized by a large, posteriorly expanded calcaneum with a single astragalar facet (unique to mylodonts, and therefore to *Glossotherium* among the

North American ground sloths), broad, flattened metatarsals, and reduced IV and V digits.

Harlan's ground sloth probably used its powerful forelimbs and claws for defense against sabertooth cats, Pleistocene lions, and packs of dire wolves. Its shaggy coat and layer of dermal ossicles were also a deterrent to predators. It has not been found in association with man at any North American site. A terminal date is 13,890 years B.P. at Rancho La Brea, but it may have survived longer in Florida (Hornsby Springs). (Brown, 1903; Cockerell, 1909; Edmund, personal communication, 1975; Hester, 1967; Hoffstetter, 1952; W. Miller, 1971; Simpson, 1945a; Stock, 1914a,b, 1917a, 1920, 1925, 1936b, 1965.)

CHAPTER ELEVEN

ORDER CARNIVORA

The carnivores have a somewhat uneven, but in many instances very good, fossil record; the order dates back to the Paleocene. Carnivores are basically predatory mammals with enlarged P^4/M_1 which function as carnassial teeth and powerful canines. Many are omnivorous, however, and some are more or less herbivorous. The order is highly diversified and is found in most kinds of continental, as well as many marine, environments. Although smaller species are often stenotopic, many carnivores are at home in varied circumstances and have a wide distribution. Holarctic species are common in this order.

Dentition and the skull and limb bones provide useful morphological characters, and species-level taxonomy is usually efficient. Many species have been intensively studied for a long time, the amount of work done being positively correlated with body size, and also with frequency of occurrence.

North American carnivores are grouped in the two suborders Pinnipedia (seals, sea lions, and walruses) and Fissipedia (land carnivores).

SUBORDER FISSIPEDIA

The Fissipedia may be divided into the infraorders Canoidea (Mustelidae, Canidae, Procyonidae, Ursidae) and Feloidea (Felidae, Hyaenidae). Many other classifications have been proposed. The land carnivores probably arose from the ancestral insectivore stock by a direct line represented by the family Miacidae, generally small, forest-dwelling animals, which appear in the Middle Paleocene. A great diversification of the Miacidae occurred in the Eocene and led to the establishment of several families that became the dominant land carnivores. Early representatives of the modern carnivore families appear in the late Eocene.

FAMILY MUSTELIDAE—WEASELS

The fossil history of the Mustelidae is poorly known since most of the members were small and forest-dwelling. The earliest known mustelids are from late Eocene deposits, and, although they are rare, martenlike animals and the highly specialized *Potamotherium* have been found in the Oligocene. By the end of the Miocene, recognizable martens, weasels, polecats, otters, badgers, honey badgers, and skunks were present. Primarily inhabitants of northern-temperate forests, mustelids invaded South America and Africa in the Pliocene. Quaternary mustelids occupy nearly all habitats, from the Arctic tundra to tropical rainforests, and have a nearly cosmopolitan distribution, except for Australia, Madagascar, and the oceanic islands.

Varying greatly in size, appearance, and habits, mustelids have a long body, small, rounded ears, and short limbs with five toes on each foot. All of them have anal glands from which they can exude or spray a vile-smelling liquid used for defense, attracting mates, and marking territory. Fine fur is characteristic of the family, and demand for pelts has reduced the numbers of many species.

Mustelids vary widely in dental adaptations. The carnassials are typically sectorial but in some groups have been secondarily modified for crushing. The number of molars is reduced to ½, and the inner lobe of the upper molar is expanded. Mustelids have a long braincase and a short rostrum. The postglenoid process partially encloses the glenoid fossa, and little lateral (and no rotary) jaw movement is possible.

Mustelids are fairly common in the North American Pleistocene, and six subfamilies—the Mustelinae, Mellivorinae, Grisoninae, Melinae, Lutrinae, and Mephitinae—are represented.

Subfamily Mustelinae

†Diluvial Fisher, *Martes diluviana* (Cope), 1889 (*Mustela diluviana; Martes parapennanti* Gidley and Gazin, 1933)

The genus *Martes* is widely distributed in the forested regions of Eurasia and North America. Martens have retained several primitive characters seen in some of the early mustelids and thus are close to the basal stock. The earliest known occurrence is in lower Miocene deposits in Europe. Sometimes included in *Mustela,* the genus is distinguished from that taxon by the presence of four premolars in both jaws and a small but well-developed metaconid and basined talonid on M_1. Never abundant, the marten's arboreal and wandering habits have further reduced the chances of finding fossil remains. Four species have been identified in the North American Pleistocene.

Martes diluviana is known from three Irvingtonian localities: Port Kennedy, Cumberland, and Conard Fissure. It belongs to the subgenus *Pekania*, distinguished from the true martens (subgenus *Martes*) by larger size and the presence of an external median rootlet on P^4. *Martes paleosinensis* (Zdansky) from the Pontian of China was ancestral. Reaching America in the Irvingtonian, this fisher was probably not directly ances-

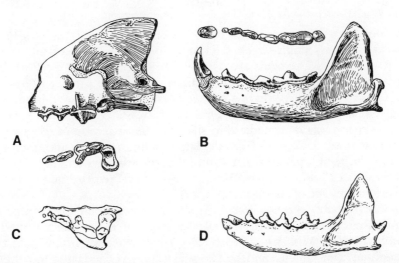

11.1 Mustelidae. *A, B, Martes diluviana*, Irvingtonian, Cumberland Cave. *A*, skull, with occlusal view of P^2–M^1; *B*, left mandible with occlusal view of dentition. *C, D, Martes nobilis*, Rancholabrean, Samwel Cave. *C*, right P^3–M^2, occlusal view; *D*, left mandible, external view. *A, B*, after Gidley and Gazin (1938); *C, D*, after Anderson (1970). 0.75×.

tral to *Martes pennanti* but was certainly its ecological forerunner. An inhabitant of wooded areas, it was probably as arboreal as the fisher. Dental specializations such as a shorter blade and larger crushing surface on the lower carnassial indicate less carnivorous habits (see fig. 11.1). About the size of a female fisher, *Martes diluviana* showed pronounced sexual dimorphism. It survived until the early Illinoian and was replaced by *Martes pennanti*. (Anderson, 1970; Brown, 1908; Cope, 1899; Gidley and Gazin, 1933, 1938; Hall, 1936.)

Fisher, *Martes pennanti* (Erxleben), 1777 (*Mustela pennanti*)

The largest extant *Martes* is the fisher, an animal found in mixed hardwood forests from British Columbia to Nova Scotia, with southward extensions into the Sierra Nevada of California, the northern Rocky Mountains, the Great Lakes region, and New England. In the late Pleistocene, its range extended as far south as Georgia (Ladds) and Arkansas (Peccary), throughout the Appalachian Mountains (New Paris No. 4, Natural Chimneys, Baker Bluff, Robinson), and west to Ohio and Missouri (Carter, Bat, and Brynjulfson). Fisher remains are turning up with increasing regularity in archeological sites in the Midwest and South. Although some of the records may represent Indian trade goods, it is becoming evident that the fisher had a much wider range in the late Wisconsinan and early Holocene than at present.

This magnificent furbearer is about the size of a fox but more slender. Although it is at home in the trees, the fisher spends a considerable amount of time on the ground. Food consists of small mammals, birds, carrion, and fruit. The fisher is one of the few animals that feeds on the porcupine, *Erethizon dorsatum*, and the presence of fishers in an area is often correlated with a decline in the porcupine population.

Martes diluviana, though closely related to the extant fisher, was probably not directly ancestral. After suffering population declines for years, fishers today are making a comeback in some areas where they have been reintroduced. (Anderson, 1970; Hall, 1936.)

†Noble Marten, *Martes nobilis* Hall, 1926 (*M. caurina nobilis*)

Martes nobilis is now known from 10 late Pleistocene faunas (Little Box Elder, Potter Creek, Samwel, Jaguar, Smith Creek, Chimney Rock, Bell, Wilson Butte, Moonshiner, Old Crow River 11A). Its remains are usually found in the lower levels of the caves in association with extinct arctic or boreal animals. At Jaguar Cave, carbon-14 dates of $10,370 \pm 350$ and $11,580 \pm 250$ years B.P. have been obtained from the charcoal hearths in the lower levels of the cave where several *Martes nobilis* bones were found.

The noble marten was intermediate in size between a female *Martes pennanti* and a male *Martes americana*. It differed from the latter in its larger size and in morphological differences in the teeth and skull. In size and proportions of the teeth, *Martes nobilis* resembled the extinct fisher, *Martes diluviana*, but lacked the external median rootlet on P^4 which is characteristic of fishers.

Martes nobilis probably reached North America from Asia in Illinoian or early Wisconsinan times and spread into the forested regions of the western United States. It survived until the late Wisconsinan/early Holocene, when it probably succumbed to competition from *Martes americana* (the two species have been found together at Chimney Rock and Bell caves), a changing climate, and perhaps man's activities. (Anderson, 1970; Hall, 1926; Kurtén and Anderson, 1972.)

Pine Marten, *Martes americana* (Turton), 1806 (*Mustela americana*)

The American pine marten is known from several late Pleistocene deposits in the eastern United States (New Paris No. 4, Natural Chimneys, Robinson, Eagle, Baker Bluff, Clark's) and has recently been identified in the upper layers of two western sites (Chimney Rock, Bell). Closely related to *Martes martes,* the European pine marten, and *Martes zibellina,* the sable, the American species probably reached North America in the early Wisconsinan and spread eastward. This population was presumably isolated in eastern North America by the ice sheet; only after the ice retreated did the species reinvade western Canada and Alaska. A later invasion from Siberia populated the West Coast, the Sierra Nevada, and the Rocky Mountains; these extant martens, *Martes americana caurina* and *Martes americana humboldtensis,* show more similarities in cranial and dental characteristics to *Martens zibellina* than to the eastern subspecies.

Pine martens inhabit dense spruce–fir forests, where they feed on rodents, other small mammals, and birds, as well as fruit, berries, and nuts. Overtrapping and an advancing civilization have reduced the population in many areas. Reintroductions have not been as successful as with *Martes pennanti,* and the pine marten does not thrive in fur farms. (Anderson, 1970; Hall, 1926, 1936.)

†Rexroad Weasel, *Mustela rexroadensis* Hibbard, 1950 (*M. gazini* Hibbard, 1958)

The genus *Mustela* includes weasels (subgenus *Mustela*), ferrets (subgenus *Putorius*), minks (subgenus *Lutreola*), and tropical weasels (subgenus *Grammogale*). Ranging across Europe, northern Africa, Asia (including Java, Sumatra, and Borneo), North America, and northern South America, these small to medium-sized mustclids are highly specialized for flesh-eating. They have three premolars and a trenchant talonid on M_1, and the metaconid on M_1 is absent (weasels and ferrets) or incipient (mink). All of them have a short face, a long, lithe body, and short legs. Three extinct species, two extinct subspecies, and six living species are recognized in the epochs covered in this book.

The Rexroad weasel, *Mustela rexroadensis,* is known only from the Blancan localities of Rexroad and Hagerman. A medium-sized weasel, it is distinguished from *Mustela frenata* by a compressed paraconid on M_1 and narrower lower premolars. The upper carnassial is similar to that of *Mustela frenata.* Both species show sexual dimorphism. It is not known if *Mustela rexroadensis* was ancestral to *Mustela frenata.* (Bjork, 1970; Hibbard, 1950.)

Long-tailed Weasel, *Mustela frenata* Lichtenstein, 1831 (*M. reliquus* Hall, 1960)

Mustela frenata is a widespread species with a long evolutionary history. The earliest records are from the pre-Nebraskan Borchers fauna and the Irvingtonian Conard Fissure fauna, and the species has been reported from more than 30 Rancholabrean faunas in California, Colorado, Florida, Georgia, Idaho, Kansas, Missouri, Nebraska, Nevada, New Mexico, Nuevo León, Ohio, Tennessee, Texas, Virginia, and Wyoming. At Moonshiner Cave in southeastern Idaho, more than 300 individuals have been recovered. The long-tailed weasel has the widest distribution of any American weasel, and its range extends from southern Canada to northern South America in almost all habitats except deserts.

Although there is some overlap in size between female *Mustela frenata* and male

Mustela erminea, they can usually be told apart by the more rugose skull, with a longer precranial region, of the former.

The ancestry of *Mustela frenata* is unknown; the species may be descended from the Blancan species, *Mustela rexroadensis.* (Hall, 1936, 1951a.)

Short-tailed Weasel, or Ermine, *Mustela erminea* Linnaeus, 1758

Remains of the short-tailed weasel have been found at Cudahy and Conard Fissure, both Irvingtonian, and in several late Wisconsinan faunas (Schulze, Klein, Crankshaft, Chimney Rock, Silver Creek, Wasden, Mooshiner, Fairbanks II). The specimens from Conard Fissure are regarded as an extinct subspecies, *Mustela erminea angustidens* Brown, which closely resembles the subspecies that lives along the edge of the tundra today (Hall, 1951a). The circumboreal ermine inhabits brushy and woody areas. Its geographic range differs from that of *Mustela rixosa* in that it extends south into the Sierra Nevada and the Rocky Mountains and is absent in the Appalachian region. Since the southern subspecies are only slightly larger than *Mustela rixosa,* direct competition between the two species probably occurred, and only one species survived in the area. In Alaska and Canada, where both species occur together, the size difference between the two is much greater.

Mustela erminea probably shows more geographic variation in size than do the other American weasels. There is size overlap with the larger *Mustela frenata* and the smaller *Mustela rixosa.* In areas where *Mustela frenata* and *Mustela erminea* occur together, they can usually be separated by the length of the postglenoidal region of the skull: in *Mustela erminea,* it is more than 46 percent of condylobasal length of the skull in males and more than 48 percent in females (Hall, 1951a). Sexual dimorphism is pronounced, with females about two-thirds the size of males.

Mustela erminea possesses typical weasel characters—a short rostrum, long postcranial region, and sectorial carnassials. Compared to the other species of weasels, the ermine's skull is more rounded, with less distinct crests. This efficient predator feeds on small mammals, especially mice, which it kills by a bite at the back of the neck.

The ancestry of *Mustela erminea* can be traced back to the late Pliocene in Europe, and the species probably reached North America in the late Blancan or early Irvingtonian. (Hall, 1936, 1951a.)

Least Weasel, *Mustela rixosa* Bangs, 1896

The least weasel was the smallest Pleistocene carnivore, and the extant animal is equally small. It is known from several late Pleistocene localities, including Natural Chimneys, New Paris No. 4, Baker Bluff, Robinson, Crankshaft, Peccary, and Moonshiner. The last three sites are outside the present range of the species, an area that includes Alaska, northern Canada, the north-central states east to Pennsylvania and south in the Appalachians to North Carolina. This mouse-sized carnivore inhabits meadows and open woods, where it feeds primarily on mice, although shrews and insects are sometimes taken. It is regarded as a rare mammal both in the field and in collections.

Taxonomic confusion still exists regarding the status of this species. Some workers (see Reichstein, 1957; Jones, 1964) regard *Mustela nivalis* and *Mustela rixosa* as only subspecifically distinct. But today in south-central Sweden, the two species exist side by side apparently without interbreeding, a condition that has persisted since the early Holocene, when *Mustela nivalis* entered the area from the south and *Mustela rixosa* came in from the east (Kurtén, 1972). Adding to the confusion are pronounced sexual dimorphism and geographical variation, and there is some overlap in size with female *Mustela erminea*. Until detailed statistical and other comparative studies are done on both the Old and New World populations, we are recognizing *Mustela rixosa* and *Mustela nivalis* as distinct species. *Mustela rixosa* is probably a highly specialized offshoot of *Mustela nivalis*. The least weasel is also known from late Pleistocene faunas in Europe. (Ewer, 1973; Hall, 1951a; J. K. Jones, 1964; Kurtén, 1968, 1972; Reichstein, 1957.)

†Melton's Mink, *Mustela meltoni* Bjork, 1974

This Blancan mustelid has been found only at Rexroad. It is a heavy-jawed, minklike animal with a metaconid crest on M_1. The premolars are crowded and have well-developed posterior cingula on P_{3-4}. Compared to that of *Mustela vison*, M_2 is greatly reduced; this advanced characteristic probably rules out *Mustela meltoni* from ancestry of the extant species. The appearance of a minklike form in the early Blancan coincides with an evolutionary explosion of microtines at that time. (Bjork, 1974.)

Mink, *Mustela vison* Schreber, 1777

Although records of *Mustela vison* extend back to the Irvingtonian (Cudahy, Cumberland, Conard Fissure), mink are uncommon in Pleistocene faunas. All of the fossil occurrences (about 20) fall within the present range of the species, an area that encompasses most of North America except desert regions. Since mink are only found along streams and lakes, their presence in a fauna is a good indicator of nearby permanent water. Amphibious in habits, mink den along stream banks and are excellent swimmers. Much of their food is aquatic and includes crayfish, fish, frogs, birds, muskrats, and other riparian mammals.

Pleistocene *Mustela vison* did not differ noticeably in size or morphology from the living animals. A slight trend toward increased size is apparent from Irvingtonian through Illinoian and Wisconsinan times. Today, the largest subspecies, *Mustela vison ingens* (Osgood), is found in Alaska.

Cranial and skeletal material of mink is sometimes confused with that of the black-footed ferret, *Mustela nigripes*, but mink have a larger inner lobe on M^1, a wider occipital region, a larger infraorbital foramen, less inflated auditory bullae, and a wider talonid and an incipient metaconid on M_1 (see fig. 11.2). The skeletons of the two animals are similar, but the mink's is usually less rugose. Both species show sexual dimorphism; females are about 10 percent smaller than males.

Mink are one of our most valuable furbearers and are raised successfully on fur farms. They are solitary and nocturnal, and wild populations flourish in many areas. (Hall, 1936; Walker, 1975.)

†Sea Mink, *Mustela macrodon* (Prentiss), 1903 (*Lutreola macrodon*)

The largest known mink, *Mustela macrodon* has been found in middens along the New England coast from Massachusetts to the Bay of Fundy. A radiocarbon date of $4,320 \pm 250$ years B.P. (Waters and Ray, 1961) was obtained from the midden at Middleborough, Massachusetts, and the sea mink probably survived until a few hundred years ago (Waters and Mack, 1962). *Mustela macrodon* shows pronounced sexual dimorphism: males were about 20 percent larger than females. It was larger than *Mustela vison ingens*, the largest extant taxon. Nothing is known about its habits. (Anderson, unpublished manuscript; Prentiss, 1903; Waters and Mack, 1962; Waters and Ray, 1961.)

152 ORDER CARNIVORA

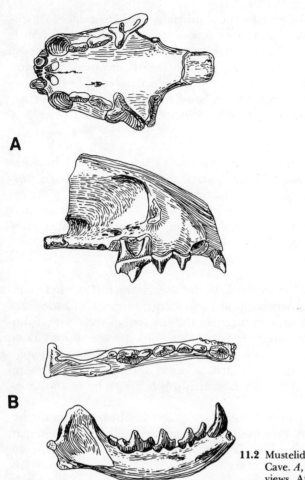

11.2 Mustelidae. *Mustela vison*, Irvingtonian, Cumberland Cave. *A*, skull; *B*, right mandible, occlusal and lateral views. After Gidley and Gazin (1938). 1.5×.

Beringian Ferret, *Mustela eversmanni* Lesson, 1827

An extinct subspecies, *Mustela eversmanni beringiae* Anderson, has been described from late Pleistocene deposits near Fairbanks. Characterized by large size and a broad facial region, this extinct ferret is larger than *Mustela eversmanni michnoi* Kashchenko, the largest living subspecies. *Mustela eversmanni* is an inhabitant of the steppe regions of Eurasia and feeds on a wide variety of small animals, especially rodents. It differs from *Mustela putorius*, the European ferret, in its larger size, pronounced postorbital constriction, relatively larger canines and carnassials, and different coloration, habits, and habitat. *Mustela eversmanni* is closely related to, and perhaps conspecific with, *Mustela nigripes*, the black-footed ferret. All of the extant ferrets are derived from *Mustela stromeri* Kormos, a small, early Middle Pleistocene species that inhabited Europe. (Anderson, 1977.)

Black-footed Ferret, *Mustela nigripes* Audubon and Bachman, 1851

The earliest reported occurrence of *Mustela nigripes* is from a late Illinoian deposit in Clay County, Nebraska, and it is recorded from Sangamonian deposits in Nebraska and at Medicine Hat. Wisconsinan occurrences include Old Crow River Loc. 65, Orr

Cave, Montana, Jaguar, Little Box Elder, Chimney Rock, Isleta, and Moore Pit. It has also been found at Moonshiner Cave, an early Holocene fauna. All of the occurrences, except Old Crow River and Moonshiner, are within the historic range of the species, an area that encompassed the Great Plains and central Rocky Mountains. The close association between this species and *Cynomys,* the prairie dog (six of the above sites contain *Cynomys*), has in recent years led to the near extermination of the black-footed ferret owing to the extensive poisoning of prairie dogs, their main prey.

A long, sinuous body, a black mask across the eyes, and black feet distinguish the animal. In contrast with *Mustela vison,* ferrets have relatively short, broad premolars, a narrow lobe on M^1, and a narrow talonid and no metaconid on M_1. In addition, ferrets have a broad palatal region and a narrow basioccipital region.

Mustela nigripes is closely related to *Mustela eversmanni,* the steppe ferret of Eurasia. Ferrets came across Beringia and spread south when conditions permitted. Apparently, they have always been rare, and today the black-footed ferret exists only as a relict population. An endangered species, *Mustela nigripes* is considered to be one of North America's rarest mammals. (Anderson, 1977; Kurtén and Anderson, 1972.)

† Martin's Weasel, *Tisisthenes parvus* Martin, 1973

Known only from the Kansan-age Java local fauna, this small mustelid is characterized by the loss of the metacone and metacone root on M^1 and the absence of P_2. Martin (1973a) believes that, although there are morphological resemblances to *Mustela,* the species' dental and orbital specializations eliminate it from that evolutionary line. (R. A. Martin, 1973a.)

† Schlosser's Wolverine, *Gulo schlosseri* Kormos, 1914 (*Gulo gidleyi* Hall, 1936)

The largest of the Mustelinae, the circumboreal wolverine is characterized by a robust build and exceptionally powerful teeth and jaws. *Gulo* is descended from *Plesiogulo,* a large Miocene and Pliocene form with less specialized dentition that inhabited Eurasia, North America, and South Africa. It, in turn, arose from a *Martes*-like ancestor in the early Miocene. The first appearance of *Gulo* is in the Kansan at Port Kennedy and in early Middle Pleistocene faunas in Europe. The European and American records are roughly contemporaneous (the American may be older), and pending further study, we are provisionally referring both to the species *Gulo schlosseri,* originally described from the Hungarian (now Romanian) site Püspökfürdö (= Betfia).

Gulo schlosseri is slightly smaller than *Gulo gulo* and has certain differences in dental proportions. *Gulo gidleyi* Hall, described from Cumberland, is distinguished from Recent wolverines by its smaller size (about 8 percent smaller in linear dimensions) and by slight differences in dental morphology. It may be transitional to *Gulo gulo,* and we tentatively include it in *Gulo schlosseri.*

A Palearctic origin of *Gulo* is suggested by the presence, in the Ruscinian of Europe and China, of an advanced form of *Plesiogulo.* However, the early appearance of *Gulo* in North America might suggest that the genus arose in the New World. (Cope, 1899; Gidley and Gazin, 1938; Hall, 1936; Kormos, 1914; Kurtén, 1970; Tobien, 1957.)

Wolverine, *Gulo gulo* (Linnaeus), 1758 (*Ursus gulo; Gulo luscus* Sabine, 1823)

Wolverines are rare in Pleistocene deposits. After their appearance in the Irvingtonian in the Port Kennedy and Cumberland faunas, they do not appear again until the Wisconsinan (Fairbanks II, Old Crow River Locs. 14N and 11A, Little Box Elder, Natural Trap, Chimney Rock, Jaguar, and Duck Point, Power Co., Idaho). Two early

Holocene sites in southeastern Idaho (Moonshiner and middle Butte caves) have yielded good samples. Circumboreal in distribution, wolverines inhabit tundra and taiga regions and today in America are found in Alaska and northern Canada; a few still survive in the Sierra Nevada, and isolated occurrences have been reported from Wyoming. Wolverines had a larger distribution within historic times, but, since they are wilderness animals, have retreated before advancing civilization.

Wolverines are overgrown martens that resemble small bears with bushy tails. Although primarily terrestrial, they can climb trees. The animals are solitary in habits and are active throughout the year. Wolverine fur is highly prized for trimming parka hoods because rime does not condense on it and thus can easily be brushed off. Like martens, wolverines feed on a wide variety of food, including mammals, birds, insect larvae, fruit, and carrion. In winter, the animals often bring down weakened caribou and moose that are stranded in the snow. Carrion forms an important part of their diet; they are known to appropriate other predators' kills and to cache what they don't eat immediately.

For mammals of their size, wolverines are unexcelled in strength. Males weigh up to 27.5 kg, with females weighing about 30 percent less. Wolverines show a gradual increase in size during Rancholabrean times. Some of the Wisconsinan animals were huge—a skull from Fairbanks has a condylobasal length of 172 mm, compared with an observed range of 130–150 mm for a Recent sample from Alaska (Anderson, 1977). Postglacial and extant wolverines are smaller. Wolverines have exceptionally strong jaws and robust teeth which enable them to crush large bones. Correlated with their strong teeth are powerful jaw and neck muscles. The sagittal crest projects well above the top of the skull and extends back beyond the level of the condyles to provide a large area of attachment for the well-developed temporalis muscles (see fig. 11.3).

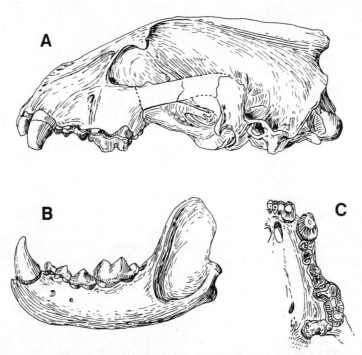

11.3 Mustelidae. *Gulo gulo*, Rancholabrean, Fairbanks. *A*, skull; *B*, left mandible, external view; *C*, occlusal view of upper dentition. After Anderson (1977). 0.5×.

Gulo gulo is descended from *Gulo schlosseri,* and the extant species is present in European faunas by the end of the Cromerian interglacial. American wolverines were formerly considered to be a distinct species, *Gulo luscus,* but Kurtén and Rausch (1959) have shown that they are only subspecifically distinct from the Eurasian population. (Anderson, 1977; Cope, 1899; Ewer, 1973; Gidley and Gazin, 1938; Hall, 1936; Kurtén, 1968, 1970; Kurtén and Anderson, 1972; Kurtén and Rausch, 1959; Walker, 1975.)

Subfamily Mellivorinae

†Voracious Flesh-eater, *Ferinestrix vorax* Bjork, 1970

This rare mustelid is known only from a mandible and a femur found in the early Blancan beds at Hagerman. Bjork (1970) believes that it is an aberrant mellivorine. Characterized by a massive jaw and robust teeth, *Ferinestrix* was a short-jawed form with crowded premolars and a wide mandibular condyle. The large carnassial, with its well-developed blade and large crushing surface, indicates both active predation and carrion-feeding habits. Proportionately larger and stronger than *Gulo, Ferinestrix* probably had similar habits. The scarcity of remains of this animal suggests that it was a solitary, far-ranging species that had a low population density. (Bjork, 1970.)

Subfamily Grisoninae

†Gazin's Grison, *Trigonictis idahoensis* (Gazin), 1934 (*Lutravus? idahoensis; Canimartes? idahoensis* Gazin, 1937; *Trigonictis kansasensis* Hibbard, 1941)

Two grisonlike animals are known from Blancan deposits. Of these, *Trigonictis idahoensis* is the largest and best known. It has been found at Hagerman, Grand View, White Bluffs, Rexroad, Sand Draw, and Broadwater; specimens from Santa Fe River IB and a post-Pliocene deposit in Charles County, Maryland are tentatively referred to this species.

Trigonictis is distinguished from all other mustelids by the structure of P^4: it is triangular in outline with a well-developed cingulum that forms a shelf extending from the protocone to the posterior part of the tooth. A long trigonid, a well-developed metaconid, and a basined talonid characterize M_1. The premolars are crowded, and no accessory cusps are present (see fig. 11.4).

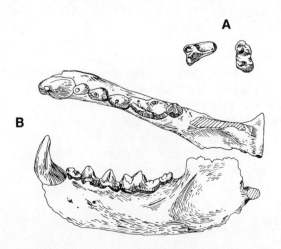

11.4 Mustelidae. *Trigonictis idahoensis,* Blancan, Hagerman. *A,* left P^4 and M^1, occlusal view; *B,* left mandible, external and occlusal views. After Bjork (1970). 0.75×.

About the size of a fisher (*Martes pennanti*), *Trigonictis idahoensis* probably preyed on ground squirrels, rabbits (*Hypolagus*), and young beavers. Apparently, this agile carnivore was a good climber and swimmer and lived in a variety of habitats.

Trigonictis is closely related to, and perhaps conspecific with, the *Pannonictis–Enhydrictis* group of galictine mustelids from Plio–Pleistocene deposits in Europe and China. A common ancestor of the group was probably the Miocene genus *Trochictis*. *Trigonictis* reached the New World in the middle Pliocene and was a relatively common Blancan mustelid. It was probably ancestral to the Neotropical genus *Galictis*. (Bjork, 1970; Gazin, 1934b; Hibbard, 1941b; Zakrzewski, 1967b.)

† Cook's Grison, *Trigonictis cookii* (Gazin), 1934 (*Lutravus? cookii; Canimartes cookii* Gazin, 1937)

A smaller species of grison, *Trigonictis cookii* has been found at Hagerman, Grand View, and Sand Draw, as well as in Pliocene deposits in Texas and California. Zakrzewski (1967b) showed that, although there is some overlap in size between the two species, *Trigonictis cookii* is a distinct species rather than a female of *Trigonictis idahoensis*.

Cook's grison was about two-thirds the size of *Trigonictis idahoensis*, or about the size of the extant *Galictis vittata* Schreber. A nearly complete skeleton was found at Hagerman; it shows many similarities to *Galictis*, though the limb bones are not as robust. Limb proportions are functionally intermediate between those of the climbing *Galictis* and the swimming *Satherium*. The two species of grisons apparently occupied different niches, with the smaller species (*T. cookii*) preying on smaller mammals such as *Cosomys* and *Pliopotamys*. At Hagerman, Bjork (1970) noted a stratigraphic separation between *Trigonictis cookii–Cosomys* (lower two-thirds of the section) and *Sminthosinus–Ophiomys* (upper third of the section). *Trigonictis cookii* is believed to be directly ancestral to *Galictis vittata*. (Bjork, 1970; Gazin, 1934b; Skinner and Hibbard, 1972; Zakrzewski, 1967b.)

† Bowler's Mouse-eater, *Sminthosinus bowleri* Bjork, 1970

Described from the upper beds at Hagerman, this small mustelid has also been identified at Broadwater. It is closely related to the *Trigonictis* group. *Sminthosinus* had a short face, crowded premolars, and well-developed carnassials. At Hagerman, its range does not overlap with *Trigonictis cookii*, and Bjork (1970) postulates that *Sminthosinus* replaced *Trigonictis cookii* ecologically. The habitat of *Sminthosinus* probably was along stream banks and in marshy areas, and the animal undoubtedly fed on the small microtine *Ophiomys*. There is a good possibility that *Sminthosinus bowleri* gave rise to the extant *Grisonella cuja*. (Bjork, 1970.)

† Cummins' Mustelid, *Canimartes cumminsi* Cope, 1893.

This poorly known, apparently monotypic genus has been found only at Blanco (late Blancan). Other specimens (e.g., *Canimartes? idahoensis, C? cookii*) were referred to this genus but have been transferred to *Trigonictis*. The genus differs from *Trigonictis* in the morphology of P^4; in *Canimartes cumminsi*, the prominent protocone is set off from the main body of the tooth by a constriction, and the hypocone is absent, as is the shelf formed by the cingulum, which extends from the protocone to the posterior part of the tooth. M^2 is apparently present. Its subfamily affinities are unknown. (Cope, 1893; Dalquest, 1975; Hibbard, 1941b.)

Subfamily Melinae

Badger, *Taxidea taxus* (Schreber), 1778 (*Ursus taxus; Taxidea sulcata* Cope, 1878; *T. robusta* Hay, 1921; *T. marylandica* Gidley and Gazin, 1933)

The stratigraphic record of the badger extends from the late Pliocene to the Recent. This fossorial mustelid is known from several Blancan faunas, including Rexroad, Deer Park, Sand Draw, Broadwater, Cita Canyon, Red Light, and Hagerman; all of these specimens have been referred to *Taxidea* sp., except those from Rexroad, which Hibbard (1941b) called *Taxidea taxus*. About the same size as the extant badger, the Blancan animal had proportionately larger teeth with heavy cingula; it is probably subspecifically distinct.

Badgers have been reported from Port Kennedy, and those from Cumberland were considered a distinct species, *Taxidea marylandica* Gidley and Gazin, until studies by Long (1964) showed the Cumberland badger to be only subspecifically distinct from the living animal. Badgers are common in Wisconsinan faunas in the West, and a few have been recovered from eastern sites (Welsh, Baker Bluff, Peccary). Badgers have also been found in unglaciated central Alaska (Fairbanks II) and in the northern Yukon (Gold Run Creek, Dominion Creek). Their presence in the East and the Far North indicates prairie or steppelike conditions in those areas and a milder climate in the Arctic. Today, *Taxidea taxus* is found in areas of friable soil from southern Canada to southern Mexico and from the Pacific Coast east to Michigan and Ohio.

Badgers reached their greatest size in the Wisconsinan, and the largest known are from Fairbanks (condylobasal length, 137.7–144.6 mm; zygomatic breadth, 90.6–100.1 mm) (see fig. 11.5) and Little Box Elder Cave (condylobasal length, 142.6 mm; zygomatic breadth, 94.9 mm; Anderson, 1977). The largest Recent badgers in Long's (1972) study were from southeastern Idaho (greatest length of skull, 126.2–139.9 mm; zygomatic breadth, 77.5–94.4 mm) and belong to the subspecies *Taxidea taxus jeffersonii* (Harlan), the northern badger, which is found throughout the western states today. Measurements of a large sample of badgers from Moonshiner Cave (early Holocene) average slightly smaller than those in Long's sample, which perhaps indicates a postglacial size reduction.

Badgers are exclusively carnivorous and feed primarily on rodents. Their teeth are adapted for crushing and are multicusped. The number of accessory cuspules on the carnassials and M^1 is highly variable. Badgers have strong jaws, and the temporalis and masseter muscles are well developed.

Although *Taxidea* shows superficial resemblances to *Meles*, the European badger, the

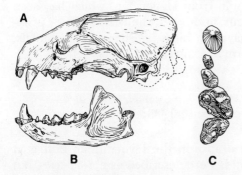

11.5 Mustelidae. *Taxidea taxus*, Rancholabrean, Fairbanks. *A*, skull; *B*, left mandible, external view; *C*, occlusal view of upper dentition. After Anderson (1977). *A, B*, 0.33×; *C*, 0.66×.

two genera are not closely related, but rather show similar adaptations to a fossorial mode of existence. In *Taxidea*, the skull is broader, P^4 is larger, M^1 is triangular, and there are only three premolars.

The probable ancestor of the American badger is *Pliotaxidea nevadensis* (Butterworth), known from Hemphillian faunas in Nevada and Oregon. It was smaller than *Taxidea* and had larger auditory bullae. The status of *Taxidea mexicana* Drescher, from supposedly middle Pliocene beds at Rincon, Chihuahua, Mexico, is uncertain. Hall (1944) referred it to the subspecies found in the area today, but this viewpoint has not been upheld by other workers. It may have been ancestral to the smaller southern badger, *Taxidea taxus berlandieri* Baird. Badgers show considerable age, sexual, individual, and geographic variation within a population, and this has led to the naming of many species and subspecies.

Badgers are generally solitary and nocturnal, and their large, deep holes are encountered more often than the inhabitants. Predator-control programs have reduced badger populations in many areas. (Anderson, 1977; Bjork, 1970; Gidley and Gazin, 1933; Hall, 1936, 1944; Hibbard, 1941b; Long, 1964, 1972; Skinner and Hibbard, 1972.)

Subfamily Lutrinae

†Blancan Otter, *Satherium piscinarium* (Leidy), 1973 (*Lutra piscinaria;*
  Satherium ingens Gazin, 1934)

Otters, the most aquatic of the mustelids, are widespread throughout much of North and South America, Europe, Africa, Asia, and the Malayan Islands. They can be roughly divided into the fish-eating otters (*Lutra, Pteroneura*) and the crab-eating otters (*Aonyx, Amblyonyx, Paraonyx,* and *Enhydra*). The latter group possibly represents two separate lineages, the small-clawed and clawless otters and the sea otters (*Enhydriodon, Enhydra*), which evolved independently from the more primitive fish-eating otters. Aquatic adaptations include a long, cylindrical body, short legs, webbed feet, a powerful tail, and a flattened skull. The teeth are extremely broad, and the carnassials are more adapted for crushing than shearing.

The Blancan otter is known from Hagerman, Grand View, Rexroad, Sand Draw, Broadwater, and Haile XVA. Larger than *Lutra canadensis,* it resembles *Pteroneura brasiliensis,* the flat-tailed otter. Skull, dental, and skeletal characters (see fig. 11.6) are not so specialized as to bar *Satherium* from the ancestry of American otters, especially *Pteroneura*.

The species is represented by abundant material from Hagerman, which has been described by Bjork (1970). An inhabitant of streams and pools, the Blancan otter probably fed upon crayfish, mussels, fish, amphibians, water birds, and riparian mammals; all of these animals have been found in fossil form at Hagerman. (Bjork, 1970.)

River Otter, *Lutra canadensis* (Schreber), 1776 (*Mustela lutra canadensis; Lutra
  rhoadsii* Cope, 1896; *L. parvicuspis* Gidley and Gazin, 1933; *L. iowa*
  Goldman, 1941)

The American river otter makes its first appearance in the Irvingtonian faunas at Port Kennedy and Cumberland. Both of these finds were described as new species, but

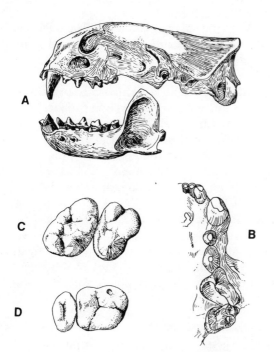

11.6 Mustelidae. *A, B, Satherium piscinarium*, Blancan, Hagerman. *A*, skull and mandible; *B*, left upper dentition. *C, D, Enhydra lutris*, Recent. *C*, right P^4–M^1; *D*, right M_{1-2}. *A, B*, after Bjork (1970); *C, D*, after Vaughan (1972). *A*, 0.66×; *B*, 1.3×; *C, D*, 1×.

later workers have referred the specimens to the extant species. Otters have been reported from late Pleistocene deposits in Arkansas, Florida, Georgia, Iowa, Missouri, Pennsylvania, and Tennessee. *Lutra iowa* Goldman was described from presumably Pleistocene beds in Wright County, Iowa; although showing some resemblances to the South American *Lutra annectens* Major, it can probably be referred to *Lutra canadensis*. Today, otters are found along inland waterways from Alaska to Florida, except in arid areas.

Aquatic in habits, otters are excellent swimmers and divers. They are long, sinuous animals with short legs, webbed feet, and a stout, tapering tail. The cheek teeth are broad, with large crushing surfaces. Otters feed on a variety of aquatic life and use their forepaws to manipulate food. Gregarious, otters are among the most playful of all mammals.

The ancestry of *Lutra canadensis* can be traced back to *Lutra licenti* Teilhard and Piveteau from the Pontian of China, and otters probably came across Beringia in the early Irvingtonian.

Recently, van Zyll de Jong (1972) proposed transferring the American river otters to the genus *Lontra* and restricting the genus *Lutra* to the Old World river otters. This conclusion was based on differences in the morphology of the baculum and on the presence of well-developed postorbital processes in the New World genus. These are probably subgeneric distinctions, however, and Sokolov (1973) believes that all river otters can be included in the genus *Lutra*. (Cope, 1896; Gidley and Gazin, 1933; Goldman, 1941; Hall, 1936; Sokolov, 1973; van Zyll de Jong, 1972.)

†Ancestral Sea Otter, *Enhydriodon* n.sp.

A few widely scattered specimens of the sea otter-like genus *Enhydriodon* have been found in late Miocene to late Pliocene deposits in southern Europe, South Africa, India, and western North America. Recent studies (Repenning, 1976) indicate that two lineages are represented. One, known from Asia and Africa, has a quadrate P^4 that shows equal development of the protocone and hypocone and a well-developed parastyle. This lineage probably culminated in *Enhydriodon sivalensis* Falconer, found in late Pliocene/early Pleistocene deposits in the Siwalik Hills, northern India. The other lineage, known from the late Miocene of Spain, Italy, and California, has a triangular P^4 that has an expanded hypocone, no parastyle, and inflated cusps. This line probably culminated in *Enhydriodon? reevei* (Newton), from the early Pleistocene of England, and *Enhydra lutris*, the living sea otter. Characteristic of the *Enhydriodon–Enhydra* group is large size (included are the largest known otters) and the presence of crushing instead of shearing-adapted carnassials, short, broad cheek teeth, and bulbous tooth cusps.

A new species of *Enhydriodon* has been described (but not named) from the late Pliocene USGS vertebrate locality M1032, Kettleman Hills, King Co., California. Blancan-age land mammals have been recovered from the same unit. The type, an isolated P^4, is distinguished from that of *Enhydriodon lluecai* Villalta and Crusafont (Late Miocene, Spain) and *Enhydriodon* cf. *lluecai* (Late Miocene, Kettleman Hills, Calif.) by its more inflated cusps, greater transverse diameter, and by the morphology of the cusps. The California specimens of *Enhydriodon* are close to the ancestral line of *Enhydra*. (Repenning, 1976.)

† Large-toothed Sea Otter, *Enhydra macrodonta* Kilmer, 1972

The only marine mustelid, the genus *Enhydra* inhabits North Pacific coastal areas. It is the only carnivore with two incisors in each half of the lower jaw. The broad, short carnassials have lost all shearing function, and the molar cusps are greatly inflated. The ancestry of *Enhydra* can be traced back to late Miocene *Enhydriodon* in California. Two species are currently recognized.

The extinct sea otter *Enhydra macrodonta* was recently described from a presumably Rancholabrean locality near Arcata, Humboldt Co., California. It is distinguished from the living species by the larger size of the posterior cheek teeth, longer tooth row, and more generalized coronoid process of the mandible. Postcranial material from the type locality shows less extreme modification by muscles than that seen in *Enhydra lutris* (Repenning, personal communication, 1976). Specimens from the Irvingtonian localities of Cape Blanco, Oregon (formerly thought to be Blancan in age) and Moonstone Beach near Eureka, California show similar characteristics and are tentatively referred to *Enhydra macrodonta*. (Kilmer, 1972; Repenning, 1976 and personal communication, 1976.)

Sea Otter, *Enhydra lutris* Linnaeus, 1758

Endemic to the North Pacific, *Enhydra lutris* ranged along coastal waters from the Aleutian Islands south to Baja California in historic times. Excessive hunting for their valuable fur decimated the population, and only an international treaty and complete protection has saved the species from extinction. Today, sea otters are making a comeback and have reestablished colonies in kelp beds along the coasts of central California, western Alaska, and the Commander and Kurile islands, but after more than 50 years of protection, they only occupy one-fifth of their former range. Remains of *Enhydra lutris* have been taken from Wisconsinan deposits at San Pedro, Newport Bay Mesa, and Santa Rosa Island. *Enhydra* sp. is reported from the Timm's Point Silt Member of the San Pedro Formation, San Pedro, California, believed to be about 1 m.y. old, and from the ?Yarmouthian Gubik Formation near Point Barrow, Alaska (northernmost record; Repenning, personal communication). Since sea ice restricts the northern distribution of sea otters by limiting their ability to dive for food (Schneider and Faro, 1975), the Point Barrow record indicates relatively ice-free conditions in the Arctic Ocean at that time (Repenning, personal communication, 1976).

Male sea otters weigh between 27 and 37 kg; females weigh about 10 kg less. These marine mustelids are long-bodied, with a blunt head and rounded, valvelike ears. The legs and tail are short, the forepaws are small, and the hind feet are webbed and flattened into broad flippers. Unlike most other marine mammals, sea otters lack an insulating layer of subcutaneous fat and instead depend on air trapped in their long, thick, luxurious underfur for protection against the cold. They feed on sea urchins, mollusks, and crustaceans and, using their chest as a table and a rock brought up from the bottom as an anvil, break open shells to obtain the contents. The lower carnassial has an enlarged metaconid which extends backward as much as half the length of the trigonid, and the talonid is as short or shorter than the trigonid (see fig. 11.6); it closely resembles the M_1 of *Enhydriodon* cf. *lluecai* from the late Miocene of California except that it is more inflated (Repenning, 1976).

Gregarious in habits, sea otters swim and float on their backs within a kilometer of the shore. Nonmigratory and unagressive, the animals need continued protection. Their ancestry can be traced back to the *Enhydriodon* that inhabited California from the late Miocene to the late Pliocene. (Leffler, 1964; W. E. Miller, 1971; Mitchell, 1966; Repenning, 1976 and personal communication; Schneider and Faro, 1975; Walker, 1975.)

Subfamily Mephitinae

†Short-jawed Skunk, *Buisnictis breviramus* (Hibbard), 1941 (*Brachyprotoma breviramus; Buisnictis meadensis* Hibbard, 1950)

Skunks, subfamily Mephitinae, are a widespread New World group whose ancestry probably extends back to the Old World Miocene genus *Miomephitis*. Their omnivorous habits are reflected in their teeth: the carnassial shear is not highly developed, and the molars have large crushing surfaces. Of the mustelids, skunks have the largest anal scent glands; the fetid secretion can be ejected a few meters and, together with the warning coloration (a distinctive black and white pattern), is used for defense. Skunks are quite common in late Pliocene and Pleistocene deposits, and six genera are recognized.

Buisnictis was a genus of small, short-faced skunks that had well-developed shearing carnassials and strong crushing molars. The stratigraphic range of the genus extends from the middle Pliocene to the early Pleistocene.

Buisnictis breviramus has been found at Rexroad and Hagerman. It was about the size of *Mustela frenata* and had crowded premolars, a small M_1 with a basined talonid and small metaconid, and a better developed protocone on P^4 than that of the Pliocene species.

The late Claude W. Hibbard collected a large series of *Buisnictis* from various sites in Meade County, Kansas, and Beaver County, Oklahoma. These specimens show the transition from the more primitive middle Pliocene species *Buisnictis schoffi* Hibbard to *Buisnictis breviramus*. Relationships to other small skunks are uncertain; the short face and crowded premolars are similar to those of *Brachyprotoma*, but the dentition is more like that of *Spilogale*. (Bjork, 1970; Hibbard, 1941b, 1950, 1954b.)

†Burrow's Short-faced Skunk, *Buisnictis burrowsi* Hibbard, 1972

A more advanced species, *Buisnictis burrowsi* has been described from Sand Draw. It was about the size of *Buisnictis breviramus* and had a four-rooted M_1 with a greatly reduced metaconid. The dentition was more sectorial than that of *Spilogale*. It probably fed on small rodents and insects. The genus became extinct in pre-Nebraskan times. (Skinner and Hibbard, 1972.)

†Rexroad Skunk, *Spilogale rexroadi* Hibbard, 1941

The earliest representative of the genus *Spilogale* is the Rexroad skunk, known from late Pliocene (Beck Ranch, Texas, Rexroad, and Blanco) faunas. It was about the size of the smallest living species, *Spilogale pygmaea* Thomas, which is found along the west coast of Mexico today. Van Gelder (1959) regards it as the most primitive living skunk. *Spilogale rexroadi* shows a more trenchant dentition than the extant species and probably had more carnivorous habits. Considered to be the most primitive member of the genus, *Spilogale rexroadi* was probably directly ancestral to later species. (Dalquest, 1972; Hibbard, 1941b; Van Gelder, 1959.)

Spotted Skunk, *Spilogale putorius* (Linnaeus), 1758 (*Viverra putorius;*
   Spilogale marylandensis Gidley and Gazin, 1933; *S. pedroensis* Gazin, 1942)

The stratigraphic range of *Spilogale putorius* extends from the pre-Nebraskan (Borchers) to the Recent. Specimens from Curtis Ranch and Cumberland were described as distinct species (*Spilogale pedroensis* and *Spilogale marylandensis*, respectively), but their "distinctive characters" can be matched in a large sample of Recent *Spilogale putorius*. These early representatives were slightly smaller than extant spotted skunks living in the same area. Spotted skunks are common in Rancholabrean faunas throughout the country, and the only record lying outside the present range is from mid-Wisconsinan deposits at Medicine Hat. An inhabitant of brushy, sparsely wooded areas and prairies, this nocturnal mammal is still common over much of its range. A distinctive black-and-white coloration and foul scent, which many people regard as more disagreeable than the scent of *Mephitis*, identify *Spilogale*.

Smallest of the extant Mephitinae, spotted skunks generally show a north-south gradation in size (the largest animals being found in the north); this trend is evident in Pleistocene samples as well. Males are about 7 percent larger than females in cranial measurements. With more trenchant carnassials than those of *Mephitis*, *Spilogale* is the most predaceous of the skunks and feeds primarily on insects and small mammals. It is considered to be an excellent mouser.

Van Gelder (1959) believes that *Spilogale putorius* originated in central Mexico from a small *Spilogale pygmaea*–like form. Although *Spilogale rexroadi* is known only from late Pliocene and early Pleistocene faunas in Texas and Kansas, its range may have extended south into Mexico and Central America; no morphological features exclude it from the ancestry of *Spilogale putorius*. Ewer (1973) regards the spotted skunk as basically a Central American genus that has been extending its range northward since the last glaciation. Unfortunately, the neomammalogists have not looked at the fossil material. To explain differences in the breeding behavior of the eastern and western forms, Ewer postulates that the populations moved north, splitting into two main branches east and west of the Rocky Mountains.

Recent studies by Mead (1968a,b) have shown that the eastern forms of *Spilogale* breed in April, with the young being born in June. In contrast, the western forms mate

ORDER CARNIVORA

in autumn and give birth in May; the delayed implantation allows the young to be born during more favorable climatic conditions and thus has survival value. Unfortunately, there have been no studies on the breeding habits of the Mexican and Central American populations of *Spilogale*. On the basis of the differences in breeding habits and concomitant physiological and cytological differences, Mead believes that the eastern and western forms are reproductively isolated and should be considered distinct species—*Spilogale putorius* east of the Rocky Mountains, and *Spilogale gracilis* in the West. (Ewer, 1973; Getz, 1960; Hall and Kelson, 1959; Mead, 1968a,b; Van Gelder, 1959.)

†Short-faced Skunk, *Brachyprotoma obtusata* Cope, 1899 (*Mephitis obtusata; Brachyprotoma pristina* Brown, 1908; *B. spelaea* Brown, 1908)

The monotypic genus *Brachyprotoma* is known from Port Kennedy, Cumberland, Conard Fissure, Frankstown, Crankshaft Pit, and Brynjulfson. These deposits range in age from early Irvingtonian to late Wisconsinan/early Holocene. Although several species have been described, the slight morphological differences separating them can probably be attributed to age differences, sexual dimorphism, and individual variation within the species. The differences *Brachyprotoma* shows from the other genera of skunks are greater than those existing among the other genera.

Brachyprotoma obtusata is characterized by a short mandible with crowded, overlapping premolars that are set obliquely in the jaw (see fig. 11.7). Owing to the crowding of the premolars, the tooth row is more curved than that of the other genera of skunks. The skull is short, with a broad frontal region and reduced postorbital processes. Only two premolars are present in the upper jaw. The canines are large compared to the size of the animal. *Brachyprotoma obtusata* was the size of a small *Spilogale*. The species is associated with a northern, or boreal, fauna at the various sites where it has been found and probably lived in a forest environment. Its dental specializations indicate a probable diet of hard-shelled insects.

Although its ancestry is unknown, *Brachyprotoma*, like *Spilogale*, was probably derived from the Eurasian Mio-Pliocene genus *Promephitis*. *Brachyprotoma* shows more resemblances to *Spilogale* than to any other skunk.

This genus has been found associated with other genera of skunks at every site, but there was apparently little competition among them. The short-faced skunk survived

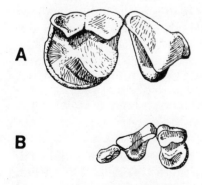

11.7 Mustelidae. *A, Conepatus leuconotus,* Recent, right P^4–M^1, occlusal view. *B, C, Brachyprotoma obtusata,* Irvingtonian, Cumberland Cave. *B,* right P^3–M^1; *C,* left C–M_2, occlusal views. *A,* after Vaughan (1972); *B, C,* after Gidley and Gazin (1938). *A,* 3×; *B, C,* 1×.

until the early Holocene; its extinction is unexplained. (Brown, 1908; Cope, 1899; Gidley and Gazin, 1938; Hall, 1936; Parmalee and Oesch, 1972.)

†Port Kennedy Skunk, *Osmotherium spelaeum* Cope, 1896 (*Mephitis fossidens* Cope, 1896; *M. orthostichus* Cope, 1896; *M. leptops* Cope, 1899; *Pelycictis lobulatus* Cope, 1896)

The genus *Osmotherium* is known only from Port Kennedy Cave (early Irvingtonian). Cope named three genera and five species of skunks from the deposit, but additional work has shown that his "species" were based on age, sexual, and individual variation within the population. Comparable variation can be found in extant populations of skunks.

Osmotherium spelaeum was about the same size as *Mephitis mephitis*. It differed from that species primarily in the morphology of the lower carnassial: the talonid is usually longer than the trigonid, the basin of the talonid is shallower, and there are from one to three accessory cusps between the bases of the entoconid and the protoconid. The lower jaw is not noticeably shortened, nor are the premolars crowded as they are in *Brachyprotoma*. The genus is closely related to *Mephitis* and may not be distinct from it. (Cope, 1896, 1899; Hall, 1936.)

Striped Skunk, *Mephitis mephitis* (Schreber), 1776 (*Viverra mephitis*)

Remains of *Mephitis* are common in Pleistocene faunas. Its habit of denning in fissures and small caves has increased the chances of its remains being preserved. The earliest record of the genus is from Broadwater (late Blancan); Irvingtonian occurrences include Inglis IA, Conard Fissure, and Coleman IIA. By the end of the Wisconsinan, the species was widespread throughout the United States. Today, striped skunks are found near water in semiopen country and brushy areas from northern Alberta to central Mexico.

Mephitis shows considerable geographic and individual variation, as well as differences related to age and sex, and this has led to the naming of several species and subspecies. Currently, two species are recognized: *Mephitis mephitis*, the striped skunk, and *Mephitis macroura*, the hooded skunk, a southern form, with a range that barely extends into the United States. All of the Pleistocene specimens are referable to *Mephitis mephitis*, and no extinct subspecies are recognized.

Mephitis is characterized by a highly arched skull which is deepest in the frontal region. The mastoid bullae are not inflated. The inferior border of the lower jaw shows a distinct step at the angle, a condition not present in the other genera of skunks. The length of P^4 is equal to or slightly less than the width of M^1.

The striped skunk, *Mephitis mephitis*, is omnivorous; insects and mice make up a large part of its diet. These house cat–sized animals are terrestrial and primarily nocturnal. They are more often smelled than seen. (Hall, 1936; Hall and Kelson, 1959.)

Hog-nosed Skunk, *Conepatus leuconotus* (Lichtenstein), 1832
(*Mephitis leuconotus*; *M. mesoleucus* Lichtenstein, 1832)

The largest member of the subfamily Mephitinae is the hog-nosed skunk, an inhabitant of semiwooded and open areas from Mexico to southern Colorado. The earliest record is from Inglis IA (early Irvingtonian). In the late Pleistocene, its range extended to Florida (Haile XIIA, Reddick, Williston) and Georgia (Ladds), as well as Texas (Schulze, Centipede, Kyle), New Mexico (Dry), and Nuevo León, Mexico (San Josecito).

The hog-nosed skunk is characterized by a large, naked, piglike snout (hence the name), long claws, and more white on the back and tail than in most specimens of *Mephitis*. Less predaceous than the other skunks, *Conepatus* feeds predominantly on insects. Its carnassials are small and poorly developed. The upper molar is huge, with a large crushing surface; it is usually longer than wide, with the lingual half of the crown displaced posteriad, and the outline of the crown is not dumbbell-shaped. The upper second premolar is usually absent, and the third upper premolar is set obliquely in the jaw. The trigonid of the lower carnassial is shorter than the talonid; this provides an increased crushing surface.

In his study of a population of *Conepatus* from Uruguay, Van Gelder (1968) has shown that there is pronounced sexual, age, and individual variation. This seems to hold true for Pleistocene and Recent North American specimens as well. (Ewer, 1973; Hall and Kelson, 1959; Ray, Olsen, and Gut, 1963; Van Gelder, 1968.)

Family Canidae—Dogs

Canids arose in the Oligocene and attained an extraordinary diversity in the Miocene, with a rich but as yet incompletely understood fossil record. In later times, the variety was much reduced through extinction of various lines, and in the Blancan only two main groups survived in North America: the true dogs and foxes and their relatives (subfamily Caninae); and the very distinct, hyenalike dogs of the *Borophagus* line, which form an important element in the Blancan fauna, and which we refer to the subfamily Simocyoninae.

The Caninae are mostly slenderly built, with digitigrade running legs; the digits are closely appressed. The jaws are long and the dentition shows little reduction. Social behavior and high intelligence are evident in many species. Although the subfamily is not much varied morphologically, differentiation on the genus and species level is great. Taxonomy is often difficult because of the conservative dentition and the temporal and spatial size variation within many species. The fossil history is mostly good, and some canids are among the most commonly found fossils. Thus, material for detailed evolutionary studies is at hand. Some canids have a wide, Holarctic distribution.

The Simocyoninae have a tendency to premolar reduction, while carnassials and anterior molars attain large size. The adaptation may be reminiscent of hyenas, as in the Borophagini, an endemic North American tribe of simocyonine dogs that survived in the Blancan.

Subfamily Simocyoninae

† Plundering Dog, *Borophagus direptor* (Matthew), 1924 (*Hyaenognathus direptor;*
  H. solus Stock, 1932; *Osteoborus hilli* Johnston, 1939;
  O. progressus Hibbard, 1944; *O. crassipinaetus* Olsen, 1956)

The genus *Borophagus* belongs to the North American hyenalike dogs, tribe Borophagini. The size of a small wolf, *Borophagus* had stout limbs with rather small and feeble feet. Its head, carried on a strong, stout neck, was very big; the muzzle was much shortened, and the skull was very broad and strongly domed. Although the anterior premolars were small, the jaws and posterior cheek teeth were powerful, especially P^4 and M_1. P_4 resembles closely a hyena premolar, and in fact, *Borophagus* was originally thought to be a member of the Hyaenidae.

The genus arose from a more primitive genus, *Osteoborus,* of the late Miocene and early Pliocene and differs from the latter in larger size, bigger canines and P^4 in relation to other teeth, more reduced M_2, shorter talonid of M_1, and a broader, more vaulted skull. Although it has generally been referred to *Osteoborus,* the species *Borophagus direptor* resembles *Borophagus diversidens* in all these key characters and so is more properly placed in this genus. It probably was derived from the early to mid-Hemphillian species *Osteoborus cyonoides* (Martin) and was somewhat smaller and more primitive than the later Blancan *Borophagus diversidens,* to which it probably gave rise.

Borophagus direptor was originally described from the upper Hemphillian of Snake Creek, Nebraska, and has subsequently been found at other sites of the same age. It survived in the earliest Blancan (Coso, Fox Canyon, Hagerman) but by Rexroad times had already been succeeded by *Borophagus diversidens.* It probably had hyenalike habits; bones apparently chewed by plundering dogs are found in Pliocene deposits. (Bjork, 1970; Hibbard, 1944, 1950; Kurtén, unpublished data; Matthew, 1924b; Stirton and VanderHoof, 1933; Stock, 1932b.)

†Bone-eating Dog, *Borophagus diversidens* Cope, 1893 (*Felis hillianus* Cope, 1893; *Hyaenognathus pachyodon* Merriam, 1903; *H. (Porthocyon?) dubius* Merriam, 1903; *Porthocyon matthewi* Freudenberg, 1910)

Remains of the bone-eating dog are not uncommon in Blancan deposits (Benson, Blanco, Broadwater, Cita Canyon, Grand View, Hudspeth, Lisco, Keefe Canyon, Red Light, Rexroad, Sand Draw, White Bluffs). There is also evidence that the species survived in the early Irvingtonian: it occurs associated with mammoth and shrub-ox at Wellsch Valley.

The species was probably derived from *Borophagus direptor,* from which it is distinguished by its larger size and more advanced characters, especially the very broad, hyenalike P_4. Clearly, it was a highly specialized predator and scavenger, relatively slow-moving but with extremely powerful jaws and teeth (see fig. 11.8). It would seem to have been a vicar of true hyenas like the genera *Hyaena* and *Crocuta* in the Old

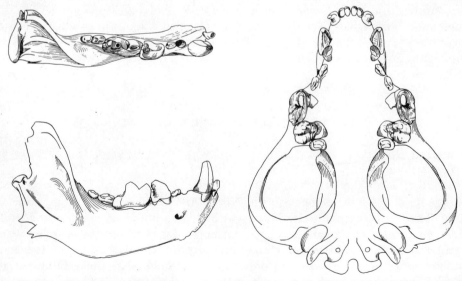

11.8 Canidae. *Borophagus diversidens,* Blancan, Blanco, skull, ventral view and right mandible, external and occlusal views. 0.33×.

World. In contrast, the contemporary American hyenid *Chasmaporthetes* had slender, slicing cheek teeth and probably did not chew bones. (Cope, 1893; Dalquest, 1969; Hibbard, 1950; Hibbard and Riggs, 1949; Kurtén, unpublished data; Matthew and Stirton, 1930; Meade, 1945; Stalker and Churcher, 1972; VanderHoof, 1936.)

Subfamily Caninae

†Johnston's Coyote, *Canis lepophagus* Johnston, 1938 (*C. arnensis* Del Campana, 1913?)

The genus *Canis,* the true dogs and wolves, comprises medium-sized to large Caninae with the full complement of three molars in each lower jaw half. The genus dates back to the late Miocene.

The species *Canis lepophagus,* which initiated the coyote evolutionary line, was very wide-ranging in the Blancan; localities include Blanco, Borchers, Broadwater, Cita Canyon, Deer Park, Donnelly Ranch, Grand View, Hagerman, Keefe Canyon, Lisco, Red Corral, Rexroad, Sand Draw, Santa Fe River IB, and White Bluffs. The same species probably also occurs at late Blancan and early Irvingtonian sites like Anita, Rock Creek, and Vallecito. The size and general characters of the skull and teeth are very similar to those of the modern coyote, but there are minor differences in the dentition. The lower premolars of *Canis lepophagus,* for instance, are somewhat narrower than those of *Canis latrans,* and the lower carnassial tends to be shorter and broader. In the limbs, the proximal segments (humerus, femur) were relatively long and the metapodial short, as compared to the modern species. This indicates a somewhat less cursorial type than the living coyote.

Johnston's coyote may well be derived from a smaller, Hemphillian species, *Canis davisi* Merriam. The size increase continued in the Blancan.

Canis lepophagus was evidently the direct ancestor of the Pleistocene and Recent coyotes. It may also be ancestral to the wolves. The Old World *Canis arnensis,* from the later Villafranchian of Europe, is indistinguishable from *Canis lepophagus,* and the two probably represent segments of a single Holarctic population which evidently originated in North America. In the late Villafranchian of Europe, it is possible to trace the sequence from such coyotelike dogs to true primitive wolves of the species *Canis etruscus* Major. This species, in turn, gave rise to *Canis lupus.* The red wolf, *Canis rufus,* probably also arose from early coyotes. Indeed, early Blancan *Canis lepophagus* may well be at the base of the radiation of all the Pleistocene *Canis* in North America.

The ecological requirements of the Blancan coyotes may have resembled those of the living species, except insofar as the different limb proportions represent an adaptive difference. (Bjork, 1970; Bonifay, 1971; Giles, 1960; Johnston, 1938; Kurtén, 1968, 1974; Nowak, 1973; Torre, 1967.)

Coyote, *Canis latrans* Say, 1823 (*C. orcutti, C. andersoni* Merriam, 1910;
   C. riviveronis Hay, 1917; *C. caneloensis* Skinner, 1942;
   C. irvingtonensis Savage, 1951)

The coyote was common and widespread in the Rancholabrean, ranging from Alaska (Fairbanks) to Pennsylvania, Florida, and Mexico; the range thus extended farther east and southeast than it does at the present day. Upwards of a hundred sites have yielded Pleistocene coyote remains. Early finds date well back into the Irvingtonian (Angus, Cumberland, Hay Springs, Irvington, Mullen).

11.9 Canidae. *A, Canis dirus. B, Canis latrans.* Both Rancholabrean, Rancho La Brea, right P^4-M^2, occlusal view. After Merriam (1912). 0.66×.

Canis latrans differs from *Canis lepophagus* in some dental proportions and especially in the relative proportions of the limb bones, which show a lengthening of the distal segments at the cost of the proximal, presumably an adaptation to faster running. The dentition is shown in figures 11.9 and 11.10. The species doubtless evolved from the Blancan form, and the numerous early Irvingtonian finds of *Canis lepophagus* may represent transitional forms.

Pleistocene coyotes, including Irvingtonian, tend to be considerably larger than living individuals, and this has resulted in the description of various extinct species and subspecies. Very late Wisconsinan finds are smaller, and there appears to be a gradual transition to the Recent form. Rancho La Brea coyotes average 5–6 percent larger in linear dimensions than Recent coyotes from the same area.

Coyotes are extremely common as fossils at sites of the trap type, such as the California tar pits and Moonshiner and Middle Butte caves, probably owing to their scavenging habits. Like other *Canis,* coyotes also take vegetarian food, showing in the southwestern United States, for instance, a great predilection for juniper berries. The species is found in many kinds of environments except dense forest.

In terms of evolutionary study, the coyote line is of great interest since it seems possible to trace it back in an unbroken sequence to the Hemphillian species *Canis davisi.* (Giles, 1960; Kurtén, 1974; Marcus, 1960; Merriam, 1912; Nowak, 1973; Savage, 1951; Slaughter, 1961a.)

†Wolf Coyote, *Canis priscolatrans* Cope, 1899 (*C. edwardii* Gazin, 1942)

A group of small wolves, or giant coyotes, from the early Irvingtonian is thought to represent the first appearance of wolflike animals in North America. Somewhat larger than the coyotes, *Canis priscolatrans* shows many of the anatomical characters of the latter but also shows such striking resemblance to *Canis rufus* that it seems probable that it represents the ancestry of that species [Nowak (1973) includes *Canis priscolatrans* in *Canis rufus*]. Because of the long temporal gap between these early Pleistocene forms and the late Rancholabrean true red wolves, however, they are here grouped as a distinct species.

Material referable to this form has been recovered from Anita, Arkalon, Curtis Ranch, Gilliland, Miñaca Mesa, Port Kennedy, and probably also Vallecito and derives from the timespan very late Blancan–early Irvingtonian. The species resembles the small European wolf *Canis etruscus* (late Villafranchian) in many respects, but since no Asian material connecting the two forms has been identified, the relationships remain obscure. The role of parallel evolution in *Canis* has yet to be clarified.

Canis priscolatrans may have arisen from Blancan *Canis lepophagus* and probably gave rise to *Canis rufus;* it may also be ancestral to the more advanced North American species *Canis armbrusteri* and *Canis dirus*. (Cope, 1899; Gazin, 1942; Kurtén, 1974; Nowak, 1973.)

Red Wolf, *Canis rufus* Audubon and Bachman, 1851 (*Lupus niger* Bartram, 1791, nom. inval.)

The status of this species is in dispute, and its fossil history is little known. However, Nowak (1973) has referred late Rancholabrean finds from 10 localities in Arkansas, Florida, Mexico, and Texas to *Canis rufus;* in addition, a number of archeological sites in the eastern United States (including Nichols Hammock) have yielded material of Holocene red wolf. The species is larger and more robust than *Canis latrans*, and somewhat smaller and more delicately built than the gray wolf, with a less reduced talonid on M_1.

Canis rufus would appear to be a little-modified survivor of the primitive early Pleistocene wolves of the *Canis priscolatrans* and *Canis etruscus* group. The time gap separating it from its probable predecessor, the early Irvingtonian wolf-coyote, is to some extent bridged by a find from the late Kansan Cudahy fauna, belonging to a wolf in this size range; unfortunately, it is only an astragalus, and so gives little information on the late Irvingtonian red wolves.

As a result of extirpation in a large part of its historical range, the red wolf now survives only as an endangered relict. (Nowak, 1973.)

†Armbruster's Wolf, *Canis armbrusteri* Gidley, 1913

The large wolf of the later Irvingtonian in North America, although closely related to both *Canis rufus* and *Canis lupus*, is now usually regarded as a distinct species. Distinctly larger than *C. rufus*, it differs from *C. lupus* in its narrower skull and somewhat more slender dentition, with a large protocone in P^4 and less reduced M_1 talonid and M2. It was smaller than *Canis dirus*, especially with regard to carnassial size (see fig. 11.10).

The main occurrences are Cumberland and Coleman IIA, both early Illinoian; the former site yielded eight skulls and numerous mandibles. In addition, late Irvingtonian fossils from Angus, Ashley River, Irvington, and the McLeod lime-rock mine near Williston, Levy Co., Florida, are here referred to this species. *Canis armbrusteri* may also be represented by early Irvingtonian material from Anita, Port Kennedy, and Rock Creek. Thus early divergence from a *Canis rufus*-like stock may be indicated.

A very similar species is the European *Canis falconeri* Major from the middle and late Villafranchian. This form may well be conspecific with *C. armbrusteri*. The earliest European record (Puebla de Valverde, Spain) is probably pre-Irvingtonian.

The extinction of *Canis armbrusteri* seems to coincide fairly closely with, and may have been hastened by, the entrance of *Canis lupus* in the New World. (Gidley and Gazin, 1938; Kurtén and Crusafont, 1977; R. A. Martin, 1974a; Nowak, 1973; Tedford, personal communication, 1975.)

Gray, or Timber, Wolf, *Canis lupus* Linnaeus, 1758 (*C. milleri* Merriam, 1912)

Although wolflike animals have been present in North America since the early Irvingtonian, it is only in the very late Irvingtonian and the Rancholabrean that we find material which can be definitely referred to the modern species. Nowak (1973) lists 56

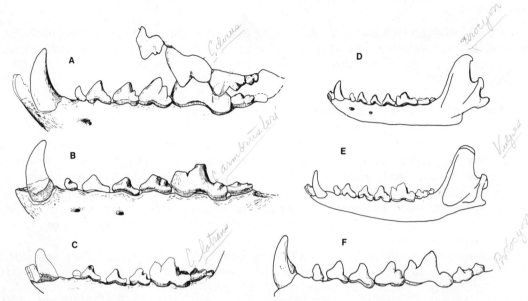

11.10 Canidae. Left mandibles, external view. *A, Canis dirus,* Rancholabrean, Rancho La Brea (showing beginning occlusion with upper teeth). *B, Canis armbrusteri,* Irvingtonian, Cumberland Cave. *C, Canis latrans,* Rancholabrean, Rancho La Brea. *D, Urocyon cinereoargenteus,* Recent. *E, Vulpes vulpes,* Recent. *F, Protocyon texanus,* Irvingtonian, Rock Creek. *A, C,* after Merriam (1912); *B,* after Gidley and Gazin (1938). All 0.5×.

Pleistocene occurrences, including the late Irvingtonian Hay Springs and Mullen; Rancholabrean records come from Alaska, Alberta, Arizona, Arkansas, California, Colorado, Georgia, Idaho, Illinois, Mexico, Michigan, Minnesota, Nebraska, Nevada, New Mexico, Oklahoma, Oregon, Pennsylvania, Saskatchewan, Texas, Virginia, Wisonsin, Wyoming, and the Yukon. In Alaska, the species was present well back in the Illinoian, and perhaps earlier.

The gray wolf is distinguished from Armbruster's wolf by its broader skull, more robust, although not necessarily longer, carnassials (with a more reduced protocone on P^4), and by the greater reduction of the M_1 talonid and M_2. On the other hand, its dentition is much less robust than that of *Canis dirus*.

Canis lupus evidently arose in the Old World, where there is a gradual sequence leading from the Villafranchian *Canis etruscus* to the gradually larger forms of the Cromerian and culminating in the typical gray wolf of the later Pleistocene. In Nowak's opinion (1973), the species made its entrance in the New World as early as the Yarmouthian, so that a Holarctic population came into being shortly after the appearance of true *Canis lupus* in Eurasia.

There are no mass occurrences like those of *Canis dirus;* a census for Rancho La Brea gave only 8 *Canis lupus* and over 1,600 dire wolves. This probably indicates important differences in habits. The gray wolf is a social, highly intelligent, and efficient predator, capable of taking a variety of game, both large and small; it is also to some extent omnivorous. It was one of the most widely distributed land mammals in the late Pleistocene. (Kurtén, 1968; Marcus, 1960; Mech, 1966, 1970; Nowak, 1973; Stock, 1929.)

†Cedazo Dog, *Canis cedazoensis* Mooser and Dalquest, 1975

Remains of a small *Canis* were described under this name from the Sangamonian deposits of Cedazo. This is only one of various records of dogs distinctly smaller than the coyote from deposits ranging in age from late Irvingtonian (Cumberland, Slaton) to late Rancholabrean (Fossil Lake). The carnassial from Cumberland shows a striking resemblance to that of the domestic dog. The relationships of the Cedazo dog are at present unresolved. A very small, coyotelike dog occurs as early as in the late Blancan of Lisco. (Dalquest, 1967; Kurtén, 1974; Lawrence, personal communication, 1971; Mooser and Dalquest, 1975b.)

Domestic Dog, *Canis familiaris* Linnaeus, 1758

Subfossil dog remains are not uncommon at archeological sites, but Pleistocene finds are very scarce. The earliest remains of dog in America were found at Jaguar Cave and have a date of 10,370 ± 350 years B.P. The material comprises dogs of two distinct size classes, which indicates that racial segregation may have taken place at this early date. If this is so, actual domestication must have been earlier. Possibly, man brought the dog with him from Eurasia; early domestication in the Old World has been suggested on the basis of fossil remains in the Magdalenian of Europe. Dog remains are also found in unstratified, perhaps Pleistocene, deposits in Alaska. *Canis familiaris* has recently been identified at Old Crow River, Loc. 11A (minimum age, 20,000 years B.P.) (B. Beebe, personal communication, 1978).

The dog is regarded as a descendant of *Canis lupus*. Interbreeding with the wild wolf population probably occurred, and in some areas interbreeding with coyote may have been important. (Beebe, personal communication, 1978; Kurtén, 1968; Kurtén and Anderson, 1972; Lawrence, 1967; Lawrence and Bossert, 1969; Musil, 1970.)

†Dire Wolf, *Canis dirus* Leidy, 1858 (*C. indianensis* Leidy, 1869;
    C. mississippiensis Allen, 1876; *C. ayersi* Sellards, 1916)

One of the most common mammalian species in the Rancholabrean, the dire wolf has been found at more than 80 sites in North America ranging from late Illinoian and Sangamonian (Butler Spring, Cragin Quarry, Reddick, Haile VIII) to late Wisconsinan. The species has not been found in the Beringian area; the northernmost occurrences are in southern Alberta. The range covers most of the United States and Mexico and extends to Peru in the south.

Equaling a large gray wolf in size, *Canis dirus* was markedly heavier of build, with a very large and broad head and sturdy limbs with relatively short lower segments. The dentition was more powerful than that of any other species of *Canis*, the carnassial teeth being, on the average, much larger than those of *Canis lupus* (see figs. 11.9, 11.10A). The braincase is relatively smaller than that of the gray wolf. The dire wolf may be referred to the subgenus *Aenocyon*, of which it is the sole known species.

The origin of the dire wolf is unknown; its seemingly unheralded appearance in the late Illinoian suggests that it originated elsewhere—in South America, for instance, where a species known as *Canis nehringi* Kraglievich shows points of resemblance. A direct origin from a North American form like *Canis armbrusteri* appears at the moment less likely since no transitional population is known.

There is a notable geographic differentiation within the dire wolf population. Californian and Mexican specimens, on the average, are smaller than those from the Great Plains, Appalachia, and Florida, and the difference in the length of limb is particularly striking.

The absence of true hyenas in the Rancholabrean may account for the presence of animals with hyenalike specializations such as the dire wolf with its very robust carnassials. A hunting–scavenging mode of life is also suggested by the extraordinary abundance of dire wolf remains in the tar pits of Rancho La Brea, where a census gives a minimum number of 1,646 individuals. Again, these mass occurrences may be a characteristic of the southwestern form. The more long-limbed eastern dire wolves presumably were adapted to the environment of the Great Plains, although they cannot have been as fleet of foot as *Canis lupus*.

Apart from the asphalt traps, most records of the dire wolf come from caves, a particularly large sample being found in San Josecito Cave. There are important cave sites in California, Kentucky, Missouri, Pennsylvania, and Florida.

The species became extinct without issue. The time of its extinction seems to tally with that of most of the North American megafauna. Terminal dates are Brynjulfson, $9,440 \pm 760$ B.P.; Rancho La Brea, $9,860 \pm 550$ B.P.; La Mirada, $10,690 \pm 360$ B.P.; and possibly Devil's Den, 8,000? B.P. (Churcher, 1959a; Galbreath, 1964; Hawksley, Reynolds, and Foley, 1973; Kurtén, in preparation; Marcus, 1960; Merriam, 1912; Miller, 1968; Nowak, 1973.)

Dhole, *Cuon alpinus* (Pallas), 1811 (*Canis alpinus*)

The genus *Cuon* today inhabits wooded areas of Asia from Siberia to Sumatra and Java. A few scattered finds record the presence of the extant species, *Cuon alpinus*, not only in the Beringian area (Fairbanks, Old Crow River, Loc. 14N) but even in the south (San Josecito Cave). The species arose in Eurasia from the middle Pleistocene *Cuon priscus* Thenius (1954). Characteristic of the evolution of the dholes is the transformation of the tubercular molars into small, trenchantly cusped structures adapted for a highly predaceous mode of life. (Harington, personal communication, 1971; Nowak, personal communication, 1971; Schütt, 1973; Thenius, 1954.)

†Troxell's Dog, *Protocyon texanus* (Troxell), 1915 (*Canis texanus*)

Dogs of the genus *Protocyon*, endemic to the New World, are known in the late Pliocene and early Pleistocene and are among the earliest known canids in South America, where they appeared in the Uquian. Approaching *Canis lupus* in size, *Protocyon* is distinguished by its short-legged build, broad and rather short-faced skull, and slender, high-cusped cheek teeth. The M_1 normally lacks a metaconid, and the talonid is short with a single trenchant cusp; the small M_2 has two sharp cusps in a longitudinal row; and M_3 is reduced or absent (see fig. 11.10F).

Protocyon was probably related to the living genus *Speothos* of South America, which is also short-legged and short-faced, although much smaller (in *Speothos*, molar reduction has gone still further, with M^2 absent). These dogs probably arose as a separate evolutionary line from early *Canis*, paralleling in their dental development the Old World hunting dogs of the genera *Cuon*, *Lycaon*, and *Xenocyon;* whether a relationship existed between the New and Old World forms is uncertain. It seems, however, that these dogs are modified Caninae, not Simocyoninae.

The North American species, *Protocyon texanus*, is known only from the early Irvingtonian of Rock Creek; the material consists of a mandible and a few limb bones. The mandible is short and deep, and a small M_3 is present [this tooth is usually absent in the contemporary (Uquian) species *Protocyon scagliarum* Kraglievich from South America].

Judging from its dental characters, *Protocyon* probably was highly predaceous, like the living bush dog, *Speothos venaticus* (Lund). The latter hunts in packs, is a good swimmer, and, despite its modest size, is able to bring down large prey. (Clutton-Brock, Corbet, and Hills, 1976; Kraglievich, 1952; Troxell, 1915a.)

†Progressive Gray Fox, *Urocyon progressus* Stevens, 1965

The gray fox genus *Urocyon* has been present in North America since Hemphillian times. Osteologically, these foxes differ from *Vulpes* in their relatively large molars, the wide separation of the lyreate temporal ridges on top of the skull, and the shape of the mandible, which has a characteristic ventral lobe in front of the angle. The characters strongly resemble those of the Old World raccoon dogs (genus *Nyctereutes*), which date back to the Ruscinian. [Although the phenetic classification suggested by Clutton-Brock, Corbet, and Hills (1976) tends to ally *Urocyon* with *Vulpes* and separate it from *Nyctereutes*, it should be noted that none of the dental and mandibular characters here considered significant was used.]

The species *Urocyon progressus* ranged through most of the Blancan (Rexroad, Red Light, and possibly Broadwater). The meager material available suggests that it was slightly larger than the modern gray fox and that the braincase was somewhat narrower. The species presumably was ancestral to *Urocyon cinereoargenteus* and may have been, like that species, an inhabitant of brushland, woodland, and forest. (Akersten, 1972; Clutton-Brock, Corbet, and Hills, 1976; Kurtén, 1968; Stevens, 1965.)

Gray Fox, *Urocyon cinereoargenteus* (Schreber), 1775 (*Canis cinereo argenteus*;
  U. atwaterensis Getz, 1960; *U. minicephalus* R. A. Martin, 1974)

The gray fox has been found at almost 40 localities, the oldest of which date back to the late Irvingtonian (Cumberland, Conard Fissure, Coleman IIA) or possibly the early Irvingtonian (Port Kennedy). A single molar from the late Blancan of Borchers was made the type of the species *Urocyon atwaterensis*, although it may be found to represent a transitional form, is here tentatively referred to the modern species. Cedazo, Haile VIIIA, and Reddick are Sangamonian records. In the Wisconsin, the gray fox was widely distributed in Arizona, Arkansas, California, Florida, Georgia, Kansas, Kentucky, Missouri, New Mexico, Tennessee, and Texas and reached its northern limit in Pennsylvania (Durham and Hartman's caves).

The late Irvingtonian gray fox from Coleman IIA had a distinctly smaller braincase, and hence more closely spaced temporal ridges, than the modern form. Taking these characters in conjunction with those of *Urocyon progressus*, which would appear to be the ancestral species, suggests that a gradual increase in brain size occurred. However, more material is needed to clarify the status of Blancan and Irvingtonian *Urocyon*.

The gray fox keeps to brushy or wooded areas and avoids open country. Omnivorous in habits, it is an excellent climber and seeks its food in trees and bushes as well as on the ground. Birds and small mammals are important food items. (Getz, 1960; R. A. Martin, 1974a; Mooser and Dalquest, 1975b.)

Island Gray Fox, *Urocyon littoralis* (Baird), 1858 (*Vulpes littoralis*)

This species, distributed in the Channel Islands, off southern California, differs from the mainland species in its smaller size, a trait often found in isolated island populations. It evidently reached the islands in the late Pleistocene. (Stock, 1943.)

Kit Fox, *Vulpes velox* (Say), 1823 (*Canis velox*)

In foxes of the genus *Vulpes*, the lower border of the mandible forms an even curve without the lobe seen in *Urocyon*; in addition, the temporal lines on the skull tend to be closer together and may unite to form a sagittal crest, the carnassial teeth are relatively larger, M_1 has a shorter talonid, and the postcarnassial teeth are smaller than the corresponding features of *Urocyon* (see fig. 11.10D,E). The Hemphillian *Canis davisi* is sometimes referred to *Vulpes* and is likely to be close to it in ancestry, but *Vulpes* is not known to be present in North America in the Blancan and early Irvingtonian. In the Old World, however, the genus is quite common from the middle Villafranchian on. This suggests that ancestral *Vulpes* migrated to the Old World in the Pliocene, became extinct in North America, and then remigrated to North America in the Pleistocene. The fact that the majority of *Vulpes* species belong to the Old World supports this argument.

The swift and kit foxes, *Vulpes macrotis* Merriam and *Vulpes velox*, are the only *Vulpes* species endemic to the New World. The differences between them may be subspecific only; in fossil material, they are not distinguishable. Fossil remains have been unearthed at upwards of 20 localities ranging in age from late Irvingtonian (Angus, Berends) and Sangamonian (Cragin Quarry) to Wisconsinan and later. Distribution in the Wisconsinan is northern and western, with records from Arizona, California, Colorado, Idaho, Kansas, Missouri, Nevada, New Mexico, Texas, and Wyoming.

Like the living form, the Pleistocene kit fox evidently was a denizen of the open country and desert, where it preyed upon small rodents, lizards, and insects. Its ancestry is unknown; phenetically, the species is very similar to the Old World species *Vulpes bengalensis* (Shaw) and *Vulpes corsac* (Linnaeus). (Clutton-Brock, Corbet, and Hills, 1976; Dalquest, Roth, and Judd, 1969.)

Red Fox, *Vulpes vulpes* (Linnaeus), 1758 [*Canis vulpes; V. palmaria* Hay, 1917; on the status of the American red fox, formerly called *Vulpes fulva* (Desmarest), 1820, see Churcher, 1959]

This species appears to be a late immigrant in North America, where it seems that no record antedates the Sangamonian, except possibly in the Far North. It has been recorded in Alaska from the Fairbanks District and at Medicine Hat in deposits of Sangamonian age; it also occurs at Old Crow.

The red fox is present at more than 25 sites of Wisconsinan age in Arkansas, California, Colorado, Idaho, Missouri, New Mexico, Tennessee, Texas, Virginia, and Wyoming. Very large samples have been found at Moonshiner and Little Box Elder caves.

Vulpes vulpes clearly originated in the Old World, where its predecessor in the Villafranchian appears to be *Vulpes alopecoides* Del Campana, a somewhat smaller animal. It is at present the most widely distributed carnivore in the world, which testifies to its extreme adaptability. It preys on a broad range of ground-living small mammals, birds, reptiles, and invertebrates and will also eat fish and amphibians.

The red fox is easily distinguished from *Vulpes velox* by its larger size.

Although the species ranged far south in the Wisconsinan, there is evidence suggesting that its range shrank toward the north with the onset of warm conditions and only recently has expanded again in response to ecological changes owing to the influence of man. (Anderson, 1968; Churcher, 1959b; Dalquest, Roth, and Judd, 1969; Guilday, 1971b; Harington, personal communication, 1976.)

Arctic Fox, *Alopex lagopus* (Linnaeus), 1758 (*Canis lagopus*)

The monotypic genus *Alopex* today inhabits the arctic regions of both the Old and the New World. The arctic fox, *Alopex lagopus*, has been found in Wisconsin-age deposits in the Yukon (Old Crow River). In Europe, the species was common in the late Pleistocene. The indication is that it originated in the Old World and is a relatively recent immigrant in the New, just as in the case of the red fox. Both species may in fact be descendants of the Villafranchian *Vulpes alopecoides*. In *Alopex*, however, the premolars are higher crowned, M_1 has a distinctly shorter talonid, and the tubercular teeth are more reduced than in *Vulpes*. The greater development of the shearing part of the dentition suggests greater dependence on animal food than in the omnivorous *Vulpes*. (Harington, personal communication, 1976.)

FAMILY PROCYONIDAE—RACCOONS

Procyonids are moderate-sized, primarily tree-living carnivores that probably share a common ancestry with the canids. Their fossil history is poorly known. The earliest identified procyonid is *Plesictis*, from early Oligocene deposits in North America and Europe. North American procyonids (subfamily Procyoninae) invaded South America in the Pliocene, and today the family reaches its greatest diversity in the Neotropics. The other subfamily, the Ailurinae, includes the living red panda (*Ailurus*) of eastern Asia and the extinct *Parailurus* of Europe and now known also from North America.

Members of the Procyonidae often have conspicuous facial markings and alternating dark and light bands on the tail. The teeth are adpated for an omnivore's diet; the premolars are not reduced, but the carnassials have lost their shearing function and are high-cusped crushing teeth. Procyonids are semiplantigrade and five-toed; the forepaw has considerable dexterity and is used in manipulating food.

SUBFAMILY PROCYONINAE

†Rexroad Raccoon, *Procyon rexroadensis* Hibbard, 1941

Raccoons, genus *Procyon*, are found from southern Canada to northern South America. One extinct and 7 extant species, 5 of them insular, are recognized. Their stratigraphic range extends back to the early Blancan.

A large raccoon, *Procyon rexroadensis* is the earliest representative of the genus. It was first described from the early Blancan Rexroad fauna. The only other record of a raccoon in the Blancan is *Procyon* sp. from Cita Canyon; although slightly smaller than the Rexroad specimens, it can probably be referred to *Procyon rexroadensis*.

The species is known only from jaw material and isolated teeth. The dentition shows a more generalized condition and is not as specialized as that of *Procyon cancrivorus*, the crab-eating raccoon of South America. The dental pattern resembles *Procyon lotor*, but the premolar–molar series is narrower. Lower-jaw measurements exceed those of the living raccoon.

Procyon rexroadensis was probably ancestral to *Procyon lotor*. Raccoons do not appear again in the Pleistocene record until the late Irvingtonian. (Hibbard, 1941b; Oelrich, 1953.)

Raccoon, *Procyon lotor* (Linnaeus), 1758 (*Ursus lotor; Procyon priscus* Leidy, 1856; *P. simus* Gidley, 1906; *P. nanus* Simpson, 1929)

The modern species first appears in late Irvingtonian deposits in Florida (Coleman IIA). By the Wisconsinan, raccoons were widespread, especially in the southern and eastern parts of the country, and their remains are common in caves and sinkholes. Today, the species occurs from Panama to Mexico, throughout most of the United States, and into southern Canada. Long winters and scarcity of water are factors limit-

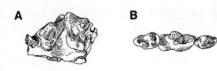
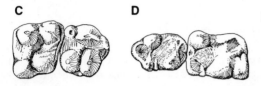

11.11 Procyonidae. *A, B, Bassariscus casei*, Blancan, Rexroad. *A*, left P³–M¹; *B*, right P₄–M₂, occlusal views. *C, D, Procyon lotor*, Recent. *C*, right P⁴–M¹; *D*, left M₁₋₂, occlusal views. *A, B*, after Hibbard (1952b), *C, D*, after Vaughan (1972). *A, B*, .66×; *C, D*, 1.5×.

ing their distribution. Raccoons are found in woody areas along streams and lakes, with highest densities reported in a mixed forest–wetlands environment.

Raccoons show considerable geographic variation in size; the largest animals are found in the colder, northern regions, and the smallest occur in the Southeast. Late Pleistocene—early Holocene specimens generally are larger than raccoons living in the corresponding areas at the present time. Raccoons from post-Wisconsinan sites in Texas were as large as those from northern California and have been referred to the extinct subspecies *Procyon lotor simus* Gidley, originally described from Cave Bear Cave in northern California. Based on slight differences in size or tooth morphology, several species of *Procyon* have been described from Pleistocene deposits; however, when large samples of Recent specimens are examined, these purported specific differences do not hold up.

A crushing type of dentition is characteristic of *Procyon;* the molars are broad, and the shearing blades of the carnassials are not well developed (see fig. 11.11). Raccoons are omnivorous and feed upon a wide variety of plant and animals; crayfish and corn are favorite foods. Since much of their food is obtained in the water, their habit of "washing" most of their food before eating it is related to the fact of "catching" it in the water, rather than being for the purpose of removing dirt. Raccoons have great forearm mobility and use their paws extensively in catching and manipulating food.

The nocturnal habits and catholic tastes of the species are factors in its successful adaptation to a changing environment. Raccoons are expanding their range, and Schneider (1973) believes that they are as numerous today as they were before the European settlement of North America. (Arata and Hutchison, 1964; Ewer, 1973; Gidley, 1906; Schneider, 1973; Simpson, 1929a; Wright and Lundelius, 1963.)

† Case's Ringtail, *Bassariscus casei* Hibbard, 1952

A slender, foxlike member of the raccoon family, the genus *Bassariscus* makes its appearance in Miocene faunas in Nebraska, Nevada, and California, and several early species are recognized. The Blancan species, *Bassariscus casei*, is probably directly ancestral to the living species, *Bassariscus astutus*. The teeth of ringtails have well-developed crushing surfaces and weak blades on the carnassials.

Known only from Rexroad, *Bassariscus casei* differs from *Bassariscus astutus* in its narrower lower premolars and molars, the more open valleys between the cusps, and the well-developed cingulum on P⁴. M¹ is triangular in shape, like that of *Bassariscus astutus*. (Hibbard, 1952b.)

†Sonoita Ringtail, *Bassariscus sonoitensis* Skinner, 1942

The Sonoita ringtail is known from two late Pleistocene faunas, Papago Springs and San Josecito. The main character separating it from *Bassariscus astutus* is the shape of M^1 which is rectangular in *Bassariscus sonoitensis* and triangular in *Bassariscus astutus*. In addition, the upper premolars are heavier in the extinct species. The Sonoita ringtail is known only from skull and postcranial material. At other sites, ringtail material (referred to *Bassariscus astutus*) consists almost entirely of lower jaws, so the material is not comparable. Skinner (1942) believed that the tooth characteristics of *Bassariscus sonoitensis* are intermediate between those of *Bassariscus astutus* and *Bassariscus sumichrasti* (Saussure), the Central American species. (Skinner, 1942.)

Ringtail, *Bassariscus astutus* (Lichtenstein), 1830 (*Bassaris astutus*)

This extant species is known from some 14 late Wisconsinan faunas in California, Nevada, New Mexico, and Texas. Today, it inhabits rocky cliffs and canyons from central Mexico and Baja California to southwestern Oregon and southwestern Wyoming.

Ringtails are extremely agile and well adapted to live in a rocky environment. They have short legs and five curved, semirectractile claws on each foot. Trapp (1972) has shown that the hind foot can rotate at least 180°, permitting rapid head-first descent of trees, cliffs, and rocky ledges. Ringtails are excellent climbers and, using their long bushy tails for balance, can leap from point to point. Omnivorous in habits, ringtails feed on rodents, cottontails, birds, insects, fruit, nuts, and carrion. They are usually found near a source of water.

The ancestry of *Bassariscus astutus* can be traced back to the Blancan species, *Bassariscus casei*. Closely related to the ringtail is the cacomistle, *Bassariscus* (*Jentinkia*) *sumichrasti*, from southern Mexico and Central America, an animal larger and more arboreal than the ringtail. (Trapp, 1972; Walker, 1975.)

Subfamily Ailurinae

†English Panda, *Parailurus anglicus* (Dawkins), 1888 (*Ailurus anglicus*)

The genus *Parailurus* is closely related to the living red panda of western China, Burma, Sikkim, and Nepal (genus *Ailurus*), but differs from these animals in its somewhat larger size and more progressive molarization of the premolars. The affinities of the pandas are somewhat uncertain; the *Ailurus–Parailurus* group is usually assigned to the Procyonidae, whereas the giant panda (*Ailuropoda*) is usually regarded as an ursid.

The North American record of *Parailurus* is based on a single M^1 from the early Blancan Taunton fauna, and, because the tooth is comparable to that of the late Ruscinian European species, *Parailurus anglicus*, the specimen is tentatively referred to that species (Tedford and Gustafson, 1977). A somewhat larger form, *Parailurus hungaricus*, occurs in the early Villafranchian, but the distinction may be subspecific only. In any case, the known temporal range of *Parailurus* in Europe is relatively short (between approximately 4 and 3 m.y. B.P.) and the Taunton occurrence may well be within these limits. The Old World was presumably the center of panda evolution, and the Nearctic record must be explained by immigration from Asia.

The living red panda (*Ailurus fulgens*) is a relatively small (head and body length ca.

60 cm), very strikingly colored animal with long, soft fur and a bushy tail. It dwells in alpine forests at altitudes of 2,000–4,000 m and is mainly arboreal in habits. Red pandas are omnivorous, and their diet includes bamboo shoots, leaves, fruits, fungi, grass, small rodents, and insects. (Dawkins, 1888; Tedford and Gustafson, 1977; Walker, 1975.)

FAMILY URSIDAE—BEARS

The bear family is characterized by a trend toward large size and, with some secondary exceptions, omnivory or even herbivory. The molars are enlarged and tubercular, and the premolars tend to reduction. The gait is plantigrade, and the feet are five-toed. Although some extinct carnivores (*Amphicyon,* for instance) are now often regarded as ursids, the living bears are probably derived from the Holarctic Miocene genus *Ursavus.* The Tertiary history of the bears is incompletely known, but the Pleistocene record is good, and in some cases excellent.

Two subfamilies are represented in the North American Pleistocene. The endemic New World Tremarctinae, surving in the present-day Andean bear, comprises three species of the genera *Tremarctos* and *Arctodus* (with additional species in South America). A good osteological character is the presence of a double masseteric fossa in the mandible. The presence in M_1 of an accessory external cusp at the junction of trigonid and talonid is also diagnostic.

The Ursinae comprise the living North American bears and their Pleistocene forerunners. The origin of this subfamily is in the Old World, where it has a long history. These bears have a single masseteric fossa in the mandible. The skull is longer and narrower and (with the exception of *Ursus maritimus*) the molars are more elongate than in the Tremarctinae. Although black bears immigrated to North America as early as the Blancan, arrival of the other ursine bears was comparatively late.

Dental and osteological characters give good taxonomic criteria, although sexual dimorphism and great evolutionary plasticity in size may complicate the picture.

SUBFAMILY TREMARCTINAE

†Florida Cave Bear, *Tremarctos floridanus* (Gidley), 1928 (*Arctodus floridanus; T. mexicanus* Stock, 1950)

The genus *Tremarctos* comprises relatively small-toothed bears, usually displaying the full complement of anterior premolars (as in other tremarctines) and a well-developed double masseteric fossa. The main cusps of M_1 form a characteristic alternating, elongate W-pattern.

Tremarctos floridanus is known from the late Blancan/Irvingtonian (Grand View, Vallecito) and may have been in existence in the early Blancan, for a humerus from Hagerman is indistinguishable from Rancholabrean material of the species. Numerous remains of the species have been found in Rancholabrean cave, sinkhole, and river deposits in Florida, and the species also occurred along the Gulf Coast (Ingleside) to New Mexico (Burnet) and Mexico (San Josecito). In the Rancholabrean, its range extended north into Tennessee (Saltpeter) and Georgia (Ladds). Thus its late Pleistocene range seems to have been restricted to the southern part of the continent. In Florida, the species survived into the Holocene; skeletons from the Devil's Den sinkhole may be

11.12 Ursidae. Distribution of *Tremarctos floridanus*. ■ = Blancan; ▲ = Irvingtonian; ● = Rancholabrean.

only 8,000 years old. Blancan and Pleistocene occurrences are shown in figure 11.12.

Tremarctos floridanus is distinguished from the living South American *Tremarctos ornatus* Cuvier by its much larger size and heavier proportions, a tendency toward reduction of the premolars and elongation of the hind molars, doming of the forehead, and elongation of the neck, a barrellike body, heavy limbs with a long humerus and femur, and shortened paws. Sexual dimorphism is pronounced. In these characters, the species is similar to the European cave bear, *Ursus spelaeus* Rosenmüller and Heinroth.

The ancestry of the species is unknown but may well lie within *Plionarctos,* a poorly known tremarctine genus from the Hemphillian.

The peculiar characters of this species, like those of the convergent European cave bear, point to an almost exclusively herbivorous existence. This is suggested by the increase in the occluding surfaces of the grinding molars and the orientation of the musculature acting upon them (leading to the development of a domed forehead) and the heavy construction of the body and limbs. However, although many *Tremarctos floridanus* fossils come from caves or sinkholes, there is no instance of mass occurrences comparable to those in European bear caves.

The Florida cave bear is often found associated with black bears (*Ursus americanus*) and so presumably differed from that species ecologically, perhaps in its more herbivorous mode of life. The cause of its extinction is unknown. (Bjork, 1970; Guilday and Irving, 1967; Kurtén, 1966; Stock, 1950.)

†Lesser Short-faced Bear, *Arctodus pristinus* Leidy, 1854 (*Ursus haplodon* Cope, 1895)

The genus *Arctodus* is distinguished from *Tremarctos* by its larger and higher-crowned teeth. The carnassials are relatively larger, the molars less elongate but very broad (especially in some South American species), and the canines are big and powerful.

Arctodus pristinus is only known from a few localities in eastern North America: Ashley River, Coleman IIA, Cumberland, Port Kennedy, Sebastian Canal, Waccasassa, and also Grapevine Cave, West Virginia (fig. 11.13). Its known history thus ranges from the early Kansan to the Wisconsinan, when it seems to have been a relict confined to Florida.

This is a primitive species of the genus; the face is only moderately shortened, and the teeth are smaller and narrower than those of *Arctodus simus*. The sparse remains of limb bones also suggest that the species had a smaller stature and less specialized characters than *Arctodus simus*. It is possible that *Arctodus pristinus* represents a relict form, close to the origin of the genus and confined to the eastern United States. (Guilday, 1971b; Kurtén, 1967a; Webb, 1974a.)

†Giant Short-faced Bear, *Arctodus simus* (Cope), 1879 (*Arctotherium simum; A. californicum* Merriam, 1911; *A. yukonense* Lambe, 1911; *Dinarctotherium merriami* Barbour, 1916)

A widespread species with a long evolutionary history, *Arctodus simus* is distinguished from *Arctodus pristinus* by its larger size, bigger, broader, and more crowded teeth, more shortened face, and relatively longer limbs. Irvingtonian sites include Rock Creek, Irvington, Gordon, and Hay Springs (see fig. 11.13); the species' recorded history thus appears to begin in the Aftonian. This bear is also known from more than 30 sites of Rancholabrean age in Alabama, Alaska, Arizona, California, Idaho, Iowa, Kansas, Mexico (Cedazo, Tequixquiac), Missouri, Montana, Nebraska, Oregon, Pennsylvania, Saskatchewan, Texas, Wyoming, and the Yukon. The Southeast, however, is devoid of finds, and the absence of this species in Florida is particularly striking since the Florida Pleistocene is rich in bears of other species.

The light, unusually long-legged and short-bodied build of this species is highly distinctive and suggests a fleetness of foot unusual in a bear. The fore- and hind feet are turned forward, in contrast to the toe-in position seen in most bears. The skull is notable for its short face and broad muzzle, which give it some semblance to that of the great cats (see fig. 11.14). Males, on the average, are 15 percent larger than females in linear dimensions. All characters indicate a highly predaceous mode of life, and *Arctodus simus* was the most powerful predator of the Pleistocene fauna of North America.

11.13 Ursidae. Distribution of *Arctodus simus* (△ = Irvingtonian; ○ = Rancholabrean) and *Arctodus pristinus* (▲ = Irvingtonian; ● = Rancholabrean).

Finds in caves are not uncommon. It may be noted that in Potter Creek Cave all of the individuals found were female (probably 8 individuals), which is suggestive of habitat selection. In other caves, both sexes have been found.

The ancestry of *Arctodus simus* is unknown, but the related *Arctodus pristinus* may possibly be ancestral. The size of the animals fluctuated, perhaps in response to Bergmann's Rule; the Aftonian (Rock Creek) and Rancholabrean bears are somewhat smaller than those of the Irvingtonian, south of the ice. In Alaska and the Yukon, gigantic forms

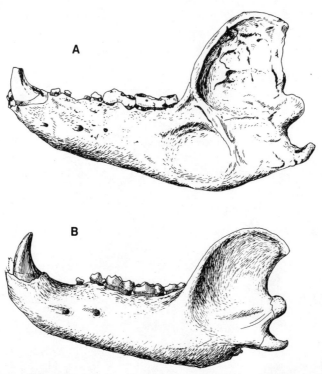

11.14 Ursidae. Left mandibles, external view. *A, Arctodus simus,* Irvingtonian, Hay Springs. *B, Ursus americanus,* Irvingtonian, Cumberland Cave. *A,* after Kurtén (1967a); *B,* after Gidley and Gazin (1938). *A,* 0.25×; *B,* 0.4×.

persisted in the late Pleistocene. The species itself persisted to the late Wisonsinan. It is possible that competition from invading brown and grizzly bears (*Ursus arctos*) may have played a role in the extinction of the species; however, the two species have been found in direct association only at one site south of the ice (Little Box Elder Cave). The two species coexisted in the Alaska–Yukon refugium during the Rancholabrean. A terminal date (Lubbock Lake) is 12,650 ± 350 years B.P.; the Little Box Elder Cave record is probably of about the same age. (Harington and Clulow, 1973; Kurtén, 1967a; Kurtén and Anderson, 1974; Merriam and Stock, 1925.)

Subfamily Ursinae

†Primitive Black Bear, *Ursus abstrusus* Bjork, 1970

The genus *Ursus,* which arose in the Old World in the early Pliocene, comprises four living and several extinct species. Bears of this genus have an elongate, narrow skull, no pre-masseteric fossa, reduced or absent anterior premolars, and elongate posterior molars.

This small bear is found in the early (Hagerman, White Bluffs) and late Blancan (Cita Canyon). It is incompletely known, but appears to be closely related to, and perhaps conspecific with, the small, ancestral black bear of the Old World, *Ursus minimus* Devèze and Bouillet, from the Ruscinian and earliest Villafranchian. This suggests that there existed in Blancan times a Holarctic population of primitive black bears which gradually gave rise to the Pleistocene and living American and Asiatic black bears. The Blancan form is distinguished from *Ursus americanus* by its smaller size. (Bjork, 1970; Johnston and Savage, 1955.)

Black Bear, *Ursus americanus* Pallas, 1780 (*U. amplidens* Leidy, 1853; *U. vitabilis* Gidley, 1913; *Euarctos optimus* J. R. Schultz, 1938)

Fossil finds of the black bear range stratigraphically from the early Irvingtonian (Port Kennedy) to the Recent. Late Irvingtonian records include Conard Fissure, Cumberland, and Trout. There are over 85 Rancholabrean sites scattered over most of the continent: Alaska, Arizona, Arkansas, California, Florida, Georgia, Idaho, Illinois, Indiana, Iowa, Kentucky, Maryland, Mexico (San Josecito), Mississippi, Missouri, New Mexico, Northwest Territories, Ohio, Oklahoma, Ontario, Pennsylvania, South Carolina, Tennessee, Texas, Virginia, and West Virginia. *Ursus americanus* is by far the most commonly found ursid in the late Pleistocene of North America.

A surviving species, the black bear still has a wide range, with strongholds in more sparsely populated areas. It differs from *Ursus arctos* (grizzly and brown bears) in its smaller size. The length of the hindmost upper molar M^2 is an especially useful character, although it does not discriminate absolutely; in the largest known fossil black bear, the M^2 is 34 mm long (30 mm in living specimens), and very few grizzlies have such a short M^2 (see fig. 11.15). The anterior premolars are normally less reduced than those of *Ursus arctos*, and the interorbital width of the skull tends to be greater than that of comparably sized specimens of *Ursus arctos*. The mandible is illustrated in figure 11.14.

The closest living relative of the American black bear is the Old World species *Ursus thibetanus* Cuvier. Both species are probably derived from an earlier Holarctic population represented in North America by *Ursus abstrusus*. The early form found in Port Kennedy is quite small and primitive. The late Irvingtonian form, although larger, retains some primitive characters (relatively large carnassial and small hind molars).

The size increase continued in the Rancholabrean. Wisconsinan black bears were in many cases as large as present-day grizzlies and have, in fact, been mistaken for such, with the result that there are erroneous reports of Wisconsinan *Ursus arctos* south of the

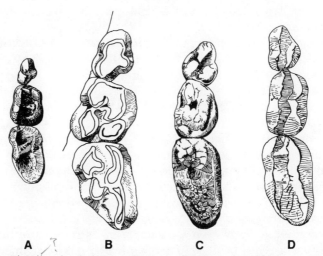

11.15 Ursidae. Right P^4–M^4, occlusal views. *A, Tremarctos floridanus*, Rancholabrean, Melbourne. *B, Arctodus simus*, Irvingtonian, Hay Springs. *C, Ursus arctos*, Rancholabrean, Welsh Cave. *D, Ursus americanus*, Irvingtonian, Cumberland Cave. *A*, after Kurtén (1966); *B*, after Kurtén (1967a); *C*, after Guilday (1968a); *D*, after Gidley and Gazin (1938). *A–C*, 0.66×. *D*, 1×.

ice (Churcher and Morgan, 1976; Stovall and Johnston, 1935). In the Holocene, there was a marked decrease in size in this species, a phenomenon also seen in many other species of large mammals. The present-day black bears of Florida, which still retain something of the giant stature of their Pleistocene ancestors, are an exception.

In feeding habits, the black bear is an omnivore, and the Pleistocene history of the species shows a gradual reduction of the shearing anterior cheek teeth and an increase in the grinding surface of the molars.

Like its Asiatic sibling, the American black bear seems not averse to denning in caves, and both species are commonly found as fossils in these locations. The most remarkable mass occurrence of *Ursus americanus* in the fossil state is in Cumberland Cave, where remains of more than 50 individuals have been found. This cave was probably a trap site, however. Other fairly large samples come from Reddick, Samwel, and Potter Creek. The species is relatively rare in asphalt deposits. (Churcher and Morgan, 1976; Gidley and Gazin, 1938; Harlow, 1962; Kurtén, 1963a, 1966, and unpublished data; J. R. Schultz, 1938; Stovall and Johnston, 1935.)

Brown and Grizzly Bears, *Ursus arctos* Linnaeus, 1758 (*U. horribilis* Ord, 1815; *U. procerus* G. S. Miller, 1899)

Some 232 Recent and 39 fossil "species" and "subspecies" (list in Erdbrink, 1953) have been proposed for this taxon—a waste of systematic effort which, as far as we know, is unparalleled.

In the Old World, the history of this species extends back to the middle Pleistocene; the ancestral form is the Villafranchian *Ursus etruscus* Cuvier. Alaskan finds date well back into the Wisconsinan, and perhaps earlier, and the species occurs together with *Arctodus simus* in this area. South of the Wisconsinan ice, there is no reliable record of the species before the last phase of the glaciation, when *Ursus arctos* migrated into the area, extending its range far beyond that known in historical times. Localities include Pit 10, Rancho La Brea (early Holocene?), Polecat Creek and Moonshiner (early Holocene), Overpeck, Butler Co., Ohio, Little Box Elder and Organ-Hedricks caves (late Wisconsinan), Welsh Cave (12,950 ± 550 B.P.), Lake Simcoe, Ontario (11,700 ± 250 B.P.), Jaguar Cave (10,370 ± 350 B.P. and possibly also late Wisconsinan), and Schulze Cave (9,680 ± 700 B.P.).

Ursus arctos is larger, on the average, than *Ursus americanus* and has longer and more slender metapodials. The teeth have a more complicated occlusal pattern, and the posterior molars are more elongate. The anterior premolars are much reduced, and often some, especially P2, are missing (see fig. 11.16). There is much local and temporal variation, especially in size. Local differentiation in the New World to some extent reflects that in the source area in northeastern Asia. The large, broad-skulled brown bears of Kodiak and Afognak islands, Alaska (*U. arctos middendorffi* Merriam), resemble the broad-skulled bears of Kamchatka. The large, narrow-skulled brown bears occurring from the Alaskan Peninsula along the coast into British Columbia (*U. arctos dalli* Merriam) and the smaller, narrow-skulled grizzly bears in the remaining part of the range (*U. arctos horribilis*) show closer affinities with the main Siberian population of narrow-skulled *Ursus arctos*.

Ursus arctos succeeded *Arctodus simus* over much of its range and may have been in competition with that species, being of a more carnivorous bent than the black bear. *Ursus arctos* only rarely dens in caves, and fossil finds are rare. (Erdbrink, 1953; Guilday, 1968a; Guilday, Hamilton, and McCrady, 1971; Kurtén, 1960, 1968, 1973d, and unpublished data; G. S. Miller, 1899.)

ORDER CARNIVORA

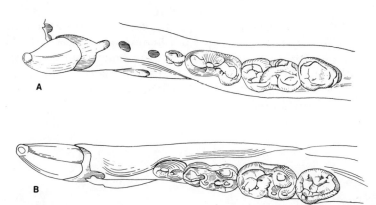

11.16 Ursidae. Mandibular dentitions, occlusal view. *A, Arctodus simus,* Rancholabrean, composite from Ester Creek and Engineer Creek, Alaska, left mandible. *B, Ursus arctos,* Rancholabrean, Goldstream, Alaska, right mandible. Both 0.5×.

Polar Bear, *Ursus maritimus* Phipps, 1774

Fossils of the polar bear are extremely rare. Fragmentary remains have been found in a lava cave in Bogoslof Hill, St. Paul, Pribilof Islands, Alaska; although of uncertain age, they are probably very recent. In the Old World, the fossil history of the species extends back to the beginning of the Weichselian.

The polar bear, a semiaquatic northern Arctic mammal, is the only obligate carnivore among living bears. Its teeth show similarities to those of *Ursus arctos,* but the cusps are higher and more trenchant, and the molars are reduced in size. The skull is long and narrow. The species may have evolved from mid-Pleistocene *Ursus arctos.* (Kurtén, 1964; Ray, 1971.)

Family Felidae—Cats

Cats appeared in North America and Eurasia at about the Eocene–Oligocene boundary. They have a fairly good fossil history, at least as far as the larger species are concerned; the fossil record of the smaller felids is very spotty. The family early became specialized for a predatory mode of life; the canine and carnassial teeth are much enlarged, whereas the postcarnassial teeth have tended to become reduced and disappear.

There are two distinct adaptive types, which have been classified in two subfamilies: the Machairodontinae and the Felinae. In the Machairodontinae, the upper canines are enlarged and flattened, and lower canines reduced. In the Felinae, or "true" cats, the canines are conical and subequal in size. The adaptive variety within both groups is great, and intermediate forms are also known.

Pleistocene American felids range in form from small animals to great creatures surpassing the living tigers and lions in size. Among the extinct Machairodontinae, three groups may be distinguished. The dirktooths and sabertooths (tribe Smilodontini), characterized by extremely elongate upper canines, died out in Eurasia in the middle Pleistocene but persisted in great numbers in the Americas up to the end of the Pleistocene. The scimitar-tooths (tribe Homotheriini), with shorter and more flattened sabers, survived to the late Pleistocene in Europe as well as in North America. The third group, the false sabertooths (*Dinofelis*), departed less from the structural type of the true cats; these animals became extinct in both hemispheres at an early date.

The conical-toothed cats (subfamily Felinae) of the American Pleistocene include ancestors and relatives of the living great and small cats such as pumas, jaguars, and lynxes. A notable Pleistocene immigrant is the lion, which has the distinction of having been the most widespread land-mammal species of all times before the advent of man.

Subfamily Machairodontinae

†Western Dirktooth, *Megantereon hesperus* (Gazin), 1933
 (*Machairodus hesperus*)

The dirktooths, genus *Megantereon*, are evidently ancestral to the Pleistocene *Smilodon* and display similar characters. The sabers are very long and only slightly recurved, and the incisors form a transverse row. The neck is long and powerful, the limbs are short and very sturdy, and the tail is short. The genus differs from *Smilodon* in its smaller size and in the presence of a functional P_3, less robust sabers (which lack serrated edges), and a fully developed saber sheath, or dependent mandibular flange.

This group probably originated in the Old World, and *Megantereon orientalis* (Kittl) from the late Miocene is its earliest representative there.

Records of western dirktooth span the entire Blancan (Hagerman, Broadwater, Rexroad (?), Santa Fe IB, Haile XVA) and indicate a wide distribution. The species is closely related to the Eurasian *Megantereon megantereon* (Croizet and Jobert), and future study may indicate specific identity. A related form, "*Smilodontidion*" *riggii* Kraglievich, existed in South America.

The dirktooth, which was about the size of a puma, probably preyed on relatively slow-moving, thick-skinned animals, which were stabbed with the upper canines, the lower jaw being swung back in an arc of some 95° to free the points of the teeth. It can be shown that the resulting stab will not penetrate very deep but will run parallel to the surface of the victim's body, causing extensive damage to the subcutaneous structures. (Gazin, 1933b; Kurtén, 1963b, 1968; Schultz and Martin, 1970a.)

†Gracile Sabertooth, *Smilodon gracilis* Cope, 1880 (*Uncia mercerii* Cope, 1895)

This incompletely known species from the early Irvingtonian may be a connecting link between *Megantereon* and the later *Smilodon* since it has characters of both. The type locality is Port Kennedy, but fragmentary finds of a similar nature have been unearthed at Inglis IA, Vallecito, and Valsequillo; the South American "*Smilodontidion*" may also be a member of this group. The dependent flange of the mandible is more reduced than that of *Megantereon*, but the canine teeth, like those of the dirktooth, are not serrated. In overall size, the species was intermediate between *Megantereon hesperus* and *Smilodon fatalis*. (Cope, 1899; Kurtén, 1963b.)

†Sabertooth, *Smilodon fatalis* (Leidy), 1868 (*Felis* (*Trucifelis*) *fatalis*;
 Machaerodus [sic] *floridanus* Leidy, 1889; *S. californicus* Bovard, 1907;
 Smilodontopsis troglodytes Brown, 1908; *S. conardi* Brown, 1908; *Smilodon
 nebraskensis* Matthew, 1918; *S. trinitiensis* Slaughter, 1960)

The sabertooth resembled its ancestor the dirktooth but was larger, attaining a size equal to that of a lion. P_3 is reduced or absent. The sabers are enormously developed, and the mandibular flanges are reduced so that the sabers protrude well beneath the

ORDER CARNIVORA

11.17 Felidae. Distribution of *Smilodon fatalis* (△ = Irvingtonian; ○ = Rancholabrean) and *Homotherium serum* (▲ = Irvingtonian; ● = Rancholabrean).

chin when the jaws are closed. The edges of the sabers have a fine serration not seen in *Megantereon*.

Early records of this species date from the later Irvingtonian (Conard Fissure, Gordon, Hay Springs, Irvington, Rushville; see fig. 11.17). The species is known from more than 40 Rancholabrean sites distributed over Alberta, Arkansas, California, Florida, Idaho, Louisiana, Mexico, Nebraska, New Mexico, Oregon, Tennessee, Texas, and Utah; it also ranged into western South America (Talara, Peru). The famous collection from Rancho La Brea consists of thousands of individuals; only the dire wolf is present in greater numbers. It follows that this is one of the best-known Pleistocene mammals. The taxonomy has been much discussed; we consider the numerous North American taxa of sabertooth (with the exception of *S. gracilis*) to have, at most, subspecific standing. In Brazil and Argentina, a closely related but apparently distinct species, *Smilodon populator* Lund (*S. neogaeus, nomen nudum*), has been found; this species was still larger and had much shortened distal limb segments.

The habits of the species have been variously interpreted, but the consensus seems to be that *Smilodon fatalis* preyed on large, slow-footed animals which were stabbed in vital areas (the neck or belly) to produce heavy bleeding and death (see fig. 11.18). The adaptation is similar to that of the dirktooth. However, the canines were also used against other carnivores and apparently also in intraspecific fights. A skull of a dire wolf penetrated by a broken *Smilodon* saber has been found, and a lesion found in the frontal bone of a *Smilodon* skull seems to have been caused by the saber of another individual. To such a predator, animals caught in the tar pits would be a powerful lure. In many instances, *Smilodon* remains from the tar show traumatic changes (broken teeth and bones, skeletal disease), which suggests that maimed individuals tended to haunt the area. Study of the age grouping in the material indicates that animals of all ages were present, including juveniles with milk teeth. On the other hand, cave finds of *Smilodon* are comparatively rare.

Popular literature sometimes features the sabertooth as an inadaptive form in which the sabers became too long to be useful to their possessors. As has been shown by Simpson (1941a), this idea is completely unfounded.

Cuts and other markings on *Smilodon* bones from Rancho La Brea may be artifactual, indicating coexistence of sabertooth and early man. A late date for *Smilodon* is 9,410 ± 155 years B.P., obtained from collagen in bone fragments from the First American Bank Site, Nashville, Tennessee. The sabertooth is also present at Devil's Den, with a suggested date of about 8,000 B.P. The cause of the extinction was probably the disappearance of important prey species. (Guilday, 1977; Merriam and Stock, 1932; G. J. Miller, 1968, 1969a,b; Simpson, 1941a; Slaughter, 1960; Webb, 1974b.)

† Idaho Sabertooth, *Ischyrosmilus ischyrus* (Merriam), 1905 (*Machaerodus ischyrus; Ischyrosmilus idahoensis* Merriam, 1918; *I. johnstoni* Mawby, 1965; *I. crusafonti* Schultz and Martin, 1970)

The genus *Ischyrosmilus* is incompletely known, and the number of valid species is difficult to ascertain. The characters employed for specific differentiation are overall size and degree of reduction of the anterior premolars, and both characters are subject to considerable temporal, individual, and sexual variation, even within a single evolving lineage.

The type specimen, a small mandible, comes from Asphalto and is probably early Blancan. *Ischyrosmilus idahoensis,* described on the basis of a very large mandible, is from the "Idaho Formation" and may be a member of the Grand View fauna (late Blancan). *Ischyrosmilus johnstoni* was described from Cita Canyon, and the material indicates considerable variation in size, although not to the extent seen in the Asphalto and Grand View specimens. *Ischyrosmilus crusafonti* from Broadwater matches the smallest Cita Canyon specimens in size; a larger individual is known from Lisco. Other records come from Sand Draw and from Channing, Texas. These specimens may represent a single lineage of a sexually dimorphic form with a tendency to size increase and gradual reduction of anterior premolars, but a definitive taxonomic evaluation may have to await the discovery of additional material. An earlier species is present in the early Pliocene Ricardo fauna of Nevada.

ORDER CARNIVORA

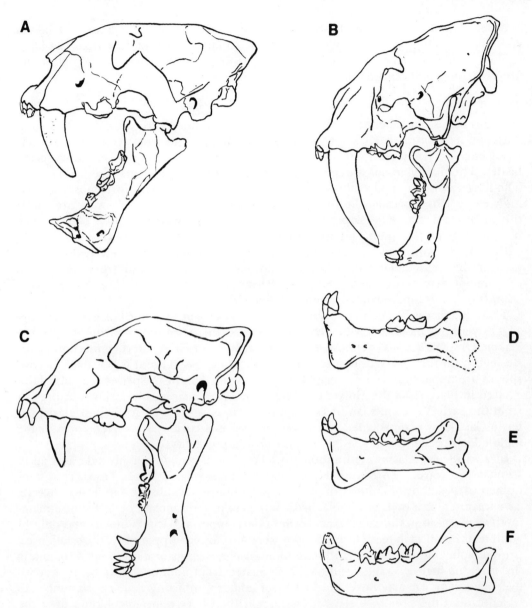

11.18 Felidae. Machairodontine skulls and jaws. *A, Ischyrosmilus* sp., Blancan, Cita Canyon. *B, Smilodon fatalis*, Rancholabrean, Rancho La Brea. *C, Homotherium serum*, Rancholabrean, Friesenhahn Cave. *D, Smilodon gracilis*, Irvingtonian, Port Kennedy Cave. *E, Megantereon hesperus*, Blancan, composite restoration. *F, Dinofelis paleoonca*, Blancan, Blanco. *A*, after Mawby (1965); *B*, after Merriam and Stock (1932); *C*, after Meade (1961). Not to scale.

These sabertooths are characterized by large, strongly curved sabers (upper canines), exceeded among Pleistocene cats only by those of *Smilodon* and much greater than those of *Homotherium;* the sabers are coarsely serrated along the edges. The mandible has a large flange serving as a protection for the sabers. The carnassials are long and slender, with serrated edges, the anterior premolars are much reduced, and the incisors form a broad curve. The limb bones are long and slender. The genus may be derived from the earlier sabertooth genus *Machairodus*, of Holarctic distribution in the late Miocene. It may also be related to *Homotherium*.

A very large, powerful predator, *Ischyrosmilus* surpassed all other Blancan felids in size. With its long, slender limbs, it clearly was much fleeter of foot than the large smilodont sabertooths of the later Pleistocene. (Kurtén, 1976; Mawby, 1965; Merriam, 1918a; Schultz and Martin, 1970a.)

†Scimitar Cat, *Homotherium serum* (Cope), 1893 (*Dinobastis serus*)

In this genus, the sabers are relatively short, very flattened and recurved (like a scimitar), and have razor-sharp, serrated edges fore and aft. The other teeth also have serrated edges. The incisors are disposed in a curve. The carnassials are thin slicing blades. The mandible has a dependent flange protecting the sabers (see fig. 11.18C). The long head was apparently carried high on a long neck. The forelimbs are strikingly elongated, especially the forearm and hand, whereas the hind limbs, are short; this indicates a markedly sloping back, which, together with the poise of the head, must have given this animal an appearance very different from that of *Smilodon*.

Early finds, which may be representative of this species or a closely related form, occur in the Irvingtonian (Inglis IA, Gilliland, Irvington); and a skull from the Blancan of Delmont may be conspecific with the Palearctic Villafranchian *Homotherium crenatidens* (Fabrini) (Cain, personal communication, 1976).

The scimitar cat is much less common than *Smilodon* in the Rancholabrean (see fig. 11.17), yet finds indicate a wide geographic range (there are records from Alaska, the Yukon, Oregon, Kansas, Tennessee, Oklahoma, and Texas). A direct connection across Beringia with the Old World species *Homotherium latidens* (Owen) is possible; the two species are apparently closely related. However, the American species may also have evolved in place from the Holarctic (?)*Homotherium crenatidens*. The genus was present from the early Villafranchian in Eurasia; in Europe, it survived to the last glaciation, but its last known representative in eastern Asia is a curiously dwarfed form in the middle Pleistocene. Descent from *Machairodus* is a possibility.

In a cave near Gassaway, Cannon Co., Tennessee, remains of juvenile and adult scimitar cats have been found. Finds made at Friesenhahn Cave, Texas, consist of numerous skeletons of animals of all ages, including small cubs; this indicates that the cave was used as a den. Strikingly, great numbers of milk molars of Jefferson's mammoth were found in the same strata [Evans (1961) reports 441 milk molars and only 14 adult teeth], and milk teeth of mastodont were also found. Apparently, the scimitar cat preyed mainly on mammoth (and mastodont) calves (see restoration in fig. 4.9), and in this case, the high leverage given by the elongated front limb may have been of adaptive value. The cat may have attacked from ambush, inflicting a quick, slashing stab which caused the prey to bleed profusely, and then have gone into hiding until the parents left the dead animal. The extinction of the scimitar cat is probably linked to that of the mammoth, its favorite prey.

The Friesenhahn occurrence dates from a late phase of the Wisconsinan. The fossils are associated with scraperlike flints which may or may not be artifacts. (Cain, personal communication, 1976; Churcher, 1966; Kurtén, 1972; Meade, 1961.)

†False Sabertooth, *Dinofelis paleoonca* (Meade), 1945 (*Panthera paleoonca*)

The genus *Dinofelis* comprises a number of Pliocene and Pleistocene species in the Old World. They are *Panthera*-sized cats with moderately enlarged upper canines. The genus became extinct in Eurasia at the end of the Villafranchian but survived into the mid-Pleistocene in Africa. *Dinofelis paleoonca* is the only known New World species.

At present, only a skull and some teeth from the type locality (Blanco) are definitely referable to this species, but some limb bones from Cita Canyon probably belong to the same species. About the size of a jaguar, the false sabertooth is distinguished by incipient sabertooth characters. The upper canines are moderately enlarged and flattened; the carnassials are elongated, comparatively slender, slicing structures; and P^2 is absent.

The New World species is considerably smaller than the gigantic *Dinofelis abeli* Zdansky from China, which is "Ruscinian" in date, and represents a comparatively advanced form. The European *Dinofelis diastemata* (Astre), of Ruscinian age, and the Indian *Dinofelis cristata* (Falconer and Cautley), from the Villafranchian Pinjaur Formation, appear to be somewhat closer to *Dinofelis paleoonca*. (Hemmer, 1973; Kurtén, 1973c; Meade, 1945.)

Subfamily Felinae

Lion, *Panthera leo* (Linnaeus), 1758 *(Felis leo; F. atrox* Leidy, 1853;
  F. imperialis Leidy, 1873)

The genus *Panthera* comprises the living great cats and their extinct relatives.

The American lion, *Panthera leo atrox,* is easily distinguished from other Nearctic Pleistocene felines by its great size and relatively long, slender limb bones. The lion ranges from the Sangamonian (American Falls, Cedazo, Cragin Quarry, Easley Ranch) to the late Wisconsinan in North America. It has been found at upwards of 40 localities in Alaska, Alberta, Arizona, California, Colorado, Florida, Idaho, Kansas, Mexico, Missouri, Nebraska, Nevada, New Mexico, Texas, Wyoming, and the Yukon, and also ranged south to Peru (Talara). Its absence in the rich Pleistocene faunas in the East and in peninsular Florida may be significant; the species probably preferred open country.

The American lion is characterized by its enormous size, rivaled only by the Eurasian cave lions *Panthera leo fossilis* (Reichenau), of the middle Pleistocene, and *Panthera leo spelaea* (Goldfuss), of the late Pleistocene. A Rancholabrean immigrant in the Nearctic, it may have descended from the mid-Pleistocene Asiatic form *Panthera leo youngi* (Pei). Late Pleistocene lions are known in Siberia and ranged east as far as the Kolyma River, and so it seems likely that a continuous population extended across Beringia during glacial phases. In Africa, the history of the lion goes back to Villafranchian times; Europe was invaded by the species in the Cromerian.

At the peak of its success, the lion ranged from Africa through Eurasia and North America into South America and appears to have been the most wide-ranging wild land-mammal species of all time.

The modern lion, unlike the other great cats, is gregarious and hunts in groups. Whether this was so in the case of the American form cannot be directly determined, but according to Hemmer (personal communication), the high degree of cephalization makes this probable. The American lion had a larger brain, relative to body size, than any of the Pleistocene or living lions of the Old World. It is found in large numbers only at trap sites; the minimum estimate for Rancho La Brea is 76 individuals.

No evolutionary change has been observed in the American lion during its span of existence; the earliest finds indicate animals of the same average size as those found later. In Europe, a gradual reduction in size occurred after the Cromerian.

Lion-hunting among Paleo-Indians is suggested by finds at Jaguar Cave, where lion remains occur among refuse radiocarbon-dated (on charcoal) at $10,370 \pm 350$ years B.P. A radiocarbon date on *Panthera leo* bone from Alaska is $22,680 \pm 300$ years B.P., which indicates that the lion was present in Beringia at the height of the Wisconsinan glaciation. It persisted in Beringia to the end of the Pleistocene, as indicated by a date of 10,370 ± 160 years B.P. from Lost Chicken Creek. Its extinction in the New World thus appears to have occurred about 10,000 years B.P. (Harington, 1971a and personal communication, 1971; Hemmer, 1974; Merriam and Stock, 1932; Whitmore and Foster, 1967; Vereshchagin, 1971.)

Jaguar, *Panthera onca* (Linnaeus), 1758 (*Felis onca; F. augustus* Leidy, 1872; *F. veronis* Hay, 1919)

The jaguar may be distinguished from the lion by its smaller size and shorter, stockier limb bones. It is present in the early Irvingtonian (Curtis Ranch, Port Kennedy), and later Irvingtonian sites include Coleman IIA, Conard Fissure, Cumberland, Delight, Gordon, Irvington, and probably the Rome Beds, in Oregon. There are over 30 Rancholabrean localities in Florida, Georgia, Kansas, Mexico, Missouri, Nebraska, Nevada, New Mexico, Oregon, South Carolina, Tennessee, and Texas; the species is also found in the Pleistocene of South America. It may be noted that pre-Wisconsinan jaguars ranged much farther north than those of the Wisconsinan; the range of the pre-Wisconsinan jaguars extended to Washington-Nebraska-Pennsylvania-Maryland, whereas that of the Wisconsinan jaguars extended only to Nevada-Kansas-Missouri-Tennessee. The present-day limit is still farther to the south, so the evidence indicates a gradual restriction of jaguar range in the Pleistocene and Holocene, even though this general trend probably was influenced by a sequence of glacial-interglacial shifts.

The earliest Irvingtonian jaguars may be conspecific with the contemporaneous (late Villafranchian to Cromerian) Palearctic jaguar *Panthera gombaszoegensis* (Kretzoi). The American form may thus be seen as a relict of what was at one time a Holarctic-Neotropical population. There is a progressive size reduction throughout this sequence in the Nearctic, and this tendency is particularly pronounced in the Holocene, where specimens from Devil's Den are intermediate in size between Wisconsinan and living jaguars. Also, there is a gradual shortening of the limbs, especially the metapodials, leading from a more generalized type (e.g., at Coleman IIA) to the characteristic jaguar form, adapted to a life in the forests, streams, and broken country.

The large Pleistocene jaguar of the Nearctic is usually called *Panthera onca augusta;* the type of the subspecies is late Irvingtonian. The Wisconsinan jaguars exceeded the living in size by 15 or 20 percent, and earlier jaguars are still larger.

The jaguar is conspicuously absent in the rich Californian faunas of the Rancholabrean, where the lion is common, yet was present in northern California in the Irvingtonian. On the other hand, the most abundant record of late Pleistocene jaguar comes from peninsular Florida and Texas and Tennessee, areas where the lion is scarce or absent. Only rarely are the two species found together (Cedazo, Cragin Quarry, San Josecito, Santa Fe River IIA).

The changes in the geographic distribution of jaguars during and after the Pleistocene may be due to environmental changes, changes in the animal's adaptation, or, most probably, both. (Guilday and McGinnis, 1972; Hemmer, 1971; Kurtén, 1965, 1973a; Simpson, 1941c.)

†Studer's Cheetah, *Acinonyx studeri* (Savage), 1960 (*Felis studeri; Uncia inexpectata* Cope, 1895?)

The genus *Acinonyx*, with the living species *Acinonyx jubatus* (Schreber) and the fossil *Acinonyx pardinensis* (Croizet and Jobert), was long thought to be endemic to the Old

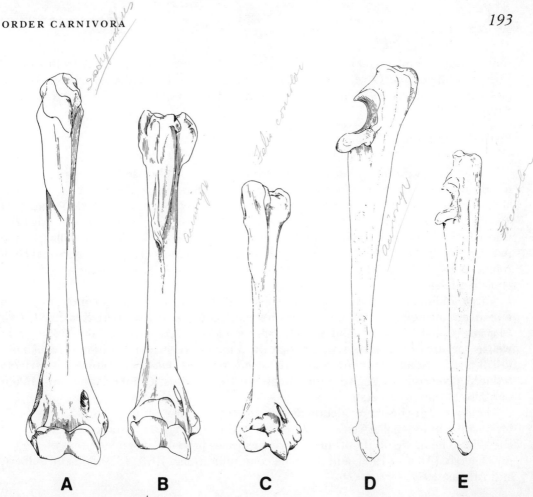

11.19 Felidae. *A–C*, right humerus, anterior view; *D–E*, left ulna, radial view. *A, Ischyrosmilus* sp., Blancan, Cita Canyon. *B, D, Acinonyx studeri*, Blancan, Cita Canyon. *C, E, Felix concolor*, Rancholabrean, San Josecito Cave. After Kurtén (1976). All 0.33×.

World. However, in recent work, still under way, Adams (personal communication, 1977) has found that at least two North American species should be referred to this genus. *Acinonyx* is distinguished from other cats mainly by characters related to a highly cursorial mode of predation, most clearly reflected in the extremely elongated and slim limb bones, light body, and small head (see fig. 11.19). Cranial and dental characters include a vaulted, broad-fronted skull, a larger outer chamber and a small inner chamber in the internal bulla, small canines, and a reduced protocone in P^4.

The American forms are referred to a distinct subgenus, *Miracinonyx*, which differs from the nominate subgenus in having a less inflated bulla and frontal sinus and fully retractile claws; in the Old World forms, the claws are not fully retractile (Adams, 1979, and personal communication, 1979). The characters of *Miracinonyx* could be regarded as primitive, and so would suggest that cheetahs originally arose in the New World. The apparently close relationship to pumas, an endemic New World group of cats, supports this conclusion. In the Old World, the first *Acinonyx* appear at the beginning of the Villafranchian. It is possible that a common ancestor lived in North America in Hemphillian times.

Acinonyx studeri is a large form, close to the Old World *Acinonyx pardinensis* in size. It is known from the Blancan (Cita Canyon) but may have survived in the very early Irvingtonian (Curtis Ranch, Gilliland, Port Kennedy). Late Irvingtonian finds from Conard Fissure, Cumberland Cave, and Mullen may be this species or the American cheetah (see below). (Adams, 1979 and personal communications, 1977, 1979; Kurtén, 1976; Savage, 1960.)

† American Cheetah, *Acinonyx trumani* (Orr), 1969 (*Felis trumani;*
  Felis longicrus Brown, 1908?; *Smilodontopsis mooreheadi* Hay, 1920?)

This species was slightly smaller than *Acinonyx studeri* and close to the living cheetah in size but shares the characters of the subgenus *Miracinonyx*. The type specimen, a skull from Crypt Cave, Pershing Co., Nevada, has been radiocarbon dated at 19,705 ± 650 years B.P. Extensive skeletal material of the same species has recently been unearthed in Natural Trap Cave in north-central Wyoming (Martin, Gilbert, and Adams, 1977) and is currently being studied by D. B. Adams (1979, and personal communications, 1977, 1979; Martin, Gilbert, and Adams, 1977); the material is of Rancholabrean age.

Three additional species, based on isolated P^4 with reduced protocone, have been tentatively included in the synonymies above: *Uncia inexpectata* (Port Kennedy), *Felis longicrus* (Conard Fissure), and *Smilodontopsis mooreheadi* (Cavetown). This character, as well as the size of the teeth, suggests that the animals are cheetahs; however, as Adams (1979 and personal communication, 1977) has pointed out, some pumas also have a reduced protocone in P^4. Additional material from Ladds, Cumberland, and Mullen may also pertain to cheetah.

The American cheetah evidently descended from *Acinonyx studeri*, from which it differs mainly in its smaller size. A similar size reduction may be observed in Old World cheetahs. The mode of life of the American species probably resembled that of the living cheetah. (Adams, 1979, and personal communications, 1977, 1979; Martin, Gilbert, and Adams, 1977; Orr, 1969.)

Puma, Cougar, or Mountain Lion, *Felis concolor* Linnaeus, 1758 (*F. hawveri*
  Stock, 1918; *F. daggetti* Merriam, 1918; *F. bituminosa* Merriam and
  Stock, 1932)

The genus *Felis sensu lato*, in the view of some authorities (e.g., Van Gelder, 1977), includes most of the living cats (including *Panthera* and *Lynx*). Other workers have had the genus comprise the "small cats," with the puma (subgenus *Puma*) as the largest member, and still others prefer to recognize a large number of genera.

Puma (*Felis concolor*) has been recorded from about 30 sites of Rancholabrean age in Arizona, California, Colorado, Florida, Idaho, Kansas, Kentucky, Mexico, Missouri, Nevada, New Mexico, Texas, and Wyoming. The earliest finds are probably from the Sangamonian (Cragin Quarry, Reddick); Irvingtonian records are uncertain. At the present day, the puma is one of the most wide-ranging carnivore species in the world, with a distribution extending from British Columbia to Patagonia. A large fossil sample comes from San Josecito. There are also numerous specimens from Rancho La Brea. Other records are based on one or a few specimens, most of them from caves and fissure fillings.

D. B. Adams (personal communication, 1979) has suggested that the puma and *Acinonyx* have a common origin, and *Acinonyx studeri* does have a number of pumalike characters (Savage, 1960). However, the evolution of the subgenus *Puma* is largely unknown. Fossil puma is present in South America, but its history there remains to be worked out.

The Rancholabrean pumas have canines smaller than those of Recent animals (a character reminiscent of *Acinonyx*) and are on average, larger than animals found in the same area today. In living pumas, average size is correlated with geographic latitude, size increasing to the north and south of the equator, and there is evidence of a similar cline in the Rancholabrean.

Although highly eurytopic, the puma prefers forest, especially in the mountains, and tallgrass prairie. Originally, its habits seem to have been mostly diurnal, another possible indication of a cheetahlike ancestor; civilization has tended to make it more nocturnal in habits. It is a good tree and broken-country climber and takes its prey (ranging in size from rodents to mule deer and wapiti) mainly by ambush.

Frequent cave-denning is indicated by the large number of finds in caves. At a few sites, the remains may be animals killed by man. (Adams, personal communication, 1979; Kurtén, 1973b, 1976; Simpson, 1941c; Savage, 1960; Van Gelder, 1977.)

† Lake Cat, *Felis lacustris* Gazin, 1933 (*F. rexroadensis* Stephens, 1959)

This species, well represented at Hagerman, the type locality (early Blancan), has also been found at Blanco, Rexroad, Curtis Ranch, and Cita Canyon and thus existed in the earliest Irvingtonian. A lynx-sized cat, it shows some resemblance to pumas as well as lynxes. The great reduction of P^2 (which is absent in lynxes) may be noted, as well as the relatively high, compressed shape of the premolars. The limb bones do not have the distal elongation typical of the lynxes, but the same is true of *Lynx issiodorensis*. On the other hand, a relationship to the pumas cannot be excluded. (Bjork, 1970; Gazin, 1933b; Stephens, 1959; Werdelin, personal communication, 1979.)

Ocelot, *Felis pardalis* Linnaeus, 1758

The species was recorded at Reddick and thus ranged to Florida in the Sangamonian. Its present range is mainly Neotropical but extends into Texas and Arizona. The ocelot is a medium-sized, forest-living cat. Of about the same size as *Lynx rufus*, it differs from the latter in its relatively larger, differently proportioned premolars. (Ray, Olsen, and Gut, 1963.)

† River Cat, *Felis amnicola* Gillette, 1976

A small cat in the jaguarundi–margay size range, this species has been reported at Aucilla River, Ichetucknee, Rock Springs, Merritt Island, Melbourne, and Waccasassa River, all in Florida, and, with less certainty, from Ladds in Georgia. All the localities are Rancholabrean in age. Much of the material had previously been referred to *Felis yagouaroundi*, which has a similar, rather deep and robust mandible, but the dentition differs from that of the jaguarundi in the proportionately taller lower canine, the great length of P_4, and the disposition of the principal cusps of the lower cheek teeth in a straight line. These characters are more like those found in the margay, which, although somewhat smaller, may be the closest living relative of the river cat. (Gillette, 1976; Ray, 1964b, 1967.)

Jaguarundi, *Felis yagouaroundi* Lacépède, 1808

Jaguarundi is found in the fossil state in the late Rancholabrean of Mexico (San Josecito Cave) and Texas (Schulze Cave). Its present range extends north into the southernmost part of Texas. Like the ocelot, it is a mainly Neotropical form but is somewhat less arboreal in habits. In size, it is intermediate between the bobcat and the margay. An early Irvingtonian form at Port Kennedy Cave (referred to *Felis eyra* by

Cope, 1899) may represent an earlier invasion of jaguarundi or a related form. (Cope, 1899; Dalquest, Roth, and Judd, 1969.)

Margay, *Felis wiedii* Schinz, 1821

A subfossil specimen of *Felis wiedii* has been reported from a shell midden in a tidal marsh by Sabine River, Orange Co., Texas; it is dated at $4,400 \pm 300$ years B.P. This record is about 600 km east of the present limit of the species, which barely extends into Texas, and so indicates a Holocene extension of the range along the Gulf Coast. Osteologically, the margay resembles the jaguarundi but is considerably smaller on the average; its range is mainly Neotropical. (Eddleman and Akersten, 1966.)

†Issoire Lynx, *Lynx issiodorensis* (Croizet and Jobert), 1828 *(Felis issiodorensis)*

The lynxes, often regarded as a subgenus of *Felis*, are characterized osteologically by the absence of P^2, a short tail, rather high compressed premolars, and, in more progressive species, a tendency to shortening of the face and body and lengthening of the limbs. Possible ancestral forms are known in the Hemphillian, the Ruscinian of Europe, and the Pliocene of Africa.

The Issoire lynx, well documented over the entire Villafranchian in the Palearctic, was a rather primitive lynx with a comparatively long face and body and short limbs (see fig. 11.20). The population was Holarctic in the early Irvingtonian, as shown by a record from Mullen I; the Nearctic form is classified as a distinct subspecies, *Lynx issiodorensis kurteni* Schultz and Martin. The species is probably ancestral to the living northern lynxes, the Nearctic *Lynx canadensis* and the Palearctic *Lynx lynx* (Linnaeus). In the Palearctic lynx lineage, a metaconid evolved on M_1 during the Villafranchian; the evolution of this character was evidently much slower in the Nearctic lineage, for it is lacking in the Irvingtonian and less expressed in modern *Lynx canadensis* than in *Lynx lynx*. (Kurtén, 1968; Savage, 1960; Schultz and Martin, 1972; Werdelin, personal communication, 1979.)

Canada Lynx, *Lynx canadensis* Kerr, 1792

The fossil history of this species is little known. It has been recorded from Sangamonian deposits at Medicine Hat and Silver Creek and from the Wisconsinan of Alaska, Idaho, Wyoming, and the Yukon.

The Canada lynx is closely related to the Palearctic *Lynx lynx*, and the two are sometimes regarded as conspecific. However, differences in size, osteology, and external characters, as well as the absence of transitional forms in the Beringia area, indicate that the species are distinct. *Lynx canadensis* probably evolved from the Nearctic *Lynx issiodorensis* population in a lineage which gradually diverged from that of *Lynx lynx;* parallel trends in the two lineages include size reduction, shortening of face and body, and lengthening of limbs, as well as the appearance of a small metaconid on M_1. However, size reduction has gone further in the Nearctic species, whereas the metaconid-talonid complex of M_1 is more advanced in the Palearctic.

The Canada lynx is a denizen of the northern forests. It is highly dependent on the snowshoe hare (*Lepus americanus*) as a food source; the populations of the two species fluctuate in close rapport. The species differs from *Lynx rufus* in its slightly larger size and in morphological details such as the position of the palatinal foramina and the more compressed P^3. (Kurtén, 1965; Merriam and Stock, 1932.)

11.20 Felidae. Restoration of *Lynx issiodorensis,* Irvingtonian, Mullen.

Bobcat, *Lynx rufus* (Schreber), 1777 (*Felis rufa; F. calcaratus* Cope, 1899)

In contrast to the lynx, the bobcat is one of the most commonly found mammals in the Pleistocene deposits. The oldest record, from Cita Canyon, is late Blancan, and there are several Irvingtonian finds (Coleman IIA, Conard Fissure, Port Kennedy, Punta Gorda, 111 Ranch in Arizona, and Trout Cave). Rancholabrean sites total over 60 and are distributed in Arizona, Arkansas, California, Colorado, Florida, Georgia, Idaho, Kentucky, Mexico, Missouri, Nevada, New Mexico, Pennsylvania, Tennessee, Texas, Virginia, West Virginia, and Wyoming.

Irvingtonian bobcats, which may be known under Cope's name, *Lynx rufus calcaratus,* are, in general, somewhat larger than Rancholabrean and Recent specimens; thus there seems to have been a gradual reduction in size. However, aberrant local trends may be noted; an example is provided by the very large size, comparable to that of *Lynx canadensis,* reached by *Lynx rufus koakudsi* (Kurtén), the Sangamonian subspecies of Florida, but this seems to have been a local phenomenon, suggesting some degree of isolation. In general, the bobcat is distinguished from the lynx by its somewhat smaller dimensions, as well as the broader P^3, larger paracone in P^4, and other morphological details (see fig. 11.21).

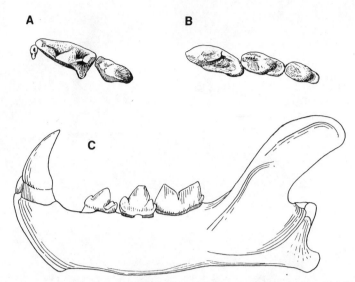

11.21 Felidae. *A, B, Lynx rufus,* Recent. *A,* right P^3–M^1; *B,* left P_3–M_1, occlusal views. *C, Felis concolor,* Rancholabrean, San Josecito Cave, left mandible, external view. *A, B,* after Vaughan (1972). *A, B,* 1.5×; *C,* 0.66×.

The bobcat is one of the most eurytopic carnivores and inhabits most kinds of environments from dense forest to desert, although it is absent at high altitudes and extreme latitudes. Food habits vary greatly; rodents often make up one-half or more of the total diet, and a good proportion of lagomorphs is added in many areas. Locally, even mule deer and, especially, white-tailed deer may be taken. (Kurtén, 1965; Merriam and Stock, 1932; Savage, 1960; Young, 1958.)

Family Hyaenidae—Hyenas

Hyenas are not now indigenous to America, and early descriptions of fossil Nearctic hyenas, made by Hay (1921) and Stirton and Christian (1940), were received with doubt. However, later discoveries have fully vindicated the conclusions formed by these authors. Finds are rare, in contrast to the situation in the Old World, where the family has a good fossil record, beginning in the Miocene.

Hyaenids arose from viverrid ancestors, probably in the early Miocene. Predominant trends in their evolution are adaptations to a predatory or scavenging modes of life. The postcarnassial teeth tend to reduction, and the carnassials form elongate, shearing blades. In the bone-eating hyaenids, exemplified by the living *Hyaena* and *Crocuta,* the premolars evolved into powerful, conical, bone-smashing structures. Many extinct hyaenids, however, had more slender, sharp-edged premolars, somewhat like those of the cats, and thus diverged less from their viverrid ancestors.

In the New World, the ecological role of the bone-eating hyenas was played by canids such as *Borophagus* and *Canis dirus.* The only hyaenid known to reach the Nearctic belonged to the slender-toothed group.

† American Hunting Hyena, *Chasmaporthetes ossifragus* Hay, 1921
  (*Ailuraena johnstoni* Stirton and Christian, 1940)

The genus *Chasmaporthetes* comprises predaceous hyenas with long, slender limbs and compressed, sharp-edged cheek teeth; M_1 has a short talonid with a single, trenchant cusp. It probably arose from Miocene Old World forms like *Thalassictis* or *Lycyaena*. The earliest known representative is the Ruscinian *Chasmaporthetes borissiaki* (Khomenko). The American hyena is found throughout the Blancan. The earliest finds may date from the very late Hemphillian, and the latest are very early Irvingtonian. Localities include Anita, Blanco, Cita Canyon, Comosi (Arizona), Goleta (Michoacán, Mexico), Inglis IA, Miñaca (Chihuahua, Mexico), and Rexroad. The known distribution seems to indicate that the species was a southern form, but there is the possibility that a milk molar from Hagerman belongs to the same species.

The American form resembles very closely the Eurasian Villafranchian *Chasmaporthetes lunensis* (Del Campana) in dental as well as limb bone structure, and although the American form has slightly more robust teeth and jaws on the average, the two may be conspecific (the mandible of *C. ossifragus* is shown in figure 11.22). The genus also existed in Africa. Old World forms have been referred to the genera *Lycyaena* and *Euryboas* (or even *Hyaena*); however, *Lycyaena* is more primitive, and *Euryboas* is a synonym of *Chasmaporthetes*.

The characters of *Chasmaporthetes* indicate a cheetahlike predator. Attention has been called to the association of *Chasmaporthetes* and *Paenemarmota* at several localities, which might be indicative of a prey–predator relationship. (Beaumont, 1967; Hay, 1921; Kurtén, 1968 and unpublished data; Repenning, 1962; Schaub, 1941; Stirton and Christian, 1941.)

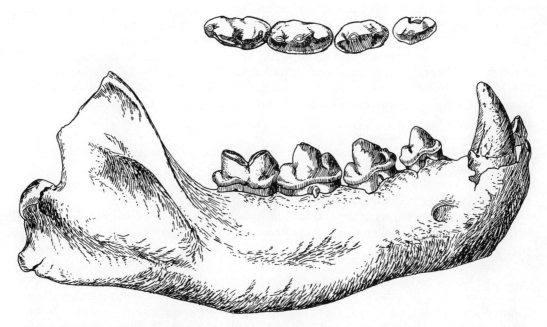

11.22 Hyaenidae. *Chasmaporthetes ossifragus*, Blancan, Cita Canyon, right mandible, external view, and occlusal view of lower dentition. After Stirton and Christian (1940). ca. 0.5×.

Suborder Pinnipedia

Seals, sea lions, and walruses are marine carnivores that arose from an arctoid carnivore stock in the temperate waters of the North Atlantic and North Pacific oceans in the mid-Tertiary. Recent discoveries of early pinnipeds (with many specimens still under preparation and study), from West Coast deposits in particular, are clarifying their early history, but classification within this suborder is still in a state of flux. Two superfamilies, the Otarioidea (the fur seals, sea lions, and walruses) and the Phocoidea (the true seals), are recognized. The main difference between them is that the otarioids can turn their hind limbs forward to help support the body in terrestrial locomotion whereas the phocoids cannot and must wriggle and hunch themselves along when traveling on land. Pinnipeds are characterized by large size (compared to the land carnivores), an insulating layer of blubber, distal limb segments modified into flippers, with cartilaginous extensions of the digits, and homodont cheek teeth. In addition, visual, auditory, circulatory, and respiratory systems are greatly modified for a marine life, and olfaction is reduced.

Family Odobenidae—Walruses

About 22 m.y. ago, a group of terrestrial, ursid canoid carnivores gave rise to the Enaliarctidae, a family that lived in warming North Pacific seas and evolved into the desmatophocids, odobenids (walruses), and otariids (fur seals and sea lions).

Walruses evolved from the Enaliarctidae about 14 m.y. ago in mid-northern latitudes and warm-temperate Pacific seas. By about 8 m.y. ago, they were the most diverse of the otarioids, with two subfamilies and many genera. Of these subfamilies, the Dusignathinae, characterized by equal enlargement of the upper and lower canines, remained in the North Pacific Basin and became extinct about 4 m.y. ago. The other subfamily, the Odobeninae, characterized by enlargement of the upper canines and reduction of the lower canines, moved south to subtropical and tropical waters and through the Central American Seaway into the Caribbean and then spread north into the North Atlantic. Here the modern walrus, *Odobenus rosmarus*, evolved; it became adapted to Arctic waters and reinvaded the North Pacific through the Arctic Ocean less than 1 m.y. ago.

Odobenids have a large, thick, heavy body, a rounded head, reduced or absent supraorbital processes, weak, platelike paroccipital processes, large auditory ossicles, a vaulted, bony palate, a deep mandibular symphysis, enlarged canines, and single-rooted cheek teeth. The extinct dusignathines had both pelagic and bottom-feeding habits, whereas the odobenines were (and are) bottom feeders.

Of all the odobenid species, only *Odobenus rosmarus* survives, and its populations are declining in number. (Repenning, 1975; Repenning and Tedford, 1977.)

Walrus, *Odobenus rosmarus* (Linnaeus), 1758 (*Phoca rosmarus*)

The monotypic genus *Odobenus* is now confined to the northern polar regions but had a much wider distribution in the Pleistocene. More than 100 specimens have been found along the Atlantic Coast, from Canada south to Cape Hatteras, North Carolina (southernmost record). Reports of *Odobenus* south of Cape Hatteras may represent odobenids from the Yorktown Formation (Pliocene) or earlier formations and should be viewed with suspicion (Ray, personal communication, 1976). Fourteen specimens of

Odobenus have been found in the eastern approaches to the Champlain Sea, where large herds probably congregated in late Wisconsinan–early Holocene times. Two walrus bones found in the northern part of eastern Michigan are artifacts and were probably brought inland from the shores of the Champlain Sea or the Atlantic Coast (Harington, 1977). The oldest record of *Odobenus* from the North Pacific is from the Kokolik River, northeast of Point Lay; it is probably less than 300,000 years old (Repenning and Tedford, 1977).Other Alaskan occurrences are from Sangamonian and Wisconsinan deposits near Barrow and on the Seward Peninsula. The southernmost Pacific occurrence is a specimen dredged up off San Francisco; it is probably late Pleistocene in age (Repenning, personal communication).

In the sixteenth century, walruses ranged along the coasts of Nova Scotia and England, and in the eighteenth century, they were observed around the Gulf of St. Lawrence. Man's activities have driven them steadily northward, and today walruses are only found near the edge of the polar ice in the Arctic Ocean. They migrate south in winter, often riding on ice floes.

A thick, swollen body, a short head and neck, small eyes, a blunt muzzle covered with coarse bristles, wrinkled, nearly hairless skin, long, oarlike fore flippers, and the absence of external pinnae and a tail characterize the walrus. The outstanding feature is the long, downward-pointing, ever-growing tusks (upper canines), which may reach a length of 100 cm in males and 60 cm in females (see fig. 11.23). Only the tip of the tusk is covered with enamel; the crown consists entirely of dentine (ivory), which is valued highly for carving. The single-rooted cheek teeth are bluntly conical or flat. Walruses feed on mollusks, but the cheek teeth are probably not used to crush the shells. Instead, the mollusk is held between the lips, and the strong tongue, in combination with

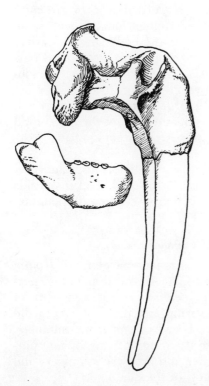

11.23 Odobenidae. *Odobenus rosmarus*, Recent, skull and mandible. After Vaughan (1972). 0.12×.

the cylinderlike mouth formed by the vaulted palate, acts as a piston to suck out the animal. Interestingly, the bony vaulted palate is a more ancient character than either the peg teeth or the tusks. At present, the only known function of the cheek teeth is to "chatter" when they are banged together underwater. Largest of the living otarioids, male walruses reach a length of 3–3.7 m and weigh more than 1 ton; females are about one-third smaller.

Gregarious in habits, walruses are found on isolated rocky coasts and ice floes. Locomotion on land is slow and clumsy; in the water, the animals use all four limbs for propulsion and are good swimmers. Walruses are important to the Eskimo economy, but excessive hunting by man has greatly reduced their numbers and range. (Harington, 1977; Ray, personal communication, 1977; Repenning, personal communication, 1974; Repenning and Tedford, 1977; Scheffer, 1958; Walker, 1975.)

Family Otariidae—Fur Seals and Sea Lions

About 12 m.y. ago, the last of the enaliarctids gave rise to the Otariidae, an extremely conservative family that evolved slowly until about 3 m.y. ago, when a trend toward single-rooted cheek teeth and taxonomic diversity began. This trend continues today, with fur seals and sea lions now attaining their maximum diversity. Fur seals are the first recognizable otariids, and the lineage leading toward *Callorhinus* has been found in deposits about 6 m.y. old. A separate sea lion lineage appeared about 3 m.y. ago on both sides of the North Pacific. From their origin in the North Pacific, otariids spread southward, reaching the Southern Hemisphere about 5 m.y. ago. Today, they are found along the shores of the Pacific, the South Atlantic, and the southern Indian Ocean.

Otariids on land are the most agile of the pinnipeds; nevertheless, they are completely adapted to a marine life. The body is slender and rather elongate, with small cartilaginous external ears and long (more than one-fourth the body length), oarlike fore flippers whose surfaces are naked and leathery. The pelage is plain colored. Characteristic of otariids are prominent supraorbital processes (especially in males), a knoblike paroccipital process joined to the mastoid process by a crest, frontals that penetrate the nasals, and small auditory ossicles. The canines are large, pointed, and recurved; the homodont cheek teeth are simple and sharply pointed, and some of the molars still retain double roots. Skeletal adaptations include a progressive shortening of the proximal parts of the limbs.

Early in their history, otariids developed pronounced sexual dimorphism and gregarious habits; evidence of this has been found in deposits 9 m.y. old. Today, otariids congregate in huge rookeries, especially during the breeding season. (Repenning, 1975; Repenning, Ray, and Grigorescu, 1979; Walker, 1975.)

Northern Fur Seal, *Callorhinus ursinus* (Linnaeus), 1758 (*Phoca ursina*)

The only known Pleistocene occurrence of *Callorhinus,* a monotypic genus, is from a bluff north of Kougechuck Creek on the Seward Peninsula, Alaska (Repenning, personal communication, 1976). The migratory fur seal (*Callorhinus ursinus*) ranges from Japan to southern California during the winter and spring and then returns to the Pribilof Islands in the Bering Sea, where the animals breed in enormous numbers. These migrations, sometimes covering 10,000 km, are the longest made by any pinniped. The animals spend most of their life in the water and are agile swimmers and expert divers. Fish, squid, and crustaceans make up most of the diet.

Sexual dimorphism is pronounced; adult males measure 1.9–2.1 m in length, weigh 180–300 kg, and are dark brown in color; females measure 1.5–1.7 m in length, weigh 36–68 kg, and are slate gray in color. *Callorhinus* is distinguished from the related *Arctocephalus* by its shorter rostrum (which gives it a puglike face), reduced premaxillary, less abrupt facial angle, smaller teeth, and more elongated fore flippers, which lack fur on the dorsal surface.

The northern fur seal has been hunted commercially since 1786. A treaty in 1910 abolished pelagic hunting, which had greatly decimated the population. The population now numbers between 1.5 and 1.9 million, and the animals are fully protected on the Pribilof Islands. (Repenning, Peterson, and Hubbs, 1971; Repenning, personal communication, 1976; Walker, 1975.)

Steller's Sea Lion, *Eumetopias jubata* (Schreber), 1776 (*Phoca jubata*)

Fossil remains of *Eumetopias* are known only from Alaska (North Point and Lincoln Bight on St. Paul Island, possibly Sangamonian in age, and Shishimaref Inlet on the Seward Peninsula, where specimens were found associated with *Mammuthus* and *Bison* (Repenning, personal communication, 1976). A subarctic and cool-temperate species, *Eumetopias jubata* is found today in the North Pacific and the Bering Sea, where it congregates in colonies along the coast.

This sea lion is the largest member of the Otariidae. An adult male is about 3.5 m in length and weighs 1,100 kg; females are much smaller (length, 2.7 m; weight, 350 kg). The species is distinguished by its large size and heavy muzzle and head (see fig. 11.24). It feeds on squid and fish. (Repenning, personal communication, 1976; Walker, 1975.)

California Sea Lion, *Zalophus californianus* (Lesson), 1828 (*Otaria californiana*)

The only published Pleistocene record of *Zalophus* is from the late Pleistocene Palos Verde Sand (lumber yard locality in San Pedro), Los Angeles Co., Newport Lagoon,

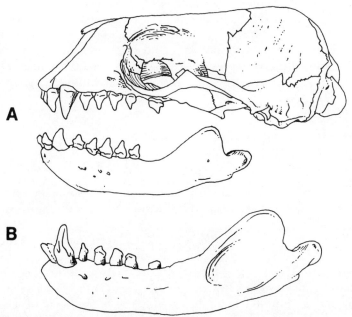

11.24 Otariidae. *A, Eumetopias jubatus,* Recent, skull and mandible. *B, Zalophus californicus,* Recent, mandible. Both after Hall and Kelson (1959). Both 0.33×.

and San Nicholas Island, all in California. Three isolated populations of this temperate to subtropical species inhabit the West Coast of North America, the Galapagos Islands, and the southern Sea of Japan. These gregarious, playful animals are often seen performing in circuses and zoos.

Both sexes have broader, heavier muzzles than *Callorhinus ursinus*. Males weigh up to 280 kg and are up to 2.35 m in length; a large female weighs about 90 kg and reaches a length of 1.8 m. The pelage is brown when dry, shiny black when wet. The species feeds on cephalopods and fish. (Kellogg, 1927; Repenning, personal communication, 1974; Walker, 1975.)

Family Phocidae—True Seals

Originating in the North Atlantic more than 15 m.y. ago, phocoids arose from an otterlike mustelid. The group underwent major evolutionary development on both sides of the North Atlantic, spread into the Paratethys (the forerunner of the Black and Caspian seas) about 13 m.y. ago, and entered the South Pacific about 5 m.y. ago. Two subfamilies are recognized: the Phocinae, found in northern waters, and the Monachinae, found in tropical and Antarctic seas.

Phocoids are earless seals that are highly specialized for an aquatic life. A thick, subcutaneous layer of blubber gives the body a fusiform shape. The fore flippers are short and well furred, and the hind flippers, though useless on land, serve as the primary propulsive organs in the water. The pelage of some phocine seals is mottled or banded. Unlike the single-rooted, single-cusped cheek teeth of otariids, phocoid cheek teeth are usually double-rooted and have accessory cusps.

The two subfamilies are derived from a common phocoid ancestor. One of the earliest known monachine seals is *Monotherium? wymani* (Leidy), found in Miocene beds in Virginia, and monachine fossils are abundant in 5-m.y.-old deposits near Cape Hatteras, North Carolina. A possible Caribbean–Gulf of Mexico center of diversification is indicated. The earliest phocine seals, genus *Leptophoca*, are known from Miocene deposits in Maryland, Virginia, and Belgium. Phocine seals are more abundant in the European Atlantic record than in the American, and several species evolved in isolation in the Paratethys. Today, the geographical ranges of the two subfamilies, with the exception of the monachine elephant seals, are distinct. (Repenning, Ray, and Grigorescu, 1979; Vaughan, 1972.)

Subfamily Phocinae

Harbor Seal, *Phoca vitulina* Linnaeus, 1758

The genus *Phoca* is circumboreal in distribution, and some endemic species are found in land-locked lakes in Eurasia. The genus includes the subgenera *Pusa*, *Histriophoca*, and *Pagophilus*.

The earliest known occurrence of *Phoca vitulina* is from the Port Orford Formation, Cape Blanco, Oregon, an early Pleistocene deposit less than 2 m.y. old (Repenning, Ray, and Grigorescu, 1979). Late Pleistocene records include specimens from the Palos Verdes Sand and Playa del Rey in Los Angeles County, California (Barnes and Mitchell, 1975) and a slumped bank below the sea bluffs west of Deering, Alaska (Repenning, personal communication, 1970). Today, harbor seals are found along the seacoasts of the Northern Hemisphere south as far as 30°N. They often ascend large rivers to follow runs of fish.

This small seal (length 1.3–2.0 m) usually has a spotted pelage. The cheek teeth are relatively large and set close together. The nasal bones are wider than those of *Phoca hispida* and are parallel-sided instead of flaring. Harbor seals feed on fish, mollusks, and crustaceans. Since they are difficult to capture, Eskimos seldom hunt them. Their main enemies are killer whales and polar bears.

Phoca sp. has been reported from several sites on the Seward Peninsula, and a specimen belonging to either *Phoca fasciata* or *Phoca vitulina*, found at the Barrow Ice Cellar in the Barrow area 10.6 m belowground, has been dated at 28,000 B.P. (Repenning, personal communication). (Barnes and Mitchell, 1975; Repenning, Ray, and Grigorescu, 1979; Walker, 1975.)

Ringed Seal, *Phoca hispida* Schreber, 1775

The commonest and most widely distributed Arctic seal, *Phoca hispida*, is found in the north-polar regions near the ice edges. Its range is limited to areas with the fast ice on which the seal breeds. Although essentially marine, the ringed seal also inhabits freshwater lakes, including Saimaa in Finland and Ladoga in Russia. There are Pleistocene records from Alaska (between Deering and Cape Deceit, middle Pleistocene to Holocene; north shore of the Nuglungnugtuk Estuary on the Seward Peninsula, Sangamonian–Wisconsinan) and the inland Champlain Sea (Hull, Quebec; Ottawa, Ontario; and Brandon, Vermont; all less than 12,000 years old). The specimen from Hull, a nearly complete skeleton found *in situ*, suggests that fast ice existed near the western margin of the Champlain Sea. Another much older specimen, a radius, found in the Malaspina District, Alaska, has been tentatively referred to the subgenus *Pusa* (Repenning, personal communication, 1970).

Smallest of the pinnipeds, the ringed seal has an average length of 1.4 m and an average weight of 68 kg (females are smaller). The face is rounded and catlike, and the dark gray back is patterned with conspicuous white rings. The cheek teeth are smaller and less crowded than those of *Phoca vitulina*, and the nasal bones are narrow posteriorly and flare out anteriorly. The ringed seal feeds on shrimplike crustaceans and fish. An abundant species, it is hunted by Eskimos and is important to the Arctic economy.

The fossil history of the subgenus *Pusa* may extend back to *Phoca pontica* and *Phoca pannonica*, endemic Paratethyan seals of 10 to 13 m.y. ago. The radius from Malaspina, Alaska, may represent an early invasion of *Pusa* into the Arctic Ocean from Paratethys; this probably occurred about 3 m.y. ago, when the Arctic Ocean was relatively ice free. Other details of the history and dispersal of *Pusa* are unknown (Repenning, Ray, and Grigorescu, 1979). (Harington, 1977; Harington and Sergeant, 1972; Repenning, personal communication, 1970; Repenning, Ray, and Grigorescu, 1979.)

Harp Seal, *Phoca groenlandica* Erxleben, 1777

Remains of the harp seal have been taken from Champlain Sea deposits at Green's Creek, Ontario, and several other reports of *Phoca* sp. from the same general area are probably referable to *Phoca groenlandica* (Harington, 1977). This deep-sea species is associated with drifting pack ice in the North Atlantic and Arctic oceans. In summer, large numbers of harp seals regularly migrate to the High Arctic. By late winter, the American population returns to its breeding areas in the Gulf of St. Lawrence and Newfoundland, where the young are born on the pack ice.

The male harp seal has a grayish pelage with a dark face and a harp, or horseshoe-shaped band, on the back; markings on the female are indistinct or absent. The animals weigh about 135 kg and reach a length of 1.5 m. The posterior margin of the palate is broadly U-shaped, and the long, narrow nasals taper gradually (anteriorly to posteriorly). The animals feed on macroplankton and fish. The newborn white pups are heavily exploited by commercial sealers. (Harington, 1977; Peterson, 1966.)

Bearded Seal, *Erignathus barbatus* (Erxleben), 1777 *(Phoca barbata)*

Specimens of the bearded seal (*Erignathus,* a monotypic genus) have been identified from middle Pleistocene deposits between Deering and Cape Deceit, Alaska, two Wisconsinan deposits on the Seward Peninsula (Repenning, personal communication, 1970), and Champlain Sea deposits near Finch, Ontario (Harington, 1977). Associated with shifting sea ice, on which it rests and breeds, the living animal inhabits shallow coastal waters of the Arctic Ocean. North American populations are sometimes found as far south as Newfoundland.

A large seal, *Erignathus barbatus* may reach a length of 3 m and weighs between 225 and 410 kg; females are smaller. A long, white tuft of stiff vibrissae on each side of the muzzle gives this seal its common name. The fore flippers are square and spadelike, with the third digit longer than the others. The large skull has a short, wedge-shaped rostrum, a short, deep jugal, and a widely flaring posterior margin of the palate (fig. 11.25). The relatively small teeth are often worn down until all that remains are the two roots, which continue to function as teeth or are lost entirely. Mollusks and other invertebrates make up a large part of the diet of this bottom-feeding seal. Its thick hide, fat, and flesh are valued by the Eskimos. (Harington, 1977; Repenning, personal communication, 1970; Walker, 1975.)

Gray Seal, *Halichoerus grypus* (Fabricus), 1791 *(Phoca grypus)*

The monotypic genus *Halichoerus* is restricted to the North Atlantic, and in North America it breeds regularly as far south as Sable Island, Nova Scotia; a marginal breeding colony exists on the small islets off Nantucket Island, Massachusetts.

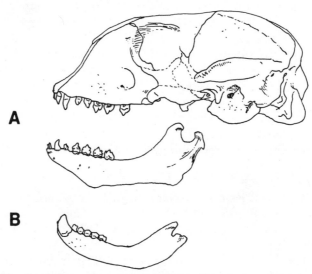

11.25 Phocidae. *A, Erignathus barbatus,* Recent, skull and mandible. *B, Cystophora cristata,* Recent, mandible. Both after Hall and Kelson (1959). Both 0.33×.

Remains of the gray seal have been found in Pleistocene deposits at the Womack gravel pit near Norfolk, Virginia, and at Edisto Beach, South Carolina. In addition, subfossil and midden remains are common in southern New England. The specimens found in the Womack gravel pit include the jaw of a suckling pup, which indicates that a breeding colony existed there.

These large seals (the males have a length of 1.6–3.6 m and weigh 158–360 kg) have a horselike head with a long snout, a broad muzzle, and high-set nasals. Gregarious in habits, gray seals live in rough waters around cliffs; in winter, they migrate up estuaries and inlets. Their food consists of herring, sculpin, and cuttlefish. They have little economic importance. (Peterson, 1966; Ray et al., 1968.)

SUBFAMILY MONACHINAE

†Caribbean Monk Seal, *Monachus tropicalis* (Gray), 1850 *(Phoca tropicalis)*

Most tropical of the pinnipeds is the genus *Monachus* (subfamily Monachinae), whose 3 species (1 now extinct) inhabit the warm waters of the Mediterranean and Black seas, the areas around the Hawaiian Islands, and formerly the Caribbean Sea.

The Caribbean monk seal, now extinct, formerly ranged along the shores of the Caribbean Sea and the Gulf of Mexico. During the Pleistocene, it was apparently restricted to southern localities (Melbourne) in the western Atlantic. These seals were pursued by members of Columbus' crew on Alta Vela, south of Haiti, in 1494, and in the years following were hunted intensively for food and oil. By 1885, they were on the verge of extinction; the last reported sighting was off the coast of Jamaica in 1952.

These lethargic seals ranged in length from 2.1–2.4 m, with females slightly smaller than males. Nonmigratory in habits, they fed on slow-moving reef fish. (Ray, 1961; Walker, 1975.)

Hooded Seal, *Cystophora cristata* (Erxleben), 1777 *(Phoca cristata)*

Today, the monotypic genus *Cystophora* is restricted to the western North Atlantic and Arctic oceans and regularly breeds on the pack ice near Newfoundland and the Gulf of St. Lawrence. The only Pleistocene record of the hooded seal is a tibia found at Plattsburgh, New York, in what was the Champlain Sea.

Male hooded seals have a distinctive inflatable pouch on the tip of the snout. Ranging in length from 2.1 to 2.4 m and weighing about 315 kg, the males are significantly larger than the females. The relatively heavy teeth are crowded close together in the anterior part of the jaw, and the upper molar, and sometimes P_4, is double-rooted. The species' fossil history is unknown. (Harington, 1977; Peterson, 1966.)

Northern Elephant Seal, *Mirounga angustirostris* (Gill), 1866
  (Macrorhinus angustirostris)

The genus *Mirounga* is found in both the North Pacific and the South Pacific (from 35°S to the Antarctic). It may be derived from *Callophoca* of the Atlantic Pliocene (Repenning, Ray, and Grigorescu, 1979). Two extant species are recognized: *Mirounga leonina* Linnaeus, the southern elephant seal, found in subantarctic waters, and *Mirounga angustirostris,* the northern elephant seal. The latter is found from the Channel Islands (off the southern California coast) south to San Benito Island (off central Baja). It ranges north to Point Reyes, California and wanderers have been reported much farther north.

The only Pleistocene record of the northern elephant seal is from Newport Bay Mesa. The possibility exists that this specimen belongs to the southern species, since it is believed that the Southern Hemisphere elephant seals expanded their range northward in the Pleistocene (Barnes and Mitchell, 1975.)

Largest of the pinnipeds, the northern elephant seal bull ranges in length from 4.5 to 6.5 m and weighs up to 3.5 tons; sexual dimorphism is pronounced, and females are considerably smaller. The common name is based on the large size and proboscislike snout, which in males may reach a length of 38 cm. The tympanic bullae are relatively large, thick-walled, and partly inflated. There is a pronounced depression in the nasal bones. Clumsy and slow-moving on land, elephant seals are strong swimmers and good divers. They feed on fish and cuttlefish. The northern elephant seal was nearly exterminated by commercial sealers; now these gregarious, phlegmatic mammals are fully protected and may even take over some California beaches. (Barnes and Mitchell, 1975; W. Miller, 1971; Repenning, Ray, and Grigorescu, 1979; Walker, 1975.)

CHAPTER TWELVE

ORDER RODENTIA

The earliest rodents appear in the late Paleocene of North America. The Rodentia is now the most diversified of all mammalian orders and is distributed over all the continents except Antarctica.

Rodents have one chisel-shaped, rootless incisor in each jaw half. In basic adaptation, the order may be considered omnivorous, but many rodents are purely vegetarian. The cheek teeth, primarily low-crowned, are hypsodont and even rootless in some groups. Arboreal, terrestrial, subterranean, and aquatic habits are represented. Although rodents are found in most environments on land and in freshwater, the habitat of most species is restricted; rodent fossils may thus be important paleoecological indicators.

Species-level taxonomy is difficult because of the common occurrence of sibling species that look alike but are nonetheless reproductively isolated. As an example, the living pocket gopher *Thomomys talpoides* has been shown, by means of karyotype studies and electrophoretic analysis of enzyme systems, to be a complex of numerous species (Thaeler and Nevo, 1973). Such speciation, which obviously creates a vast evolutionary potential, is not as yet detectable in fossil material. If there were subsequent morphological divergence, it could of course be studied in the same manner as for other mammals.

Major classification of rodents is at present uncertain. A majority of North American rodents may be referred to the cosmopolitan suborder Myomorpha (Cricetidae, Zapodidae). The Caviomorpha, of South American origin, includes the Erethizontidae and Hydrochoeridae. Other families are treated as *incertae sedis*. Study of Pleistocene rodents is still in its early stages.

FAMILY APLODONTIDAE—SEWELLELS

The family contains a single extant genus and species, *Aplodontia rufa,* an animal considered to be the most primitive living rodent. The species still retains the primitive zygomasseteric structure, with the masseter muscle limited to the ventral surface of the zygoma. The family reached its zenith in the Tertiary, when at least two lineages and a number of genera inhabited North America. The prosciurines are probably ancestral to the aplodontids.

Sewellel, *Aplodontia rufa* (Rafinesque), 1817 (*Anisonyx rufa*)

Fossil remains of the sewellel, or "mountain beaver," are known only from the Wisconsinan of Samwel, Potter Creek, and Hawver caves in northern California. The living animal inhabits the humid forests of the Pacific Northwest, from southern British Columbia to San Francisco Bay.

Resembling a tailless muskrat, this stocky, short-legged rodent is about the size of a small house cat. The teeth are quite hypsodont, and the lower molars show a reduced trigonid and an enlarged talonid (see fig. 12.1). There are five digits on each foot, and the well-developed claws on the front feet are used in digging. Sewellels are fossorial

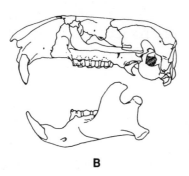

12.1 Aplodontia. *Aplodontia rufa*, Recent. *A*, right M^{1-2}; *B*, skull and mandible. *A*, after Vaughan (1972); *B*, after Hall and Kelson (1959). *A*, ca. 3×; *B*, 0.66×.

and dig elaborate burrows and tunnel systems close to the surface. Almost any green plant material serves as food, and in winter, the animals feed on bark and spruce needles. Sewellels are seldom seen, since they are nocturnal and usually solitary. (Rensberger, 1975; Walker, 1975.)

Family Sciuridae—Squirrels

This widespread family first appears in middle Oligocene deposits in North America; it reached the Old World in the late Oligocene. Ground squirrels, tree squirrels, and flying squirrels are identifiable by the Miocene. Today, sciurids occupy terrestrial and arboreal niches throughout much of the world, being absent only in the Australian region, Madagascar, southern South America, the polar regions, and some Old World deserts.

Relatively unspecialized, sciurid skulls have well-developed postorbital processes, a flattened zygomatic plate, and a small infraorbital foramen. The occlusal pattern of the low-crowned, rooted cheek teeth features transverse ridges. The limb bones, too, are unspecialized, and there is no reduction of freedom at the elbow, wrist, or ankle joints. Five digits, usually having sharp claws, are present. In the semifossorial forms, the forelimbs are stronger, and the tail is shorter.

Sciurids are common in Blancan and Pleistocene deposits, and many species are recognized.

†Giant Marmot, *Paenemarmota barbouri* Hibbard and Schultz, 1948
    (*Marmota mexicana* Wilson, 1948)

As far as is known, *Paenemarmota* is a monotypic genus. Characters of the astragalus and calcaneum of *Paenemarmota* appear to indicate a closer relationship to *Marmota* than to *Spermophilus* or *Cynomys*. A number of earlier Hemphillian species which have been tentatively ascribed to *Marmota* [e.g., *M. nevadensis* (Kellogg), *M. sawrockensis* Hibbard, and *M. oregonensis* Shotwell] seem to be closer to the ancestry of *Paenemarmota* than to *Marmota*.

Paenemarmota barbouri, an animal nearly as large as a beaver, was the largest of the ground-dwelling squirrels. Its stratigraphic range extends from the Hemphillian to the late Blancan, and specimens have been found from Mexico (in Michoacán and Chihuahua) to Washington (White Bluffs), as well as at Hagerman, Blanco, Broadwater, and Fox Canyon. It was mistakenly identified from Anita by Lindsay and Tessman (1974) (Zakrzewski, personal communication, 1976).

The species differs from *Marmota* in its larger size, massive build, and dental details.

ORDER RODENTIA

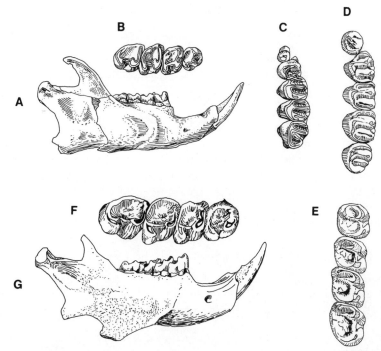

12.2 Sciuridae. *A–C, Spermophilus boothi*, Blancan, Sand Draw. *A*, right mandible, external view; *B*, right lower, and *C*, left upper cheek teeth, occlusal views. *D, E, Paenemarmota barbouri*, Blancan, Rexroad. *D*, left upper, and *E*, right lower cheek teeth, occlusal views. *F, G, Cynomys hibbardi*, Blancan, White Rock. *F*, right lower cheek teeth, occlusal view; *G*, right mandible, external view. *A–C*, after Skinner and Hibbard (1972); *D, E*, after Repenning (1962); *F, G*, after Eshelman (1975). *A*, 0.88×; *B, C*, 1.3×; *D, E*, 0.83×; *F*, 3×; *G*, 1.6×.

For example, P^3 has a double-cusped "protoloph" followed by a distinct valley, and the base of the lower incisor extends farther back than in the true marmots (see fig. 12.2).

This species is associated with the hyenid *Chasmaporthetes ossifragus* at most sites, and it is possible that there was a special predator–prey relationship between the two. The association may also reflect biotope preference since the cheetahlike hyenid probably favored open ground. (Black, 1963; Lindsay and Tessman, 1974; Repenning, 1962; Zakrzewski, 1969, and personal communication, 1976.)

Woodchuck, or Groundhog, *Marmota monax* Linnaeus, 1758

The stratigraphic range of the extant Holarctic genus *Marmota* extends back to the Clarendonian in North America, the Villafranchian in China, and the late Middle Pleistocene in Europe. A Nearctic origin is indicated. Black (1963) has suggested the Hemphillian *Marmota minor* (Kellogg) as a possible ancestor for the later taxa since it is structurally intermediate between the Clarendonian *Marmota vetus* (Marsh) and the Recent *Marmota monax*. Subsequent evolution has involved a general increase in size, enlargement of P_4, modification of the molars, and probably additional fossorial specializations. These large, stocky, short-legged, ground-dwelling rodents are common over much of their range today.

Marmota monax has been identified at some 25 Pleistocene localities in Arkansas, Georgia, Illinois, Kentucky, Maryland, Missouri, Pennsylvania, Tennessee, Virginia, and West Virginia. The earliest occurrences are late Irvingtonian (Conard Fissure,

Cumberland, Trout). An inhabitant of open woods, brushy and rocky areas, and clearings where well-drained soil is available for burrowing, the woodchuck is found in the eastern half of the United States south to Alabama and from Labrador to central Alaska. Diurnal in habits, it feeds on succulent plants and hibernates from October to February. (Black, 1963; Guilday, 1962b; Guilday, Hamilton, and McCrady, 1969; Guilday, Martin, and McCrady, 1964; Repenning, 1967a.)

†Arizona Marmot, *Marmota arizonae* Hay, 1921

Hay (1921) described a partial marmot skull found in the late Blancan Anita fauna as a new species and noted that it was similar to *Marmota flaviventris* but differed from it in having a narrower, more rounded snout and broader nasal processes of the premaxillaries. If the species is valid, it may be ancestral to *Marmota flaviventris* (Zakrzewski, personal communication, 1976). Marmots are not found in Arizona today. (Hay, 1921, Zakrzewski, personal communication, 1976.)

Yellow-bellied Marmot, *Marmota flaviventris* (Audubon and Bachman), 1841
  (*Arctomys flaviventer*)

A western species, the yellow-bellied marmot is found from central California east to the foothills of Colorado and the mountains of northern New Mexico (from lowland valleys to about 3,636 m). It is known from about 20 late Wisconsinan sites in Arizona, California, Colorado, Idaho, Nevada, New Mexico, and Wyoming. Finds of *Marmota flaviventris* south of the present range (at Papago Springs, Rampart, Isleta, and Dry caves, for example) are probably indicative of pluvial conditions in these areas since a limiting factor in marmot distribution is sufficient early spring moisture to sustain luxuriant plant growth all summer.

Slightly smaller than *Marmota monax,* this species has a head and body length of 355–480 mm and a tail length of 114–230 mm and weighs 2.5–5.0 kg. Their extensive burrows are usually located near a large rock, which they use as a lookout post. *Marmota* sp. has been reported from San Josecito and Boney Spring. (Harris, 1970.)

†Howell's Ground Squirrel, *Spermophilus howelli* (Hibbard), 1941
  (*Citellus howelli*)

Ground squirrels are terrestrial, semifossorial sciurids that are widely distributed in North America and Eurasia. Found in a variety of habitats, from the Arctic to the desert, *Spermophilus* displays great thermoregulatory flexibility, and many species hibernate or aestivate to escape climatic extremes. Ground squirrels have been known under several generic names, including *Citellus,* an Oken name now considered invalid, and *Otospermophilus* and *Callospermophilus,* now viewed as subgenera; only *Spermophilus* and *Ammospermophilus* are now recognized by most workers. The subgenera *Spermophilus, Ictidomys, Poliocitellus, Otospermophilus,* and *Callospermophilus,* all represented in the North American Pleistocene, differ from each other in details of dentition, cranial characters, and baculum (see Hall and Kelson, 1959). Their fossil history extends back to the middle Miocene, and remains of ground squirrels are common in many Blancan and Pleistocene faunas.

Spermophilus howelli has been identified at Fox Canyon, Rexroad 3, Blanco, and Hagerman. It is distinguished from other Blancan ground squirrels by its smaller size; it is slightly larger than Recent *Spermophilus mexicanus.* The protoconid and metaconid are separated by a deep, broad notch, and the hypoconid on P_4 is smaller than of

Recent species of *Spermophilus*. *Spermophilus howelli* shows closer resemblances to the subgenus *Callospermophilus* than to the subgenus *Otospermophilus*, and it may be near the point where the two subgenera separated. *Spermophilus howelli* and *Spermophilus rexroadensis* are more similar to each other than to any Recent ground squirrel. (Hall and Kelson, 1959; Hibbard, 1941a; Vaughan, 1972; Zakrzewski, 1969.)

† Rexroad Ground Squirrel, *Spermophilus rexroadensis* (Hibbard), 1941
   (*Citellus rexroadensis*)

Known from the type locality, Rexroad 3, and from Fox Canyon, this large ground squirrel was the size of *Spermophilus columbianus*. Size and characteristics of the ramus distinguish it from all other Great Plains ground squirrels. (Hibbard, 1941a.)

† Benson Ground Squirrel, *Spermophilus bensoni* (Gidley), 1922
   (*Citellus bensoni*)

Known from the early Blancan Benson fauna and the middle Irvingtonian Irvington fauna, this poorly known species was slightly smaller than *Spermophilus beecheyi*. According to Gazin (1942), it has a smaller talon on M^3, and the metaconule is closer to the metacone on P^4 and M^1 than in other species of the subgenus *Otospermophilus*. The lower teeth are similar to those of *Otospermophilus*, and the species probably belongs to that subgenus. (Gazin, 1942.)

† Melton's Ground Squirrel, *Spermophilus meltoni* Hibbard, 1972

Spermophilus meltoni, a Blancan species found only at Sand Draw, was intermediate in size between *Spermophilus franklini* and *Spermophilus mexicanus*. P^3 is more than one-third the size of P^4, and the protocones of P^4 and M^1 are not broadened or joined by a parastyle. (Skinner and Hibbard, 1972.)

† Johnson's Ground Squirrel, *Spermophilus johnsoni* Hibbard, 1972

This extinct species is known only from Sand Draw. It was about the size of *Spermophilus parryii* but had a deeper jaw and wider teeth. Differences from *Spermophilus boothi* include a shorter jaw, slightly higher-crowned teeth, a larger parastyle on P^4, and larger upper premolars. This species may represent a specialized line of ground squirrels. (Skinner and Hibbard, 1972.)

† Booth's Ground Squirrel, *Spermophilus boothi* Hibbard, 1972

The largest known ground squirrel in the subgenus *Otospermophilus*, this extinct species has been found at Sand Draw and White Rock. It has a broad rostrum and low-crowned cheek teeth similar to those of *Spermophilus variegatus*. The parastyles of P^4–M^2 do not change direction to join the protocones as in *Spermophilus magheei* and *Spermophilus johnsoni*. (Eshelman, 1975; Skinner and Hibbard, 1972.)

† Maghee's Ground Squirrel, *Spermophilus magheei* (Strain), 1966
   (*Citellus magheei*)

Known only from the late Blancan Hudspeth fauna, this species was about twice as large as *Spermophilus meadensis* and three times as large as *Spermophilus mexicanus*. As in *Spermophilus johnsoni*, the parastyle of P^4 abruptly changes direction to join the protocone. The upper molars are wider than long, and P^4–M^2 are moderately high-crowned. P_4 is subtriangular, and M_2 is subquadrate. (Strain, 1966.)

†Finlay Ground Squirrel, *Spermophilus finlayensis* (Strain), 1966
 (*Citellus finlayensis*)

This large ground squirrel has been reported only from Hudspeth. It was twice as large as *Spermophilus tridecemlineatus* or *Spermophilus spilosoma*. P_4 is subtriangular, and M_2 is slightly rhombohedral and narrowed posteriorly. The species may belong to the subgenus *Otospermophilus*. (Strain, 1966.)

†Cragin's Ground Squirrel, *Spermophilus cragini* (Hibbard), 1941
 (*Citellus cragini*)

Known only from the pre-Nebraskan Borchers fauna, *Spermophilus cragini* is distinguished from other fossil and Recent ground squirrels by the size and character of M^3, which most closely resembles that of *Spermophilus columbianus*, although the parastyle is higher lingually and joins the protocone near the top of the crown. Cragin's ground squirrel was larger than Franklin's ground squirrel, which is found in the region of Borchers today. (Hibbard, 1941d.)

†Anita Ground Squirrel, *Spermophilus tuitus* (Hay), 1921 (*Citellus tuitus*)

Hay (1921) described this medium-sized ground squirrel from the late Blancan Anita fauna. He noted that the upper tooth rows coverge backward and that the palate is narrow. (Hay, 1921.)

†Taylor's Ground Squirrel, *Spermophilus taylori* (Hay), 1921 (*Citellus taylori*)

A doubtful species, *Spermophilus taylori* is known from a single specimen collected somewhere in the vicinity of San Diego, Texas, by one of Cope's collectors. Hay (1921) said that it resembled *Spermophilus townsendii*. (Hay, 1921.)

†Meade Ground Squirrel, *Spermophilus meadensis* (Hibbard), 1941
 (*Citellus meadensis*)

Originally described from the Borchers fauna, this species has been tentatively identified at Wellsch Valley. It was nearly the size of *Spermophilus mexicanus*. On P_4, the protoconid and paraconid are well developed and separated by a deep sulcus or groove. The hypoconid of P_4 is larger than that of *Spermophilus tridecemlineatus*. (Hibbard, 1941d.)

†Cochise Ground Squirrel, *Spermophilus cochisei* (Gidley), 1922
 (*Citellus cochisei*)

Found only at Curtis Ranch, this poorly known species was about the size of *Spermophilus columbianus*. The teeth differ from those of *Spermophilus bensoni* in being more compressed anteroposteriorly, with more emphasis on the development of the transverse crests. (Gazin, 1942.)

Richardson's Ground Squirrel, *Spermophilus richardsonii* (Sabine), 1822
 (*Arctomys richardsonii*)

The earliest record of this extant species is from Cudahy (late Kansan), and specimens have been recorded from the Illinoian Sandahl and Doby Springs faunas, as well as Rancholabrean sites in Alberta, Colorado, Idaho, Kansas, New Mexico, South Dakota, and Wyoming. The Kansas and New Mexico localities are south of the present range. This medium-sized ground squirrel is often called "picket pin" from its habit of sitting upright. (Semken, 1966.)

Townsend's Ground Squirrel, *Spermophilus townsendii* Bachman, 1839

Found in the fossil state at Jaguar, Moonshiner, and Wasden, this small species inhabits grassland and sagebrush areas of western Oregon, southern Idaho, western Utah, and Nevada. A colonial species, it feeds on green plants and seeds. (Burt and Grossenheider, 1976.)

Arctic Ground Squirrel, *Spermophilus parryii* (Richardson), 1825
(*Arctomys parryii*)

Bones, seed caches, and preserved nests (some containing mummies) of the Arctic ground squirrel are common in the muck deposits around Fairbanks. The species appears to have been abundant throughout the Yukon–Tanana uplands in the late Pleistocene but is not found in the area today. An inhabitant of tundra and brushy meadows of the Far North, *Spermophilus parryii* is the only ground squirrel in the area. To survive in this harsh environment, it hibernates about 7 months of the year.

Fossil skulls have slightly wider zygomatic arches and a relatively greater mastoid breadth than those of modern specimens, but teeth of the fossils are indistinguishable from those of Recent animals. With a head and body length of 215 to 350 mm and a tail length of 75 to 150 mm, *Spermophilus parryii* is one of the largest species of ground squirrels. Radiocarbon dates of 14,760 ± 850 and 14,510 ± 450 years B.P. have been obtained on nest materials and coprolites, respectively (Péwé, 1975). The species' extirpation in interior Alaska was probably due to the disappearance of its preferred habitat at the end of the Wisconsinan glaciation. (Guthrie, 1968b; Hill, 1942; Péwé, 1975.)

Columbian Ground Squirrel, *Spermophilus columbianus* (Ord), 1815
(*Arctomys Columbianus*)

Fossils of this species have been identified only at Wasden. An inhabitant of meadows and forest edges in the interior Pacific Northwest, this colonial species feeds on green vegetation. It has a head and body length of 254–305 mm and a tail length of 76–127 mm and weighs 340–812 g. (Burt, 1976; Guilday, 1969.)

Uinta Ground Squirrel, *Spermophilus armatus* Kennicott, 1863

The only Pleistocene record of this species is from Silver Creek. A colonial species, it inhabits meadows and field edges up to elevations of 2,438 m in eastern Idaho, southwestern Montana, western Wyoming, and north-central Utah. It is about the size of *Spermophilus richardsonii* (head and body length, about 225 mm; weight, 284–425 g). (Burt and Grossenheider, 1976; W. Miller, 1976.)

Thirteen-lined Ground Squirrel, *Spermophilus tridecemlineatus* (Mitchell), 1821
(*Sciurus tridecemlineatus*)

In stratigraphic range, this species extends from the late Irvingtonian (Cudahy and Cumberland) to the Recent; it has been identified in 28 faunas (fig. 12.3). *Spermophilus tridecemlineatus* has the greatest geographic range of any of the ground squirrels and is found throughout the shortgrass prairies. During the Pleistocene, it ranged far eastward, and remains have been recovered at New Paris No. 4, Bootlegger Sink, Cumberland, Eagle, Trout, Natural Chimneys, Clark's, Welsh, Guy Wilson, Baker Bluff, Robinson, Conard Fissure, and Peccary. Its presence in the East is indicative of a former semiprairie or parkland environment in that area since the thirteen-lined ground squirrel is not found in forested regions.

A small species (head and body length, 115–165 mm; tail length, 63–133 mm), the

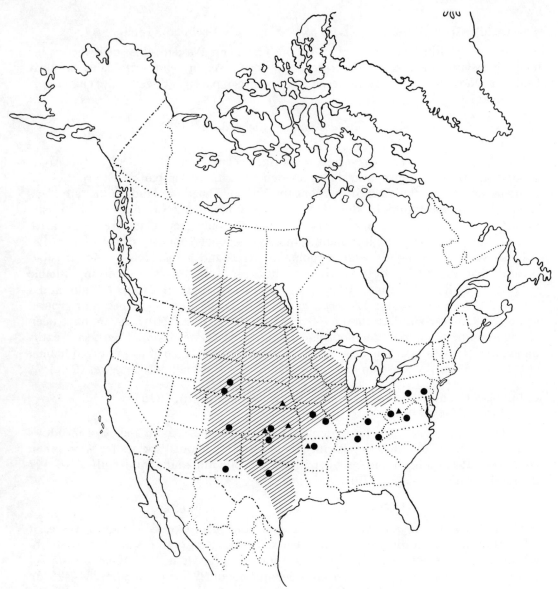

12.3 Sciuridae. Distribution of *Spermophilus tridecemlineatus*. ▲ = Irvingtonian; ● = Rancholabrean; *hatching* = Recent.

thirteen-lined ground squirrel gets its name from 13 whitish stripes on its back and sides; some of the stripes may be broken into rows of spots. It feeds on seeds, insects, and meat (deer mice, meadow voles, etc.). This solitary species hibernates about 6 months of the year. (Guilday, 1962b; Guilday, Hamilton, and McCrady, 1969, 1971; Guilday, Martin, and McCrady, 1964.)

Spotted Ground Squirrel, *Spermophilus spilosoma* Bennett, 1833

The earliest record is from Slaton (late Irvingtonian), and this species has been identified at Centipede, Damp, Easley Ranch, Isleta, Schulze, and San Josecito. Extant spotted ground squirrels live in areas of open forests and scattered brush from central Mexico north to North Dakota and west to central Arizona. This small species (head and body length, 125–152 mm; tail length, 57–90 mm) is active throughout the year. It has smaller jaws, shorter tooth rows, and smaller teeth than *Spermophilus tridecemlineatus,* but the two species are difficult to separate. (Dalquest, Roth, and Judd, 1969.)

Mexican Ground Squirrel, *Spermophilus mexicanus* (Erxleben), 1777
(*Sciurus mexicanus*)

There are a few Texan (Schulze, Howard Ranch, Centipede, Damp) fossil records of this medium-sized species (head and body length, 171–190 mm; tail length, 114–127 mm). Living populations inhabit the grassland, mesquite, and cactus country of southeastern New Mexico, the western half of Texas, and eastern Mexico. (Dalquest, Roth, and Judd, 1969.)

California Ground Squirrel, *Spermophilus beecheyi* (Richardson), 1823
(*Arctomys beecheyi*)

Fossils of this chiefly Californian species have been found at Costeau Pit, Emery Borrow Pit, Hawver, Newport Bay Mesa, Potter Creek, and Samwel. A medium-sized species, it lives in pastures and on slopes with scattered trees. Diurnal and colonial, it feeds on plants, insects, birds, and eggs and is regarded as a pest by farmers. (W. Miller, 1971.)

Rock Squirrel, *Spermophilus variegatus* (Erxleben), 1777 (*Sciurus variegatus*)

Late Rancholabrean records of this species include Burnet and Papago Springs (New Mexico), Centipede, Damp, and Williams (Texas), Little Box Elder (Wyoming), and McKittrick and Rancho La Brea (California); the last three sites are outside the present range. Living in rocky canyons and on boulder-strewn slopes, the rock squirrel is the largest ground squirrel in its range, an area that includes much of the Southwest. It is diurnal and solitary and climbs nearly as well as *Sciurus.* (Burt and Grossenheider, 1976.)

Franklin's Ground Squirrel, *Spermophilus franklinii* (Sabine), 1822
(*Arctomys franklinii*)

Pleistocene records of *Spermophilus franklinii* are few, and specimens have been identified only at Angus, Ben Franklin, Brynjulfson, Cudahy, and Jones. The record from Ben Franklin is outside the present range of the species, an area that extends from central Alberta to western Indiana and south to southern Kansas. This medium-sized, colonial species lives in tall grass or herbs along the borders of fields or the edges of forests. (Burt and Grossenheider, 1976.)

Golden-mantled Ground Squirrel, *Spermophilus lateralis* (Say), 1823
(*Sciurus lateralis*)

Found in the mountainous areas of the West, this chipmunklike ground squirrel has been found in the fossil state at Bell, Chimney Rock, Horned Owl, Jaguar, Little Box Elder, Moonshiner, and Ventana caves. A medium-sized species, the golden-mantled ground squirrel feeds on seeds, fruit, insects, meat, and eggs. It hibernates over the winter. (Burt and Grossenheider, 1976.)

White-tailed Antelope Squirrel, *Ammospermophilus leucurus* (Merriam), 1889

The genus *Ammospermophilus* has low-crowned cheek teeth, and the metaloph on M^1 and M^2 does not join the protocone as it does in the genus *Spermophilus*. These ground squirrels are found in the southwestern United States and northern Mexico.

Known in the fossil state only from Centipede Cave, the white-tailed antelope squirrel is found in the desert and foothill regions of the West. A solitary species, it is active throughout the year and is able to tolerate high air and soil temperatures. This small rodent feeds on seeds, insects, and meat and stores food in its burrow. Like *Dipodomys merriami*, *Ammospermophilus leucurus* produces concentrated urine, an adaptation which conserves water. (Vaughan, 1972.)

† Hibbard's Prairie Dog, *Cynomys hibbardi* Eshelman, 1975

Prairie dogs (genus *Cynomys*) are plump, oversized ground squirrels that have a flattened head, low, rounded ears, short legs, and a short, slender tail. *Cynomys* has higher-crowned teeth than other Nearctic Sciuridae. Colonial and semifossorial, prairie dogs are found in the prairies and mountain valleys of the West. Feeding primarily on forbs, they are regarded as pests by ranchers, and their numbers have been greatly reduced. Two subgenera of prairie dogs are recognized. The subgenus *Leucocrossuromys* [including *Cynomys leucurus*, *C. gunnisoni*, and *C. parvidens* (not known from Pleistocene faunas)] is distinguished from the subgenus *Cynomys* (including *C. ludovicianus* and *C. mexicanus*) by the presence of a stylid on the basin of M_3 that abuts the ectolophid and with wear joins it, thus dividing the anterior–posterior valley.

Prairie dogs are a late Pliocene offshoot of the spermophile line and first appear in the fossil record in the late Blancan. A number of extinct species have been named, and other specimens have been referred to living species; a comparative study of Blancan and Pleistocene prairie dogs is badly needed. *Cynomys meadensis* Hibbard (1956b), described from Deer Park was thought to be the earliest known prairie dog; however, these specimens are now believed to be intrusive since the preservation is different from other specimens in the Deer Park fauna and are considered to be immature *Cynomys* cf. *ludovicianus* rather than a member of the pre-Nebraskan Deer Park fauna (L. Martin, personal communication, 1977).

Cynomys hibbardi is known only from White Rock. It is distinguished from other members of the genus by the prominent anterolabial curve formed by the protolophid of the lower molars and the less developed metalophid on M_3. This small prairie dog was about the size of *Spermophilus parryii*. Eshelman (1975) has placed it in the subgenus *Cynomys* (which has the ectolophid of M_3 separated from the talonid basin by an anterior–posterior valley) and suggested that it may represent a taxon near the base of the lineage where prairie dogs split from ground squirrels. (Eshelman, 1975; Hibbard, 1956b; L. Martin, personal communication, 1977; Semken, 1966; Walker, 1975.)

† Old Prairie Dog, *Cynomys vetus* Hibbard, 1942

First described from a probable late Blancan deposit in Jewell County, Kansas, this species has also been reported from White Rock. It is distinguished from other early prairie dogs by its small size. It differs from the extant *Cynomys mexicanus* in its smaller size, more circular P^3, smaller infraorbital foramen, and less robust infraorbital process of the maxillary. Dalquest (1967) believes that *Cynomys vetus* is directly ancestral to *Cynomys gunnisoni* and related forms. (Dalquest, 1967; Eshelman, 1975; Hibbard, 1942.)

† Niobrara Prairie Dog, *Cynomys niobrarius* Hay, 1921

The type locality of this poorly known species is Hay Springs, and specimens have also been reported from Angus, Mullen II, and Rushville, all Irvingtonian in age. Hay (1921) believed that the species was allied to *Cynomys leucurus* but probably was larger and had a longer tooth row and a broader, deeper groove for the anterior branch of the masseteric muscle. Dalquest (1967) thinks that both *Cynomys niobrarius* and *Cynomys spispiza* Green are conspecific with *Cynomys ludovicianus,* but other workers (see Eshelman, 1975) question this assignment. (Dalquest, 1967; Eshelman, 1975; Hay, 1921.)

Black-tailed Prairie Dog, *Cynomys ludovicianus* (Ord), 1815 (*Arctomys ludoviciana*)

The stratigraphic range of this species extends from the Irvingtonian [Holloman, Medicine Hat (Yarmouthian), Butler Springs, Sandahl] to the Recent, and specimens have been identified at some 22 localities in Alberta, Arizona, Kansas, Nebraska, New Mexico, Oklahoma, and Texas. Of these, Medicine Hat (Sangamonian and Wisconsinan), Hand Hills, Clear Creek, Freisenhahn, Ingleside, and Ventana are outside the present range of the species, the shortgrass prairies from southern Saskatchewan to northern Mexico.

This species is distinguished from other prairie dogs by its larger teeth, longer tooth rows, and relatively long occlusal length of P_4–M_2 in relation to M_3. Thousands of black-tailed prairie dogs may inhabit a single "town" covering a vast area. Their bare, crater-shaped mounds are distinctive. The animals do not hibernate but are inactive during periods of inclement weather. The black-footed ferret is their chief predator. (Semken, 1966; Storer, 1975.)

White-tailed Prairie Dog, *Cynomys leucurus* Merriam, 1890

This species is poorly known in Pleistocene faunas. It has been reported from Horned Owl and Rainbow Beach, and material from Little Box Elder may also be referable. *Cynomys spispiza* Green from the Wisconsinan of South Dakota has been referred to this species (see Black, 1963), but Dalquest (1967) believes that it is more closely allied to *Cynomys ludovicianus*. The white-tailed prairie dog is found in intermontane valleys in Wyoming, northern Colorado, northwestern Utah, and southern Montana. *Cynomys leucurus* is the largest member of the subgenus *Leucocrossuromys* and is nearly as large as *Cynomys ludovicianus* but has a shorter tail. White-tailed prairie dogs live in fairly small colonies and hibernate about 4 months of the year. (Black, 1963; Clark, Hoffmann, and Nadler, 1971; Dalquest, 1967; Green, 1963; Semken, 1966.)

Gunnison's Prairie Dog, *Cynomys gunnisoni* (Baird), 1858 (*Spermophilus gunnisoni*)

The most primitive extant prairie dog is *Cynomys gunnisoni,* now an inhabitant of high mountain valleys and plateaus of the southern Rocky Mountains. It has been reported from Sandahl, Kentuck, Isleta, Williams, Bell, and Mesa de Maya. Dalquest (1967) referred the specimens from Sandahl to *Cynomys vetus,* but this has been questioned (Eshelman, 1975).

Smaller than *Cynomys leucurus,* the species has a short tail and small auditory bullae, and the maxillary arm of the zygoma is broad and flaring. Gunnison's prairie dog is inactive during the winter and probably hibernates. The burrow systems and mounds of this species are more similar to those of *Spermophilus* than those of other prairie dogs. Recent studies by Pizzimenti and Hoffmann (1973) have shown that the diploid chromosome number is $2n = 40$ for *Cynomys gunnisoni,* compared to $2n = 50$ for the other species of *Cynomys*. The lower number is characteristic of ground squirrels, and these authors believe that *Cynomys gunnisoni* is closely related to *Spermophilus*. Like the other species of prairie dogs, *Cynomys gunnisoni* is considered to be an agricultural pest and has been the target of extensive poisoning campaigns. It is now on the endangered-species list. (Dalquest, 1967; Eshelman, 1975; Pizzimenti and Hoffmann, 1973.)

†Noblest Chipmunk, *Tamias aristus* Ray, 1965

Chipmunks are small, active, diurnal, ground-dwelling squirrels that have facial stripes and carry their tail erect when running; they comprise two genera, *Tamias* and *Eutamias*. *Tamias* is distinguished from *Eutamias* by the presence of a single upper premolar in each half of the jaw.

Known only from the type locality, Ladds (Sangamonian), *Tamias aristus* is distinguished from *Tamias striatus* by much larger size—almost all the cranial and dental dimensions are 10 to 30 percent greater than those of the largest extant representatives. Morphologically, the two species are similar, and *Tamias aristus* may merely be an extinct giant subspecies. (Ray, 1965.)

Eastern Chipmunk, *Tamias striatus* (Linnaeus), 1758 (*Sciurus striatus*)

The single extant species, *Tamias striatus* is found in deciduous forests and brushy areas in eastern United States and southeastern Canada. It has been identified at 18 Pleistocene faunas in Arkansas, Georgia, Kentucky, Maryland, Missouri, Pennsylvania, Tennessee, Texas, and Virginia. Its stratigraphic range extends back to the late Irvingtonian (Cumberland). The species is larger than *Eutamias*, with a head and body length of 127–152 mm and a tail length of 76–102 mm. Like *Eutamias*, the eastern chipmunk has unlined cheek pouches, stores food, and remains in a torpid state during periods of inclement weather. (Burt and Grossenheider, 1976; Ray, 1965.)

Least Chipmunk, *Eutamias minimus* (Bachman), 1839 (*Tamias minimus*)

Eutamias, one of the two genera of chipmunks, differs from *Tamias* in having two upper premolars in each half of the jaw.

Remains of *Eutamias minimus*, the least chipmunk, have been identified at Little Box Elder, Moonshiner, Silver Creek, and Wasden caves. Recently, the species has been recognized in three Appalachian faunas, Clark's Cave, Baker Bluff, and Back Creek Cave No. 2 (Virginia) (Guilday, Parmalee, and Hamilton, 1977). Its stratigraphic range extends back to the late Sangamonian. *Eutamias minimus* is the smallest and most variable of the chipmunks and has the widest range, both geographically and altitudinally. An inhabitant of mountain glades, coniferous and mixed forests, and sagebrush deserts, it is found today over much of the West and in Canada east to Quebec. (Other species of *Eutamias* are found as far south as northern Mexico and in northern Asia.) Although primarily ground-dwelling, chipmunks are expert climbers. These extremely active, diurnal animals feed on plants, seeds, nuts, and insects.

Eutamias sp. has been reported from Arroyo Seco, Carpinteria, Horned Owl, Jaguar and Potter Creek. (Burt and Grossenheider, 1976; Guilday, Parmalee, and Hamilton, 1977.)

Fox Squirrel, *Sciurus niger* Linnaeus, 1758

Tree squirrels, genus *Sciurus*, inhabit deciduous, coniferous, and tropical forests throughout Europe, Asia, and the New World from southern Canada to northern Argentina. About 55 Recent species are recognized.

Tree squirrels are poorly represented as fossils because of their arboreal habits. Their fossil history extends back to the late Oligocene in Europe, the middle Miocene in North America, and the Pleistocene in South America. Four species of *Sciurus* have been identified in Pleistocene deposits, and *Sciurus* sp. is reported from some 15 sites, the oldest of which are Irvingtonian (Inglis IA and Trout).

Sciurus niger inhabits open hardwood and pine forests in the eastern half of the United States and has been introduced on the West Coast. It has been identified at several Wisconsinan sites, including Brynjulfson, Clear Creek, Devil's Den, Hill-Shuler, Kyle, Peccary, Robinson, and Savage caves. Largest of the eastern tree squirrels, it has a head and body length of 255–380 mm and a tail length of 230–355 mm. *Sciurus niger* has 22 teeth. For squirrels feed on nuts, acorns, seeds, fungi, bird's eggs, and cambium. They spend considerable time on the ground, foraging and burying nuts. (Burt and Grossenheider, 1976; Walker, 1975.)

Gray Squirrel, *Sciurus carolinensis* Gmelin, 1788

The primarily arboreal gray squirrel is found in hardwood forests, forest edges, and along stream bottoms in the eastern United States and southern Canada. It has been identified in 18 Pleistocene faunas dating back to the late Irvingtonian (Coleman IIA) in Florida, Kentucky, Maryland, Missouri, Pennsylvania, Tennessee, Texas, and Virginia. Smaller than the fox squirrel, this species has a head and body 200–255 mm in length and a tail equally long. (Burt and Grossenheider, 1976.)

Arizona Gray Squirrel, *Sciurus arizonensis* Coues, 1867

This species is similar to *Sciurus carolinensis* but has a longer tail. It is found in oak and pine forests in southwestern Arizona and northern Sonora, Mexico. Material from Blackwater Draw (late Wisconsinan) has been referred to *Sciurus arizonensis*. (Slaughter, 1964.)

Allen's Squirrel, *Sciurus alleni* Nelson, 1898

The only Pleistocene record of this Mexican squirrel is from the late Wisconsinan fauna of San Josecito. The species lives in oak, pine, and madrona forests in the states of Tamaulipas, Coahuila, Nuevo León, and San Luis Potosi, at elevations of 666–2,424 m. It is characterized by a single premolar in each half of the jaw. Its slightly smaller size distinguishes it from the closely related *Sciurus oculatus* Peters, which it resembles in color and proportions. (Hall and Kelson, 1959; Jakway, 1958.)

Red Squirrel, *Tamiasciurus hudsonicus* (Erxleben), 1777 (*Sciurus vulgaris hudsonicus; Sciurus tenuidens* Hay, 1920)

Two species of red squirrels, genus *Tamiasciurus*, inhabit North American forests. *Tamiasciurus* is smaller than *Sciurus*.

The smallest tree-dwelling squirrel in its range, *Tamiasciurus hudsonicus* lives in coniferous and mixed hardwood forests of the northern United States from Alaska to the East Coast, with extensions south into the Rocky Mountains and the Appalachians. Its fossil record extends back to the Irvingtonian (Conard Fissure, Cumberland, Trout), and specimens have been identified in 13 Wisconsinan faunas in Kentucky, Maryland, Missouri, Pennsylvania, Tennessee, Virginia, and West Virginia. An extinct subspecies, *Tamiasciurus hudsonicus tenuidens* (Hay) (see Guilday, Martin, and McCrady, 1964), characterized by a larger size and more robust build, has been found at Cavetown, Cumberland, Natural Chimneys, New Paris No. 4, and Whitesburg (type locality). The presence of red squirrel in a fauna may indicate a cooler climate. (Guilday, Martin, and McCrady, 1964.)

Douglas' Squirrel, *Tamiasciurus douglasii* (Bachman), 1839 (*Sciurus douglasii*)

A noisy inhabitant of the coniferous forests of the Pacific Northwest and the Sierra Nevada, *Tamiasciurus douglasii* is known from two Rancholabrean faunas, Potter Creek and Samwel caves. It is slightly smaller than the red squrrel and has a head and body length of 152–178 mm and a tail length of 121–127 mm. Both species nest in tree cavities and store vast quantities of nuts. (Burt and Grossenheider, 1976.)

† Webb's Flying Squirrel, *Cryptopterus webbi* Robertson, 1976 (*Petauria* sp. Webb, 1974)

The genus *Cryptopterus* includes large, extinct flying squirrels that inhabited warm, forested regions of Eurasia in the Miocene and Pliocene. The genus has recently been recognized in the Blancan of Florida.

Known only from the type locality, Haile XVA (late Blancan), *Cryptopterus webbi* is most similar to the late Pliocene species *Cryptopterus tobieni* from the lignites of Wolfersheim–Wetterau in West Germany. It is the largest known flying squirrel in the New World. *Cryptopterus webbi* has a low-crowned, subovate M_3 that is slightly tapered toward the posterior end and greatly expanded anterolingually. A unique characteristic of this species is the presence of a vestigial hypoconulid on M_3. The enamel of the tooth is not finely crenulated.

Flying squirrels underwent most of their evolutionary development in Eurasia, and the most complete fossil record of the group is found in Europe. *Cryptopterus webbi* probably reached North America from the Old World via Beringia in Hemphillian or early Blancan times. The presence of this forest-dwelling gliding squirrel in western Germany and Florida suggests that there was a continuous band of forest extending from Eurasia into eastern North America during that time. (Robertson, 1976)

Southern Flying Squirrel, *Glaucomys volans* (Linnaeus), 1758 ([*Mus*] *volans*)

Flying squirrels of the genus *Glaucomys* are small, agile, nocturnal mammals that can glide from tree to tree by means of an outstretched membrane extending along the sides of the body. They have long limbs and 22 teeth and feed on seeds, nuts, insects, and birds' eggs. *Glaucomys* first appears in the Irvingtonian (Coleman IIA, Cumberland), and two extant species are recognized. Flying squirrels are abundant in faunas in which activities of nocturnal raptors account for much of the bone accumulation.

Glaucomys volans is found in deciduous and mixed-hardwood forests in the eastern United States north as far as the Canadian border and in isolated areas in Mexico and Guatemala. It has been identified in 18 Pleistocene faunas in Florida, Missouri, Pennsylvania, Tennessee, Virginia, and West Virginia. Its stratigraphic range extends back to the late Irvingtonian (Trout). Late Pleistocene specimens from the Appalachians are larger than individuals from extant local populations. The ranges of *Glaucomys volans* and *Glaucomys sabrinus* overlap in the Great Lakes region and in the Appalachians, and the species have been found together at Baker Bluff, Clark's, Natural Chimneys, New Paris No. 4, and Robinson caves. (Guilday, 1962b; Guilday, Hamilton, and McCrady, 1969; Guilday, Martin, and McCrady, 1964; Guilday, Parmalee, and Hamilton, 1977.)

Northern Flying Squirrel, *Glaucomys sabrinus* (Shaw), 1801 (*Sciurus Sabrinus*)

Remains of the northern flying squirrel have been found at several late Wisconsinan localities including Baker Bluff, Bootlegger Sink, Clark's, Eagle, Natural Chimneys,

New Paris No. 4, Peccary, Potter Creek, Robinson, and Samwel caves. The species inhabits coniferous and mixed forests from southern Alaska south across Canada into the Great Lakes region and the northeastern states, with extensions south into the Sierra Nevada, Idaho, Utah, Wyoming, and the Appalachian Mountains.

Glaucomys sabrinus is larger (head and body length, 140–162 mm; tail length, 110–140 mm) than *Glaucomys volans*. Recent specimens from the eastern United States are the smallest, with size increasing to the west and north. Specimens from Clark's, New Paris No. 4, and Robinson caves are larger than individuals from extant local populations. (Burt and Grossenheider, 1976; Guilday, 1962b; Guilday, Hamilton, and McCrady, 1969; Guilday, Martin, and McCrady, 1964; Guilday, Parmalee, and Hamilton, 1977.)

FAMILY GEOMYIDAE—POCKET GOPHERS

This is an endemic New World family, the history of which begins in the Clarendonian (late Miocene). The geomyids are fossorial, stockily built rodents with strong, curved, front claws, small eyes and ears, and large, external, fur-lined cheek pouches that open on either side of the mouth. The tail is short, sparsely haired, and sensitive to touch.

Evolutionary trends in geomyid dentition include increasing hypsodonty, loss of roots, and development of dentine tracts in the cheek teeth (presence of such tracts leads to interruption of the enamel on the wearing surface of the tooth). These trends proceed at different rates in different lineages, but by the beginning of the Blancan, most geomyids had acquired ever-growing, rootless teeth.

Owing to their burrowing habits, pocket gophers are not uncommon in the fossil state. Their somewhat sedentary life results in considerable stability of geographic ranges, as well as differentiation into many local subspecies and species. Many "species" are in fact complexes of reproductively isolated but morphologically indistinguishable populations, a situation which makes modern taxonomy difficult and fossil taxonomy even more so.

Adaptation of such local populations to different conditions ensures that some form of pocket gopher is met with in most environments. There are no true arctic forms, however, and only one Recent species, *Thomomys talpoides*, has an appreciable range in Canada.

†Gidley's Pocket Gopher, *Thomomys gidleyi* R. Wilson, 1933

This genus has forefeet less specialized for digging than *Geomys*. There is no interruption of the enamel of the cheek teeth. The upper incisors are smooth, apart from a minute groove near the inner edge in front. The genus appears at the beginning of the Blancan, and its ancestry is not precisely known.

The species *Thomomys gidleyi* has been found in the Blancan deposits of Hagerman and White Bluffs. It is a small form, near the size of the living subspecies *Thomomys talpoides quadratus* Merriam (found in eastern Oregon and adjacent areas). Its mandible is characterized by a shallow temporal fossa and a well-developed valley between the ascending ramus and the capsular process (see fig. 12.4) There are resemblances to two Recent species, *Thomomys bottae* and *Thomomys talpoides*, and *Thomomys gidleyi* could be ancestral to either or both. (Akersten, 1973; R. W. Wilson, 1933a: Zakrzewski, 1969.)

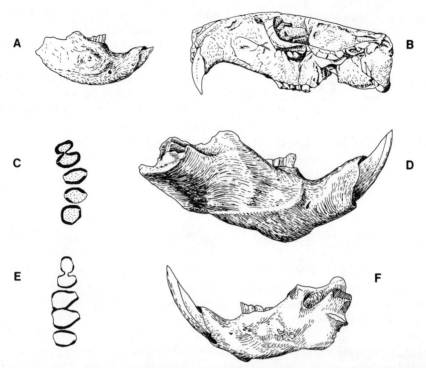

12.4 Geomyidae. *A, Nerterogeomys smithi*, Blancan, Fox Canyon, right mandible, external view. *B, Geomys garbanii*, Irvingtonian, Vallecito, skull. *C, Nerterogeomys persimilis*, Irvingtonian, Curtis Ranch, right P_4–M_3, occlusal view. *D, Geomys jacobi*, Blancan, Rexroad, right mandible, external view. *E, F, Thomomys gidleyi*, Blancan, Hagerman. *E*, left P_4–M_3, occlusal view; *F*, left mandible, external view. *A, C, D*, after Hibbard (1967); *B*, after White and Downs (1961); *E, F*, after Zakrzewski (1969). *A, F*, 2×; *B*, 1×; *C*, 5×; *D*, 2.4×; *E*, 4×.

† Potomac Pocket Gopher, *Thomomys potomacensis* (Gidley and Gazin), 1933
(*Plesiothomomys potomacensis*)

Found at Cumberland and Trout caves, this species is a member of the Irvingtonian fauna of the Appalachian region. It is comparatively large, close to *Thomomys talpoides* in size. Characters cited as distinctive include a very robust and deep mandible with a nearly straight masseteric crest, lower molars oval in cross section and less compressed lingually than those of other *Thomomys*, and a buccal surface not flattened or grooved. Whether *Pleisothomomys* is valid as a subgenus is uncertain; its dental characters may occur in extant species as well.

The species is closely related to the somewhat younger *Thomomys orientalis*. These eastern forms are extinct, and the genus is not now present in eastern North America. (Gidley and Gazin, 1938; Russell, 1968; Zakrzewski, personal communication, 1976.)

† Eastern Pocket Gopher, *Thomomys orientalis* Simpson, 1928

Found only at one site, Sabertooth Cave in Florida, this species is clearly allied to *Thomomys potomacensis* and belongs to the same group (subgenus *Plesiothomomys?*). The age is Rancholabrean and may be Sangamonian. The species is markedly smaller than the Appalachian form. (Gidley and Gazin, 1938; Simpson, 1928.)

Valley Pocket Gopher, *Thomomys bottae* (Eydoux and Gervais), 1836
(*Oryctomys bottae*)

Fossil remains of this species have been found in Rancholabrean deposits in California (several localities), Nevada (Glendale), New Mexico (Burnet, Dry, and Isleta caves), and Texas (Klein Cave). The oldest record seems to be that from Newport Bay Mesa, dated at 103,000 years B.P. Klein Cave in southwestern Texas is slightly outside the present-day range of the species.

Thomomys bottae is highly variable in size and other characters, and its distinction from *Thomomys umbrinus*, which replaces it to the south, is uncertain. Northern, valley forms are larger than those of the southern deserts. Color varies locally from almost white to nearly black. Habitat preference is loamy soil in valleys and montane meadows, also sandy and rocky deserts. Osteological characters include a rounded anterior prism on P^4. A sphenoidal fissure is present.

Thomomys gidleyi of the Blancan is a plausible ancestor. Irvingtonian *Thomomys*, not identified as to species, are known from Irvington, Vallecito, and Cudahy. (W. E. Miller, 1971; Roth, 1972.)

Pygmy Pocket Gopher, *Thomomys umbrinus* (Richardson), 1829 (*Geomys umbrinus*)

Finds from the late Rancholabrean of San Josecito Cave, Nuevo León, and Schulze Cave, Texas, have been referred to this species. The latter site is slightly east of the present-day range.

Thomomys umbrinus is distinguished by its small size but in other features is similar to *Thomomys bottae* and may be a derivative of that species. Hall and Kelson (1959) consider the two species synonymous, and, in fact, a number of fossil finds within the range of *Thomomys bottae* have been referred to *Thomomys umbrinus* (Potter Creek, Samwel, Burnet caves). (Dalquest, Roth and Judd, 1969; Hall and Kelson, 1959.)

Townsend's Pocket Gopher, *Thomomys townsendii* (Bachman), 1839 (*Geomys townsendii*)

Fossil records of this species come from Sangamonian (American Falls) and Wisconsinan (Fossil Lake, Rainbow Beach) sites, well outside of its present-day, very restricted relict range, which consists of a few river valleys in northern Nevada and neighboring areas. The fossil evidence thus suggests a very considerable shrinkage in the range of this species since Rancholabrean times.

With sphenoidal fissure present and characters of P^4 rather like those of the *Thomomys bottae–T. umbrinus* group, *T. townsendii* although somewhat larger on the average, is evidently a close relative of these species. (Allison 1966; Elftman, 1931; White, personal communication, 1975.)

Northern Pocket Gopher, *Thomomys talpoides* (Richardson), 1828 (*Cricetus talpoides*)

The earliest fossil find attributed to this species comes from Sangamonian beds at Medicine Hat. Late Wisconsinan records from the northwestern United States (Jaguar, Wasden, Silver Creek, Bell caves) are within the present-day range of the species. However, there is also a record from Dry Cave (southeastern New Mexico), which is well south of the range, and even tentative identifications as far afield as Howard Ranch (Texas) and Brynjulfson (Missouri). These are well beyond the limits of any present-day *Thomomys* and so indicate a considerable range extension to the south and east in Wisconsinan times.

This species, which, on the average, is intermediate in size between *Thomomys bottae* and *Thomomys umbrinus*, has a robust skull with broad zygomatic arches. There is no sphenoidal fissure. The anterior prism of P^4 is triangular.

A prairie form in the north, it is found in more montane situations (alpine meadows, open pine forests) farther south. The burrowing habits of this species are considered to contribute to soil and water conservation.

Descent from *Thomomys gidleyi* is possible. (Guilday and Adam, 1967; Harris, 1970; Parmalee and Oesch, 1972.)

†Sinclair's Mazama Gopher, *Thomomys microdon* Sinclair, 1905

A comparatively small form similar to the living *Thomomys mazama* Merriam, this species has been found in Wisconsinan-age deposits from Potter Creek and Samwel caves in northern California. The species is stated to differ from the living form by virtue of its short and broad rostrum and the presence of a prominent ridge lateral to the alveolus of the upper incisor. The fossil form may be a subspecies of the modern mazama pocket gopher. The latter species occurs in mountain meadows in western Washington and Oregon and northern California. (Kellogg, 1912; Sinclair, 1905.)

†Zakrzewski's Pygmy Pocket Gopher, *Pliogeomys parvus* Zakrzewski, 1969

The genus *Pliogeomys* comprises pocket gophers in which the cheek teeth are still rooted and the upper incisors, as in *Geomys*, are bisulcate (i.e., have two grooves); it is very similar to *Nerterogeomys* of the Blancan, and earlier forms of *Pliogeomys* are probably ancestral to that genus.

Pliogeomys parvus, found at Hagerman, is the last-known representative of the genus and the only Blancan geomyid with rooted teeth. Its marginal geographic position, compared with contemporary and later *Nerterogeomys* and *Geomys*, is seen as an example of Matthew's (1939) rule, which states that species on the periphery of the range tend to retain primitive characteristics. (Matthew, 1939; Zakrzewski, 1969.)

†Smith's Pocket Gopher, *Nerterogeomys smithi* (Hibbard), 1967 (*Geomys smithi*)

The genus *Nerterogeomys* is characterized by a forward projection (the so-called heteromyid projection) of the masseteric crest in the mandible, beneath which the mental foramen is situated. The masseteric crest has a ventral and medial, but no dorsal, branch. *Nerterogeomys* is an extinct line of geomyids which ranges from early Blancan to early Irvingtonian. The cheek teeth were evergrowing; the ancestral Hemphillian *Pliogeomys* had rooted teeth. A transitional form, *Pliogeomys transitionalis*, is known from the latest Hemphillian.

Nerterogeomys smithi from Fox Canyon may be the earliest *Nerterogeomys*. It is a small gopher, somewhat smaller than the Recent *Geomys bursarius texensis*. The pit for the temporal muscle between M_3 and the ascending ramus is less developed than that of *Nerterogeomys minor*, which is the same size as *N. smithi*. (Akersten, 1973; Hibbard, 1967.)

†Small Pocket Gopher, *Nerterogeomys minor* (Gidley), 1922 (*Geomys minor*)

This species, from Rexroad and Benson, is similar in size to *Nerterogeomys smithi*, but the temporal pit is more developed. (Gazin, 1942; Hibbard, 1967; Zakrzewski, 1969.)

†Early Plains Gopher, *Nerterogeomys paenebursarius* (Strain), 1966 (*Geomys paenebursarius*)

This species, from Hudspeth and Red Light in western Texas, is probably late Blancan in age. A long diastema, relatively short tooth row, absence of enamel on the hind wall of P^4, and square reentrants of P_4 are among the characters of this species, which is about the same size as an average individual of *Geomys bursarius*. (Akersten, 1972; Strain, 1966.)

†Hay's Pocket Gopher, *Nerterogeomys persimilis* (Hay), 1927 (*Geomys persimilis; G. parvidens* Gidley, 1922, non *G. parvidens* Brown, 1908)

The type species of *Nerterogeomys*, this small pocket gopher is close to *Geomys bursarius* in size but has a somewhat shorter face and smaller incisors; the mental foramen is slightly more anterior in position than that of other *Nerterogeomys*. It is known from the earliest Irvingtonian of Curtis Ranch and is thus the latest known *Nerterogeomys*. Unlike *Nerterogeomys paenebursarius*, it retains enamel on the posterior wall of P^4. (Gazin, 1942; Hibbard, 1967.)

†Adams's Pocket Gopher, *Geomys adamsi* Hibbard, 1967

The genus *Geomys* is thought to have arisen from the Hemphillian *Pliosaccomys*, which may also be close to the ancestry of the line leading to *Nerterogeomys;* a possible common ancestor is known from the late Clarendonian. *Geomys* has rootless cheek teeth in which dentine tracts develop to form an interrupted enamel pattern. The mental foramen is in front of the masseteric ridge, which loses the heteromyid projection and develops a dorsal branch.

Geomys adamsi from the Fox Canyon local fauna is the earliest *Geomys*. A small species, it is inferior in size to *Nerterogeomys smithi* from the same site and is characterized by the very shallow temporal pit in the mandible. (Hibbard, 1967.)

†Jacob's Pocket Gopher, *Geomys jacobi* Hibbard, 1967

This species, from the Rexroad local fauna, has a temporal pit in the mandible slightly deeper than that of its predecessor *Geomys adamsi* from Fox Canyon but shallower than that of *Geomys quinni*. It was also somewhat smaller than *Geomys quinni*. (Hibbard, 1967.)

†Quinn's Pocket Gopher, *Geomys quinni* McGrew, 1944

This late Blancan species is known from Broadwater, Deer Park, Lisco, and Sand Draw. A relatively small form, it is larger than *Geomys jacobi* and has a deeper temporal pit. The interruption of the enamel pattern of the premolars is delayed, a primitive character. Also primitive is the shape of the premolars, which have divergent reentrant folds (as a result of wear, this character is lost at an early stage in the life history of modern *Geomys* but persists in old individuals in some extinct species). The very narrow rostrum is distinctive. (Skinner and Hibbard, 1972; White and Downs, 1961.)

†Garbani's Pocket Gopher, *Geomys garbanii* White and Downs, 1961

A large sample of this species has been recovered from the Irvingtonian Vallecito beds. It is a large species, surpassing in size the average individual of *Geomys bursarius*, and is distinguished by its broad, deep rostrum, relatively short diastema, and large tympanic bullae. As in *Geomys quinni,* a delayed interruption of the enamel pattern in the premolars may still be observed. Such is not the case in later *Geomys,* in which the interruption comes early in life.

The living genus *Zygogeomys* of Mexico seems to retain some characters of *Geomys garbanii* and is thought to be descended from that species. (White and Downs, 1961; Zakrzewski, 1969.)

†Tobin Pocket Gopher, *Geomys tobinensis* Hibbard, 1944

This species is known from the later Blancan (Sanders) and Irvingtonian (Cudahy) of Kansas, and Paulson (1961) has pointed out some differences between the samples from the two sites. In evolutionary development, it is similar to the larger *Geomys garbanii;* in both species, there is a delayed interruption of the premolar enamel pattern during wear. In this respect, *Geomys tobinensis* appears to be slightly more advanced than *Geomys garbanii.* It also differs from that species in its deeper temporal pit. (Paulson, 1961; White and Downs, 1961.)

Plains Pocket Gopher, *Geomys bursarius* (Shaw), 1800 (*Mus bursarius; Geomys bisulcatus* Marsh, 1871; *G. parvidens* Brown, 1908)

The species *Geomys bursarius* has been found in deposits ranging from the late Irvingtonian (Kentuck, Mullen II, Conard Fissure) to the late Pleistocene and Holocene. Upwards of 30 sites in Arkansas, Colorado, Illinois, Kansas, Kentucky, Missouri, Nebraska, Oklahoma and Texas have yielded fossils of the plains pocket gopher. These sites are mostly within the present range, with the notable exception of Welsh Cave, Kentucky, which is well east of the modern limit of the species. The Welsh Cave fauna dates from about 13,000 years B.P. At that time, the area was a prairie; only later was it invaded by deciduous forest, which does not support *Geomys.*

The early form from Arkansas described as *Geomys parvidens* is regarded as a subspecies of *Geomys bursarius* by White and Downs (1961). It is small, but some present-day subspecies are no larger. The living species shows a size cline, with size increasing from south to north.

With a body length of 190–350 mm, this species is one of the largest pocket gophers. The upper incisors are bisulcate, and the premolar pattern is interrupted. The temporal pit is deeper than that of *Geomys tobinensis* or *Geomys quinni,* either of which could be ancestral.

The species inhabits the grassy plains, where it is a burrower, and nests in tunnels or in mounds built on the surface. It feeds on roots, bulbs, and stems of the indigenous vegetation and may become a pest in cultivated fields. The cheek pouches are used to transport food to subterranean stores.

Material of *Geomys,* not determined as to species, has been found at many other Rancholabrean sites. A record from Centipede Cave is outside the range of *Geomys bursarius* and may represent another species, perhaps the local *Geomys personatus* True. (Brown, 1908; Guilday, Hamilton, and McCrady, 1971; Lundelius, 1963; White and Downs, 1961.)

Southeastern Pocket Gopher, *Geomys pinetis* Rafinesque, 1817

This species has a limited present-day distribution in Alabama, Georgia, and Florida. All of the fossil finds come from Florida, where specimens have been found at more than 15 sites, ranging in age from late Irvingtonian (Coleman IIA) and Sangamonian

(Reddick) to late Wisconsinan and Holocene. The southernmost localities are Vero, on the east coast, and Seminole Field, on the west; both sites are within the modern range of the species.

A somewhat smaller form than *Geomys bursarius*, the species has a long diastema, a comparatively broad and shallow rostrum, and, typically, hourglass-shaped nasal bones as seen from above. Its habitat is pinewoods and fields.

Although this species is now considered distinct from *Geomys bursarius*, Guilday, Hamilton, and McCrady (1971) point to the possibility that the two species may have formed a continuous prairie population in Wisconsinan times and that the present dichotomy is due to extinction of interjacent and transitional segments of that population (of which the Welsh Cave *Geomys* sp. could have been a member). Much more material is needed, however, to substantiate such a hypothesis. (Guilday, Hamilton, and McCrady, 1971; R. A. Martin, 1974a; Weigel, 1962; White and Downs, 1961.)

† Benson Pocket Gopher, *Cratogeomys bensoni* Gidley, 1922

Cratogeomys is distinguished from other genera of pocket gophers in having a single median groove on each upper incisor.

The Benson pocket gopher is only known from the early Blancan of Benson. It was comparable in size to the living *Cratogeomys castanops*. It is placed in this genus because the upper incisors carry only one groove. However, it has been noted that there are also traces of an inner groove in the Benson form, although not comparable to that in *Geomys;* this suggests an earlier common ancestry with *Geomys*, and Akersten (1973) has suggested that *Cratogeomys* arose by way of the late Hemphillian *Parapliosaccomys* from earlier Hemphillian *Pliosaccomys*, both of which are also ancestral to *Geomys*.

Compared with *Cratogeomys castanops*, the Benson species has somewhat narrower cheek teeth and a more slenderly built mandible. (Akersten, 1973; Gazin, 1942; Russell, 1968.)

Mexican Pocket Gopher, *Cratogeomys castanops* (Baird), 1852 *(Pseudostoma castanops)*

Fossil records of this species come from western Texas (Williams Cave), New Mexico (Burnet Cave), and Mexico (San Josecito Cave, etc.), within its present range; all are of Wisconsinan age. It is a relatively large form, measuring some 210–320 mm in length, and has single-grooved upper incisors. Its habitat is sandy soil. (Cushing, 1945.)

† Giant Hispid Gopher, *Heterogeomys onerosus* Russell, 1960

Heterogeomys has unisulcate upper incisors with a deep groove, located to the inner side of the median line.

Heterogeomys onerosus has been found only at San Josecito Cave, Nuevo León, Mexico, and is the only known fossil occurrence of the genus. It is outside the modern range of the genus, which is from southern Tamaulipas into Guatemala. Presence of this southern form could indicate that part of the Rancholabrean deposit in San Josecito Cave is of Sangamonian age.

The fossil species is distinguished from the living *Heterogeomys hispidus* (Le Conte) by its somewhat larger size and deeper, more robust mandible. This could be an infraspecific clinal difference. On the other hand, the fossil form could be ancestral to the living, in which case a late Wisconsinan–Holocene size reduction must be postulated. (Russell, 1960.)

Family Heteromyidae—Kangaroo Rats

The New World Heteromyidae have a stratigraphic range extending from the lower Oligocene to the Recent in North America and in the Recent in northern South America. They are closely related to the pocket gophers (Geomyidae).

The five extant genera vary considerably in external appearance, but all have a long tail and invertible, fur-lined cheek pouches. *Dipodomys* and *Microdipodops* are saltatorial and have small forelimbs and powerful hind limbs; in *Perognathus, Liomys,* and *Heteromys,* the limbs are of nearly equal size. Heteromyid skulls are thin-walled, with threadlike zygomatic arches and greatly inflated auditory bullae. The incisors are thin and compressed, and the cheek teeth have a simplified crown pattern.

Heteromyids are superbly adapted to live in arid regions. Physiological adaptations of the desert-living species include the ability to obtain water by oxidation of food and to conserve water by concentration of urine and feces and reduction of evaporative loss from the skin and respiratory passages. Species that inhabit semiarid regions do not show these extreme specializations. Heteromyids are nocturnal and feed primarily on seeds, which are often stored in their deep burrows.

Heteromyids became adapted to desert living in the Pliocene, when much of their habitat was arid. Three subfamilies (Dipodomyinae, Perognathinae, and Heteromyinae) and five genera are recognized in the North American Blancan and Pleistocene.

Subfamily Dipodomyinae

†Central Kangaroo Rat, *Prodipodomys centralis* (Hibbard), 1941 (*Liomys centralis; Prodipodomys rexroadensis* Hibbard, 1954)

The fossil record of the Dipodomyinae (*Prodipodomys, Etadonomys, Dipodomys*) extends back to the Pliocene in western North America. *Prodipodomys,* a Blancan and Irvingtonian genus, was formerly thought to be ancestral to *Dipodomys.* It has rooted cheek teeth.

Prodipodomys centralis is known from Rexroad, Sand Draw, Donnelly Ranch, and Blanco. It is larger than *Prodipodomys kansensis* Hibbard from middle Pliocene deposits in Kansas and smaller than *Prodipodomys idahoensis* from Hagerman. The teeth are moderately hypsodont and lack dentine tracts. The tips of the nasals extend beyond the deeply grooved incisors (see fig. 12.5). At Sand Draw, the animal was part of the upland, or dry plains, community.

Prodipodomys centralis is thought to be a generalized form that persisted after the development of the more advanced *Prodipodomys idahoensis.* Instead of being directly ancestral to *Dipodomys, Prodipodomys* was probably a collateral lineage. (Hibbard, 1962; Skinner and Hibbard, 1972.)

†Idaho Kangaroo Rat, *Prodipodomys idahoensis* Hibbard, 1962

Originally described from Hagerman, this species has also been identified at California Wash. The Idaho kangaroo rat has rooted premolars and molars and slight dentine tracts developed along the sides of the teeth. The teeth are more hypsodont than those of *Prodipodomys centralis. Prodipodomys idahoensis* is the most advanced species of the genus.

Prodipodomys has also been reported from Cita Canyon, Red Light, Benson, White

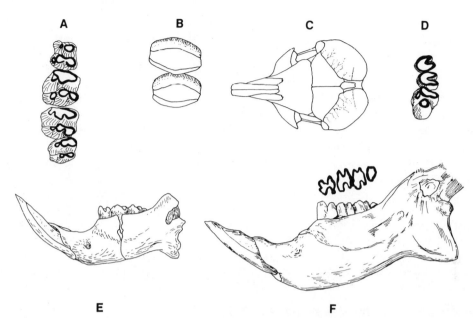

12.5 Heteromyidae. *A, E, Perognathus maldei,* Blancan, Hagerman. *A,* left P_4-M_3, occlusal view; *E,* left mandible, external view. *B, C, Dipodomys merriami,* Recent. *B,* right M^{1-2}, occlusal view; *C,* skull, dorsal view. *D, F, Prodipodomys centralis,* Blancan, Sand Draw. *D,* left M^{1-3}, occlusal view; *F,* left mandible, external view, with occlusal view of P_4-M_3. *A, E,* after Zakrzewski (1969); *B, C,* after Vaughan (1972); *D, F,* after Skinner and Hibbard (1972). *A,* 8×; *B,* ca. 10×; *C,* 1×; *E, D, F,* 4×.

Rock, Sanders, Java, Vallecito Creek, and Arroyo Seco, but the material has not been identified as to species. (Hibbard, 1962; Zakrzewski, 1969.)

†Tihen's Kangaroo Rat, *Etadonomys tiheni* Hibbard, 1943

A problematical monotypic genus, *Etadonomys* was described from the Borchers fauna and is also present in the Arroyo Seco and Vallecito Creek faunas. It is distinguished primarily by the position of the temporal fossa on the lower jaw. The teeth are indistinguishable from those of *Prodipodomys. Etadonomys tiheni* was about the size of *Dipodomys ordii.* (Hibbard, 1943b; Zakrzewski, personal communication, 1976.)

†Gidley's Kangaroo Rat, *Dipodomys gidleyi* Wood, 1935

The kangaroo rat genus *Dipodomys* comprises the largest heteromyids and is characterized by a large head, compact body, small forelimbs, large, powerful hind limbs, and a long, proplike, tufted tail. The enormous auditory bullae, whose volume exceeds that of the braincase, are formed by the mastoid and tympanic bones and are adapted to amplify low-intensity sounds (see fig. 12.5C). A keen sense of hearing enables the animals to avoid predators. With a size that ranges from that of a large mouse to that of a rat, kangaroo rats feed primarily on seeds; they seldom drink water. These saltatory rodents inhabit arid and semiarid regions west of the Missouri River, from southern Canada to south-central Mexico. Their ancestry may extend back to *Eodipodomys,* a recently described dipodomyine from a Clarendonian deposit in Nebraska that shows many resemblances to *Dipodomys* (Voorhies, 1975).

The only known extinct taxon, *Dipodomys gidleyi* was described from Curtis Ranch (early Irvingtonian). It is distinguished in part by the protolophid of P_4 being narrower, but longer, than the metalophid, and the M_3 is not greatly reduced. *Dipodomys* has been reported from the Blancan (Broadwater, Grand View, Arroyo Seco) and the Irvingtonian (Vallecito Creek, Tusker, Angus), but the material has not been identified as to species. (Vaughan, 1972; Voorhies, 1975; Wood, 1935.)

Ord's Kangaroo Rat, *Dipodomys ordii* Woodhouse, 1853

The most widely distributed kangaroo rat is *Dipodomys ordii,* which inhabits the Great Basin, Rocky Mountains, and Great Plains. The earliest records of this species are late Illinoian (Butler Springs, Slaton), and specimens have been found at Jinglebob, Schulze, Howard Ranch, Klein, Isleta, and Wasden. A small species, it has a head and body length of 100–115 mm and a tail length of 127–152 mm. The incisors are rounded, not flat across. (Dalquest, Roth, and Judd, 1969.)

Giant Kangaroo Rat, *Dipodomys ingens* (Merriam), 1904 *(Perodipus ingens)*

Found today only in the Central Valley of California, the giant kangaroo rat apparently had an equally limited Pleistocene distribution, for it has been identified only at McKittrick. It is one of the largest species, with a head and body length of 140–153 mm and a tail length of 175–205 mm. Unlike the other species, it eats mostly green vegetation. (Burt and Grossenheider, 1976.)

Agile Kangaroo Rat, *Dipodomys agilis* Gambel, 1848

This is the common kangaroo rat of coastal southern California; it has been found in the fossil state at Newport Bay Mesa and Rancho La Brea. It prefers gravelly or sandy soils on slopes, washes, and open chaparral areas. (Burt and Grossenheider, 1976.)

Bannertail Kangaroo Rat, *Dipodomys spectabilis* Merriam, 1890

This extant species inhabits the arid and semiarid grasslands of Arizona, New Mexico, and northern Mexico. The only fossil occurrences are from Isleta and Dry caves. A large species, the bannertail kangaroo rat is characterized by its conspicuous, white-tipped tail. It builds large mounds and stores vast quantities of seeds. (Burt and Grossenheider, 1976.)

Merriam's Kangaroo Rat, *Dipodomys merriami* Mearns, 1890

This species inhabits the low desert regions of the southwestern United States, Baja California, and Mexico; the only reported Pleistocene occurrence is from Isleta Cave. It is the smallest kangaroo rat, with a head and body length of 102 mm and a tail length of 127–160 mm. (Burt and Grossenheider, 1976.)

Subfamily Perognathinae

† Rexroad Pocket Mouse, *Perognathus rexroadensis* Hibbard, 1950

Pocket mice (subfamily Perognathinae, genus *Perognathus*) are small heteromyids with hind legs not much longer than the forelimbs. They inhabit the arid and semiarid regions of southwestern Canada, the western half of the United States, northern Mexico, and Baja California; all of the Blancan and Pleistocene occurrences are within the present range. Pocket mice can go for long periods without drinking, and their body is physiologically adapted to conserve water. The fossil history of *Perognathus* extends back to the Miocene.

Known only from the Blancan Fox Canyon and Blanco faunas, *Perognathus rexroadensis* is larger than *Perognathus gidleyi* and differs from it in the morphology of P_4. It differs from the ancestral *Perognathus mclaughlini* Hibbard from the late Hemphillian Saw

Rock fauna in its larger size and the larger posterior loph on M_3. In size and morphology, the teeth are similar to those of *Perognathus hispidus,* and the latter is probably a descendant species. An evolutionary sequence starting with *Perognathus mclaughlini* and ending with *Perognathus hispidus* is thus suggested. (Hibbard, 1950.)

† Gidley's Pocket Mouse, *Perognathus gidleyi* Hibbard, 1941

A Blancan species, *Perognathus gidleyi* is present in the Rexroad and Borchers faunas. It is larger than *Perognathus pearlettensis.* The lower teeth have deeper reentrant angles and valleys than specimens with comparable wear from Recent species. (Hibbard, 1941b.)

† Large Pocket Mouse, *Perognathus magnus* Zakrzewski, 1969

Known only from Hagerman, this large species was about the size of *Perognathus hispidus* but differed from it and other pocket mice in the circular enamel pattern on the occlusal surface of M^3. *Perognathus magnus* does not appear to be closely related to any other species of pocket mouse. (Zakrzewski, 1969.)

† Malde's Pocket Mouse, *Perognathus maldei* Zakrzewski, 1969

Another species found only at Hagerman is *Perognathus maldei.* Medium-sized, it was about the size of *Perognathus parvus* but differed from it in having a larger, more quadrate P_4. The H pattern on the molars is not as strongly developed as in *Perognathus magnus. Perognathus maldei* closely resembles the extant *Perognathus parvus* and was probably ancestral to it. (Zakrzewski, 1969.)

† Pearlette Pocket Mouse, *Perognathus pearlettensis* Hibbard, 1941

This small, late Blancan pocket mouse is known from Fox Canyon, Sanders, Borchers (type locality), White Rock, and Blanco; it has also been reported from a Clarendonian deposit in Oklahoma. It is distinguished from *Perognathus flavus* by its slightly larger size, heavier incisors, greater depth of ramus below the posterior edge of M_2, and more robust condyloid process.

Other Blancan localities containing *Perognathus* include Red Light, Cita Canyon, Benson, and White Rock; the material has not been identified as to species. (Hibbard, 1941d.)

California Pocket Mouse, *Perognathus californicus* Merriam, 1889

An inhabitant today of chaparral and live oak–covered slopes from central California south into Baja California, *Perognathus californicus* is known from the late Pleistocene of Newport Bay Mesa, Costeau Pit, and Rancho La Brea. A medium-sized species, it has a head and body length of 79–90 mm and tail length of 100–150 mm. It has spinelike hairs on its rump. (W. Miller, 1971.)

Silky Pocket Mouse, *Perognathus flavus* Baird, 1855

The only Pleistocene occurrence of this small species is from Isleta Cave. The silky pocket mouse lives on the shortgrass prairie, from Arizona east to western Nebraska and south into Mexico. (Harris and Findley, 1964.)

San Joaquin Pocket Mouse, *Perognathus inornatus* Merriam, 1889

This species is endemic to the San Joaquin and Sacramento valleys of California, where it is found in dry, open, grassy or weedy areas. The only Pleistocene record of *Perognathus inornatus* is from McKittrick. (Burt and Grossenheider, 1976.)

Rock Pocket Mouse, *Perognathus intermedius* Merriam, 1889

This extant species was identified at Isleta. It lives on rocky slopes and old lava flows with sparse vegetation in Arizona, south-central New Mexico, and northern Mexico. The rock pocket mouse is about the size of *Perognathus californicus* but has a shorter tail. (Harris and Findley, 1964.)

Great Basin Pocket Mouse, *Perognathus parvus* (Peale), 1848 (*Cricetodipus parvus*)

Remains of this extant species have been found at Wasden and Moonshiner. *Perognathus parvus* inhabits the sagebrush and chaparral country of the Great Basin. It has a head and body length of 63–76 mm and a tail length of 82–102 mm. *Perognathus maldei* was probably ancestral. (Burt and Grossenheider, 1976.)

Hispid Pocket Mouse, *Perognathus hispidus* Baird, 1858

The hispid pocket mouse has been found at some 20 Pleistocene localities, the earliest of which are late Irvingtonian (Berends, Sandahl, Slaton). A common species, it is found today in areas of sparse vegetation and friable soil across the shortgrass prairies. It is a fairly large species, with a head and body length of 115–127 mm and a tail length of 90–115 mm. Kennerly (1956) found no morphlogical differences between the mandibles of a large sample (>300) of specimens from Friesenhahn Cave and those of Recent animals. (Kennerly, 1956; Semken, 1966.)

Merriam's Pocket Mouse, *Perognathus merriami* J. A. Allen, 1892

This small species (head and body length, 55–70 mm; tail length, 38–51 mm) is found on the open plains of central and eastern New Mexico, the western half of Texas, and northern Mexico. It has been identified at Longhorn Caverns and Schulze Cave. At Schulze, more than 1,000 specimens have been found, probably representing the remains of owl kills. The species shows wider environmental tolerances than other species of pocket mice. (Dalquest, Roth, and Judd, 1969.)

Little Pocket Mouse, *Perognathus longimembris* (Coues), 1875 (*Otognosis longimembris*)

An inhabitant of the southwestern deserts, the little pocket mouse has only been reported in the fossil state from Glendale (late Pleistocene). It is about the size of Merriam's pocket mouse but has a longer tail. It feeds primarily on small seeds. (Burt and Grossenheider, 1976.)

Apache Pocket Mouse, *Perognathus apache* Merriam, 1889

The only reported Pleistocene occurrence of this species is from Papago Springs. This tiny species inhabits semiarid and arid regions with sparse cover in northeastern Arizona, southeastern Utah, southwestern Colorado, and the western half of Mexico. (Burt and Grossenheider, 1976.)

SUBFAMILY HETEROMYINAE

Mexican Spiny Pocket Mouse, *Liomys irroratus* (Gray), 1868 (*Heteromys irroratus*)

The genus *Liomys* is found from the Lower Rio Grande Valley in southern Texas to northern South America. Mouselike in appearance, spiny pocket mice are covered with the coarse, stiff, flattened guard hairs that give the animals their name.

Liomys irroratus is found mainly in the Sonoran zone of central Mexico; the only Pleistocene record is from San Josecito. The species has a robust skull, a slender rostrum, widely spreading zygomatic arches, and moderately inflated auditory bullae. Food consists of seeds and tender herbaceous plants. (Hall and Kelson, 1959.)

Family Castoridae—Beavers

Beavers are large, semiaquatic rodents best known for their tree-cutting and dam-building activities. Their fossil history extends back to the early Oligocene in North America, the late Oligocene in Europe, and the early Miocene in Asia. Restricted to the Northern Hemisphere, beavers are found along stream banks and the shores of lakes and marshes.

Beavers have high-crowned cheek teeth, and a well-developed fourth premolar is present in both jaws. The infraorbital foramen is not enlarged, but there is a large channel below the orbit. The origin of the family is uncertain.

The Castoridae are represented by two groups in the North American Blancan and Pleistocene. The first of these, the extinct castorid group, includes *Dipoides*, *Paradipoides*, *Procastoroides*, and *Castoroides* and is characterized by ever-growing cheek teeth, an S-shaped occlusal pattern on the molars, and convex-faced incisors. The second group, which includes the modern beaver, genus *Castor*, is characterized by semiflattened, chisellike incisors.

† Rexroad Beaver, *Dipoides rexroadensis* Hibbard and Riggs, 1949

Dipoides, a poorly known genus of early beavers, has been found in Pliocene deposits in Eurasia and has a stratigraphic range extending from the Hemphillian to the late Blancan in North America.

Dipoides rexroadensis has been identified at Rexroad 3 (type locality), Keefe Canyon, Wendell Fox, and Sand Draw. The large cheek teeth are prismatic and hypsodont, and the molars show an S-shaped occlusal pattern. The species is probably a descendant of *Dipodes wilsoni* Hibbard from the Hemphillian Saw Rock local fauna in Kansas. (Woodburne, 1961.)

† Intermediate Beaver, *Dipoides intermedius* Zakrzewski, 1969

Known only from Hagerman, this early Blancan species is intermediate in size between *Dipoides rexroadensis* and *Procastoroides sweeti*. It is distinguished from other species of *Dipoides* by its large size and the complete metastria on M^3 (see fig. 12.6C). Characteristics of the teeth and foot bones indicate that *Dipoides intermedius* was close to the stock that gave rise to *Procastoroides*. (Zakrzewski, 1969.)

† Stovall's Beaver, *Paradipoides stovalli* Rinker and Hibbard, 1952

The genus *Paradipoides* has been reported from Berends, Cumberland, and (with less certainty) Mt. Scott. The taxonomic status of the genus has been questioned (see Martin, 1969; Zakrzewski, 1969); the taxon may be congeneric with *Dipoides*.

The type of *Paradipoides stovalli*, described from Berends, was based on an immature individual, and many of its characters are duplicated in immature specimens of *Castoroides*. The species was about the size of *Castor canadensis*. The Cumberland specimen, an isolated molar, belongs to a small beaver about the size of *Dipoides wilsoni;* it is tentatively referred to *Paradipoides*. (R.A. Martin, 1969; Zakrzewski, 1969 and personal communication, 1976.)

†Sweet's Beaver, *Procastoroides sweeti* Barbour and Schultz, 1937 (*Eocastoroides lanei* Hibbard, 1937)

Procastoroides, a large beaver about two-thirds the size of *Castoroides,* is known from the Blancan and early Irvingtonian of North America, where two species are recognized.

Procastoroides sweeti is a Great Plains species first described from Broadwater; it has also been found at Sand Draw, Mullen I, Delmont, Rexroad, Deer Park, Dixon, White Rock, and Red Corral (Texas). The absence of crenulations on the enamel of the incisors is a distinguishing character (see fig. 12.6A). Absence of these crenulations was one of the reasons that led Woodburne (1961) to believe that *Procastoroides* was not ancestral to *Castoroides* and was too specialized to be transitional to *Castor.* (Woodburne, 1961.)

†Idaho Beaver, *Procastoroides idahoensis* Shotwell, 1970

This species has been found only at Jackass Butte in the Grand View local fauna (very late Blancan). Unlike the condition in *Procastoroides sweeti,* crenulations are present on the enamel of the incisors. These longitudinal grooves, along with characteristics of the occlusal pattern of the cheek teeth, indicate that *Procastoroides* was ancestral to *Castoroides,* as Shotwell (1970) has suggested. (Shotwell, 1970; Zakrzewski, personal communication, 1976.)

†Giant Beaver, *Castoroides ohioensis* Foster, 1838 (*C. nebrascensis* Barbour, 1931; *Burosor efforsorius* Starrett, 1956)

Castoroides, an animal about the size of a black bear, was the largest rodent in North America during the Pleistocene. *Procastoroides* is probably ancestral to *Castoroides.* A somewhat smaller but otherwise similar animal, *Trogontherium,* is known from the Pleistocene of Europe. Only one species, *Castoroides ohioensis,* is now recognized (Zakrzewski, personal communication, 1976).

With a temporal range extending from the Blancan (Santa Fe River IB) to the late Wisconsinan, *Castoroides ohioensis* has been reported from some 30 local faunas, as well as hundreds of isolated sites. Although its remains are known from Alaska south to Florida and from Nebraska east, the giant beaver was most abundant in the region south of the Great Lakes, in what is now Indiana and Illinois. Most of the material is fragmentary, but a nearly complete skeleton was found in Randolph County, Indiana, in 1889 (mounted at Earlham College).

An inhabitant of lakes and ponds bordered by swamps, *Castoroides ohioensis* was more similar in habits to *Ondatra* than to *Castor.* There is no evidence that the giant beaver built dams or felled trees. It had shorter legs than *Castor,* small front feet, and large hind feet that were probably webbed. The tail was relatively longer and narrower than that of *Castor.* The overall length of this giant swamp rodent was about 2.5 m, and the animal weighed between 150 and 200 kg. Clumsy on land, *Castoroides* was probably a powerful swimmer.

The enormous, convex incisors (extending 100 mm beyond the gum line) are longitudinally fluted with enamel on the anterior and labial surfaces. The tips of the incisors are blunt and rounded, quite unlike the sharp, straight-edged incisors of *Castor* (see fig. 12.6). The cheek teeth consist of a number of flattened tubes of enamel enclosing dentine; the tubes are held together by cement and resemble those of the capybara (*Hydrochoerus*). The teeth were used in cutting off and grinding up the coarse swamp vegetation on which the giant beaver fed. At Boney Spring 8 mastodont (*Mammut*

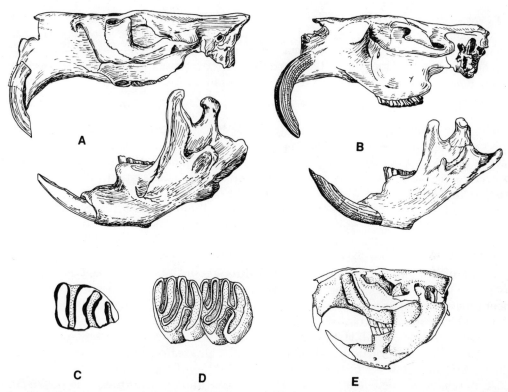

12.6 Castoridae. *A, Procastoroides sweeti,* Blancan, Broadwater, skull and mandible. *B, Castoroides ohioensis,* Rancholabrean, St. Paul, Ramsey Co., Minnesota, skull and mandible. *C, Dipoides intermedius,* Blancan, Hagerman, right M^3, occlusal view. *D, E, Castor canadensis,* Recent. *D,* right M^{1-2}, occlusal view; *E,* skull and mandible. *A,* after Barbour and Schultz (1937); *B,* after Erickson (1962); *C,* after Zakrzewski (1969); *D, E,* after Vaughan (1972). *A,* 0.33×; *B, E,* 0.25×; *C, D,* 2×.

americanum) cheek teeth show tooth marks probably made by the blunt, lower incisors of *Castoroides,* which has been identified at the site (Saunders, 1977).

Extinction was probably due to competition with *Castor* and reduction and disappearance of the species' preferred habitat. (Erickson, 1962; R. A. Martin, 1969; Moore, 1890; Powell, 1948; Saunders, 1977; Stirton, 1965; Zakrzewski, personal communication, 1976.)

†Kellogg's Beaver, *Castor californicus* Kellogg, 1911

The genus *Castor* formerly inhabited the forested regions of the Northern Hemisphere, but populations are now greatly reduced, especially in Eurasia. Two extant species, *Castor fiber* Linnaeus in Eurasia and *Castor canadensis* in North America, are recognized. *Castor fiber* has a good fossil record, extending back to the Villafranchian in Europe.

Three species of *Castor* are known from the North American Pleistocene. Although both extinct species are larger than the extant one, the primary difference among them appears to be a change in the length of the striae along the sides of the teeth. Unfortunately, no detailed comparison has ever been made of the material assigned to the extinct species; it is possible that such a study might show that the taxa are conspecific or that some of the assignments are incorrect (Zakrzewski, personal communication, 1976).

Castor californicus was first described from the Kettleman Hills, Fresno, County, California, and has also been reported from Hagerman (an extensive collection), Sand Point, Idaho, Powell County, Montana (late Pliocene or early Pleistocene), and Mullen I. The striae on the sides of the teeth are shorter than those of *Castor accessor*. (Kellogg, 1911; Rasmussen, 1974; Zakrzewski, personal communication, 1976.)

† Hay's Beaver, *Castor accessor* Hay, 1927

The type locality of this late Blancan–late Irvingtonian species is Grand View, and specimens have also been found at Angus and Mullen II. The striae on the sides of the teeth are longer than those of *Castor californicus* but shorter than those of the living beaver, *Castor canadensis*. *Castor accessor* has been found in association with *Castoroides* at Mullen and Angus.

Castor sp. has been reported from White Bluffs, White Rock, Mt. Scott, Java, Rushville, Gordon, Berends, Fossil Lake, and American Falls. (Shotwell, 1970.)

Beaver, *Castor canadensis* Kuhl, 1820

A widespread and fairly common species throughout the Pleistocene, *Castor canadensis* continues to flourish along waterways in North America, with the exception of peninsular Florida, desert areas, and Arctic shores. The earliest record is late Blancan (Haile XV); Irvingtonian occurrences include Port Kennedy, Trout, Cumberland, Conard Fissure, and Medicine Hat (D–E), and the species has been reported from some 45 Rancholabrean sites, including several in Florida, an area outside the animal's present range. Remains of fossil beaver dams have been found in many areas. [Kaye (1962) believes that beaver activities in ponds stir up the sediments, thus upsetting pollen stratigraphy.]

Beavers are large (weighing from 9 to 32 kg), thickset rodents that have short legs and a broad, scaly tail. The small eyes are equipped with a nictating membrane, and the nostrils and ear openings are valvular and can be closed during submersion. The hind feet are webbed, and the second and third claws on each foot are split to form a comb for grooming the thick, luxurious fur. With their many aquatic adaptations, beavers naturally are good swimmers and divers.

The robust skull has a long external auditory meatus that is surrounded by a tubular extension of the bullae. Beavers have huge, chisellike incisors and high-crowned cheek teeth that have a complex occlusal pattern with transverse folds (see fig. 12.6E).

Beavers can modify their environment by cutting down trees and damming streams. Using their upper incisors as levers, they fell trees by a gnawing action of the lower incisors. A downed tree is cut into sections and dragged to the water. In northern and montane areas, beavers build elaborate lodges with underwater chambers out of mud and sticks; in the Southwest and Midwest, they dig burrows in the banks of streams and lakes. Beavers feed on bark, cambium, small twigs, leaves, and roots of aspen, poplar, birch, willow, and alder.

Remains of *Castoroides ohioensis* and *Castor canadensis* have been found together at Sandahl, Berends, Carter, Hartman, Ichetucknee River, Santa Fe River IIA, Old Crow River, and Fairbanks, and competition with *Castor canadensis* was probably a factor in the extinction of the giant beaver. Beavers have been trapped for their valuable fur for millennia, and their flesh and perineal gland secretion, castoreum, are highly prized, yet beavers continue to thrive in many areas. (Kaye, 1962; Vaughan, 1972; Walker, 1975.)

Family Cricetidae—Hamsters and Voles

Although early members of this family appeared well back in the Tertiary, its real evolutionary proliferation, which has made it richer in species than any other rodent family now present in North America, came in Blancan and Pleistocene times. Within this timespan, more than 130 species have been recorded from fossil remains. These species are evenly distributed between the two subfamilies Cricetinae and Microtinae, which are regarded by some authors as distinct families.

The Cricetinae comprise, besides the Old World hamsters, the native cricetine rats and mice of North America. Varied in external appearance, these animals usually have a long tail and pointed muzzle. The molars are basically tubercular, with cusps arranged in two longitudinal rows; trends in evolution include an increase in hypsodonty and, occasionally (e.g., *Neotoma*), a tendency toward a volelike arrangement of alternating triangular prisms (see fig. 12.7).

The Microtinae (Arvicolinae of some authors)—the voles, muskrats, water rats, and lemmings—are shorter-tailed than the cricetines and have a rounded muzzle. The molars are prismatic, and the flat occlusal surface shows triangles and loops separated by reentrant angles; the number of triangles tends to increase in many lineages. Another common evolutionary trend is the development of dentine tracts, or lines, where enamel is absent, extending up from the enamel base; these may take the shape of inverted V's or U's. With molar wear, the tracts are exposed as interruptions of the enamel pattern on the occlusal face. Gradual appearance of cement in the reentrants can be observed in some lineages. There is also a trend toward increasing crown height and the development of rootless, evergrowing molars. An intermediate evolutionary stage, in which molars are rootless in young individuals and rooted in adults, is characteristic of some species. Such trends occur in parallel fashion in various evolving lineages.

By virtue of their abundance and diversity, the cricetid rodents are rapidly becoming one of the most important stratigraphic tools in the study of the Pleistocene.

Subfamily Cricetinae

Marsh Rice Rat, *Oryzomys palustris* (Harlan), 1837 (*O. fossilis* Hibbard, 1955)

The rice rats are found mainly in South America. North American fossil and Recent forms belong to the genus *Oryzomys*, which is characterized by well-developed palatal pits, internal nares positioned well behind the tooth rows, and molars with main cusps directly opposite each other in two longitudinal rows.

The marsh rice rat is the only species found north of Mexico. The fossil form is regarded as a distinct subspecies, *Oryzomys palustris fossilis*. It is known from Illinoian and Sangamonian deposits (Mt. Scott, Reddick, Jinglebob) and occurs in Wisconsinan deposits at some 20 localities distributed in Florida, Georgia, Missouri, and Texas. Some of these records are north (Brynjulfson Cave) and west (Uvalde County, Schulze Cave, Easley Ranch, Howard Ranch) of the present-day range of the species, and the earlier occurrences in Kansas are likewise outside the range.

The marsh rice rat prefers lush meadows and humid thickets close to water, although drier areas may be inhabited if dense cover is available. It consumes shellfish as well as plant matter. (Dalquest, 1965; Dalquest, Roth, and Judd, 1969; Hibbard, 1955c, 1963a.)

†Wetmore's Harvest Mouse, *Reithrodontomys wetmorei* Hibbard, 1952

The harvest mice, genus *Reithrodontomys*, are among the smallest cricetines. The history of the genus begins in the early Blancan. The upper incisors have a deep median groove, and the molars are tubercled, with cusps diagonally placed in two longitudinal rows.

Known only from the early Blancan of Fox Canyon, *Reithrodontomys wetmorei* was about the size of *R. megalotis* but had a more generalized, primitive occlusal pattern. Ectostylids are present on the lower molars. Hibbard has suggested that it was a member of the *megalotis* group and lived in lowland areas. (Hibbard, 1952a.)

†Rexroad Harvest Mouse, *Reithrodontomys rexroadensis* Hibbard, 1952

This species, like *Reithrodontomys wetmorei*, is known only from Fox Canyon. It is smaller than the latter species and lacks the accessory ectostylids on the lower molars. Its size suggests that it inhabited upland areas. (Hibbard, 1952a.)

†Meadow Harvest Mouse, *Reithrodontomys pratincola* Hibbard, 1941

Known from the late Blancan of Kansas (Borchers, White Rock), this species is very small, even smaller than the *Reithrodontomys montanus*, which now occurs in the same area. (Hibbard, 1941d.)

†Moore's Harvest Mouse, *Reithrodontomys moorei* (Hibbard), 1944
   (*Cudahyomys moorei*)

Known only from the deposits of Kansan age at Cudahy, this species is smaller than the related Recent species *Reithrodontomys humulis* and *Reithrodontomys montanus*, which, like the fossil specimens, possess a labial shelf on M_1 and M_2. In the opinion of Paulson (1961), however, M_3 is too reduced and the cusp pattern of M_1–M_2 too advanced for Moore's harvest mouse to be ancestral to living forms. (Hibbard, 1944; Paulson, 1961.)

Plains Harvest Mouse, *Reithrodontomys montanus* (Baird), 1855
   (*Reithrodon montanus*)

The earliest record of this species comes from the Illinoian Sandahl fauna; specimens have also been found in the Sangamonian Jinglebob fauna in Kansas. Later records come from Easley Ranch, Klein Cave, Longhorn Cavern, and Schulze Cave, all in Texas. These sites are within or close to the present-day range of the species.

Reithrodontomys montanus is smaller than *Reithrodontomys megalotis* and has a characteristic C-pattern on M_3. It is a plains and upland form, not found in wet or marshy areas, and prefers short, dense grass. (Dalquest, Roth, and Judd, 1969; Hibbard, 1955c; Semken, 1966.)

Eastern Harvest Mouse, *Reithrodontomys humulis* (Audubon and Bachman),
   1841 (*Mus humulis*)

Fossils of this species ranging from the late Irvingtonian (Coleman IIA) through the Rancholabrean (Arredondo, Haile XI, Reddick, Vero) have been found in Florida. The smallest of all living North American harvest mice, it inhabits wet grasslands and marshes. It is now confined to the southeastern United States, with a range extending west along the Gulf Coast into eastern Texas. Apart from its small size, the species is characterized by recumbent external cusps and labial shelves on M_1–M_2 and an M_3 less reduced than that of *Reithrodontomys montanus* or *Reithrodontomys moorei* (see figs. 12.7 and 12.8).

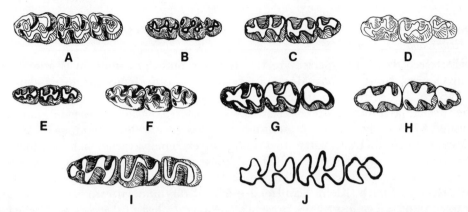

12.7 Cricetidae. Cricetine left M_{1-3}, occlusal views. *A, Oryzomys palustris,* Rancholabrean, Jinglebob. *B, Reithrodontomys humulis,* Irvingtonian, Kanopolis. *C, Peromyscus cragini,* Irvingtonian, Cudahy. *D, Peromyscus progressus,* Rancholabrean, Cragin Quarry. *E, Baiomys rexroadensis,* Blancan, Rexroad. *F, Bensonomys eliasi,* Blancan, Sanders. *G, Symmetrodontomys simplicidens,* Blancan, Rexroad. *H, Onychomys fossilis,* Blancan, Borchers. *I, Sigmodon medius,* Blancan, Borchers. *J, Neotoma micropus,* Rancholabrean, Cragin Quarry. *A,* after Hibbard (1955c); *B, D, J,* after Hibbard and Taylor (1960); *C,* after Hibbard (1944); *E, G, H, I,* after Hibbard (1941a). All 7×.

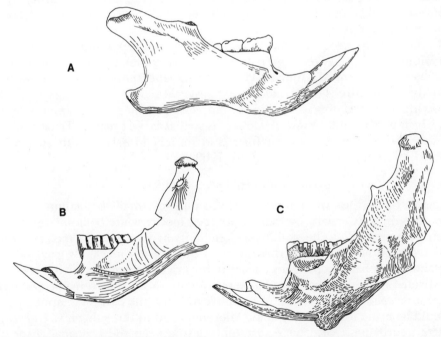

12.8 Cricetidae. Cricetine and microtine mandibles, external views. *A, Reithrodontomys humulis,* Irvingtonian, Kanopolis, right mandible. *B, Ophiomys meadensis,* Blancan, Sanders, left mandible. *C, Pliophenacomys finneyi,* Blancan, Fox Canyon, left mandible. *A,* after Hibbard et al., (1978); *B,* after Hibbard (1956b); *C,* after Hibbard and Zakrzewski (1972). All 3×.

A fossil sample from Costeau Pit, probably of late Illinoian or Sangamonian age, was tentatively referred to this species by W. Miller (1971). The 31-specimen sample is adequate, but in view of the great separation, the identification, as Miller notes, remains problematical; if the record is substantiated, it would indicate that the species was distributed over a great part of the continent in earlier times, (R. A. Martin, 1974a; W. E. Miller, 1971.)

Western Harvest Mouse, *Reithrodontomys megalotis* (Baird), 1858
(*Reithrodon megalotis*)

Earliest records of this species are late Illinoian and Sangamonian (Butler Spring, Cragin Quarry, Newport Bay Mesa). Later records come from California, Missouri, New Mexico, Texas, and San Josecito Cave, in Nuevo León, Mexico.

The species has a tooth pattern resembling that of *Reithrodontomys montanus* but is distinguished by its large size. Its present-day range is very wide, extending from southern Mexico to southern Canada. It is found in extremely varied habitats, including desert, brushland, open woodland and glades, and subarctic mountain meadows. Dense forests are, however, avoided. Some Wisconsinan-age Texan localities (Easley Ranch, Howard Ranch, Klein, Schulze) are east of the modern range, in a hiatus difficult to account for, unless by the presence of *Reithrodontomys fulvescens*, with which the species is not generally sympatric (Semken, personal communication). (Dalquest, Roth, and Judd, 1969; W. E. Miller, 1971; Parmalee, Oesch, and Guilday, 1969; G. E. Schultz, 1969; Semken, personal communication, 1976.)

Fulvous Harvest Mouse, *Reithrodontomys fulvescens* Allen, 1894

There are four Texan fossil records of this species; the earliest is probably early Illinoian (Slaton), although the identification is tentative. Wisconsinan records are from Klein and Schulze caves and Easley Ranch.

Like *Reithrodontomys megalotis*, this is a large species, and separation of the two taxa is difficult without well-preserved material. A distinctive character of the fulvous harvest mouse is the S-pattern of M_3, which is due to a lingual extension of the posterior fold of the tooth; this feature is obliterated by extensive wear.

The occurrence at Slaton is outside of, but close to, the present-day range of the species, which now includes most of Mexico as well as large areas in Texas and neighboring states. It is thus a southern form and prefers grassland with sparse brush. (Dalquest, 1967; Dalquest, Roth, and Judd, 1969.)

†Kansas Mouse, *Peromyscus kansasensis* Hibbard, 1941

The genus *Peromyscus*, white-footed mice, comprises small to medium-sized cricetines, characterized externally by their white feet, mostly white bellies, large ears, and long, usually haired tails. The skull has a long, narrow rostrum, the molars are small, tuberculate, and brachydont, and the upper incisors are not markedly grooved.

The genus dates back to the Miocene. It is very rich in fossil and living species and may be divided into a number of subgenera. Among those known in the Quaternary, the subgenus *Haplomylomys* is the most primitive and also the earliest to appear. In fact, a case could be made for including some Miocene forms in this subgenus. *Haplomylomys* is characterized by absence of accessory tubercles between the principal cusps of the molars. Tubercles are present in the two other subgenera, *Peromyscus* and *Podomys*.

Like most other late Tertiary (including Blancan) *Peromyscus*, the Kansas mouse is in the subgenus *Haplomylomys*. It is known from the early (Rexroad) and late (Sand Draw) Blancan. It is distinguished by the broad valleys between the molar cusps. The length of the series M_1–M_3 is 4.1 mm. (Hibbard, 1941a; Skinner and Hibbard, 1972.)

†Baumgartner's Mouse, *Peromyscus baumgartneri* Hibbard, 1954

Recorded from the Fox Canyon and Rexroad faunas, this species, a member of the subgenus *Haplomylomys*, is one of the smallest of the known Pliocene *Peromyscus* (M_1–M_3,

3.5 mm). It differs from the equally small *Peromyscus cragini* in the shape of M_1, which is broader and lacks an anterior groove. (Hibbard, 1954a.)

†Irvington Mouse, *Peromyscus irvingtonensis* Savage, 1951

Known only from the type locality, Irvington, this species is also a member of the subgenus *Haplomylomys;* it is somewhat larger and has more massive teeth than *Peromyscus kansasensis*. M_3 is reduced, but much less so than in modern representatives of the subgenus. The length M_1–M_3 is 5.2 mm. (Savage, 1951.)

†Cragin's Mouse, *Peromyscus cragini* Hibbard, 1944

This species was originally described from the Kansan Cudahy fauna and has also been identified at White Rock and, tentatively, Mesa de Maya and Kanopolis. It appears to be a late Blancan and Irvingtonian species of the Great Plains. Although having the characters of the subgenus *Haplomylomys*, it is much smaller than the species described above (M_1–M_3, 3.4 mm). (Paulson, 1961; Zakrzewski, personal communication, 1976.)

Canyon Mouse, *Peromyscus crinitus* (Merriam), 1891 (*Hesperomys crinitus*)

This small species has been recorded from Dry Cave in southeastern New Mexico and also, tentatively, in the Newport Bay Mesa fauna. The Dry Cave occurrence is well outside the present-day range of the species, which barely reaches the northwestern corner of New Mexico. Most of the living species of the subgenus *Haplomylomys* have small, relict ranges; the canyon mouse is an exception, being widely distributed in Nevada, Utah, and the neighboring states. It is found in arid situations such as rocky slopes, lava fields, and the like. (Harris, 1970; W. E. Miller, 1971.)

†Santa Rosa Mouse, *Peromyscus nesodytes* Wilson, 1936

An insular species, known only from Rancholabrean deposits on Santa Rosa Island, off the coast of California, this is a large form (M_1–M_3, 5.9 mm). It lacks accessory tubercles on the cheek teeth. (Hibbard, 1967; White, 1966.)

†Anacapa Mouse, *Peromyscus anyapahensis* White, 1966

This species is also known only from Rancholabrean deposits on Anacapa Island, off the Californian coast. It is similar to *Peromyscus nesodytes* but significantly smaller. White has suggested that the large size of these two insular species is indicative of an adaptive trend toward a *Neotoma*-like habitus. The woodrat (*Neotoma*) is not found on either of the islands. (White, 1966.)

†Imperfect Mouse, *Peromyscus imperfectus* Dice, 1925

The species is only known from Rancho La Brea. It is about as large as the living *Peromyscus maniculatus gambelii* of California. Small accessory tubercles are present on M^1 and M^2. The species has a skull with palatine slits proportionally larger and the shelf of the bony palate proportionally shorter than other species of the genus. (Dice, 1925; Hibbard, 1967.)

Cactus Mouse, *Peromyscus eremicus* (Baird), 1858 (*Hesperomys eremicus*)

Fossil cactus mouse is recorded from Friesenhahn Cave, close to the eastern margin of its present-day range. It is an average-sized *Peromyscus*, with high, inflated braincase. It lives in hot, desert areas. (R. A. Martin, 1968b.)

†Progressive Mouse, *Peromyscus progressus* Hibbard, 1960

This species occurs mainly in Illinoian faunas of the Great Plains (Mesa de Maya, Mt. Scott, Sandahl, Slaton) but may be present as early as the Yarmouthian (Kanopolis); it survived in the Sangamonian (Cragin Quarry). Individuals in this population range from the typical *Haplomylomys* condition without accessory tubercles to advanced forms with well-developed accessory folds (see fig. 12.7D). The species represents a transitional position between *Haplomylomys* and more advanced species of *Peromyscus*. It is a medium-sized species (M_1–M_3, 3.75 mm) and appears to be closely related to the living *Peromyscus leucopus*. The cheek teeth are, however, markedly narrower than those of the modern species. (Hibbard and Taylor, 1960; Semken, 1966; Zakrzewski, personal communication, 1976).

†Berends Mouse, *Peromyscus berendsensis* Starrett, 1956

Recorded from the Illinoian Berends and Mt. Scott faunas, this form is similar in size and dental characters to *Peromyscus progressus*, and the two species may turn out to be identical. However, both species were recorded in the Mt. Scott fauna. (Hibbard, 1963a; Semken, 1966.)

†Hagerman Mouse, *Peromyscus hagermanensis* Hibbard, 1962

Known only from the Hagerman Blancan, this species is the same size as *Peromyscus dentalis* Hill; it differs from the latter species in the lack of a deep fossa between the coronoid process and last molar of the lower jaw. A tendency to develop accessory lophs on the molars distinguishes the species from the subgenus *Haplomylomys*. M_1–M_3 is 3.82 mm. (Hibbard, 1962; Zakrzewski, 1969.)

†Cumberland Mouse, *Peromyscus cumberlandensis* Guilday and Handley, 1967

An Irvingtonian form known from the Appalachian region (Cumberland Cave, Trout Cave), this species is also present in the Ladds fauna in Georgia, which is of late Pleistocene age.

It is a large species (M_1–M_3, 5.2 mm), with accessory tubercles well developed in the upper molars. Although more advanced than the *Haplomylomys* condition, it is difficult to fit into any of the Recent subgenera because accessory cusps are not always present in the lower molars. (Guilday and Handley, 1967.)

†Cochran's Mouse, *Peromyscus cochrani* Hibbard, 1955

The subgenus *Peromyscus* (*P. maniculatus*, *P. truei*, *P. polionotus*, *P. gossypinus*, *P. leucopus*, *P. pectoralis*, *P. boylii*) is characterized by the presence of more or less well-developed accessory tubercles in the molars. This subgenus, which evolved toward the end of the Irvingtonian, tends to become dominant in the course of the Rancholabrean, while at the same time there is a decline in the previously dominant subgenus *Haplomylomys*.

An early representative of the subgenus *Peromyscus* is Cochran's mouse, which has been tentatively identified in Illinoian faunas (Doby Springs); the type comes from the Sangamonian of Jinglebob. A record from Clear Creek may extend the stratigraphic range well into the Wisconsinan. It is about the size of the living *Peromyscus leucopus* (M_1–M_3, 3.7 mm). In characters, it is intermediate between *Peromyscus* (*Haplomylomys*) *cragini* of the Kansan and the modern *Peromyscus leucopus* and *P. maniculatus*, both of which belong to the subgenus *Peromyscus*. (Hibbard, 1955c; Slaughter and Ritchie, 1963; Stephens, 1960.)

Deer Mouse, *Peromyscus maniculatus* (Wagner), 1845 (*Hesperomys maniculatus*)

The deer mouse has been found at some 25 localities of Rancholabrean age distributed in Arkansas, California, Georgia, Idaho, Kansas, Missouri, Montana, New Mexico, Pennsylvania, Tennessee, Texas, Utah, Virginia, and Wyoming. Most finds are Wisconsinan in age.

The species has an enormous geographic range, extending from Alaska over most of Canada and the United States, with the exception of the extreme north and the Southeast. Local variation in size and other characters is great, often exceeding differences between good species. However, the species tends to be smaller than *Peromyscus leucopus* in areas where both are found.

As may be gathered from its wide distribution, the deer mouse is highly adaptable and is found in extremely varied situations, including forest, grassland, and desert; local populations tend to be more strictly adapted to their environments. When found associated with *Peromyscus leucopus*, the deer mouse is restricted to open areas. (Burt and Grossenheider, 1976; Ray, 1967.)

Oldfield Mouse, *Peromyscus polionotus* (Wagner), 1843 (*Mus polionotus*)

This species, now found in the southeastern United States, has been recorded in the fossil state from peninsular Florida, at Rancholabrean sites from the Sangamonian on (Arredondo, Reddick, Vero, Haile XIB, Devil's Den). Osteologically, the species resembles *Peromyscus maniculatus*, but is noticeably smaller than sympatric forms of the latter. It inhabits sandy fields and beaches. (Webb, 1974a.)

White-footed Mouse, *Peromyscus leucopus* (Rafinesque), 1818
(*Musculus leucopus*)

The species is widely distributed in Rancholabrean deposits of Arkansas, New Mexico, Pennsylvania, Tennessee, Texas, and Virginia, all within the present-day range. In addition, there is a record from Moonshiner Cave, Idaho, which is definitely west of the range (which extends into eastern Montana and northeastern Wyoming). The Idaho record is early Holocene; other finds date from the Sangamonian (e.g., Easley Ranch) and Wisconsinan.

Peromyscus leucopus is not easy to distinguish from *Peromyscus maniculatus;* however, it tends to be larger than the deer mouse where the ranges overlap. The ancestral form may be *Peromyscus progressus*. If this is so, the Easley Ranch record would presumably postdate that of the earlier species from Cragin Quarry.

The white-footed mouse is typically an inhabitant of deciduous forests, and sometimes open areas. It does not range into the northern coniferous forest, and its presence at sites like New Paris No. 4 may be due to contamination during excavations. (Dalquest, 1962; Guilday Martin, and McCrady, 1964; Hibbard and Taylor, 1960.)

Cotton Mouse, *Peromyscus gossypinus* (Le Conte), 1853 (*Hesperomys gossypinus*)

The cotton mouse occurs in various fossil faunas of peninsular Florida from the Sangamonian onward (Reddick, Arredondo). It has also been found in the Upper Shuler Lewisville fauna in Dallas, Texas, which is west of the present-day range.

Osteologically, the cotton mouse resembles *Peromyscus maniculatus* and *Peromyscus leucopus,* but in size, especially that of the molar teeth, it tends to be larger. It is a woodland animal also found in damp surroundings, but not arid situations (which are thought to have developed gradually in Upper Shuler time).

Like other species of the subgenus *Peromyscus,* this species probably arose from ancestral forms close to *Peromyscus progressus.* (Slaughter et al., 1962; Webb, 1974a.)

Brush Mouse, *Peromyscus boylii* (Baird), 1855 *(Hesperomys boylii)*

Remains of this southwestern species have been found in California (Hawver Cave), Mexico (San Josecito Cave), and Texas (Klein and Schulze caves), all within its present-day range; none of these sites is definitively earlier than the Wisconsinan.

Osteological characters include a somewhat depressed rostrum and an inflated braincase. The upper molars are large and the accessory elements more complex than those of the sympatric *Peromyscus maniculatus* and *Peromyscus pectoralis;* M^1 and M^2 often show closed, isolated enamel islands not seen in other species of this genus.

The brush mouse is found in arid and rocky situations of the chaparral type. (Dalquest, Roth, and Judd, 1969; Jakway, 1958.)

White-ankled Mouse, *Peromyscus pectoralis* Osgood, 1904

This species has been found at three Wisconsinan sites in Texas and New Mexico (Klein, Schulze, and Dry caves).

The jaws and teeth are, on the average, slightly larger than those of *Peromyscus leucopus,* and the molars tend to be somewhat more complex in structure. The weakly developed capsular process for the lower incisor is distinctive. The same character is also found in *Peromyscus boylii,* which is, however, larger and odontologically distinct, as noted above.

Although the white-ankled mouse is distributed primarily in Mexico, it is also found in Trans-Pecos Texas and southeastern New Mexico. Absent in very arid areas, it occurs in hilly regions and chaparral, a distribution suggesting relict status from an earlier continuous population. (Dalquest, Roth, and Judd, 1969.)

Piñon Mouse, *Peromyscus truei* (Shufeldt), 1885 *(Hesperomys truei)*

The piñon mouse has been tentatively identified in a late Wisconsinan assemblage from Blackwater Draw, New Mexico. This site is close to the eastern border of the range, which extends west and south from New Mexico and Colorado. A medium-sized member of the subgenus *Peromyscus,* this species is found in rocky areas with piñon pine and juniper. (Lundelius, 1967.)

Rock Mouse, *Peromyscus difficilis* (Allen), 1891 *(Vesperimus difficilis)*

The rock mouse has been found in Wisconsinan deposits at Dry Cave; the record is within its present-day range, which extends from Colorado to Mexico. This medium-sized species is characterized by its comparatively large auditory bullae. It inhabits rocky ground, canyon walls, and cliffs. (Harris, 1970.)

† Oklahoma Mouse, *Peromyscus oklahomensis* Stephens, 1960

This species may be referred to the subgenus *Podomys,* which is distinguished from the subgenus *Peromyscus* by somewhat more hypsodont molars and by the accessory tubercles, which do not form a loop extending to the outer edge of the tooth as in the latter subgenus. The molar pattern has a low or moderate degree of complexity.

The Oklahoma mouse is a little-known Illinoian to Sangamonian species from Doby Springs and Easley Ranch. It is a large form, about the size of *Peromyscus floridanus*, from which it differs in the shape of the reentrant valleys between the molar cusps. (Dalquest, 1962; Stephens, 1960.)

Florida Mouse, *Peromyscus floridanus* (Chapman), 1889 (*Hesperomys floridanus*)

This is the only living member of the subgenus *Podomys*. Fossil representatives occur in the Pleistocene of Florida, from the early Illinoian to the Holocene (Coleman IIA, Arredondo, Reddick, Haile XI, Devil's Den). There is a suggestion of a progressive size reduction: M_1–M_3 averages 4.7 mm at Reddick and 4.4 mm in modern specimens.

The Florida mouse probably shared a common ancestor with *Peromyscus oklahomensis*, and the indication is that the subgenus *Podomys* was a more widespread group in the early Rancholabrean. Perhaps it was affected by competition from the subgenus *Peromyscus*, which entered its period of great radiation in mid-Rancholabrean times.

The habitat of *Peromyscus floridanus* is high, sandy ridges with pine or palmetto. (Webb, 1974a.)

Golden Mouse, *Ochrotomys nuttalli* (Harlan), 1832 (*Arvicola nuttalli*)

The genus *Ochrotomys* is distinguished from *Peromyscus* by the markedly compressed occlusal surface pattern of the teeth and resulting contact of enamel folds; worn teeth show subtriangular dentine islands. The enamel is relatively thicker than that of *Peromyscus*. Some authors consider *Ochrotomys* only subgenerically distinct from *Peromyscus*.

The genus is monotypic. Fossils of the golden mouse are best known from Florida, where they date from the early Illinoian to the Holocene (Coleman IIA, Arredondo, Reddick, Haile VIII and IX, Ichetucknee River, Devil's Den), but have also been found at Herculaneum, Missouri. All records are within the Recent range.

This is an arboreal species, found in forests and thickets with vines and Spanish moss. (Webb, 1974a.)

†Boreal Pygmy Mouse, *Baiomys aquilonius* Zakrzewski, 1969

Baiomys, the pygmy mouse genus, is similar to *Peromyscus* but smaller; the coronoid process of the mandible is larger and more hooked. These are the smallest of the native mice of North America.

The boreal pygmy mouse, known only from Hagerman, resembles the living *Baiomys taylori* in the lack of well-developed stylids on the lower molars (with the exception of M_1, which has an anteromedian stylid). M_3 is reduced. The species may be a collateral or a descendant of *Baiomys rexroadensis*. The length M_1–M_3 is 2.52 mm.

The species may have inhabited meadows or valley slopes. The occurrence at Hagerman is far north of other fossil or Recent records of the genus. Evidently, *Baiomys* was very widespread in the late Pliocene. At Hagerman, there is another, unnamed species of the genus that is slightly larger (M_1–M_3, 2.85 mm) and lacks the anteromedian stylid in M_1. (Zakrzewski, 1969.)

†Rexroad Pygmy Mouse, *Baiomys rexroadensis* Hibbard, 1941

This species, slightly smaller than *Baiomys aquilonius*, which it resembles, is only known from the early Blancan of Rexroad and Fox Canyon (see fig. 12.7E). (Hibbard, 1941a,b, 1950.)

†Least Pygmy Mouse, *Baiomys minimus* (Gidley), 1922 (*Peromyscus minimus*)

A small form, known only from the Blancan of Benson, this species differs from the Recent *Baiomys taylori* in its somewhat more brachydont molars; the anterior lobe of M_1 is narrow and double-cusped. (Gazin, 1942.)

†Kolb's Pygmy Mouse, *Baiomys kolbi* Hibbard, 1952

This species is known only from the Blancan Fox Canyon fauna. It is a relatively large form (M_1–M_3, 2.89 mm). The M_3 is relatively unreduced. (Hibbard, 1952a.)

†Short-faced Pygmy Mouse, *Baiomys brachygnathus* (Gidley), 1922
 (*Peromyscus brachygnathus*)

The sole known specimen, from the early Irvingtonian of Curtis Ranch, has heavily worn teeth. It is slightly larger than *Baiomys minimus*.

Baiomys sp. has been identified at California Wash and Irvington. (Downs and White, 1968; Gazin, 1942.)

Northern Pygmy Mouse, *Baiomys taylori* (Thomas), 1887 (*Hesperomys taylori*)

Fossils of this species have been found at Schulze Cave, which is within the modern range. A relative increase in numbers from the lower to upper levels may corroborate the suggestion of Packard (1960) that the species invaded this part of its range at a late date.

This is a southern form, mainly distributed in Mexico and the southeastern part of Texas. It is the only species of *Baiomys* now occurring in the United States. Apparently, the genus became extinct farther north with the onset of Irvingtonian glaciations.

The M_3 is less reduced than in the related boreal pygmy mouse, *Baiomys aquilonius*, which may indicate that the latter is not directly ancestral.

The pygmy mouse inhabits areas where there is nearby cover of dense grass or weeds. (Dalquest, Roth, and Judd, 1969; Packard, 1960.)

†Arizona Benson Mouse, *Bensonomys arizonae* (Gidley), 1922
 (*Eligmodontia arizonae*)

The dental pattern of the genus *Bensonomys* resembles that of the living South American *Eligmodontia*, but M_1 has only two roots, rather than four as in the living genus. The resemblance is probably due to parallel evolution (Skinner and Hibbard, 1972). The genus makes its first appearance in the Hemphillian (Baskin, unpublished manuscript) and does not survive into the Irvingtonian.

Bensonomys arizonae, which is known from Benson and Rexroad (Loc. G), has somewhat narrower lower incisors than *Bensonomys eliasi* and a triangular M_1. It differs from the late Blancan *Bensonomys meadensis* in the development of the masseteric crest. Length M_1–M_3 is 3.93 mm. (Baskin, unpublished manuscript; Gazin, 1942; Gidley, 1922; Hibbard, 1956b; Skinner and Hibbard, 1972.)

†Elias's Benson Mouse, *Bensonomys eliasi* (Hibbard), 1937 (*Peromyscus eliasi*)

Known from Fox Canyon and Rexroad, this species has somewhat broader lower incisors and a more rectangular M_1 than *Bensonomys arizonae*. Length M_1–M_3 is 3.82 mm (see fig. 12.7F). (Hibbard, 1956b.)

†Meade Benson Mouse, *Bensonomys meadensis* Hibbard, 1956

This species occurs in the late Blancan faunas of Sanders and Sand Draw. It is close to *Bensonomys arizonae* in size (M_1–M_3, 3.5 mm). The front end of the masseteric crest is more dorsal and posterior, relative to the mental foramen, than that of the Arizona species. (Hibbard, 1956b; Skinner and Hibbard, 1972.)

†Simple-toothed Mouse, *Symmetrodontomys simplicidens* Hibbard, 1941

Symmetrodontomys is a comparatively primitive cricetine genus with simple broad cusps, wide valleys, and poorly developed cingula on the lower molars; protoconid and metaconid are nearly opposite rather than alternating.

The species *Symmetrodontomys simplicidens* occurs at Fox Canyon, Keefe Canyon, and Rexroad. M_1–M_3 is 4.28 mm (see fig. 12.7g). (Hibbard, 1941a, 1950.)

†Benson Grasshopper Mouse, *Onychomys bensoni* Gidley, 1922

Grasshopper mice, genus *Onychomys*, are stout, short-tailed animals. They are carnivorous (eating insects, small mammals, lizards, etc.) and frequent burrows dug by other animals. The molars are more hypsodont than those of *Peromyscus*, and accessory cusps are rare. The mandible has a long coronoid process. The genus is represented in the Blancan by various forms such as *Onychomys gidleyi* in Kansas and undetermined species at Red Light, Sand Draw, and White Rock.

The Benson grasshopper mouse is only known from the type locality. It is the same size as the living *Onychomys torridus*, but the molars are less hypsodont and M_3 is less reduced. (Gazin, 1942.)

†Gidley's Grasshopper Mouse, *Onychomys gidleyi* Hibbard, 1941.

This early Blancan species occurs in the Fox Canyon and Rexroad faunas. It is the same size as *Onychomys leucogaster* (M_1–M_3, 4.0 mm), but the molars have lower cusps, endostylids more strongly developed, and the mental foramen is lower down on the side of the ramus. (Hibbard, 1941a, 1950.)

†Fossil Grasshopper Mouse, *Onychomys fossilis* Hibbard, 1941

Found in the Borchers and White Rock faunas of late Blancan age, this species resembles *Onychomys bensoni* in having an unreduced M_3 but is somewhat larger. It probably is a descendant of *Onychomys bensoni* and may be ancestral to *Onychomys leucogaster*. (Eshelman, 1975; Hibbard, 1941d.)

†San Pedro Grasshopper Mouse, *Onychomys pedroensis* Gidley, 1922

Curtis Ranch, the type locality, has yielded a sample of jaws of an animal noticeably larger and more robust, with more hypsodont molars, than the Benson form. Again, the M_3 shows less reduction than that of living *Onychomys*. The age is earliest Irvingtonian, but the species may also be present in the somewhat earlier Wolf Ranch local fauna. Records unidentified as to species come from the Irvingtonian of Cudahy and Vallecito Creek. (Gazin, 1942; Lindsay and Tessman, 1974.)

†Jinglebob Grasshopper Mouse, *Onychomys jinglebobensis* Hibbard, 1955

Originally found in the Jinglebob fauna (Sangamonian), the species has also been tentatively identified at Slaton (Illinoian). It is related to the living *Onychomys leucogaster*, but the teeth tend to be somewhat larger (M_1–M_3, 4.5 mm). The distinction from *Onychomys fossilis* is somewhat uncertain.

According to Dalquest (1967), specimens from the Borchers fauna have also been referred to *Onychomys jinglebobensis*. (Dalquest, 1967; Hibbard, 1955c.)

Northern Grasshopper Mouse, *Onychomys leucogaster* (Wied-Neuwied), 1841
  (*Hypudaeus leucogaster*)

Fossil records of this species are common from Illinoian times onward (Angus, Doby Springs, Butler Spring, Cragin Quarry, Easley Ranch) and are widely scattered in late Rancholabrean deposits. Records exist from Arizona, Idaho, Kansas, Missouri, Nebraska, New Mexico, Oklahoma, and Texas. The occurrence in Crankshaft Cave is east of the modern range.

This species has a more reduced M_3 than the fossil forms discussed above. M^3 is nearly circular in shape. The teeth are larger than those of *Onychomys torridus*, and there is little overlap in size. The species may be derived from *Onychomys fossilis*.

A wide-ranging species, the northern grasshopper mouse inhabits arid regions from northern Mexico through the Great Plains and western states into southern Canada. (Dalquest, Roth, and Judd, 1969; Eshelman, 1975.)

Southern Grasshopper Mouse, *Onychomys torridus* (Coues), 1874
  (*Hesperomys torridus*)

Late Rancholabrean records of this species come from Rancho La Brea and Isleta Cave, sites close to the present-day range, which extends from Mexico to the Northwest into parts of California and Nevada. The species is distinctly smaller than *Onychomys leucogaster*; M^1 is larger relative to skull size, and M^3 is more transverse. The preferred habitat is desert and valley floors. (Harris and Findley, 1964.)

†Intermediate Cotton Rat, *Sigmodon medius* Gidley, 1922 (*S. minor* Gidley, 1922; *S. intermedius* Hibbard, 1938; *S. hilli* Hibbard, 1941)

The cotton rats, genus *Sigmodon*, appeared in North America in the early Blancan, probably as immigrants from the south (Hershkowitz, 1966). With a modern southern distribution, *Sigmodon* is nonetheless the most common cricetine in Blancan deposits, ranging north into Nebraska and Colorado. Cotton rats are medium-sized, somewhat volelike cricetines with a robust skull and heavy supraorbital ridges. The trends toward hypsodonty and involution of enamel crests in molars parallel evolutionary developments in the voles.

Sigmodon medius is a small form having relatively low-crowned, bunodont molars with wide reentrant folds; M_1 is typically two- or three-rooted. The species appeared in the early Blancan and survived throughout this age and into the earliest Irvingtonian (Arroyo Seco, Bender, Benson, Blanco, Borchers, California Wash, Curtis Ranch, Donnelly Ranch, Layer Cake, Haile XVA, Red Corral, Rexroad, Santa Fe 1B, Sand Draw, Sanders, White Rock). Its range includes Arizona, California, Colorado, Florida, Kansas, and Nebraska. Later North American species may be descended from early Blancan populations of this species. (R. A. Martin, 1974a; Skinner and Hibbard, 1972.)

†Hudspeth Cotton Rat, *Sigmodon hudspethensis* Strain, 1966

Only known from Hudspeth, this late Blancan species has an occlusal pattern smilar to that of *Sigmodon medius*, of which it may be a descendant, but is larger and has a more strongly developed inner root on M_1. It may be ancestral to *Sigmodon curtisi* of the Irvingtonian. (R. A. Martin, 1974a.)

†Curtis Cotton Rat, *Sigmodon curtisi* Gidley, 1922

This species has been found at very late Blancan and very early Irvingtonian sites in Arizona and Colorado (California Wash, Curtis Ranch, Donnelly Ranch). It often occurs associated with *Sigmodon medius*. The Curtis cotton rat shows a degree of hypsodonty approaching that of living species, from which it differs by the presence of only three roots in M_1. It is about the same size as *Sigmodon hudspethensis*, from which it may be derived. Living species belonging to the *Sigmodon curtisi* group, with hypsodont teeth and three-rooted M_1, are found in Mexico and South America. This group probably also gave rise to the living forms with four-rooted M_1.

Irvingtonian *Sigmodon*, not determined as to species, is also found at Vallecito Creek. (R. A. Martin, 1974a.)

†Baker's Cotton Rat, *Sigmodon bakeri* Martin, 1974

This species has only been found in Florida, where it ranges from the late Irvingtonian (Coleman IIA) to the early Rancholabrean (Williston IIIA). It is the oldest known member of the *Sigmodon hispidus* species group, characterized by four well-developed roots on M_1 and relatively hypsodont molars. It is distinguished from other members of the group (*S. hispidus, S. ochrognathus, S. fulviventer,* and *S. alleni*) by the lack of an anterior cingulum on M_2 and M_3 and the somewhat aberrant M_1 anteroconid. It may be derived from *Sigmodon curtisi* and could be ancestral to *Sigmodon hispidus* and related species. (R. A. Martin, 1974a.)

Hispid Cotton Rat, *Sigmodon hispidus* Say and Ord, 1825

This species is of common occurrence in Rancholabrean deposits of Florida and Texas and is also found in Georgia, Kansas, Mexico, New Mexico, and Oklahoma. The record from Kansas, at Kentuck, is also the oldest (Kansan), but the identification is tentative. Other early records are Illinoian to Sangamonian (Coppell, Haile VIIIA, XIB, Reddick, and others). All are within the modern range of the species. The range has been extended to the north in recent years.

This is the largest living *Sigmodon* in North America and has the most hypsodont teeth. It may be derived from *Sigmodon bakeri*. The habitat is moist areas with tall grass, sedge, and weeds. (R. A. Martin, 1974a; Semken, 1966.)

Yellow-nosed Cotton Rat, *Sigmodon ochrognathus* Bailey, 1902

The species has been identified in an early Holocene fauna from Centipede Cave, which is slightly east of its present-day range. Somewhat smaller than the related *Sigmodon hispidus*, it has very similar tooth and bone morphology, and some records of hispid cotton rat may be misidentifications (R. A. Martin, 1974a). The yellow-nosed cotton rat is found in mountains and foothills, where it seeks cover in dense vegetation. (Lundelius, 1963; R. A. Martin, 1974a.)

†Rexroad Woodrat, *Neotoma quadriplicata* (Hibbard), 1941
   (*Parahodomys quadriplicatus*)

Woodrats (or packrats), genus *Neotoma*, are hairy-tailed, fine-furred cricetines that often use sticks and other plant debris to build their surface nests. The rooted molars tend to hypsodonty and resemble those of microtines but have a less complex pattern.

Apart from the type locality (Rexroad), *Neotoma quadriplicata* has been tentatively identified at Hagerman and White Bluffs. It is referred to the extinct subgenus *Paraneotoma*, in which the molars are still rather brachydont and have thick enamel; M_3 resembles that of the subgenus *Hodomys* in wearing to form an S pattern.

Length M_1–M_3 in *Neotoma quadriplicata* is 9.5 mm. A related species, *Neotoma sawrockensis* Hibbard, is found in the Saw Rock Canyon local fauna, Kansas, of late Hemphillian age. (Hibbard, 1967.)

†Taylor's Woodrat, *Neotoma taylori* Hibbard, 1967

Known from the Borchers and White Rock local faunas, of late Blancan age (around the Blancan/Irvingtonian transition), this species resembles *Neotoma quadriplicata* but is smaller and has somewhat more hypsodont molars. It may be ancestral to the living subgenus *Hodomys*, but fossil documentation is lacking. (Eshelman, 1975; Hibbard, 1967.)

†Cave Woodrat, *Neotoma spelaea* (Gidley and Gazin), 1933
  (*Parahodomys spelaeus*)

Cumberland Cave (late Irvingtonian), is the only locality. The species is in the subgenus *Parahodomys*, which differs from the subgenus *Paraneotoma* in having higher-crowned molars and thinner enamel. The M_3 has a third reentrant fold on the posterior inner lobe, whereas in the subgenera *Hodomys* and *Neotoma*, there is a corresponding fold on the anterior outer lobe. Length M_1–M_3 is 9.5 mm. (Gidley and Gazin, 1938.)

†Fossil Woodrat, *Neotoma fossilis* Gidley, 1922

This incompletely known species from the Blancan of Benson may be an early representative of the subgenus *Neotoma*; it has the open and "pocketed" condition of the posteroexternal reentrants in the lower molars seen in *Neotoma floridana*. (Gazin, 1942.)

Eastern Woodrat, *Neotoma floridana* (Ord), 1818 (*Mus floridanus; N. magister* Baird, 1857)

The species has been found at about 35 sites in Arkansas, Florida, Georgia, Kansas, Missouri, Pennsylvania, Tennessee, Texas, Virginia, and West Virginia. The earliest records are late Irvingtonian (Cumberland Kanopolis, Mt. Scott, Trout). Fossil finds are within or close to the present-day range.

This is a large member of the subgenus *Neotoma*. Its habitat varies considerably, from cliffs in the Northeast to swamps and palmetto in the Southeast and yucca and cacti in the West. (Gidley and Gazin, 1938; Dalquest, Roth, and Judd, 1969; Patton, 1963.)

Southern Plains Woodrat, *Neotoma micropus* Baird, 1855

There are a few records of this species in Texas; the earliest come from the Sangamonian (Coppell, Easley Ranch). Intermediate in size between the large *Neotoma floridana* and the small *Neotoma albigula*, the species is similar to both morphologically. It inhabits brushlands, valleys, and plains in the western half of Texas and adjacent areas to the north and northwest. (Dalquest, Roth, and Judd 1969.)

Whitethroat Woodrat, *Neotoma albigula* Hartley, 1894

Fossils of this species have been found in San Josecito and Isleta caves and a few sites in Texas, the earliest record being Illinoian at Slaton. All these localities are within the present-day range. The range is sympatric with that of *Neotoma micropus* in western

Texas, and since only slight differences separate the two species, identification of fragmentary material is difficult. The species occurs in mesquite brushland and on rocky cliffs. (Dalquest, Roth, and Judd, 1969.)

Desert Woodrat, *Neotoma lepida* Thomas, 1893

This western species occurs in late Rancholabrean deposits in California (Emery Borrow Pit, McKittrick), Nevada (Glendale), and New Mexico (Burnet Cave). These sites are within or close to the modern range. M^1 has a shallow anterointernal angle. This is mainly a desert form. (J.R. Schultz, 1938.)

Mexican Woodrat, *Neotoma mexicana* Baird, 1855

Records of this species come from New Mexico (Burnet and Papago Springs caves, and probably Dry Cave) and Texas (Centipede Cave); the last-mentioned locality is slightly outside of the modern range. All records are Wisconsinan to Holocene. The species has a deep anterointernal reentrant on M^1. It inhabits mountains, rocks, and cliffs. (Schultz and Howard, 1935; Wells and Jorgensen, 1964.)

Dusky-footed Woodrat, *Neotoma fuscipes* Baird, 1858

A chiefly Californian species, with marginal distribution in Oregon and Baja California, this woodrat has been found in the fossil state at McKittrick, Newport Bay Mesa, San Pedro, and Hawver Cave. The earliest record dates from the Sangamonian. The species resembles *Neotoma cinerea*, but M_1 differs from that of the latter species in the lesser degree of constriction of the anterior lobe. Found along streams in woods and chaparral, this species builds conspicuous stick houses. (C. Stock, 1918.)

Bushytail Woodrat, *Neotoma cinerea* (Ord), 1815 (*Mus cinereus*; *Teonoma spelaea* Sinclair, 1905)

Fossils of this species are found in late Rancholabrean and Holocene deposits in Wyoming, Idaho, Colorado, New Mexico, and California. The records are within the present-day range. M^3 has a closed anterior triangle and two confluent posterior loops. The species inhabits mountain slopes and pinewoods and nests in fissures or under logs.

Neotoma middens preserved in dry rock shelters in the Southwest have provided a wealth of information on changing vegetation during climatic shifts in the late Pleistocene. Material from more than 130 middens has been radiocarbon dated (from a few hundred to more than 40,000 years B.P.) and the preserved vegetation identified (Wells, 1976). The acquisitive habits of woodrats account for the large quantities of plant and animal macrofossils; the material has often been impregnated with woodrat urine, which acts as a preservative. Of the species involved (*Neotoma cinerea*, *N. lepida*, *N. albigula*, and *N. mexicana*), *Neotoma cinerea* is an almost obligate cliff- and cave-dweller and has left behind an outstanding record of Quaternary vegetation in its abundant and well-preserved middens. (Kellogg, 1912; Wells, 1976.)

Subfamily Microtinae

†Grass-eating Furrowtooth Vole, *Ogmodontomys poaphagus* Hibbard, 1941

The genus *Ogmodontomys* is one of the basal groups in the Pliocene radiation of voles with rooted molars and dates back to the Hemphillian; a still earlier member of this group in North America is the Hemphillian *Prosomys*. Early members of *Ogmodontomys* may have given rise to several other genera, e.g., the closely related *Cosomys*, in the Blancan. In *Ogmodontomys*, M^3 is usually three-rooted (it is two-rooted in *Cosomys*). The upper incisor is faintly grooved.

The species *Ogmodontomys poaphagus* may be derived from the smaller *Ogmodontomys sawrockensis* from the terminal Hemphillian of Saw Rock Canyon. *Ogmodontomys poaphagus transitionalis* Zakrzewski from Fox Canyon is transitional. Within the Blancan species, recorded also from Rexroad, Bender, Deer Park, and Sand Draw, a gradual increase in hypsodonty can be observed, and dentine tracts appear on M_1 in some populations.

This vole may have had aquatic tendencies; with the arrival of the muskrat *Pliopotamys* in the Great Plains, *Ogmodontomys* disappears, owing perhaps to competition. (Hibbard, 1941a; Skinner, and Hibbard, 1972; Zakrzewski, 1967.)

†Prime Coso Vole, *Cosomys primus* Wilson, 1932

The genus *Cosomys* comprises voles with rooted cheek teeth and three or four alternating triangles on M_1; in these characters, the genus resembles the Eurasian *Mimomys*, butt the resemblance is probably due to parallel evolution. The genus also resembles *Ogmodontomys* in many characters, including presence of an accessory prism fold on the anterior loop of M_1 (see fig. 12.10B), a character which persists longer in life in individuals of *Cosomys*. The genus probably arose from earlier Pliocene *Ogmodontomys*. There are no grooves on the upper incisors.

Cosomys primus is very common in the early Blancan Hagerman fauna, where it outnumbers all other mammals; it is also known from Coso and Buttonwillow, California. Within the Hagerman complex, its abundance tends to vary inversely with that of the muskrat *Pliopotamys;* both existed in the marsh–meadow community, but *Cosomys* was eventually replaced by *Pliopotamys*. Local variation in the number of alternating triangles in M_1 may be of adaptive significance. (R. W. Wilson, 1932; Zakrzewski, 1969.)

†Taylor's Snake River Vole, *Ophiomys taylori* (Hibbard), 1959 (*Nebraskomys taylori; Pliophenacomys idahoensis* Hibbard, 1959)

The genus *Ophiomys* comprises a group of primitive voles having rooted teeth without cement or labial dentine tracts; there are three to five alternating triangles on M_1. The genus is distinguished from the related *Pliophenacomys* by a lesser degree of closure of triangles on M_1 and weaker development of the dentine tracts. Its ancestry may lie within the genus *Ogmodontomys*.

The species *Ophiomys taylori* has a considerable stratigraphic range within the Glenns Ferry Formation of Idaho. Length M_1–M_3 is 5.7 mm. The species is variable in M_1 pattern, having from three to five alternating triangles; three triangles are dominant in the lower part of the section (Hagerman), whereas stratigraphically higher localities show a progressive increase in the count, which finally stabilizes at five. (Hibbard and Zakrzewski, 1967.)

†Small Snake River Vole, *Ophiomys parvus* (Wilson), 1933 (*Mimomys? parvus*)

This species occurs in the Grand View fauna (late Blancan). M_1 has five alternating triangles, and the lingual root of M^1 is less developed than that of *Ophiomys taylori*. Length M_1–M_3 is 5.5 mm. The dentine tracts are higher than those of *Ophiomys taylori* or *Ophiomys meadensis* (see fig. 12.10C). The species probably arose from *Ophiomys taylori*. (Hibbard and Zakrzewski, 1967.)

12.9 Cricetidae. Development of dentine tracts in microtine left M_1, external view. *A, Ophiomys fricki*, Blancan, Sand Draw. *B, Pliophenacomys primaevus*, Blancan, Sand Draw. *C, Ondatra nebracensis*, Irvingtonian, Kanopolis. *A, B*, after Skinner and Hibbard (1972); *C*, after Hibbard et al. (1978). *A, B*, 4×; *C*, 2×.

A B C

†Meade Snake River Vole, *Ophiomys meadensis* (Hibbard), 1956
(*Pliophenacomys meadensis*)

A southern species, *Ophiomys meadensis* is known from the Blancan of Sanders, Dixon, and White Rock. It is the largest known *Ophiomys* but is less advanced with regard to dentine tracts than *Ophiomys parvus* (see fig. 12.8). The ancestry of this species is not known. (Hibbard, 1956b; Hibbard and Zakrzewski, 1967.)

†Magill's Snake River Vole, *Ophiomys magilli* Hibbard, 1972

Recorded from the late Blancan of Sand Draw, this species was larger than *Ophiomys taylori* and had higher-crowned teeth. (Skinner and Hibbard, 1972.)

†Frick's Snake River Vole, *Ophiomys fricki* Hibbard, 1972

Like the previous species, *Ophiomys fricki* is known only from Sand Draw (late Blancan). It was somewhat smaller than *Ophiomys magilli* and had five alternating triangles in M_1 (see fig. 12.9A). (Skinner and Hibbard, 1972.)

†Finney's Vole, *Pliophenacomys finneyi* Hibbard and Zakrzewski, 1972

The *Pliophenacomys* voles have teeth with closed roots and lacking cement. M_1 has five alternating, closed or nearly closed triangles and conspicuous dentine tracts on both sides of the tooth. These features, as well as the closure of the triangles and the higher position of the mental foramen, distinguish the genus from the collateral *Ophiomys* (see figs. 12.8, 12.9, 12.10D).

Finney's vole, *Pliophenacomys finneyi*, is represented by a large sample from Fox Canyon (very early Blancan). It is smaller than other species of the genus (M_1–M_3, 6.26

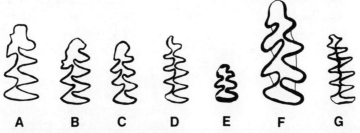

12.10 Cricetidae. Various microtine left M_1, occlusal view. *A, Ogmodontomys poaphagus*, Blancan, Rexroad. *B, Cosomys primus*, Blancan, Hagerman. *C, Ophiomys parvus*, Blancan, Grand View. *D, Pliophenacomys finneyi*, Blancan, Fox Canyon. *E, Atopomys salvelinus*, Irvingtonian, Trout Cave. *F, Clethrionomys gapperi*, cf. Illinoian alluvial fill, southwestern Ellis Co., Kansas. *G, Pliolemmus antiquus*, Blancan, Sanders. Thin lines in *F* denote cement. *A*, after Hibbard (1941a); *B*, after Zakrzewski (1969); *C, G*, after Hibbard (1956b); *D*, after Hibbard and Zakrzewski (1972); *E*, after Zakrzewski (1975); *F*, after Zakrzewski (1971). All 6×.

mm), and the teeth are lower-crowned, with less developed dentine tracts. The species is probably ancestral to *Pliophenacomys primaevus*. (Hibbard and Zakrzewski, 1972.)

† Primeval Vole, *Pliophenacomys primaevus* (Hibbard), 1938 (*Phenacomys primaevus*)

With a wide stratigraphic range in the Blancan, this species is found at Rexroad, Sand Draw, and probably White Rock. It is absent, however, at the Rexroad type locality, perhaps owing to competition from *Sigmodon*. The species is more advanced than *Pliophenacomys finneyi*, from which it differs by larger average size and somewhat higher molar crowns and better-developed dentine tracts (see fig. 12.9B). It is regarded as a descendant of the Fox Canyon species. (Hibbard, 1938; Hibbard and Zakrzewski, 1972.)

† Osborn's Vole, *Pliophenacomys osborni* L. D. Martin, 1972

This is a still more advanced species, found at Mullen I, Dixon, and White Rock (late Blancan–early Irvingtonian). It is distinguished from *Pliophenacomys primaevus* by its higher tooth crowns and dentine tracts and may be a progressive descendant of that species. On the other hand, the characters of the species are closer to those of *Pliomys deeringensis* (see below) than those of the *Pliophenacomys* species previously described. The species is provisionally retained in *Pliophenacomys*. (Eshelman, 1975; L. D. Martin, 1972.)

† Deering Vole, *Pliomys deeringensis* Guthrie and Matthews, 1971

The genus *Pliomys*, which is characterized by closed roots, five enamel triangles in M_1, and absence of cement in the reentrant molar folds, comprises primitive moles known mostly from the Old World, where they (and the related genus *Dolomys*) are common in the Pliocene and early Pleistocene.

The Deering species, from the early Irvingtonian Cape Deceit fauna, has a comparatively simple dental pattern, with poorly developed dentine tracts, for a *Pliomys* and may represent a form that evolved in isolation from the main Eurasian branches of the genus. It is, however, very similar to Osborn's vole (which see), and the two presumably belong to the same genus. (Eshelman, 1975; Guthrie and Matthews, 1971.)

† Rexroad Nebraska Vole, *Nebraskomys rexroadensis* Hibbard, 1970

The genus *Nebraskomys* comprises small voles with rooted teeth; M_1 has three alternating triangles and an anterior loop, and the mental foramen is just in front of M_1.

The early Blancan species *Nebraskomys rexroadensis* occurs only at the Rexroad type locality. It has shorter dentine tracts and less confluent alternating triangles than the later *Nebraskomys mcgrewi*. (Hibbard, 1970a.)

† McGrew's Nebraska Vole, *Nebraskomys mcgrewi* Hibbard, 1957

Found at Dixon, Sand Draw, and White Rock, this late Blancan species has higher dentine tracts and more confluent alternating triangles than the early Blancan *Nebraskomys rexroadensis*, from which it is probably descended. (Hibbard, 1957, 1970a; Skinner and Hibbard, 1972.)

† Balcones Vole, *Atopomys texensis* Patton, 1965

The genus *Atopomys* comprises small voles with rooted and cementless teeth and well-developed dentine tracts; M_1 has a posterior loop, three alternating or almost confluent triangles, and a relatively simple anterior loop. It has been suggested that the genus evolved from *Nebraskomys,* but since this would presuppose a reduction in M_1 length and in loop development of M^3 (a reversal of the common pattern of microtine evolution), the resemblance is probably due to parallel evolution.

The Balcones vole is known only from Fyllan Cave, Balcones Trail, Austin, Texas (probably Kansan in age). It differs from the other known species, *Atopomys salvelinus,* in its less-developed dentine tracts and the somewhat more complex anterior loop on M_1. (Patton, 1965; Zakrzewski, 1975b.)

†Trout Cave Vole, *Atopomys salvelinus* Zakrzewski, 1975

Remains of this species occur in the late Irvingtonian of the Appalachian region (Trout and Cumberland caves). It is distinguished from *Atopomys texensis* by its more developed dentine tracts, simpler anterior loop, and deeper second internal reentrant on M_1 (see fig. 12.10E). (Zakrzewski, 1975b.)

†Monahan's Vole, *Mimomys monahani* L. D. Martin, 1972

Voles of the genus *Mimomys* were the dominant microtines in Europe in the Ruscinian and Villafranchian. These primitive voles have cheek teeth with closed roots. *Mimomys* was ancestral to *Arvicola,* the Old World water voles.

Monahan's vole, found only at Mullen I (probably Irvingtonian), differs from *Cosomys, Ogmodontomys,* and other primitive American voles in its more hypsodont molars, high dentine tracts, and cement in the reentrant folds and has for this reason been referred to the Old World genus *Mimomys,* which shows these characters. Cheek teeth of juveniles lack roots, but these form with increasing age. This juvenile characteristic corresponds to the evolutionary stage seen in advanced *Mimomys.* Reference to the genus *Mimomys* is tentative, however, and *Mimomys monahani* might be a descendant of *Cosomys.* The animal was about the size of *Microtus pennsylvanicus.* (L. D. Martin, 1972.)

Boreal Redback Vole, *Clethrionomys gapperi* (Vigors), 1830 (*Arvicola gapperi*)

Redback voles, genus *Clethrionomys,* have molars that are rooted in adults; cement occurs in the reentrants (see fig. 12.10F). The inner reentrants in the lower molars are little if at all deeper than the outer. The lower incisors pass from the lingual to the labial side of the molars between the bases of M_2 and M_3 and extend far back toward the base of the condyle. The genus is Holarctic, and most species have a northern distribution.

Clethrionomys gapperi is the only species found in the Pleistocene of North America. Specimens have been found at about 20 sites, in Arkansas, Kansas, Kentucky, Michigan, Missouri, Nebraska, Pennsylvania, Tennessee, Virginia, and Wyoming. The oldest records are late Irvingtonian (Angus, Duck Creek, Trout Cave); the others are Rancholabrean. The species resembles the living *Clethrionomys rutilus* in dental pattern but can be distinguished by its more completely ossified hard palate. It presumably represents an early immigration from the Old World, where the genus is present from the earliest Pleistocene. Length M_1–M_3 is 5.04 mm (New Paris No. 4). It inhabits coniferous, deciduous, or mixed forests, preferably damp. Its distribution is similar to that of the heather vole, except that *Clethrionomys* ranges farther south in the northern United States and in the Rockies and Appalachians. Many of the fossil occurrences are south of the present distribution. An occurrence of *Clethrionomys,* not assigned to species, has been reported from central Alaska. (Guilday, 1971b; Guilday, Martin, and McCrady, 1964; Repenning, Hopkins, and Rubin, 1964; Zakrzewski and Maxfield, 1971.)

†Ancient Lemming Vole, *Pliolemmus antiquus* Hibbard, 1938

Voles of the genus *Pliolemmus* are the first voles with ever-growing, rootless teeth to appear in North America. The M_1 has seven alternating triangles (see fig. 12.10G). In spite of this advanced character, the teeth have no cement (there are well-developed dentine tracts, however).

Pliolemmus antiquus is the sole known species. It ranges from the early (Bender fauna, Rexroad Formation) to the late Blancan (Dixon, Sanders, Deer Park, Sand Draw). The species is most common at the Nebraskan site (Sand Draw), and it has been suggested that southern Kansas was the southern limit of the range. The ancestry of *Pliolemmus* is not known. (Skinner and Hibbard, 1972.)

Heather, or Spruce, Vole, *Phenacomys intermedius* Merriam, 1889

The genus *Phenacomys* is confined to the New World and probably evolved in North America. The cheek teeth are rooted, lack cement, and have a distinctive occlusal pattern.

Phenacomys intermedius, which has a wide distribution in Canada and the mountainous regions of the western United States, has been found in Wisconsinan and early Holocene deposits in Arkansas, Idaho, Iowa, Kentucky, Pennsylvania, Tennessee, Utah, Virginia, West Virginia, and Wyoming. Many of these records are far south of its modern range, the extremes being Arkansas (Peccary), and Tennessee (Baker Bluff, Guy Wilson). In addition, the genus *Phenacomys* occurs in pre-Wisconsinan faunas of "cold" aspect, like Cudahy (though not at the Cudahy type locality), Cumberland, and Trout caves.

Some more ancient forms originally regarded as *Phenacomys* are now placed in *Pliophenacomys*, which is not considered to be closely related to *Phenacomys*. The heather vole is now found in a variety of habitats, including pine and spruce forests, grassy and rocky mountain slopes, and tundra. (Guilday, 1971b; Guilday and Parmalee, 1972; Guilday, Martin, and McCrady, 1964.)

†Kormos's Steppe Vole, *Microtus pliocaenicus* (Kormos), 1933 (*Allophaiomys pliocaenicus*)

At the present time, *Microtus* is richer in species than any other vole genus. Molars are rootless, with cement in the reentrant angles, and the outer and inner reentrants are more nearly equal in size than those of *Phenacomys*. The upper incisors are not grooved, and the roots of the lower incisors extend far behind the molars on their outer side. Eurasia was the evolutionary center for *Microtus*. Van Der Meulen (1978) believes that there were two different migrations of *Microtus* from Eurasia to North America: the earlier took place more than 1.2 m.y. ago and is represented by the Wathena and Kentuck forms; the second immigration occurred just over 0.7 m.y. ago and is represented by *Microtus paroperarius, M. meadensis,* and *M. cumberlandensis*. The subgenera *Allophaiomys, Microtus, Stenocranius, Pedomys,* and *Pitymys* are represented in the North American Pleistocene.

The earliest forms are referred to the subgenus *Allophaiomys,* in which M_1 has three closed and two confluent triangles and anterior and posterior loops. M_3 has a posterior loop, two confluent triangles, and a small anterior loop.

The species *Microtus pliocaenicus* is common in the early Pleistocene of Europe. A small vole from the Java fauna (Irvingtonian) has been referred to the European species, as has some material from Kentuck and from Fyllan Cave in Texas. Other specimens from Kentuck and Wathena, a site in northeastern Kansas, are believed to represent a possibly new species, *Microtus (Allophaiomys)* sp. (Van Der Meulen, 1978). These specimens were originally assigned to *Microtus llanensis*. The question of whether these taxa are conspecific with the European form or represent a parallel development requires additional study. (Einsohn, 1971; R. A. Martin, 1975; Van Der Meulen, 1972, 1978; Zakrzewski, 1975.)

†Cape Deceit Vole, *Microtus deceitensis* Guthrie and Matthews, 1971

The subgenus *Microtus* is normally characterized by five closed triangles in M_1 (the number varies), three transverse loops and no triangles in M_3, and three closed triangles in M^3.

The Cape Deceit vole from the Irvingtonian Cape Deceit and Wellsch Valley faunas has a more primitive dental pattern than that of *Microtus paroperarius*, although more advanced than the *Allophaiomys* condition described above. M^3 has two closed triangles, M_1 never more than four, and M_3 has the second loop bisected by the posterior labial reentrant. Its combination of dental characters is unique, and the species probably represents either an early evolutionary stage or a side branch of the genus. (Guthrie and Matthews, 1971; Van Der Meulen, 1978.)

†Hibbard's Tundra Vole, *Microtus paroperarius* Hibbard, 1944

This late Irvingtonian species was a member of the Cudahy local fauna and has also been identified at Cumberland Cave, Conard Fissure, Mullen I, and Vera. The species shows variability in M_1, which has four or five closed triangles (see fig. 12.11). Length of M_1–M_3 is 5.96 mm. (Hibbard, 1944; L. D. Martin, 1972; Paulson, 1961; Van Der Meulen, 1978.)

†Meade Vole, *Microtus meadensis* (Hibbard), 1944 (*Pitymys meadensis*)

Numerous remains of this species have been found at Cudahy (late Kansan). The molars are *Microtus*-like. On M_1, the posterior loop, the first three triangles, and the anterior loop are closed; the labial reentrants are angular. Length of M_1–M_3 is 6.0 mm. *Microtus meadensis* is larger than *M. cumberlandensis* and shows close resemblances to *M. arvalidens* Kretzoi, a middle Pleistocene species in Europe. (Paulson, 1961; Van Der Meulen, 1978.)

Meadow Vole, *Microtus pennsylvanicus* (Ord), 1815 (*Mus pennsylvanicus*)

This species appears in late Irvingtonian faunas (Angus, Berends, Hay Springs, Kanopolis, Mullen I, Rezabek) and has been identified at about 40 Rancholabrean sites in Arkansas, Florida, Idaho, Iowa, Kansas, Kentucky, Missouri, Ohio, Oklahoma, Pennsylvania, Tennessee, Texas, Virginia, West Virginia, and Wyoming. Many sites are far south of the present range. Occurrences in Texas (several sites) and Florida (Arredondo IA, Devil's Den, Waccasassa River, Withlacoochee River) are especially noteworthy and make the meadow vole the most widely distributed microtine in the Pleistocene (and Recent) of North America. This testifies to the species' great adaptability, which is also evident from the variety of situations—forests, grasslands, stream borders, and swamps—in which it is found today.

In this species, M^2 has five sections, M_1 has five to seven closed triangles, and M_1–M_3 is about 6 mm (see fig. 12.11). It is closely related to (in the view of some authors, conspecific with) the European field vole, *Microtus agrestis*. That species, and its sibling *Microtus arvalis*, date back to the middle Pleistocene.

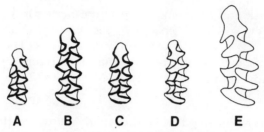

12.11 Cricetidae. *Microtus*, left M_1, occlusal view. *A, Microtus paroperarius*, Irvingtonian, Cudahy. *B, Microtus pennsylvanicus*, Irvingtonian, Hay Springs. *C, Microtus llanensis*, Irvingtonian, Cudahy. *D, Microtus meadensis*, Irvingtonian, Cudahy. *E, Microtus aratai*, Irvingtonian, Coleman IIA. Thin lines denote cement. A, C, D, after Paulson (1961); B, after Hibbard (1956b); E, after R. A. Martin (1974a). All 6×.

The earliest known meadow voles, from Kanopolis, are also the smallest, and may merit subspecific distinction. This form has only five closed triangles in M_1, a condition also found in most Illinoian and Sangamonian specimens. The percentage of specimens with a higher number of triangles (six or seven) tends to increase over time, reaching a high in Recent populations of the plains, with populations in the eastern forested region having five as the modal number. (Cuffey, Johnson, and Rasmussen, 1964; Dalquest, 1962; Guilday, 1971b; Hibbard, 1955c; Slaughter and Hoover, 1961; Semken, 1966; Zakrzewski, personal communication.)

Mountain Vole, *Microtus montanus* (Peale), 1848 (*Arvicola montana*)

Remains of the mountain vole have been taken at seven Rancholabrean sites in Colorado, Idaho, and Wyoming, all within the present range; a record from Fossil Lake, and an uncertain record from Glendale (which may be *M. californicus*), are somewhat outside the present distribution. In this species, M^2 has four closed triangles, and the incisive foramina are constricted posteriorly. The species mainly occurs in mountain valleys. (Guilday and Adam, 1967.)

California Vole, *Microtus californicus* (Peale), 1848 (*Arvicola californica*)

This species occurs at a number of late Rancholabrean sites in California, ranging from the Los Angeles area in the south to the Shasta County caves in the north; in addition, there are records from Nevada (Tule Springs and Glendale; the latter may be *M. montanus*), east of the modern range. The incisive foramina of this species are wide and not constricted. Marsh–meadow and grassy slopes are common habitats. (W. E. Miller, 1971.)

Long-tailed Vole, *Microtus longicaudus* (Merriam), 1888 (*Arvicola longicaudus*)

A tentative record of this species from American Falls may be pre-Wisconsinan (although the specimen may be *M. pennsylvanicus*); other records are Wisconsinan (Chimney Rock, Dry, Little Box Elder, Moonshiner caves) or later. All sites are within or close to the present range. The shape of the incisive foramina is diagnostic: the borders are parallel or taper gradually posteriorly. The species ranges from stream banks to mountain meadows. (Anderson, 1968.)

Mexican Vole, *Microtus mexicanus* (Saussure), 1861 (*Arvicola mexicana*)

Rancholabrean specimens of the Mexican vole have been found at Blackwater Draw and Burnet, Dry, Papago Springs, and San Josecito caves, all within or slightly outside of the modern range. The incisive foramina are short, broad, and truncated posteriorly. The habitat is mountain meadows and park forests in dry situations. (Schultz and Howard, 1935.)

Rock Vole, *Microtus chrotorrhinus* (Miller), 1894 (*Arvicola chrotorrhinus*)

A northeastern species, the rock vole appears in the late Irvingtonian (Trout Cave) and occurs in Wisconsinan or early Holocene deposits at Baker Bluff, Bootlegger Sink, Clark's Cave, Eagle Cave, Natural Chimneys, and New Paris No. 4. The records are within or some distance outside of the Appalachian portion of the present range. M^3 normally has triangles 1–2 confluent, 3 isolated, and 4–5 confluent; the incisive foramina are short and broad. A south–north size cline, with size increasing to the south, may be observed in the living population; the New Paris No. 4 sample is the same size as modern specimens from Labrador–Quebec. The species inhabits rocky woodlands. (Guilday, 1971b; Guilday, Martin, and McCrady, 1964.)

Yellow-cheeked Vole, *Microtus xanthognathus* (Leach), 1815
(*Arvicola xanthognatha*)

This species has been found at 12 sites in the eastern United States, distributed in Arkansas, Illinois, Iowa, Kentucky, Missouri, Pennsylvania, Tennessee, Virginia, and West Virginia and dating from the Wisconsinan or early Holocene. It is also recorded from the Rancholabrean of Alaska (Fairbanks). At New Paris No. 4, it is the most common terrestrial mammal in the deposit, with numbers decreasing in the upper levels. A tundra and boreal swamp- and forest animal, its present distribution is restricted to parts of Alaska and northwestern Canada.

Odontologically, the species resembles the meadow and rock voles but is much larger (length M_1–M_3 is 7.53 mm at New Paris No. 4). There are three closed triangles in M^3, and cement is absent in the third lateral reentrant of this tooth. The incisive foramina are long and narrow. (Guilday, 1971b,; Guilday, Martin, and McCrady, 1964; Hallberg, Semken, and Carson, 1974.)

Singing Vole, *Microtus miurus* Osgood, 1901

This species belongs to the subgenus *Stenocranius*, which is characterized by an unusually narrow skull with a sharp median crest. Fossil records come from the Fairbanks area and Old Crow River and are Rancholabrean in age. At Fairbanks, it is found in every major stratigraphic unit, from Illinoian to terminal Wisconsinan. Although found in adjacent areas of Alaska, Yukon Territory, and western Northwest Territories, the species is not found in the Yukon-Tanana upland today, having been replaced by *Microtus pennsylvanicus* and *Microtus longicaudus*, which are believed to be early Holocene immigrants. An inhabitant of dry, open tundra, this vocal vole feeds on forbs; it is active throughout the year and and caches large amounts of food for winter use.

The Nearctic form is closely related to the Holarctic *Microtus oeconomus* Pallas and is often considered to be conspecific with it, but the habitat requirements of the two species are different (*M. oeconomus* lives in damp tundra areas), and *Microtus miurus* has five closed triangles on M^1 instead of four. The presence of a simple trefoil on M_1 distinguishes the species from *Microtus pennsylvanicus*. (Guilday, Martin, and McCrady, 1964; Guthrie, 1968b; Harington, 1970a; Rausch, 1964.)

†Guilday's Vole, *Microtus guildayi* Van Der Meulen, 1978 [*M. speothen*
(Cope), 1871 (*Arvicola speothen; A. tetradelta* Cope, 1871); *M. involutus*

(Cope), 1871 (*Arvicola involuta*); *Microtus dideltus* (Cope), 1871 (*Arvicola didelta; A. sigmodus,* Cope, 1871); all *nomina dubia*]

A number of *Microtus*-like voles have been referred to the subgenera *Pedomys* and *Pitymys*, which are very similar odontologically. In his recent biometric study of *Microtus,* Van Der Meulen (1978) recognizes both subgenera on the basis of the morphology of M_1, and traces the *Pedomys* evolutionary lineage from *Microtus (Allophaiomys)* sp. from Wathena and Kentuck (the earliest known North American form) to *Microtus guildayi* to *Microtus llanensis* to the extant *Microtus ochrogaster.*

There is a nomenclatural problem with the voles (*M. speothen, M. dideltus,* and *M. involutus*) that Cope (1871) described from Port Kennedy Cave. The material is poorly preserved, and comparisons with other species are nearly impossible. Hibbard (1955a) looked at the material and reduced several species to synonymy. Guilday (personal communication, 1976) believes that all of Cope's microtine species should be disregarded. He noted that the type of *Microtus speothen* "is a hunk of mud with a tooth impression on it." Van Der Meulen (1978) observed that the molars on the holotype of *Microtus involutus* are missing and that the M_1's from Cumberland Cave assigned to this species (see Gidley and Gazin, 1938) actually include three different species, *Microtus guildayi, Microtus cumberlandensis,* and *Microtus paroperarius.* Van Der Meulen believes that *Microtus involutus* is a *nomen dubium*. We concur with the views of Guilday and Van Der Meulen and believe that the species *Microtus speothen, Microtus involutus,* and *Microtus dideltus* should be *nomina dubia*.

Microtus guildayi is known only from the type locality, Cumberland Cave (middle Irvingtonian). The species resembles *Microtus ochrogaster* in having three triangles on M_1 and two triangles on M^3. In *Microtus ochrogaster,* the anterior cap of of M_1 bears two well-developed salient angles and shallow reentrant angles; these characters are lacking or poorly developed in *Microtus guildayi.* The two central triangles in M^3 are usually communicating in *Microtus guildayi* but separated in *Microtus ochrogaster.* (Gidley and Gazin, 1938; Guilday, personal communication, 1976; Hibbard, 1955a; Van Der Meulen, 1978.)

† Llano Vole, *Microtus llanensis* Hibbard, 1944

The Llano vole, a member of the *Pedomys* lineage, is known from Cudahy (type locality), Conard Fissure, and Vera. M_1 is somewhat larger than that of *Microtus guildayi,* and the length of M_1–M_3 is 5.9 mm. (Paulson, 1961; Semken, 1966; Van Der Meulen, 1978.)

Prairie Vole, *Microtus ochrogaster* (Wagner), 1842 (*Hypudaeus ochrogaster*)

The fossil record begins with the late Irvingtonian (Angus, possibly Rezabek); in addition, there are 25 or more Rancholabrean sites in Arkansas, Colorado, Kansas, Missouri, New Mexico, Oklahoma, Texas, and Wyoming. The distribution of the finds may indicate an earlier continuous range connecting the Gulf Coast population of today (sometimes regarded as a distinct species, *M. ludovicianus* Bailey) with the main Great Plains population. The teeth resemble those of *Microtus pinetorum* in having a constricted or divided middle loop. As the common name indicates, the species inhabits open prairie. Length of M_1–M_3 is 6.0 mm (Jinglebob). (Hibbard, 1955c; Patton, 1963; G. E. Schultz, 1967; Slaughter and Hoover, 1963.)

†Cumberland Pine Vole, *Microtus cumberlandensis* (Van Der Meulen), 1978
(*Pitymys cumberlandensis*)

The remaining species of *Microtus* are referred to the subgenus *Pitymys*, which is often regarded as a distinct genus. The dentition of *Pitymys* shows thick, little differentiated enamel, more or less rounded salient angles, and rather narrow reentrants (compared with the rather thin, well-differentiated enamel, pointed salient angles, and wide reentrants in *Microtus*); M_1 has three triangles.

Microtus cumberlandensis, an Irvingtonian species known from Cumberland Cave and Conard Fissure, is the earliest-known representative of the subgenus *Pitymys* and is ancestral to the extant species, *Microtus pinetorum*. The Cumberland pine vole is more primitive than the living species in having virtually undifferentiated enamel, a less anterior cap on M_1, and an unreduced M^3. (Van Der Meulen, 1978.)

†Arata's Pine Vole, *Microtus aratai* (R. A. Martin), 1974 (*Pitymys aratai*)

This vole from the late Irvingtonian of Coleman IIA is larger than any other known *Pitymys*, to which it is referred on the basis of the configuration of M_1 (see fig. 12.11). The length of M_1–M_3 is 8.0 mm. (R. A. Martin, 1974a.)

†Hibbard's Pine Vole, *Microtus hibbardi* (Holman), 1959 (*Pitymys hibbardi*)

An early Rancholabrean species from Florida, *Microtus hibbardi* has been found at Willston IIIA and Bradenton. It exceeds Recent pine voles in size; the length of M_1–M_3 is 6.9 mm. (Holman, 1959b.)

†McNown's Pine Vole, *Microtus mcnowni* (Hibbard), 1937 (*Pitymys mcnowni*)

A species of about the same size as *Microtus hibbardi* (length M_1–M_3 is 7.0 mm), it differs in having broader salient angles in M_1, with rounded rather than angular apices. It comes from a Rancholabrean deposit in Brown County, Kansas. (Hibbard, 1937.)

Pine Vole, *Microtus pinetorum* (Le Conte), 1830 (*Psammomys pinetorum*)

This species has been found at more than 20 Rancholabrean sites in Arkansas, Florida, Georgia, Missouri, Pennsylvania, Tennessee, Texas, Virginia, and West Virginia. The oldest record is Sangamonian (Reddick). The finds are within or close to the modern range. In the New Paris No. 4 fauna, *Microtus pinetorum* is very rare and is the only microtine found that does not occur north of the southern edge of the Canadian life zone. Today, the species is the most common microtine found in the area of the cave.

Dentitions of *Microtus pinetorum* are easily confused with those of *Microtus ochrogaster* (which see). The species is somewhat smaller than *Microtus hibbardi*. *Microtus cumberlandensis* was probably ancestral to *M. pinetorum*. The living pine vole of northern Florida is sometimes considered a distinct species, *Microtus parvulus* (Howell); it is smaller than the form of the main population. The species is common in the Eastern deciduous forest and in the pine forests of the South. (Guilday, Martin, and McCrady, 1964; Martin and Webb, 1974; Patton, 1963.)

Sagebrush Vole, *Lagurus curtatus* (Cope), 1868 (*Arvicola curtata*)

The steppe or sagebrush lemming genus, *Lagurus*, is mainly distributed in the Old World, where its history begins in the Villafranchian. These voles have ever-growing cheek teeth with widely open folds, lacking cement; in the Old World, they are the earliest voles with rootless teeth. The genus evidently is a comparatively recent immigrant in the New World, for the finds (all of which belong to the living species) date back only to the Wisconsinan. However, the New World species also has been referred to a separate genus (*Lemmiscus*).

Lagurus curtatus is recorded from Jaguar, Moonshiner, Wasden (Idaho), Bell, Horned Owl, Little Box Elder (Wyoming), Warm Springs (Montana), and Isleta and Dry caves (New Mexico); the last-mentioned location is the only site outside (to the east) of the present range. As the common name implies, the species is found in arid situations, where sagebrush may form the staple food. (Davis, 1939; Guilday and Adam, 1967.)

† Guilday's Water Rat, *Proneofiber guildayi* Hibbard and Dalquest, 1973

The genus *Proneofiber*, with only one known species, resembles the living *Neofiber*, to which it is presumably ancestral; unlike that genus, however, adults have rooted cheek teeth. As in *Neofiber*, cement and high dentine tracts are present.

Proneofiber guildayi, from the early Irvingtonian (Aftonian?) of Gilliland, is about the same size as the primitive muskrat *Ondatra annectens*. It may be derived from a primitive microtine like *Pliopotamys*. (Hibbard and Dalquest, 1973; Zakrzewski, 1974.)

† Diluvial Water Rat, *Neofiber diluvianus* (Cope), 1896 (*Microtus diluvianus;*
  Schistodelta sulcatus Cope, 1899)

Although water rats, genus *Neofiber*, live in bogs and are good swimmers, they are less specialized for an aquatic life than is *Ondatra*. The cheek teeth are ever-growing, with cement in the reentrant valleys; the triangles in M_3 are tightly closed. Dentine tracts are well developed. The genus probably evolved from *Proneofiber*.

Neofiber diluvianus, known only from Port Kennedy, has upper molars with narrower alternating triangles a more pointed apex, and wider lingual reentrant angles than those of *Neofiber alleni;* however, the status of the species is somewhat uncertain. (Hibbard, 1955a; Hibbard and Dalquest, 1973).

† Leonard's Water Rat, *Neofiber leonardi* Hibbard, 1943

The stratigraphic range of this species, which is known from Rezabek, Kanopolis, Slaton, Trout, and McLeod (Levy Co., Fla.) includes the Yarmouthian and early Illinoian. It differs from *Neofiber diluvianus* in the larger size of the first alternating triangle of M^1 and its broader apex. Most M_3's in this species have two labial reentrant angles instead of one as in *Neofiber alleni*. However, the modern condition occurs as an occasional variant in the fossil species. Thus an early Illinoian transition between the two species is suggested. (Frazier, 1977; Guilday, personal communication, 1976; Hibbard, 1943a, 1955a; Hibbard and Dalquest, 1973.)

Florida Water Rat, *Neofiber alleni* True, 1884

This living species, which probably arose from the Irvingtonian *Neofiber leonardi*, first appears in the late Irvingtonian Coleman IIA fauna. At the end of the Irvingtonian, *Neofiber* evidently became extinct over most of its range, surviving only in Florida and Georgia. Fossils of the modern species have been found at more than 15 Rancholabrean and Holocene sites in Florida and at Ladds in northern Georgia.

The Florida water rat is dependent on permanent bodies of water that support the emergent vegetation on which it feeds. It constructs feeding platforms in shallow water

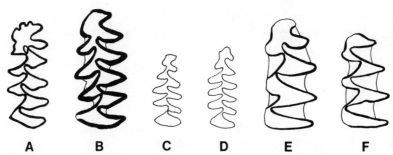

12.12 Cricetidae. Left M_1, occlusal view, in muskrats, lemmings, and bog lemmings. *A, Pliopotamys meadensis,* Blancan, Dixon. *B, Ondatra annectens,* Irvingtonian, Cudahy. *C, Predicrostonyx hopkinsi,* Irvingtonian, Cape Deceit. *D, Dicrostonyx torquatus,* Recent. *E, Synaptomys vetus,* Blancan, Grand View. *F, Synaptomys cooperi,* Recent. Thin lines denote cement. *A,* after Hibbard (1956b); *B,* after Paulson (1961); *C, D,* after Guthrie and Matthews (1971); *E, F,* after Zakrzewski (1972). All 6×.

but builds its nests in stumps, mangroves, or open savanna. Frazier (1977) believes that cold and aridity were (and are) limiting factors to its distribution. It is not presently sympatric with *Ondatra.* (Frazier, 1977; R. A. Martin, 1974a; Webb, 1974a.)

†Pygmy Muskrat, *Pliopotamys minor* (Wilson), 1933 (*Ondatra idahoensis minor*)

The evolution of the muskrat genus *Ondatra* can be traced in some detail to the genus *Pliopotamys,* which is distinguished by its small size and lack of cement and absence, or weak development, of lingual dentine tracts in the teeth. *Pliopotamys* may have arisen from an early member of the genus *Ogmodontomys* or from an ancestor common to both. *Pliopotamys* may also be ancestral to *Proneofiber.*

Pliopotamys minor is common (at least 380 individuals) in the early Blancan Hagerman fauna, where it has been found at 77 collecting localities ranging through 105 m of the deposits. Its abundance increases markedly in the upper deposits.

The species differs from *Pliopotamys meadensis* in the absence of lingual dentine tracts on the cheek teeth and the occlusal pattern of M_1. (Hibbard, 1959; Hibbard and Zakrzewski, 1967; Nelson and Semken, 1970; R. W. Wilson, 1933a; Zakrzewski, 1969.)

†Meade Muskrat, *Pliopotamys meadensis* Hibbard, 1938

This species was taken at late Blancan–early Irvingtonian localities (Dixon, Mullen I, Sanders). As befits its stratigraphic position, it is somewhat advanced over *Pliopotamys minor,* as evidenced by its slightly larger size and the presence of small lingual dentine tracts (see fig. 12.12). It may be a descendant of *Pliopotamys minor* and be ancestral to *Ondatra.* However, an unnamed, somewhat more advanced form has been discovered in an early Blancan deposit near Benson, Arizona, and correlated with the Benson fauna. (Hibbard, 1956b; L. D. Martin, 1972; Zakrzewski, 1969, 1974.)

†Idaho Muskrat, *Ondatra idahoensis* Wilson, 1933

The genus *Ondatra* is distinguished from its predecessor *Pliopotamys* by its larger size, the presence of cement in the reentrant angles, and the development of dentine tracts, which produce an interrupted enamel pattern in well-worn cheek teeth.

The earliest known member of the genus is *Ondatra idahoensis,* found in transitional late Blancan to early Irvingtonian deposits (Grand View, Borchers, White Rock, Curtis Ranch). It is one of the smallest species in the genus, and the molars have less cement (none reticulated) and less well-developed dentine tracts than those of other *Ondatra;* as in most fossil muskrats, M^1 is three-rooted.

An evolutionary line may be traced from this species to the living *Ondatra zibethicus*, with *Ondatra annectens* as intermediate. The trend is one of increasing size, cement deposition on molar surfaces, and development of the dentine tracts. (Hibbard, 1959; Nelson and Semken, 1970; Semken, 1966; R. W. Wilson, 1933a.)

†Cope's Muskrat, *Ondatra hiatidens* (Cope), 1871 (*Anaptogonia hiatidens;*
   Sycium cloacinum Cope, 1899)

Smaller than *Ondatra idahoensis*, this species is distinguished by its better-developed dentine tracts. It has been found in Port Kennedy Cave (Irvingtonian). Whether it is on the main line of *Ondatra* evolution is uncertain. (Hibbard, 1955a.)

†Brown's Muskrat, *Ondatra annectens* (Brown), 1908 (*Fiber annectens; O.
   kansasensis* Hibbard, 1944)

This Irvingtonian species, originally described from Conard Fissure, has been identified at Cudahy, Cumberland, Java, Trout, and Vera, and may be present at Hay Springs. It thus appears to range from the Kansan to the Yarmouthian. It continues the trend of increasing body size and height of dentine tracts. The species probably arose from *Ondatra idahoensis* and may be ancestral to *Ondatra nebracensis* (see fig. 12.12) (Nelson and Semken, 1970; Semken, 1966; Stephens, 1960.)

†Nebraska Muskrat, *Ondatra nebracensis* (Hollister), 1911 (*Fiber nebracensis; O.
   triradicatus* Starrett, 1956)

Very late Irvingtonian (Yarmouthian to early Illinoian) muskrats are placed in this species, intermediate in characters between *Ondatra annectens* and *Ondatra zibethicus*. It has been found at Angus, Berends, Doby Springs, Gordon, Hay Springs, Kentuck, Mullen I, and Rushville. Presence of two distinct size groups at Hay Springs suggests heterochrony or, alternatively, the coexistence of two distinct species.

Apart from smaller size, the species is distinguished from *Ondatra zibethicus* by higher frequency of three-rooted M^1. In the living form, most young individuals have only two roots in this tooth, but a third root tends to develop with age (see fig. 12.9).

Ondatra nebracensis may form a link in the evolution from *Ondatra annectens* to the modern muskrat. (Nelson and Semken, 1970; Semken, 1966; Starrett, 1956.)

Muskrat, *Ondatra zibethicus* (Linnaeus), 1766 (*Castor zibethicus; Fiber oregonus*
   Hollister, 1911)

Fossil remains of this species are very common and have been reported from about 50 Rancholabrean sites in Alberta, Arkansas, Florida, Georgia, Idaho, Illinois, Kansas, Kentucky, Louisiana, Maryland, Missouri, Montana, Nevada, New Mexico, Ohio, Oklahoma, Oregon, Pennsylvania, South Carolina, Tennessee, Texas, Virginia, West Virginia, Wyoming, and the Yukon. The earliest finds are late Illinoian in age (Mt. Scott) and are somewhat smaller, on the average, than modern specimens. The trends of size increase, cement deposition on the surface of molars, and development of dentine tracts, with interruption of the enamel pattern in the worn cheek teeth, are most pronounced in this species. The species may be derived from *Ondatra nebracensis*, but as noted above, the possibility of multiple muskrat lineages in the Irvingtonian cannot be ruled out.

The muskrat is the most aquatic microtine; it has partially webbed feet, and the scaly, nearly hairless tail is flattened laterally and is used as a rudder. The muskrat spends most of its time in marshes, lakes, or along stream banks; it builds marsh houses or burrows in banks with underwater entrances and feeds off water plants, mollusks, and aquatic vertebrates. Muskrats are valuable fur-bearers. (Nelson and Semken, 1970; Semken, 1966; Starrett, 1956; Stephens, 1960; Walker, 1975.)

Brown Lemming, *Lemmus sibiricus* (Kerr), 1792 (*Mus lemmus sibiricus*)

The genus *Lemmus* is mainly distributed in the Palearctic, where it presumably originated; early finds date back to the Günz. The genus is characterized by a robust skull, ungrooved upper incisors, cheek teeth with cement and comparatively few loops, and diverging tooth rows. As in *Dicrostonyx* and *Synaptomys,* the lower incisor passes to the lingual side of the cheek teeth. Although the American brown lemming is often regarded as a separate species, *Lemmus trimucronatus* (Richardson), we follow Macpherson (1965), Ognev (1947), and Rausch (1953) in assigning it to the same species as the Siberian form.

Fossil remains of the brown lemming have been found at Cape Deceit (Irvingtonian) and in the Fairbanks District (Rancholabrean). The modern range comprises part of Alaska and northern Canada, where the species inhabits tundra and alpine meadows. (Guthrie and Matthews, 1971; Macpherson, 1965; Ognev, 1947; Rausch, 1953.)

† Hopkins's Lemming, *Predicrostonyx hopkinsi* Guthrie and Matthews, 1971

The monotypic genus *Predicrostonyx* is considered ancestral to *Dicrostonyx*.

Good samples of the entire dentition of this species have been found at the Irvingtonian Cape Deceit site in Alaska. This primitive lemming differs from *Dicrostonyx* in the simplicity of its molar crowns (the lower molars lack anterior buds and the upper molars lack posterior buds) (see fig. 12.12). Its habitat was probably treeless tundra. (Guthrie and Matthews, 1971.)

Collared Lemming, *Dicrostonyx torquatus* (Pallas), 1799 (*Mus torquatus*)

The cementless, many-looped cheek teeth (M_1 has seven closed triangles) set off the genus *Dicrostonyx* from other lemmings (see fig. 12.12). The genus is Holarctic; *Predicrostonyx* may represent an early stage in its evolution. In Europe, the genus appears in the Günz.

This species can be distinguished from the closely related *Dicrostonyx hudsonius* by the presence of accessory cusps in M^1, M^2, and M_3. Fossil finds come from the Fairbanks District (Alaska), Jaguar Cave (Idaho), and Bell and Little Box Elder caves (Wyoming), which suggests that the species inhabited the tundra and alpine habitats around the western part of the Wisconsinan ice sheet. It is now found on the open tundra in the extreme north of the continent, from Alaska to the Hudson Bay and Greenland. The existence of several subspecies with partly disjunct distributions suggests fragmentation during the last glaciation. (Guilday, 1968b; Guilday and Doutt, 1961; Macpherson, 1965; Rausch and Rausch, 1972.)

Hudson Bay Collared Lemming, *Dicrostonyx hudsonius* (Pallas), 1778 (*Mus hudsonius*)

Differing from *Dicrostonyx torquatus* in the absence of accessory molar cusps, this species, with a present-day distribution east of Hudson Bay, has been found in Wisconsinan deposits at New Paris No. 4 in Pennsylvania. It thus appears that the species lived in the tundra close to the eastern part of the Wisconsinan ice sheet.

It is possible that in Wisconsinan times there was a continuous cline from *Dicrostonyx torquatus* in the West to *Dicrostonyx hudsonius* in the East, and that this cline was interrupted when the population shifted to the north at the end of the glaciation. The Ungava form (*D. hudsonius*) would then have become isolated in its present habitat. This presupposes a continuous tundra belt south of the Wisconsinan ice and necessitates a very late date for speciation. If there was a separate eastern tundra enclave, the species could be much older. (Guilday, 1963.)

† Rinker's Bog Lemming, *Synaptomys rinkeri* Hibbard, 1956

Unlike the Arctic lemming genera *Lemmus* and *Dicrostonyx,* the bog lemming *Synaptomys* is now endemic to North America, where its history begins in the Blancan. The origin of this genus appears to be in the Old World, however; a very primitive form occurs in the middle Villafranchian of Poland. In addition to the internal position of the lower incisors, a feature common to all the lemmings, this genus has a relatively simple dental pattern, grooved upper incisors, and less divergent tooth rows than *Lemmus*. In the evolution of the genus, the external triangles of the lower molars and the internal triangles of the upper molars have undergone a reduction. The European *Synaptomys europaeus* Kowalski represents a very early stage in this process and is referred to a distinct subgenus, *Praesynaptomys*. Of the currently recognized American subgenera, *Synaptomys* is the oldest presently known. In this subgenus, the first two triangles in the lower molars are closed, and the external reentrant angles are deep.

Synaptomys rinkeri belongs to the nominate subgenus. Known from the late Blancan of Dixon and White Rock, it was about the size of a big *Synaptomys cooperi;* the occlusal pattern of M_1 and M_2 is interrupted, as in the living species, to which the Blancan form may be ancestral. (Hibbard, 1956b; Kowalski, 1977.)

† Bunker's Bog Lemming, *Synaptomys bunkeri* Hibbard, 1939

Known only from a late Pleistocene deposit in Beaver County, Oklahoma, this species is a member of the subgenus *Synaptomys* and is distinguished by its small size. (Hibbard, 1939b.)

Southern Bog Lemming, *Synaptomys cooperi* Baird, 1858

Fossil remains of this species have been found at 23 localities in Arkansas, Georgia, Kansas, Maryland, Missouri, Oklahoma, Pennsylvania, Tennessee, Texas, Virginia, and West Virginia; there is also a record from San Josecito Cave in Mexico. The earliest are late Irvingtonian (Cumberland, Trout). Southern and western localities are well beyond the modern range of the species, which extends west to the Kansas–Colorado state line (where there is a relict population in a spring) and south to the 36th and 37th parallel. At some sites, the species replaces *Synaptomys borealis* in the late Wisconsinan. With the characters of the subgenus *Synaptomys*, this species differs from the very closely related *Synaptomys australis* in its smaller size (see fig. 12.12).

A cline with increasing size to the south has been observed within the modern population. The appearance of very large individuals in an otherwise homogeneous population (e.g., at Ladds) may signify shifts in the cline, within a heterochronous assemblage.

The species may be a descendant of *Synaptomys rinkeri*. Large southern variants of *Synaptomys cooperi* probably gave rise to *Synaptomys australis*, which may be only subspecifically distinct.

Today, the species inhabits bogs and damp meadows. (Gidley and Gazin, 1938; Guilday, Martin, and McCrady, 1964; Ray, 1967.)

†Florida Bog Lemming, *Synaptomys australis* Simpson, 1928

This form is mainly known from the Florida Pleistocene, where it occurs in several Sangamonian and Wisconsinan assemblages, and may have persisted to about 8,000 years B.P. (Devil's Den). It occurs in latest Illinoian to Wisconsinan deposits in Texas and Kansas (Clear Creek, Easley Ranch, Sims Bayou, Jinglebob, Mt. Scott) and has also been tentatively identified at Crankshaft Cave. The latter specimen may be a large *Synaptomys cooperi*.

Morphologically, *Synaptomys australis* is similar to *Synaptomys cooperi*, from which it differs only in being up to 35 percent larger. It may have been a large subspecies of *Synaptomys cooperi* representing a southern extension of the cline. Whether it was reproductively isolated is uncertain. The presence of two size groups at Ladds might point to such isolation, but these finds may also represent different time planes in the deposit. (Olsen, 1958; Ray, 1967; Simpson, 1928.)

†Old Bog Lemming, *Synaptomys vetus* Wilson, 1933

The extinct subgenus *Metaxyomys* has a combination of characters found in the other two subgenera: the first two alternating triangles in M_1 are confluent as in *Mictomys*, and the external reentrant angles are well developed as in *Synaptomys* (see fig. 12.12). Cement may be present or absent in the posteroexternal reentrant; it is absent in *Mictomys*, and present in *Synaptomys*.

Synaptomys vetus from the late Blancan Grand View fauna is one of the earliest known members of this subgenus. The anterior loop of M_1 is smaller than that of other *Metaxyomys*. The posterior loop of M^3 is ellipsoidal and lacks enamel on its posterior face. (Zakrzewski, 1972.)

†Landes's Bog Lemming, *Synaptomys landesi* Hibbard, 1954

This species, from the late Blancan of Borchers, differs from *Synaptomys vetus* in the following characters: the anterior loop of M_1 is larger, with shallow internal and external reentrants, and the posterior loop of M_3 is triangular, with enamel on its posterior face. It is the largest member of the subgenus *Metaxyomys*. (Zakrzewski, 1972.)

†Anza Bog Lemming, *Synaptomys anzaensis* Zakrzewski, 1972

The latest occurring member of the subgenus *Metaxyomys*, this species is found in the Irvingtonian of Vallecito. A characteristic feature is the large, volelike anterior loop of M_1, with deep internal and external reentrants. (Zakrzewski, 1972.)

†Melton's Bog Lemming, *Synaptomys meltoni* Paulson, 1961

In the subgenus *Mictomys*, the lower molars have confluent triangles and weakly developed external reentrant angles. The species *Synaptomys meltoni* is known from Sappa and Cudahy and thus belongs to the later Irvingtonian. It is close in size to the living *Synaptomys borealis*, but the teeth are larger; in the lower molars, the enamel on the anterior walls of the triangles is noticeably thicker than that on the posterior walls. Like the living form, it is regarded as a cold-climate indicator. (Paulson, 1961; Schultz and Martin, 1970b.)

†Kansas Bog Lemming, *Synaptomys kansasensis* Hibbard, 1952

This species is known from the Kentuck deposits (probably Kansan) of McPherson County, Kansas. It resembles *Synaptomys meltoni* but is distinctly larger, and the capsular process for the lower incisor is more posterior in position. The two species *Synaptomys meltoni* and *Synaptomys kansasensis* may represent an extinct branch of the subgenus *Mictomys*, not directly ancestral to the living northern bog lemming. (Semken, 1966.)

Northern Bog Lemming, *Synaptomys borealis* (Richardson), 1828 (*Arvicola borealis*)

Fossils of this extant species have been recovered from nine Wisconsinan sites in Missouri, Pennsylvania, Tennessee, Virginia, and northern Arkansas. All are far south of the present range of the species, which is mainly Canadian, extending to northernmost Washington, Idaho, Montana, and northern New England. It is thus an important cold-climate indicator.

The species differs from the *S. meltoni*–*S. kansasensis* group in the thickness of enamel on the molars, which is the same on the anterior and posterior walls of the triangles. It inhabits boreal forests, muskeg, and moist alpine environments. (Guilday, 1971b; Guilday, Hamilton, and McCrady, 1969; Guilday, Martin, and McCrady, 1964.)

Family Zapodidae—Jumping Mice

In addition to the living North American genera *Zapus* (also found in China) and *Napaeozapus*, this small family includes the Eurasian Sicistinae. The Zapodidae have a long but spotty Holarctic history in the Tertiary, beginning in the Eocene. The North American jumping mice have large hind feet and long tails; some of them have an almost froglike mode of progression. In addition to vegetable food, insects and spiders are eaten. The upper incisors are grooved, and the molars are tubercular to lophodont and rooted; P^4 may be present but is small. Trends in evolution in the Pleistocene relate to the occlusal pattern of the cheek teeth.

†Rinker's Jumping Mouse, *Zapus rinkeri* Hibbard, 1951

The history of the genus begins with *Zapus rinkeri*, from the early Blancan of Fox Canyon. *Zapus* may have arisen from the earlier Pliocene *Pliozapus* (which has been tentatively recorded from the late Blancan of Broadwater). The teeth are comparatively high-crowned, and the lower molars have a well-developed anteroconid, usually with an anteromedian fold in front. P^4 is present (see fig. 12.13).

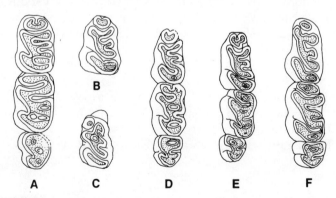

12.13 Zapodidae. Left lower dentitions, occlusal views: *A, D, E* are M_{1-3}; *B, C* are M_1. *A, Zapus rinkeri,* Blancan, Rexroad. *B, Zapus sandersi,* Blancan, Rexroad. *C, Zapus burti,* Blancan, Borchers. *D, Zapus hudsonius,* Rancholabrean, Jinglebob. *E, Zapus hudsonius,* Recent. *F, Zapus princeps,* Recent. After Klingener (1963). All 10×.

The Fox Canyon species, *Zapus rinkeri*, had a massive jaw and broad molars; it probably represents an extinct lineage. (Klingener, 1963.)

† Sanders' Jumping Mouse, *Zapus sandersi* Hibbard, 1956

With a long stratigraphic range extending from the early Blancan (Rexroad) to the late Blancan and Irvingtonian (Sanders, White Rock, Sappa, Java, Cudahy, and perhaps Cita Canyon and Port Kennedy), this species contains two successional subspecies, the early *Zapus sandersi rexroadensis* Klingener and the later type subspecies. The mandible is less massive than that of *Zapus rinkeri* but heavier than that of *Zapus princeps;* the anteroconid of M_1 is smaller than that of *Zapus rinkeri* or *Zapus hudsonius* but larger than that of *Zapus princeps* (see fig. 12.13) An increase in the depth of the anteriomedian fold may be noted over the course of time. The species is probably ancestral to *Zapus hudsonius*. (Klingener, 1963.)

Meadow Jumping Mouse, *Zapus hudsonius* (Zimmermann), 1780 (*Dipus hudsonius; Z. adamsi* Hibbard, 1955)

Early records of this species come from the late Irvingtonian (Angus, Cumberland, Sandahl). A series of successional subspecies continues the evolutionary lineage initiated in *Zapus sandersi*, with trends including a progressive deepening of the anteromedian fold in M_1 (see fig. 12.13). The early subspecies *Zapus hudsonius transitionalis* Klingener lived in Illinoian and early Sangamonian times (Doby Springs, Mt. Scott), whereas *Zapus hudsonius adamsi* is present later on in the Sangamonian (Jinglebob). Other Rancholabrean (mostly Wisconsinan) records come from Arkansas, Georgia, Kansas, Pennsylvania, Tennessee, Virginia, and West Virginia. Some of these sites are slightly outside the current range of the species, which extends to eastern Kansas, northeastern Oklahoma, and northern Georgia.

The species occurs in varied environments, preferring low meadows for feeding. Like other jumping mice, it is mainly nocturnal in habits. (Klingener, 1963.)

† Burt's Jumping Mouse, *Zapus burti* Hibbard, 1941

Known only from the late Blancan Borchers fauna, this species has broad teeth and in some characters resembles the living *Zapus princeps*, to which it may be ancestral; however, some dental peculiarities may indicate that there is no direct connection between the two (see fig. 12.13). (Hibbard, 1941d; Klingener, 1963.)

Western Jumping Mouse, *Zapus princeps* Allen, 1893

The only fossil record of this species is from Schulze Cave, with an approximate age of 10,000 years B.P. The area of the present range closest to this site is central New Mexico, where the species inhabits humid meadows and other moist areas with dense ground cover. The species has a very shallow anteromedian fold in M_1 (see fig. 12.13). (Dalquest, Roth, and Judd, 1969.)

Woodland Jumping Mouse, *Napaeozapus insignis* (Miller), 1891 (*Zapus insignis*)

The genus *Napaeozapus* lacks P^4, which is present in *Zapus*. Fossils are rare; the earliest record, tentatively referred to the modern species, comes from Cumberland Cave and is thus Irvingtonian. Other finds, of late Rancholabrean age, come from Baker Bluff, Boney Spring, Bootlegger Sink, Clark's Cave, Natural Chimneys, New Paris No. 4, and Robinson Cave. The Missouri record (Boney Spring) is well outside the modern range. Late Wisconsinan specimens from New Paris No. 4 average 10 percent larger than modern specimens from the central Appalachians, which suggests a positive "Bergmann's response" and indicates that the animals were adapted to boreal conditions, as are the populations living today in the Hudsonian/Canadian life-zones in Canada.

Like other jumping mice, this species prefers damp environments. It occurs in forested or brushy areas. (Gidley and Gazin, 1938; Guilday, Martin, and McCrady, 1964.)

Family Erethizontidae—New World Porcupines

The erethizontid porcupines are native to the Americas and first appear in the fossil record in the late Pliocene or early Pleistocene in North America with *Coendou cascoensis* White. Two genera, *Coendou* and *Erethizon,* are represented in the North American Pleistocene, and *Coendou* probably gave rise to *Erethizon* in the Irvingtonian. The two genera resemble each other morphologically but are distinguished by differences in functional morphology of mastication and adaptation to an arboreal (*Coendou*) or arboreal–terrestrial (*Erethizon*) mode of life.

Four extinct North American species are recognized, and erethizontid material from Port Kennedy, San Josecito, and Aguascalientes is referred to *Coendou* sp. The presence of *Coendou* in a fauna probably indicates an Irvingtonian age, except in Mexico, where the genus has survived to the present day.

†Deep-jawed Coendou, *Coendou bathygnathum* (Wilson), 1935 (*Erethizon bathygnathum*)

Living representatives of the genus *Coendou* inhabit tropical forests from Mexico south to Brazil and Bolivia. Adapted to an arboreal existence, the animals have a long, spineless, prehensile tail, specialized hands and feet for climbing, and a laterally directed field of vision. The orthodont incisors are used in cutting the leaves and tender shoots on which they feed. The upper tooth row is subparallel, and a projection of the longitudinal axis of the lower cheek tooth row would pass laterad to the lower incisor.

Known only from the very late Blancan locality of Jackass Butte, Grand View local fauna, *Coendou bathygnathum* is the largest reported fossil erethizontid. It is characterized by a deep, massive ramus and large teeth. White (1968) referred the species to the genus *Coendou* because a projection of the longitudinal axis of the cheek tooth row would pass laterad to the alveolus of the lower incisor, as it does in *Coendou* but not in *Erethizon.* (White, 1968, 1970; R. W. Wilson, 1935.)

†Stirton's Coendou, *Coendou stirtoni* White, 1968

This species is known only from Vallecito Creek, an Irvingtonian deposit in the Anza Borrego Desert State Park. About the size of an adult *Erethizon,* it has large teeth, with P^4 as large as M^1. The lower jaw is not as massive as that of *Coendou bathygnathum.* White (1968) believes that the species probably lacked a prehensile tail and that its habits were similar to those of *Erethizon.* (White, 1968, 1970.)

†Appalachian Coendou, *Coendou cumberlandicus* White, 1970

White (1970) regards this species as structurally intermediate between *Coendou* and *Erethizon.* The upper cheek tooth rows are subparallel as in *Coendou,* but the incisors, the longitudinal axis of the lower cheek tooth row, and auditory bullae show characteristics of both genera. It is known from Cumberland and Trout caves, both late Irvingtonian in age. (White, 1970.)

Porcupine, *Erethizon dorsatum* (Linnaeus), 1758 (*Hystrix dorsata*)

Erethizon first appears in late Irvingtonian deposits (Coleman IIA, Conard Fissure) and has been found in about 40 faunas in Alaska, Alberta, Arkansas, California, Colorado, Florida, Idaho, Kentucky, Maryland, Missouri, New Mexico, Pennsylvania, Tennessee, Texas, Utah, Virginia, and Wyoming. Several occurrences (Seminole Field, Waccasassa River, Cherokee, Brynjulfson, Crankshaft) are outside the present range of the species, an area that includes tundra, timbered and brushy tracts, rangelands, and deserts from Alaska and northern Canada to northern Mexico. It is absent today from the southeastern and south-central United States, areas that were inhabited by porcupines in the late Quaternary.

Porcupines are covered with thick, sharp, barbed, loosely attached quills interspersed with stiff guard hairs. The underparts lack quills, which makes the animal vulnerable to attacks by fishers, bobcats, cougars, and other predators. In some areas, fishers (*Martes pennanti*) are used to control porcupine populations.

These clumsy, slow-moving animals have a robust body, a small head, short legs, a vestigial prehallux, and a stout, spiny, nonprehensile tail. They weigh from 3.5 to 7.0 kg. The skull is robust, with a deep rostrum and large infraorbital foramen. White (1968, 1970) studied the functional aspects of their mastication and found that the upper cheek tooth rows converged markedly anteriad and that the longitudinal axis of the lower cheek tooth rows is lingual to the lower incisor (see fig. 12.14.) This results in a more efficient chewing mechanism than that of *Coendou*. The proodont incisors have a scraping, rather than cutting, function. Porcupines feed on the cambium layer of certain trees, as well as on leaves, buds, twigs, and young conifer needles, and will gnaw on anything with a trace of salt.

White (1970) believed that *Coendou* gave rise to *Erethizon;* the chief evidence for this is the intermediate nature of the population at Cumberland Cave. If this is the case, then *Erethizon* evolved in temperate North America and should be considered as having northern rather than southern affinities. (Frazier, personal communication, 1975; Ray and Lipps, 1970; Walker, 1975; White, 1968, 1970; C. A. Woods, 1973; Zakrzewski, personal communication, 1975.)

A

B

12.14 Erethizontidae and Hydrochoeridae. *A, Erethizon dorsatum*, Recent, right M^{1-2}. *B, Hydrochoerus holmesi*, Rancholabrean, Sabertooth Cave, M_3. *A*, after Vaughan (1972); *B*, after Simpson (1928). *A*, ca. 4×; *B*, 1×.

Family Hydrochoeridae—Capybaras

Capybaras are well represented in the Pliocene and Pleistocene of South America and in the Pleistocene of North America, Central America, and the West Indies. Today, *Hydrochoerus hydrochoeris* inhabits the northern half of South America, and *Hydrochoerus isthmius* is found in Panama east of the canal. Their fossil history extends back to the early Pliocene, and the family probably had a common ancestry with the Caviidae.

Capybaras are the largest living rodent; *Hydrochoerus hydrochoeris* weighs up to 50 kg. With a robust build, capybaras have short limbs and a large head with a deep rostrum and a truncated snout. White, slightly grooved incisors are characteristic. The evergrowing cheek teeth are made up of transverse lamellae joined together by cement. Semiaquatic in habit, capybaras live along borders of marshes and banks of streams, where they feed on succulent plants. They are good swimmers and divers and seek shelter in the water when pursued.

† Holmes's Capybara, *Hydrochoerus holmesi* Simpson, 1928

Originally described from Sabertooth Cave, this species has also been found at Inglis IA (earliest occurrence), Coleman IIA, Reddick, Seminole Field, Vero, Melbourne, and Santa Fe River IIA. The cheek teeth are larger than those of a large extant capybara but much smaller than those of *Neochoerus pinckneyi* (see fig. 12.14.) The lower incisor is smaller and less compressed anteroposteriorly than that of the living species.

The present distribution of *Hydrochoerus* suggests that the extinct species must have lived in North America when winters, even in Florida, were warmer than they are now and, in addition, that extensive aquatic habitats must have existed even in interglacial times. (Simpson, 1928.)

† Pinckney's Capybara, *Neochoerus pinckneyi* (Hay), 1923 (*Hydrochoerus pinckneyi*)

From the abundant material found at Sinton, Hay (1926) named a new genus of capybara, *Neochoerus,* and referred to it the type material found at Charleston, South Carolina. *Neochoerus pinckneyi* has also been identified at Bradenton, Ichetucknee River, Aucilla River, and Melbourne. The undescribed capybara material from the Irvingtonian Tusker fauna may represent a new genus (Lance, 1966.)

Neochoerus pinckneyi is characterized by large size. It was about 40 percent larger, with teeth some 30 percent larger, than robust Recent capybaras. The frontal region of the skull is broad and nearly flat, and the postorbital processes are almost twice as wide at the contact with the supraoccipitals as those of living species. The coronoid process is reduced.

The phylogenetic relationship between *Neochoerus* and *Hydrochoerus* is unknown at this time. (Hay, 1926; Lance, 1966; Simpson, 1930b; Zakrzewski, personal communication, 1976.)

CHAPTER THIRTEEN

ORDER LAGOMORPHA

Lagomorphs, like rodents, have a dentition characterized by chisellike, ever-growing incisors, which are separated from the cheek teeth by a long diastema. In some old classifications, this group appears as a suborder, Duplicidentata, of the Rodentia, but this arrangement is no longer considered valid.

In contrast to the Rodentia, which have only one incisor in each jaw half, the lagomorphs have two upper incisors. The first incisor has an anterior longitudinal groove; the second incisor, which is tiny, is situated just to the rear of the first. The cheek teeth are prism-shaped and wear to produce oval or rhomboidal occlusal surfaces, which may display reentrant angles resulting from flexures of the enamel. Early in the history of the lagomorphs, the cheek teeth became hypsodont and rootless. In the hind limb, the fibula is fused to the tibia over half or more of its length and articulates with the protarsus.

Unlike the Rodentia, which are often omnivorous, all lagomorphs are herbivores. They are also comparatively uniform, whereas the rodents are extremely rich in species and adaptive types. All of the Pleistocene lagomorphs of North America belong to one of the two extant families, the Ochotonidae and the Leporidae.

The order is native to all continents except Australia and Antarctica. It has a fairly good fossil record, but the taxonomy, especially on the species level, is difficult owing to the conservativeness and interspecific overlap of dental characters.

FAMILY OCHOTONIDAE—PIKAS

Pikas are small lagomorphs with short hind legs. In the cheek tooth row, M^3 is absent, and M_3 is reduced to a single prism (or, in some extinct forms, is absent.) The mental foramen is situated beneath the last lower molar, the nasals are wider anteriorly than posteriorly, and the frontals lack postorbital processes. The family appeared in the middle Tertiary and has a Holarctic distribution. All of the Pleistocene pikas of North America belong to the sole living genus, *Ochotona*. Fossils are rare. A form described by Cope (1871) from Port Kennedy Cave under the name *Praotherium palatinum* may be an ochotonid, but its status is uncertain.

†Wharton's Pika, *Ochotona whartoni* Guthrie and Matthews, 1971

In the genus *Ochotona*, M_3 is present as a single prism. The columns of the lower molars have an angular internal surface, and P_3 has more than one reentrant. The width of the columns is more than twice that of the connecting necks. The genus dates back to the Hemphillian in Eurasia; the American forms are immigrants.

An early Pleistocene form is Wharton's pika from the Irvingtonian of Cape Deceit; the species has also been found in the Gold Hill Cut at Fairbanks and at Old Crow Loc. 14N. *Ochotona whartoni* is distinguished from other North American *Ochotona* by its much larger size. Species of about the same size are known from the early and middle Pleistocene of China (*O. complicidens* Teilhard and the living *O. koslowi* Buchner), and it is possible that *Ochotona whartoni* belongs to one of these, size being the only distinctive character of its dentition. Its habitat may have been tundra and steppe, as in some Eurasian pikas, rather than the scree slopes of the living *Ochotona princeps*. (Guthrie and Matthews, 1971.)

Pika, *Ochotona princeps* (Richardson), 1828 (*Lepus princeps*)

The living North American pika belongs to the subgenus *Pika*, in which the palatal and incisive foramina of the upper jaw are separated rather than confluent as in the subgenus *Ochotona*. Both subgenera are present in Eurasia.

Late Irvingtonian records of *Ochotona* sp. come from Cumberland, Rapps, and Trout caves, which are very far from the present range of the genus. Rancholabrean records from Bell, Chimney Rock, Horned Owl, Jaguar, Little Box Elder, and Smith Creek caves are within or close to the range of the living *Ochotona princeps*, which has a patchy distribution from British Columbia to California, in the Rocky Mountains from Alberta to northern New Mexico, and in parts of Utah and Nevada.

The pika is a diurnal feeder on grass and herbs and lives on mountain scree slopes near the timberline, where it seeks shelter in the talus interstices. Pikas are often called "haymakers" for their habit of cutting, sun-drying, and storing plant materials for use as winter food. The ancestry of the species is presumably Eurasian. (Gidley and Gazin, 1938; Guilday and Adam, 1967; Walker, 1975.)

Collared Pika, *Ochotona collaris* (Nelson), 1893 (*Lagomys collaris*)

A mummified pika carcass was found in a hydraulic mining pit 15 m below the surface at Chatanika, 48 km north of Fairbanks, and dung pellets of pika overlain by more than 15 m of loess were recovered at Wilber Creek in interior Alaska. Guthrie (1973) referred the specimens to *Ochotona collaris*, the species found in the Yukon–Tanana upland and the Alaska Range today. Additional remains of the species have been reported from Fairbanks (Wisconsinan) and Old Crow River Loc. 44 (Sangamonian). Guthrie (1973) believes that speciation took place during the Wisconsinan when populations of *Ochotona princeps* were isolated by the ice sheets. (Guthrie, 1973; Harington, 1971c.)

Family Leporidae—Hares and Rabbits

The leporid family, which dates back to the Eocene, is native to all continents except Australia and Antarctica. Modern leporids have long ears and hind legs. The postorbital processes are well-developed, the nasals are not wider anteriorly than posteriorly, and the mental foramen is in a "normal" position. Dental characters include the presence of M^3 (in most species) and a bilobed M_3. Of the three subfamilies known, one (the Palaeolaginae) died out before the Blancan; one (the Archaeolaginae) survived into the early Pleistocene, and one (the Leporinae) is extant.

The Archaeolaginae differ from the Leporinae in the structure of their cheek teeth, especially P_3. The archaeolagine P_3 is rather simply built, with two persistent external reentrant folds, of which the posterior is deeper than the anterior and usually extends less than halfway across the occlusal surface of the tooth. In the Leporinae, the pattern tends to be more complicated, with additional reentrants, enamel lakes, and crenulation of enamel. Presence of cement increases masticatory efficiency.

Study of fossil leporids is made difficult by the overlap of specific and generic characters and the spottiness of the record.

Subfamily Archaeolaginae

†Hagerman Rabbit, *Hypolagus limnetus* Gazin, 1934

The genus *Hypolagus* has a relatively short and broad P_3 with a well-developed anteroexternal reentrant. The upper molars have reentrants with coarsely folded enamel borders. Cement is well developed on the teeth. Species are poorly characterized.

Hypolagus sp. has been recorded from various Blancan localities including Blanco, Borchers, Broadwater, Lisco, White Bluffs, and Sand Draw.

Hypolagus limnetus occurs in the early Blancan Hagerman fauna, and closely similar forms, perhaps the same species, are known from Benson and from the early Pliocene of California, Nevada, and South Dakota. It is a small form, agreeing in size with the living *Sylvilagus nuttalli*. The rostrum is relatively short, the cranial part elongate; P^2 has two unequal reentrants, P_3 a relatively deep anterior reentrant. (Dice, 1917; Gazin, 1934a; Hibbard, 1969.)

†Ancient Rabbit, *Hypolagus vetus* (Kellogg), 1910 (*Lepus vetus*)

A large species, *Hypolagus vetus* was originally described from the early Pliocene of Nevada; it, or a closely related species, survived at Hagerman (early Blancan), where it is the most common leporid, greatly outnumbering *Hypolagus limnetus*. It is especially common at the lower levels. The rostrum is longer than in *Hypolagus limnetus*, and P^2 has a well-developed anterior reentrant and an internal groove (see fig. 13.1). (Gazin, 1934a; Hibbard, 1969.)

†Royal Rabbit, *Hypolagus regalis* Hibbard, 1939

This species from the early Blancan Rexroad fauna has also been tentatively identified from Layer Cake and Arroyo Seco, as well as Cita Canyon, and so appears to

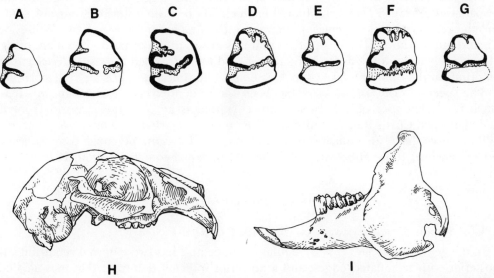

13.1 Leporidae. *A–G*, left P_3, occlusal view. *A, Hypolagus vetus,* Blancan, Hagerman. *B, Pratilepus vagus,* Blancan, Hagerman. *C, Aluralagus bensonensis,* Blancan, Benson. *D, Nekrolagus progressus,* Blancan, Rexroad. *E, Lepus americanus,* Recent. *F, Sylvilagus aquaticus,* Recent. *G, Sylvilagus floridanus,* Recent. *H, Hypolagus vetus,* Blancan, Hagerman, skull. *I, Pratilepus vagus,* Blancan, Hagerman, left mandible, external view. *A, B, H, I,* after Hibbard (1969); *C,* after Downey (1968); *D–G,* after Hibbard (1963). *A–G,* 5×; *H, I,* ca. 0.8×.

have survived to the late Blancan. It is a large species. The anterior reentrant of P_3 is wide, the posterior narrow; both have crenulated enamel. (Downs and White, 1968; Hibbard, 1939a).

†Furlong's Rabbit, *Hypolagus furlongi* Gazin, 1934

The species occurs at various localities, probably of very late Blancan age, of the Grand View local fauna. It is a small species, like *Hypolagus limnetus*, but the rostrum appears to have been shorter; P_3 is triangular in outline, with a shallow anteroexternal fold. P^2 has a deep, crenulated anterior reentrant and a shallow anteroexternal groove. (Gazin, 1934a.)

†Tusker Rabbit, *Hypolagus arizonensis* Downey, 1962

Only known from the Blancan Tusker fauna, this species is in the size range of *Hypolagus furlongi* and *Hypolagus limnetus*. It has deep external reentrants on P_3, the posterior one extending about two-thirds of the way across the tooth; the anterior reentrant is directed obliquely forward. (Downey, 1962).

†Brown's Rabbit, *Hypolagus browni* (Hay), 1921 (*Brachylagus browni*)

Again, this is a small species, close to *Hypolagus limnetus* in size; however, the ascending ramus of the lower jaw rises less steeply, and the condyle is lower in position. In some specimens, P_3 has a posterior enamel lake in the projection of the posterior enamel fold, a character seen also in some *Hypolagus regalis* specimens. It is known only from Anita, where it occurs in association with specimens determined as *Lepus*. This may suggest a very late Blancan age. (Gazin, 1934a; Hay, 1921.)

†Small Rabbit, *Notolagus lepusculus* (Hibbard), 1939 (*Dicea lepuscula*)

The genus *Notolagus*, which may date back to the early Pliocene, has a large anterointernal reentrant on P_3 with crenulated enamel. There is no anterior reentrant, and in this the genus resembles *Pratilepus*, *Aluralagus*, *Brachylagus* and *Romerolagus*. White (personal communication, 1975) considers that this group of genera arose from an *Alilepus*-like ancestor without the anterior reentrant, and he is disposed to classify *Notolagus* as a leporine. In the opinion of Dawson (personal communication), it is an archaeolagine.

The species *Notolagus lepusculus* has been found at Rexroad (early Blancan). It is a small form, and probably became extinct without issue. A larger leporid from the Gidley Pocket, Curtis Ranch, which has been tentatively identified as *Notolagus* cf. *velox* Wilson, does have an anterior reentrant on P_3. (Dawson, personal communication, 1976; Downey, 1968; Hibbard, 1939a; White, personal communication, 1975).

Subfamily Leporinae

†Plains Rabbit, *Pratilepus vagus* (Gazin), 1934 (*Alilepus? vagus*)

In the genus *Pratilepus*, P_3 is triangular in shape and has two external reentrants, the posterior with crenulated edges, and a posterior internal reentrant that may also take the shape of an enamel lake. As in *Notolagus*, there is no anterior reentrant. P^2 has a single anterior reentrant. The skull is larger and broader than that of *Hypolagus* (see fig. 13.1).

Pratilepus vagus is known only from the early Blancan Hagerman fauna, where it is next to *Hypolagus vetus* (among the leporids) in abundance. The lower molars have a simple pattern without crenulations on the edges of the reentrants. (Gazin, 1934a; Hibbard, 1969.)

†Kansas Rabbit, *Pratilepus kansasensis* Hibbard, 1939

This species, from the early Blancan Rexroad fauna, differs from *Pratilepus vagus* in the strong crenulation of the posterior borders of the reentrants on the lower molars. The reentrants of the upper molars have wavy borders, which produces a repeated trefoil pattern. *Aluralagus* probably descended from this or some other species of *Pratilepus*. (Hibbard, 1939a, 1969; White, personal communication.)

†Benson Rabbit, *Aluralagus bensonensis* (Gazin), 1942 (*Sylvilagus bensonensis*)

The pattern of P_3 in *Aluralagus* may have been derived from that in *Pratilepus* by connection of the enamel lake with the posterior reentrant, which thus crosses almost the entire occlusal surface; the borders are crenulated. The anteroexternal reentrant is fairly deep and consists of a series of folds. There is no anterior reentrant. The posterior borders of the reentrants on P_4–M_1 are crenulated.

The species *Aluralagus bensonensis* occurs at a number of levels in the early Blancan Benson fauna. No descendants are known. (Downey, 1968.)

Pygmy Rabbit, *Brachylagus idahoensis* (Merriam), 1891 (*Lepus idahoensis*)

In this genus, as in the probably related *Pratilepus–Aluralagus* group, P_3 lacks an anterior reentrant. The posteroexternal reentrant extends almost across the tooth.

The only known species, *Brachylagus idahoensis* has a restricted distribution, mainly in the states of Idaho, Nevada, Oregon, and Utah. Fossils come from the Rancholabrean of Jaguar, Middle Butte, Moonshiner, Rainbow Beach, Silver Creek, Tule Springs, and Wasden, sites within or close to the present range. The pygmy rabbit is a small form, with short ears and short hind legs, and digs its own burrows in areas with dense sagebrush, which forms its main food. (Hibbard, 1963b; Mawby, 1967; White, personal communication, 1975.)

†Progressive Rabbit, *Nekrolagus progressus* (Hibbard), 1939 (*Pediolagus progressus*)

The genus *Nekrolagus*, which initiated a series of leporids with a well-developed anterior reentrant fold on P_3, was presumably derived from an *Alilepus*-like ancestor with this character in the Pliocene. There is an anteroexternal and a posteroexternal reentrant. The posteroexternal reentrant extends more than halfway across the tooth and has a crenulated posterior border and a cement-filled enamel lake lingual to it. In some specimens, the reentrant and the lake unite to form a *Sylvilagus*-like pattern. Thus *Nekrolagus* may occupy a pivotal position in the evolution of the modern genera *Lepus* and *Sylvilagus*.

Nekrolagus sp. has been reported from some early Blancan faunas like Layer Cake, White Bluffs, and the Boo Boo fauna of Cochise County, Arizona; the species *Nekrolagus progressus* occurs at Rexroad (early Blancan). (Hibbard, 1939a, 1963b.)

Eastern Cottontail, *Sylvilagus floridanus* (Allen), 1890 (*Lepus sylvilagus floridanus*)

In the genus *Sylvilagus*, the young are born naked, blind, and helpless. Cottontail rabbits live in burrows but do not dig their own. The genus ranges from southern Canada to Argentina and Paraguay. In *Sylvilagus*, as in *Lepus*, the posteroexternal reentrant of P_3 extends across the tooth, except for the internal wall; in early forms, the *Nekrolagus* pattern may persist as a rare variant (Borchers). Distinguishing between *Lepus* and *Sylvilagus* on the basis of dental material is difficult; according to White (personal communication, 1975), probably no more than 75 percent of the specimens, at best, are determinable. Undetermined *Sylvilagus* spp. have been recorded from numerous localities, including late Blancan (Broadwater, Haile XVA), Irvingtonian (Gilliland, Cudahy, Java, Rezabek, Vallecito), and about 50 Rancholabrean sites.

Sylvilagus floridanus has been tentatively identified in the early Irvingtonian (Curtis Ranch, Port Kennedy) and is present in the late Irvingtonian (Conard Fissure, Mullen). Rancholabrean records, numbering more than 30, come from Alberta, Florida, Illinois, Missouri, New Mexico, Pennsylvania, South Carolina, and Texas; the Alberta record (Medicine Hat, Sangamonian) is well outside the present range but within that of the living *Sylvilagus nuttalli*. Many of the most prolific sites, with large numbers of remains, are "death trap" type caves with vertical shafts, for which this species seems to have an affinity.

Sylvilagus floridanus has the largest geographic range of any species of *Sylvilagus*, and the zonal range extends from the Canadian Life-zone into the Tropical Life-zone. The species shows considerable geographic variation in the size and shape of the skull. (Dalquest, Roth, and Judd, 1969; Hall and Kelson, 1959; Parmalee, Oesch, and Guilday, 1969; Stalker and Churcher, 1970.)

Mountain Cottontail, *Sylvilagus nuttalli* (Bachman), 1837 (*Lepus nuttalli*)

Only a few remains have been referred to this species, which has been recorded from the Wisconsinan of Dry Cave, Jaguar Cave, and perhaps Wasden. However, some remains ascribed to *Sylvilagus floridanus* may in fact be mountain cottontail. *Sylvilagus nuttalli* is now a fairly widespread northwestern species, occurring in mountains, thickets, sagebrush, and forests. It differs from *Sylvilagus floridanus* in having a more slender rostrum and larger external auditory meatus, and the tip of the supraorbital process projects free from the skull. (Guilday and Adam, 1967.)

New England Cottontail, *Sylvilagus transitionalis* (Bangs), 1895 (*Lepus sylvilagus transitionalis*)

A few remains of this species have been found at Clark's Cave and Ladds. The obsolete anterior supraorbital process and the slender, free posterior supraorbital process distinguish it from other eastern species of rabbits. The sites are close to the present range. The species is found in mountains, open forests, and brushy areas. (Dawson, personal communication; Ray, 1967.)

Desert Cottontail, *Sylvilagus audubonii* (Baird), 1858 (*Lepus audubonii*)

This southwestern species occurs at about 15 sites in Arizona, California, Mexico, New Mexico, and Texas; some date back to the Sangamonian (Cedazo). Osteological characters include inflated bullae, a prominent supraorbital process, and long hind legs. The species inhabits open plains, foothills, and valleys with grass, sagebrush, and juniper. (W. E. Miller, 1971; Mooser and Dalquest, 1975b.)

Brush Rabbit, *Sylvilagus bachmani* (Waterhouse), 1839 (*Lepus bachmani*)

The brush rabbit, now occurring along the West Coast from Oregon into Baja California, has been found in fossil form at Costeau, McKittrick, Newport Bay Mesa, Rancho La Brea, and San Pedro. All the records are Rancholabrean, the earliest being

Sangamonian. It is a small species, occurring in dense chaparral, where it makes tunnellike runways. (W. E. Miller, 1971.)

Marsh Rabbit, *Sylvilagus palustris* (Bachman), 1837 (*Lepus palustris*)

The marsh rabbit has been found at several sites in Florida, the earliest of which is Sangamonian (Reddick). The hind feet are short. The species inhabits wetlands and hummocks, probably not over 150 m in altitude, and thus may be an important indicator of ancient sea level. (Ray, 1958; Webb, 1974a.)

†Pygmy Marsh Rabbit, *Sylvilagus leonensis* Cushing, 1945 (*S. palustrellus* Gazin, 1950)

Pygmy rabbits possibly representing the genus *Sylvilagus* have been reported from San Josecito, Melbourne, and Vero (Wisconsinan). These finds have been insufficiently studied, and two species may be represented. (Cushing, 1945; Gazin, 1950; Sellards, 1916a.)

Swamp Rabbit, *Sylvilagus aquaticus* (Bachman), 1837 (*Lepus aquaticus*)

Fossil remains of this marsh and wetland species have been found at Herculaneum, Missouri (Wisconsinan), which is close to the present range. It is a large species and a good swimmer. (Olson, 1940.)

†Benjamin's Hare, *Lepus benjamini* Hay, 1921

The genus *Lepus* comprises the true hares, in which the young are born fully furred, with the eyes open. Hares do not dig or occupy burrows. The genus ranges across Eurasia south to Java and Sumatra, over most of Africa, and in North America from southern Canada to central Mexico. It includes the largest living lagomorphs. The hind limbs are long and powerful. On P_3, the posteroexternal reentrant extends across the tooth almost to the inner wall. The anterior reentrant is weakly developed. Distinguishing this genus from *Sylvilagus* is often difficult in fossil material, and species distinctions are uncertain unless much of the skeleton is recovered.

Early forms referred to *Lepus* sp. have been identified well back in the Irvingtonian (Angus, Bautista, Java, Rushville, Vallecito) and may have a so-called pro-*Lepus* P_3 pattern with a somewhat less developed posteroexternal reentrant. *Lepus* has also been found at sites regarded as Blancan (Tusker). These records include *Lepus benjamini* from Anita, the sole known locality of the species. The affinities and status of this species are uncertain.

There are also upwards of 30 records of *Lepus* sp. from scattered Rancholabrean deposits. (Hay, 1921; White, personal communication, 1975.)

Antelope Jack Rabbit, *Lepus alleni* Mearns, 1890 (*L. giganteus* Brown, 1908)

The antelope jack rabbit has a long history in North America, beginning in the early Irvingtonian (Inglis IA); other records come from Coleman IIA (late Irvingtonian) and Burnet (late Wisconsinan). It may be the same species that has been reported under the name *Lepus giganteus* from the Irvingtonian of Conard Fissure and Mullen. It is a large species, with relatively small P_3 and a long, slender diastemal part of the mandible. The Pleistocene form may be larger, on the average, than living representatives.

The sites in Florida are very far from the present range of the species, which is now restricted to desert, grassland, and mesquite in western Mexico and adjoining parts of Arizona and New Mexico. It thus appears to be a relict of a population that was widespread in the Irvingtonian. (Brown, 1908; R. A. Martin, 1974a; Schultz and Howard, 1935.)

Black-tailed Jack Rabbit, *Lepus californicus* Gray, 1837

This species is the most commonly found fossil *Lepus* in North America, with tentative early records from the early Irvingtonian or even earlier (Curtis Ranch, possibly Borchers). It also occurs at Slaton (late Irvingtonian) and at more than 20 Rancholabrean sites in Arizona, California, Mexico, Nevada, and Texas. The species occurs in the same area today; its habitat is prairie and desert. It is distinctly shorter-limbed than *Lepus alleni* or *Lepus townsendii*. (Dalquest, Roth, and Judd, 1969; Patton, 1963.)

White-tailed Jack Rabbit, *Lepus townsendii* Bachman, 1839

The earliest fossil record of this species comes from Mullen II (late Irvingtonian). Rancholabrean fossils, most of them only tentatively referred to this species, are from American Falls, Burnet, Dry, Jaguar, Little Box Elder, Medicine Hat (Sangamonian and mid-Wisconsinan), and Schulze. The sites in New Mexico and Texas are south of the present range, which extends approximately to the Colorado–New Mexico state line.

Intermediate in size between *Lepus arcticus* and *Lepus americanus,* this species is a plains form with nocturnal habits. (Dalquest, Roth, and Judd, 1969; Schultz and Howard, 1935.)

Snowshoe Rabbit, *Lepus americanus* Erxleben, 1777

This northern form occurs in the late Irvingtonian faunas of Cumberland and Conard Fissure and has been identified at several Wisconsinan localities (Bat, Crankshaft, Frankstown, Herculaneum, Jaguar, Natural Chimneys, New Paris No. 4, Samwel, Welsh). Although the species now ranges quite far south in the Appalachians and Rocky Mountains and in California, the records from Arkansas, Missouri, and Illinois are well south of its modern range and probably indicate the existence of climate colder than that at present.

The snowshoe rabbit is a relatively small species; the basal length of the skull is less than 67 mm, and the first upper incisors describe an arc with a radius less than 9.6 mm. The ears are relatively small, the feet large. It is an inhabitant of swamps, forests, and mountains. (Guilday, 1962b; Guilday, Martin, and McCrady, 1964; Peterson, 1926.)

Arctic Hare, *Lepus arcticus* Ross, 1819

The relationship between the Arctic hares of North America (*Lepus arcticus* and *Lepus othus* Merriam) and those of Eurasia (*Lepus timidus* Linnaeus) is apparently close. Fossil records come from Old Crow Locs. 14N and 44 (an area in which *Lepus othus* now occurs). These records are Sangamonian and Wisconsinan. The European species *Lepus timidus* dates back to the Last Interglacial.

The Arctic hare is distinguished from *Lepus americanus* by its larger size. It is an inhabitant of the northern tundra. (Hall, 1951b; Harington, 1971c; Kurtén, 1968.)

CHAPTER FOURTEEN

Order Perissodactyla

This order, which arose in the late Paleocene, has a good fossil record. It comprises medium-sized to large herbivores in which the axis of the foot passes along the third digit. Side toes tend to reduction, typically to a three-toed or one-toed condition. A highly varied and successful order in the early and middle Tertiary, the Perissodactyla were well past their apogee in Blancan and Pleistocene times; only two families, the Tapiridae and the Equidae, survived in North America. The former belongs to the suborder Ceratomorpha, the latter to the Hippomorpha. A third suborder, the Ancylopoda, became extinct in North America in the Miocene.

Species-level taxonomy in perissodactyls is notoriously difficult because of the inconstancy of many characters and the large number of fragmentary finds. The literature is widely scattered, and much of it is inconclusive.

The perissodactyl family Rhinocerotidae (suborder Ceratomorpha), which became extinct in North America in the Pliocene, survived to the late Pleistocene in northeastern Asia and is extant in southern Asia and parts of Africa. With so many other cold-adapted Eurasian mammals entering Alaska during the Ice Age, the absence of the woolly rhinoceros, *Coelodonta antiquitatis* (Blumenbach), is particularly striking.

Family Equidae—Horses

The equid family has long served as one of the best examples of evolution to be demonstrated by paleontology (Simpson, 1961; Stirton, 1940). From their beginnings in the late Paleocene, various equid lineages can be traced through geologic time. The center of equid evolution lay in North America throughout the Tertiary, but several migrations led to the population of the Old World (and, later, South America) by various kinds of equids, which then underwent further differentiation.

Evolutionary trends in the Equidae include size increase (often reversed, however), reduction of the number of digits (ultimately, to one, much strengthened), increase of relative and absolute brain size, a change from tubercular to lophodont cheek teeth, molarization of premolars, progressive hypsodonty, changes in the structure of skull and jaws, and development of a rigid, straight backbone. By Blancan times, most of these changes had occurred, and one-toed horses of basically modern aspect were already present, together with hypsodont three-toed equids. Blancan and Pleistocene horses were essentially grazers.

The fossil record of equine horses is excellent. The prismatic cheek teeth and massive limb bones are very durable and tend to preserve well. Nonetheless, taxonomy of Pleistocene horses is in a state of confusion. The reasons for this are biological as well as historical. Important among the former are the great variability of the cheek teeth, which may produce striking differences in the occlusal surface even within a single species, and the pronounced changes that occur as the teeth wear down. The practice of many early workers of erecting new species based on incomplete dental material and intraspecifically and intraindividually variable characters has led to a proliferation of named "species," especially of the genus *Equus sensu lato* (more than 40 species have been proposed for the Blancan and Pleistocene of North America). The resulting impasse is illustrated by the fact that we have counted approximately 100 local faunas in which the genus is represented by *Equus* sp. A situation encountered at many sites is the presence of three size groups—a large horse, a medium-sized horse, and a small horse (Whitmore, personal communication, 1971).

A modern revision of the late Cenozoic Equidae has long been needed. Walter W. Dalquest most generously placed the results of his then unpublished preliminary revision at our disposal (published, 1978), and much of the review that follows is to be credited to him. The major aims of the revision have been to (a) determine the true biological species, (b) assign valid names to them, and (c) place other names in synonymy.

In view of the fragmentary nature of the types of many of the earlier-named species, Savage (1951) has suggested that such names be considered indeterminate *nomina vana*. Dalquest (1967), however, fears that such treatment would lead to instability and believes that additional material and more discriminating study will permit eventual revision of the Pleistocene Equidae by conventional taxonomic methods. Important contributions to this end have been provided by the work of Skinner and Hibbard (1972), Mooser and Dalquest (1975b), Dalquest (1978), and others.

†Beck Ranch Gazelle-Horse, *Nannippus beckensis* Dalquest and Donovan, 1973

The genus *Nannippus* ranged from the Clarendonian, where it is poorly represented, through the Hemphillian and Blancan. It comprises small to tiny horses with three functional toes on each foot. The limbs were slender and elongated, the teeth hypsodont, and the protocones of the upper cheek teeth isolated, except on P^2. These are the last representatives in North America of the three-toed grazing horses, or hipparions. The genus can be distinguished from the related *Neohipparion* (which did not survive in the Blancan) by the shape of the ectoflexid (the median enamel fold between the protoconid and the hypoconid) in the lower molars: in *Nannippus*, the ectoflexid penetrates deep into the isthmus between the metaconid and metastylid, whereas in *Neohipparion* (and in horse premolars generally), it is shallow.

A tiny *Nannippus*, which is unnamed at present, is known from several Blancan sites (Beck Ranch, Sanders, Red Light) and is probably derived from the Hemphillian *Nannippus minor* (Sellards) of Florida or *Nannippus aztecus* Mooser from Guanajuato, Mexico.

Nannippus beckensis is known only from the early Blancan Beck Ranch. It is a moderately large form with an elongated muzzle and rather high-crowned cheek teeth; parastylids occur in most of the lower cheek teeth. (Dalquest and Donovan, 1973.)

†Gazelle-Horse, *Nannippus phlegon* (Hay), 1899 (*Equus phlegon; E. minutus* Cope, 1893; *Hipparion cragini* Hay, 1917)

Most Blancan fossils identified as belonging to this genus have been referred to *Nannippus phlegon*, which was for more than 70 years the only accepted Blancan species of the genus. The holotype, from Blanco, is an isolated tooth, of little value for comparative purposes. Thus, in Dalquest's opinion, records of *Nannippus phlegon* from other localities are uncertain; some may be *Nannippus beckensis*, and some may belong to undescribed species.

Material referred to *Nannippus phlegon* or *Nannippus* sp. has been reported from numerous localities in Arizona, Florida, Kansas, Nebraska, and Texas. The Blancan *Nannippus* thus has a central, southern, and southeastern distribution, and it may be significant that the genus apparently became extinct in the San Pedro Valley of Arizona about 2.5 m.y. B.P. but survived about half a million years longer elsewhere.

Nannippus phlegon is currently seen as a highly advanced species, differing from *Nannippus beckensis* in having a shorter muzzle and more hypsodont teeth, as well as in the lack of parastylids in all lower cheek teeth (see fig. 14.1). (Dalquest, 1978; Hibbard, 1941a; Lindsay, Johnson, and Opdyke, 1975; Stirton, 1940.)

†American Zebra, *Equus simplicidens* Cope, 1892 (*E. proversus* Merriam, 1916; *E. idahoensis* Merriam, 1918; *Pliohippus francescana* Frick, 1921; *Plesippus shoshonensis* Gidley, 1921)

The genus *Equus sensu lato* comprises all the living equids—horses, zebras, and asses—and their predecessors in the Pleistocene and late Pliocene. Various classifications, with division into genera and subgenera, have been proposed. In Dalquest's opinion, the genus *Equus sensu stricto* (true horses and zebras) may be derived from the Hemphillian *Dinohippus;* in both genera, the ectoflexid of the lower molars penetrates deeply into the metaconid–metastylid isthmus. This group would thus represent a lineage separate from that of *Astrohippus–Asinus*, which lacks this character. All of these genera have the protocone of the upper cheek teeth connected to the protoselene (in contrast to the condition in the hipparions, e.g., *Nannippus*) and may have a common origin in Miocene *Pliohippus* (see figs. 14.1–14.3). The number of digits is reduced to one on each foot.

Although recognizing that it may be found necessary to divide the *Equus* complex into two or more genera, we prefer at present to refer the various species to *Equus sensu lato*. These species fall into a number of more or less well-defined groups, some of which have received subgeneric designation.

The subgenus *Equus* (true horses) is, according to Dalquest, characterized by broad and shallow, usually U-shaped linguaflexids (the internal folds between the metaconid

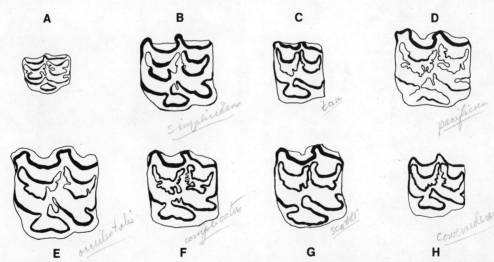

14.1 Equidae. Upper cheek teeth: *A* is P⁴; *B–H* are M¹. *A*, *Nannippus* cf. *phlegon*, Blancan. *B*, *Equus simplicidens*, Blancan, Hagerman. *C*, *Equus tau*, Irvingtonian, Lissie Formation, Texas. *D*, *Equus pacificus*, Rancholabrean, Ingleside. *E*, *Equus occidentalis*, Rancholabrean, Rancho La Brea. *F*, *Equus complicatus*, Rancholabrean, Ingleside. *G*, *Equus scotti*, Irvingtonian, Arkalon, Kansas. *H*, *Equus conversidens*, Rancholabrean, Valley of Mexico. *A, E*, after Stirton (1940); *B*, after Gazin (1936); *C, H*, after Lundelius and Stevens (1970); *D, F*, after Lundelius (1972a); *G*, after Hibbard (1953c). All 0.66×.

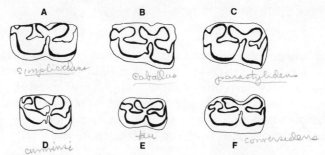

14.2 Equidae. Right M_1, occlusal view. *A, Equus simplicidens,* Blancan. *B, Equus caballus,* Recent. *C, Equus parastylidens,* Rancholabrean, Cedazo. *D, Equus cumminsii,* Blancan, Blanco. *E, Equus tau,* Irvingtonian. *F, Equus conversidens,* Rancholabrean. After Dalquest (1978). Not to scale.

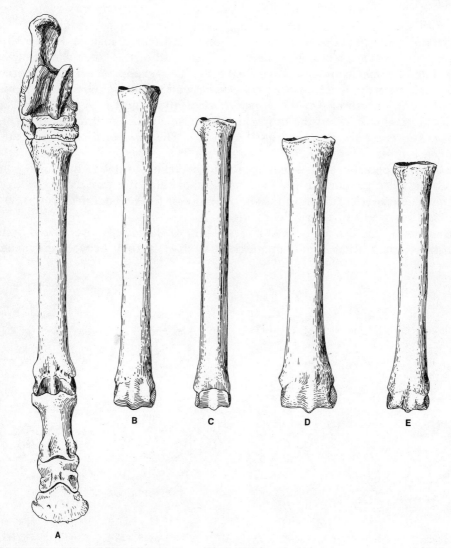

14.3 Equidae. Left pes and metatarsals. *A, Equus simplicidens,* Blancan, Hagerman, articulated pes. *B, Equus calobatus,* Irvingtonian, Arkalon, Kansas. *C, Equus tau,* Irvingtonian, Lissie Formation, Texas. *D, Equus scotti,* Irvingtonian, Arkalon, Kansas. *E, Equus conversidens,* Irvingtonian, Slaton. *A,* after Gazin (1936); *B, D,* after Hibbard (1953c); *C,* after Lundelius and Stevens (1970); *E,* after Dalquest and Hughes (1965). All 0.25×.

and metastylid of the lower cheek teeth). Although definite records from the Blancan or Pleistocene of North America do not exist, a number of specimens have been referred to *Equus caballus* Linnaeus, the Old World horse; they may be extreme variants of other species that resemble *Equus caballus*. A skull with lower jaws found on a sandbar of the Kansas River near Lawrence, Kansas, shows the characters of the subgenus *Equus* and is the type of *Equus laurentius* Hay, 1913; the specimen is permineralized, and other skulls of undoubted Pleistocene age have been found in the Kansas River under similar circumstances. Matthew (1924a) thought that the skull was from a Recent horse, however, and Dalquest (1978) inclines to the same view.

In the subgenus *Dolichohippus,* the linguaflexids are narrow and usually V-shaped; there are no parastylids in the lower cheek teeth. This subgenus includes the living Grévy's zebra of Africa. Skinner (in Skinner and Hibbard, 1972) has remarked on the close similarity of the American Blancan zebras to *Equus grevyi*. Like that species, *Equus simplicidens* has a long skull with a prominent occipital region. Among dental characters may be noted the mostly large P^1, the relatively simple enamel pattern of the cheek teeth, and the presence of "cups" (enamel folds) on the lower incisors. The limbs are slender, and the hoofs are small; the "splint bones" (metapodials of the reduced side toes) are fairly large; which is apparently a primitive character.

Dolichohippine zebras lived in North America from early Blancan (Rexroad) to late Blancan times. Sites include Blanco, Broadwater, Cita Canyon, Deer Park, Donnelly Ranch, Grand View, Hagerman (where the famous Horse Quarry has yielded numerous skeletons), Haile XVA, Hudspeth, Lisco, Rexroad, Sand Draw, Sanders, Santa Fe IB, Seger, White Bluffs, and White Rock. Whether one or more species should be recognized is not certain (Dalquest, 1978.) The species is not found in very early Blancan (late Gilbert) faunas, and Lindsay (personal communication) thinks that it appeared about 3.2 m.y. B.P. (Gazin, 1936; McGrew, 1944; Matthew, 1924a; Mooser and Dalquest, 1975b; Savage, 1951; J. R. Schultz, 1937; Shotwell, 1970; Skinner and Hibbard, 1972.)

†Mooser's Horse, *Equus parastylidens* Mooser, 1959

This large horse is only known from two lower jaw fragments from Cedazo. The linguaflexids are V-shaped as in the subgenus *Dolichohippus,* but unlike this group of zebras, the Cedazo species, which is Rancholabrean, in age, has parastylids on the lower cheek teeth (except M_3) and has been referred to a distinct subgenus, *Parastylidequus*. It is distinct from other described American equids but has characters in common with the living lowland zebra of Africa, *Equus burchelli* (Gray) (see fig. 14.2). The latter is in the subgenus *Hippotigris*. Whether this is a case of parallel evolution or represents a real relationship is uncertain at present. (Mooser and Dalquest, 1975b.)

†Cummins's Ass, *Equus cumminsii* Cope, 1893

A genus *Asinus*, separate from *Equus* and including the asses and half-asses, is recognized by many workers. A distinctive character of the lower molars is that the ectoflexids do not enter the metaconid–metastylid isthmus. This character is also seen in the Hemphillian *Astrohippus*, which Dalquest (1978) regards as ancestral to *Asinus*. If this interpretation is correct, the genera *Equus* and *Asinus* have been separate lineages since the Miocene. The great majority of American Pleistocene horses would thus belong to *Asinus* since they exhibit this character.

In the present review, we adhere to a conservative classification and use *Asinus* as a subgenus for the true donkeys and asses. In these animals, the linguaflexids of the lower cheek teeth tend to be narrow and V-shaped, as is also true of *Equus cumminsii*. The latter is a Blancan species, known from Beck Ranch, Blanco, and Seward County, Kansas. It is thus one of the earliest of the typical American nonzebrine horses. It is a small horse, with a very simple enamel pattern (see fig. 14.2); a referred phalanx is slender. Specimens from Hudspeth and Red Light, referred to *Equus cumminsii*, are regarded by Dalquest as belonging to another species. (Akersten, 1972; Dalquest, 1975, 1978; Hibbard, 1944; Strain, 1966.)

† Stilt-legged Onager, *Equus calobatus* Troxell, 1915

The stilt-legged American horses of the subgenus *Hemionus* are closely related to the Old World onagers. They range in size from quite large to extremely small, and most species can be distinguished by means of size. The earliest records are late Blancan, and the latest are Wisconsinan or early Holocene. The slender, elongated metapodials separate the hemiones from other American forms of *Equus* (see fig. 14.3). The dentition is not strongly characterized, but the cheek teeth tend to be relatively slender and very hypsodont, and the internal valleys (linguaflexids) of the lower cheek teeth are narrow and usually V-shaped.

Equus calobatus is the largest member of the group, with metatarsals that approach or exceed 300 mm in length. The length of the lower tooth row usually exceeds 180 mm. The species ranges from the late Blancan (Sand Draw, Donnelly Ranch) through the Irvingtonian [Arkalon, Rock Creek, Medicine Hat (D–E), Hay Springs, Rushville, Gordon, Slaton, Angus] and into the early Rancholabrean (Cedazo, Cragin Quarry). (Mooser and Dalquest, 1975b; Skinner and Hibbard, 1972; Troxell, 1915a.)

Onager, *Equus hemionus* Pallas, 1775 (*Onager altidens* Quinn, 1957?)

Also given the names kiang, kulan, and others, this species (or group of species) is now extant in Asia and has a Pleistocene history in Eurasia. A metatarsal from Gold Run Creek in the Yukon is similar to that of the modern Asian form; its length (267 mm) is less than that of *Equus calobatus*.

The postcranial material of *Equus altidens* from the late Pleistocene Berclair Terrace of Bee County, Texas, seems surely referable to the subgenus *Hemionus;* a metatarsal measures 283 mm and is relatively more stocky than the metatarsals of *Equus calobatus*. It is close in age to the Yukon species and may represent the same species. (Harington and Clulow, 1973; Quinn, 1957.)

† Pygmy Onager, *Equus tau* Owen, 1869 (*E. littoralis* Hay, 1913; *E. francisci* Hay, 1915; *E. achates* Hay and Cook, 1930; *E. quinni* Slaughter et al., 1962)

This is the smallest species of the subgenus *Hemionus*, as well as the smallest American *Equus*. It is known from earliest Irvingtonian (Holloman) to Wisconsinan deposits and ranged from the Valley of Mexico to Oklahoma, Texas, and Florida. It is a stilt-legged horse with extremely slender metapodials; the metatarsals are ca. 277–283 mm long and 32–38 mm wide at the proximal end (see fig. 14.3). Upper and lower cheek tooth rows usually measure less than 120 mm in length (135 mm in the type skull of *E. francisci*) (see figs. 14.1, 14.2). (Hibbard and Dalquest, 1966; Lundelius and Stevens, 1970; Quinn, 1957.)

†Giant Horse, *Equus giganteus* Gidley, 1901, and the probably related forms *Equus pacificus* Leidy, 1869 (*E. mexicanus* Hibbard, 1955); *Equus pectinatus* Cope, 1899; and *Equus crinidens* Cope, 1894

Although the species of *Equus* already discussed are fairly well understood, perhaps because most of them belong to extant subgenera, the status and relationships of the remaining species (all of which have the *Asinus*-like molar pattern defined above) are more or less obscure. We have listed and grouped them in accordance with the opinions of Dalquest (1978). The listing includes species which have been considered valid on the basis of seemingly adequate material but leaves out a number of nominal species based on inadequate material or not recently restudied; most are probably synonyms of better-known species, but this cannot be shown at present.

Some species are clearly more closely related than others and are here grouped together. Various subgeneric names have been proposed, but in most cases, there are no useful anatomic characters to justify subgeneric separation at this time.

The present grouping is mainly based on the shape of the protocone of the upper cheek teeth and the degree of complexity reached by the folding of the enamel. In the group typified by the giant horse, the protocone is short and broad, and the enamel is complexly folded. All of these horses are large in size.

Equus giganteus was described on the basis of a single upper premolar from an uncertain locality in Texas. Teeth from Borchers, Gilliland, and Holloman seem to be correctly referred to this form. These records are late Blancan and early Irvingtonian. (Hibbard and Dalquest, 1966.)

Equus pacificus may be a related form; Lundelius (1972) has noted that it differs from *Equus giganteus* in being slightly smaller, with more slender and elongate protocones on the upper teeth and a slightly less complex enamel pattern. Most records seem to be from rather late Pleistocene sites and range from California to Oregon and from Texas south to the Valley of Mexico. (Hibbard, 1955b; Lundelius, 1972a; Mooser and Dalquest, 1975b.)

Equus pectinatus comes from the early Irvingtonian of Port Kennedy, and teeth from Illinois have been referred to the same species. The available material indicates a horse smaller than *Equus pacificus* and larger than *Equus complicatus*, with short protocones on the upper teeth, including the molars, and very complexly folded enamel. (Gidley, 1901.)

Equus crinidens is known only from the late Pleistocene of the Valley of Mexico. Only the upper premolars are known; they have short and broad protocones, and it is probable, but not certain, that the molars were similar. The enamel pattern is complex. *Equus crinidens* seems to have been slightly smaller than the contemporary *Equus pacificus*. (Hibbard, 1955b.)

†Western Horse, *Equus occidentalis* Leidy, 1865 (*E. bautistensis* Frick, 1921?)

Leidy based his description on three upper teeth: a third premolar from the auriferous gravels of Tuolumne County, California, and two teeth from Asphalto (Blancan). Gidley (1901) chose the tooth from Tuolumne County as the lectotype (a tooth figured by Leidy). The fact that the teeth from Asphalto were probably of a Blancan zebra (Savage, 1951) is thus not relevant. In Dalquest's opinion, there is no reason to consider the tooth from Tuolumne County pre-Pleistocene. It is the same size as, and closely resembles, teeth from Rancho La Brea. The name *Equus occidentalis* should thus be continued to be used for the Rancho La Brea horse unless it can be proved that the lectotype tooth is not of the same species. A large sample is available from this locality, and there are additional records from California, Arizona, and Nevada.

Equus occidentalis represents a group of horses which, like the previously described giant horse group, have a short, broad protocone, but in which the enamel pattern is simple (see fig. 14.1). *Equus bautistensis,* from the Irvingtonian of Bautista Creek, is similar in size and dental characters. Although Frick listed differences between this species and the Rancho La Brea form, the two may nevertheless be conspecific. (Frick, 1921; Gidley, 1901; Savage, 1951; Stock, 1965; Willoughby, 1948.)

†Complex-toothed Horse, *Equus complicatus* Leidy, 1858

Together with *Equus fraternus,* this species represents a group of horses with slender, elongated protocones and complexly folded enamel. *Equus complicatus* was the first American Pleistocene horse to be described. It is a large form, approximately the size of *Equus scotti* but differing from the latter in the more complex enamel pattern, especially on the upper teeth, and in other details. It seems to have been the common large horse of the eastern United States during the later Pleistocene. Records extend from the Gulf Coast of Texas to Florida, South Carolina, Kentucky, and Missouri, and the time range is Irvingtonian to late Wisconsinan. (Lundelius, 1972a.)

†Brother Horse, *Equus fraternus* Leidy, 1860

The type locality is near Charleston, South Carolina, and specimens from Florida and as far west as Texas (Gulf Coast, Dallas) have been identified as *Equus fraternus.* Determinations are usually based on upper teeth that are smaller than but otherwise similar to those of *Equus complicatus;* referred lower teeth (Slaughter et al., 1962) are too fragmentary to show details. A metacarpal figured by Slaughter et al. (1962) is short and stout; a referred metatarsal figured by Quinn (1957), on the other hand, is slender and *Hemionus*-like. (Quinn, 1957; Slaughter et al., 1962.)

†Scott's Horse, *Equus scotti* Gidley, 1900 (*E. midlandensis* Quinn, 1957)

Equus scotti, like the forms that follow, belongs to a group of horses with a slender, elongated protocone and simple enamel pattern; the limbs are short and stout. This group may be referred to the subgenus *Amerhippus.* Scott's horse is known from several skeletons from the type locality, Rock Creek, and additional well-preserved material is at hand from other localities. It ranges from the late Blancan (Red Light) to the Wisconsinan but is best represented in Irvingtonian deposits (Gilliland, Holloman, Rock Creek, Punta Gorda, Medicine Hat). The size is large, approximately that of a large modern horse, and the limbs are heavy, with short and stout metapodials (see fig. 14.3). The teeth are quite hypsodont, but with a simple enamel pattern (see fig. 14.1). (Gidley, 1900; Hibbard and Dalquest, 1966; Troxell, 1915a.)

†Niobrara Horse, *Equus niobrarensis* Hay, 1913 (*E. hatcheri* Hay, 1915)

Equus niobrarensis is based on a nearly complete skull from near Grayson, Nebraska, and the type of *Equus hatcheri,* from the same area, is a skull with lower jaws. The species ranged from Texas and New Mexico north to Nebraska, and possibly even farther north and west. Most records seem to be of Irvingtonian age.

The species is similar to *Equus scotti* (see Savage, 1951) but smaller, with cheek tooth rows 15–25 mm shorter. The enamel pattern of the cheek teeth is simple, and the metapodials are short and stout. It also resembles the still smaller *Equus conversidens*. These three related species are sometimes found together. In the Gilliland local fauna, *Equus scotti* is abundant, but neither of the two other species occurs. At Slaton, however, both *Equus niobrarensis* and *Equus conversidens* are present, with *Equus scotti* absent. At Cedazo, *Equus conversidens* is common, and the other two species are absent. (Dalquest, 1967; Hibbard and Dalquest, 1966; Lundelius, 1972b; Savage, 1951.)

†Mexican Horse, *Equus conversidens* Owen, 1869 (*E. semiplicatus* Cope, 1893)

This is the smallest of the American stout-limbed Pleistocene horses and seems to have been the most common small horse in much of North America in the later Pleistocene (see fig. 14.3). It ranged from the Valley of Mexico (the type comes from this area) to Alberta and probably over much of the Great Plains area. The species apparently occurred in the early Irvingtonian (Rock Creek) and survived to early Holocene times.

The teeth resemble those of a large *Equus tau*, especially when the latter are moderately worn. In younger animals, the teeth are more slender and high-crowned, the internal valleys of the lower molars are narrower and usually V-shaped, and the upper premolars are more curved (see figs. 14.1, 14.2). All of these characters are variable, however, and separation of isolated teeth of the two species is often difficult or impossible. Published records of *Equus conversidens* should be reexamined. (Dalquest, 1967; Dalquest and Hughes, 1965; Quinn, 1957; Skinner and Hibbard, 1972.)

†Lambe's Horse, *Equus lambei* Hay, 1917

Several skulls and other skeletal material of this species have been found at Gold Run Creek, and specimens have also been found at Fairbanks, Lost Chicken Creek, and Hunker Creek; all the finds may be Wisconsinan in age. The species thus seems to have been a member of the Alaska–Yukon refugium fauna. It is a medium-sized horse with stout metapodials. (Harington and Clulow, 1973.)

†Noble Horse, *Equus excelsus* Leidy, 1858 (*Onager hibbardi* Mooser, 1959; *Onager arellanoi* Mooser, 1959; *Asinus aguascalentensis* Mooser, 1959)

This is one of the first Pleistocene horses to be described from North America; the type locality is in Nebraska. Skulls, jaws, and metapodials from Cedazo have been referred to the same species. The dentition of the Mexican specimens is distinctly smaller than that of *Equus lambei* from the Yukon, but similar morphologically; the metapodials, on the other hand, average somewhat larger than those of *Equus lambei* but are similar in proportions. The species ranged from central Mexico to Nebraska and from the Irvingtonian to the earlier part of the Rancholabrean. (Mooser and Dalquest, 1975b.)

Terminal dates of North American *Equus*

Late dates for small horses (which have been referred to *E. conversidens*) come from Jaguar Cave, 10,370 ± 350 years B.P. (where the horse is also present in younger deposits); Pashley, Alberta, ca. 8,000 years B.P.; and Beverly Pit near Edmonton, Alberta, which has been tentatively correlated with a deposit dated at 8,150 ± 100 years B.P. A large horse was also present at Jaguar Cave (10,370 ± 350 years B.P.) and survived beyond that date. (Churcher, 1968; Churcher and Stalker, 1970; Fuller and Bayrock, 1965; Kurtén and Anderson, 1972.)

Family Tapiridae—Tapirs

Tapirs are heavily-built perissodactyls of a primitive stamp, retaining four toes on the front feet and three on the hind. The family can be traced back to the Eocene, and evolutionary trends include a tendency to retraction of the nasal bones, correlated with the development of a small proboscis. The cheek teeth are brachydont, with a simple pattern of crests and additional tubercles. In modern tapirs, there is a rather unusual development of the tusks, which are formed by the lower canines and the third upper incisor, the upper canine being vestigial. Modern tapirs are browsers, feeding on soft vegetable tissues; semiaquatic in habit, they live in moist forests. Four extant species are recognized, three in Central and South America and one in Southeast Asia.

The fossil history of tapirs is rather spotty, and it is only in the Pleistocene that a fuller record is available. The conservativeness of the family is well illustrated by the fact that all Pleistocene species may be referred to the living genus *Tapirus* or a few very closely related genera. The known species are distinguished on the basis of a mosaic of osteological and odontological characters that bear little relation to zoogeography. Fossil tapirs are distinguished mostly by size, but even this criterion is sometimes difficult to apply on isolated teeth whose exact position in the tooth row may be uncertain. The degree of molarization of P^1 and P^2 appears to be a promising character for differentiating geographic populations. Another character used in distinguishing tapir species is the configuration of the sagittal crest(s) of the skull and the table enclosed between the double crests. It is unusual, however, to find material well enough preserved to show this character, which may, moreover, be affected by age of the individual.

Of the numerous finds reported only as *Tapirus* sp., most are Rancholabrean in age. However, *Tapirus* sp. has also been recorded from the Blancan of Florida (Santa Fe River IB, Haile XVA) and the Irvingtonian of Florida (Inglis IA) and Maryland (Cumberland).

Geographically, the distribution of Pleistocene tapirs in North America was limited to that part of the continent south of the glaciated area and (with a few exceptions near the Mexican border) areas which now have a mean annual precipitation of 500 mm or more (i.e., along the Pacific coast and east of 100° west longitude). Such areas now have a humid mesothermal climate, and this probably was true also of the time in which they were inhabited by tapirs (Simpson, 1945b.)

† Merriam's Tapir, *Tapirus merriami* Frick, 1921

The type and sole known specimen comes from the Irvingtonian beds of Bautista Creek. It is a lower jaw fragment with two anterior molars, which are distinguished from homologs in other tapirs by their significantly larger size. (Frick, 1921; Simpson, 1945b.)

† California Tapir, *Tapirus californicus* Merriam, 1912 (*T. haysii californicus*)

This large form (probably within the size range of *T. copei*), which closely resembles the living South American *Tapirus terrestris* Linnaeus, is only known on the basis of a few teeth from various localities in California (National City, San Diego Co.; between San Pedro and Lomito, Los Angeles Co.; Corralillos Canyon, Santa Barbara Co.; Zuma Creek) and Oregon (Cape Blanco, Curry Co.). The Zuma Creek record, and perhaps some of the others, is Rancholabrean. *Tapirus* sp. from Newport Bay Mesa (1066) and Rancho La Brea may represent the same species. Relationships with the fragmentary tapir remains from New Mexico (Dry Cave) and Arizona (Lehner Ranch, Murray

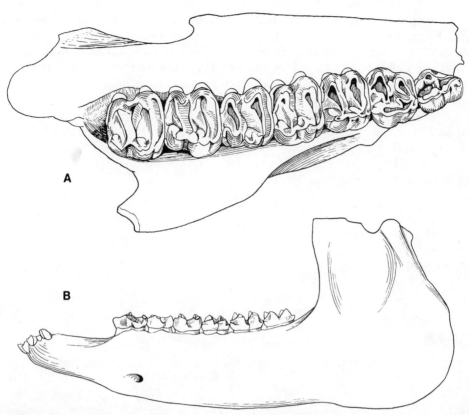

14.4 Tapiridae. *Tapirus veroensis*, Rancholabrean, Florida. *A*, left P^1–M^3, occlusal view; *B*, left mandible, external view. *A*, 0.66×; *B*, 0.33×.

Springs, Ventana) are uncertain. Little can be said about the affinities of this taxon, and more and better material is needed to determine validity as a species. (Simpson, 1945b.)

†Cope's Tapir, *Tapirus copei* Simpson, 1945 (*T. haysii* Leidy, 1860?)

Although the name *Tapirus haysii* was long used for these large tapirs, the type is indeterminate and of unknown provenance, and the name *Tapirus copei* has been used instead. The type of *Tapirus copei* is from the early Irvingtonian of Port Kennedy, but large tapirs which may belong to the same species have been reported from the late Blancan (Hudspeth, Donnelly Ranch), the Irvingtonian (Gilliland), and the Rancholabrean (Melbourne, Moore Pit, Mullen II, Natchez, Sabertooth Cave; also El Paso County, Tex., and Apollo Beach, Fla.). The species is larger than all other North American Pleistocene tapirs except *Tapirus merriami*; P^2 shows an advanced state of molarization, and P^1 is unusually large, with a greater development of the protocone. (Simpson, 1945b.)

†Vero Tapir, *Tapirus veroensis* Sellards, 1918 (*T. tennesseae* Hay, 1920; *T. excelsus* Simpson, 1945)

Most finds come from the Rancholabrean of Florida, where the species has been reported from more than 15 sites, the oldest probably being Sangamonian in age. Other records are from Georgia, Kansas (Kanopolis), Missouri, Tennessee, and Texas. In addition, the many records of *Tapirus* sp. east of 100° west longitude (Arkansas, Florida, Georgia, Kentucky, Missouri, Pennsylvania, Tennessee, Texas, Virginia) may represent this species. *Tapirus excelsus* was described on the basis of a juvenile skeleton from Enon Sink, and specimens from Crankshaft Pits and Ingleside were referred to it; however, doubts on the validity of its distinctive characters (mainly the morphology of the sagittal crest) have been voiced. Lundelius and Slaughter (1976) note that if larger samples from more localities reveal subspecific differences, the larger specimens should be known as *Tapirus veroensis excelsus*. The Vero tapir is, on the average, slightly larger than living Neotropical species, and the skull has a number of distinctive characters (see fig. 14.4). (Hay, 1920; Hibbard et al., 1978; Lundelius and Slaughter, 1976; Ray, 1964a; Sellards, 1918; Simpson, 1945b.)

CHAPTER FIFTEEN

ORDER ARTIODACTYLA

This highly varied order, which dates back to the Eocene, comprises even-toed ungulates in which the axis of the foot lies between the third and fourth digits. The most characteristic feature is the structure of the astragalus; the presence of a distal trochlea in addition to the "normal," proximal one results in a double-tarsal joint. The efficiency of this joint may have contributed to the success of the order, to which belong the majority of living ungulates.

The artiodactyls have an excellent fossil record, and a great number of families are known. All of the Pleistocene families in North America are still extant, although one of them (the Camelidae) is no longer found in the area.

There are two suborders. The Suina, or piglike artiodactyls, have molar teeth with rounded (bunodont) cusps and canine tusks; the feet are four-toed, usually with separate metapodial bones, and the stomach is simple. Here belong the peccaries and the Old World swine and hippopotami.

In the suborder Ruminantia, or cud-chewers, the cusps of the molars tend to become crescent-shaped (selenodont), the upper incisors are reduced or absent, the feet have reduced side toes or none at all, with the two central metapodials tending to fuse into a cannon bone, and there is a compound stomach. The camellike ruminants, infraorder Tylopoda, evolved in North America and later spread to other continents. All other Pleistocene ruminants are in the infraorder Pecora. Of the numerous endemic North American pecorans, only the pronghorn family survived in the Pleistocene; the other Pleistocene pecorans (deer and bovids) are of Old World origin. The species-level taxonomy of the pecorans is facilitated by horn and antler characters.

FAMILY TAYASSUIDAE—PECCARIES

Peccaries, though related to the Old World pigs, have had a long separate history. They have been found in Oligocene and Miocene deposits in the Old World, and in North America their stratigraphic range extends from the Lower Oligocene to the Recent. Peccaries differ from true pigs in having vertical upper canines, relatively short, simple molars, a fused radius–ulna, reduced side toes, and a large dorsal musk gland.

Platygonus and *Mylohyus* were the Pleistocene representatives of the family in North America. Showing wide environmental tolerances, they were found from the periglacial regions at the edge of the ice sheet to the tropics. *Platygonus compressus* was probably the most numerous medium-sized mammal during the late Pleistocene. South American species of *Platygonus* and *Mylohyus* show more primitive characters than their North American counterparts, who became increasingly specialized.

The two closely related extant species *Tayassu peccari*, the white-lipped peccary, and *Tayassu tajacu*, the collared peccary (placed in a separate genus, *Dicotyles*, by some workers; see Woodburne, 1968), are primitive compared to *Platygonus* and *Mylohyus* and show the following unspecialized characters: retention of I_3, forward position of the eye socket, a shallow basicranial axis, primitive olfactory apparatus, and possession of dew claws.

Recently, living representatives of the "extinct" Pleistocene species *Catagonus wagneri* (Rusconi) have been discovered in the Chaco of western Paraguay (Wetzel, 1977; Wetzel et al., 1975). It is less similar to the other two living species than they are to each other but shows similarities in skull morphology and dentition to *Platygonus,* to which it is more closely related. Like some of the early species of *Platygonus,* it possesses three lower incisors and a keeled mandibular symphysis; it differs from *Platygonus* in its larger teeth, longer tooth rows and shorter diastema. A relict species, it survives today in a scrub-thorn and grass refugium.

An unknown Tertiary peccary was probably ancestral to the Quaternary species. According to Guilday (personal communication, 1975) there were three groups of peccaries in the Pleistocene: *Mylohyus, Platygonus–Catagonus,* and *Tayassu.*

† Kinsey's Peccary, *Mylohyus floridanus* Kinsey, 1974

Mylohyus, a genus of long-nosed, long-legged American peccaries, has a stratigraphic range extending from the late Blancan to the late Wisconsinan or early Holocene. *Mylohyus* was derived from the Pliocene genus *Prosthenops* and differed from that genus in its larger size, more elongated skull, longer postcanine diastema, molarized premolars, and fused metatarsals. *Mylohyus* and the extant white-lipped peccary, *Tayassu pecari,* are closely related and probably had a common ancestry in *Prosthenops.*

The newly recognized species *Mylohyus floridanus* first appears in Blancan faunas in Florida (Haile XVA, Santa Fe River IB). It was about the same size as *Mylohyus nasutus* but had longer pre-and postcanine diastemas and a shorter tooth row. Kinsey (1974) believes that the Blancan species is probably ancestral to *Mylohyus nasutus.* (Kinsey, 1974.)

† Long-nosed Peccary, *Mylohyus nasutus* (Leidy), 1869 (*Dicotyles nasutus; D. fossilis* Leidy, 1860; *D. pennsylvanicus* Leidy, 1889; *Mylohyus tetragonus* Cope, 1899; *M. browni* Gidley, 1920; *M. exortivus* Gidley, 1920; *M. gidleyi* Simpson, 1929)

Although a number of post-Blancan species of *Mylohyus* have been named, the descriptions are based on fragmentary material, and the amount of intraspecific variation is unknown. We believe that only one species, *Mylohus nasutus,* occurred in the eastern and central United States in Irvingtonian (Port Kennedy, Cumberland, Conard Fissure, Angus) and Rancholabrean times. It has been identified in some 35 faunas in Arkansas, Florida, Georgia, Indiana, Iowa, Kansas, Kentucky, Maryland, Missouri, Nebraska, Oklahoma, Pennsylvania, Tennessee, Texas, Virginia, and West Virginia. This long-legged, cursorial peccary inhabited open areas, perhaps forest edges, and apparently was much less common than *Platygonus* except in Florida. At New Paris No. 4, its environment was a cold-temperate woodland.

Mylohyus nasutus has a long slender rostrum and mandible, with a long postcanine diastema in both jaws. Molarized premolars and bunodont cheek teeth characterize the dental series (see fig. 15.2). The auditory bullae are more elongate than those of *Platygonus.* Keen eyesight is indicated by the placement of large orbits high and far back on the skull.

The long-nosed peccary was about the size of a small white-tailed deer (*Odocoileus virginianus*) and had long slender limbs (see fig. 15.1). Metacarpals III and IV are slender and unfused; metatarsals III and IV are fused and less expanded at the distal end.

15.1 Tayassuidae. Restoration of *Platygonus compressus* and *Mylohyus nasutus*.

Unlike the gregarious *Platygonus*, *Mylohyus nasutus* was probably solitary and apparently did not often seek shelter in caves. Its remains are usually found singly, though at Friesenhahn Cave five individuals were found together. Lundelius (1960) believes that *Mylohyus nasutus* was the ecological equivalent of *Sus scrofa*. In the Appalachians, *Mylohyus nasutus*, *Platygonus compressus*, and *Ursus americanus* were probably thrown into competition by the disappearance of several niches at the end of the Wisconsinan; only the black bear survived.

Mylohyus nasutus has not been found in direct association with man, although both cultural materials and bones of *Mylohyus* were removed from Hartman's Cave in the 1800s. *Mylohyus nasutus* survived until late glacial or early Holocene times. (Gidley, 1920; Guilday, 1967a; Guilday, Martin, and McCrady, 1964; Johnston, 1935; Lundelius, 1960; Semken and Griggs, 1965; Simpson, 1929a; Woodburne, 1968.)

† Pearce's Peccary, *Platygonus pearcei* Gazin, 1938

The genus *Platygonus* is characterized by a moderately elongated snout, nonmolariform premolars, molars with two pairs of cusps connected by transverse crests, and side toes reduced to splints or nodules of bone. A trend in the evolutionary history of *Platygonus* is the reduction and then loss of the third lower incisor. Apparently, *Platygonus* was a gregarious animal that traveled in small herds. Ranging stratigraphically from the Blancan to the late Rancholabrean, the genus has been found from southern Canada to Mexico.

Numerous species of *Platygonus* have been described, but as is the case in many other groups, descriptions are often based on fragmentary material, incomplete comparisons with other specimens, and inadequate consideration of individual variation. It now appears that four species of *Platygonus* inhabited North America: *Platygonus pearcei* and *P. bicalcaratus*, in the Blancan and early Irvingtonian; *P. vetus*, in the late Irvingtonian and early Rancholabrean; and *P. compressus*, in the middle and late Rancholabrean.

Smallest and best known of the Blancan peccaries is *Platygonus pearcei*, first described from Hagerman and later identified at Grand View and White Bluffs. The well-preserved Hagerman material includes both cranial and skeletal remains. The species was slightly smaller and less robust than *Platygonus vetus*. The canines are relatively slender, the lower third incisor is present, and the upper teeth do not show the well-isolated cross lophs seen in the upper teeth referred to *Platygonus bicalcaratus* from Blanco. The skull is larger and more elongate than that of *Platygonus compressus*. The metacarpals are not fused. The metatarsals generally are fused, but not to the extent seen in *Platygonus vetus*.

Woodburne (1968) believes that this long-snouted species with widely flaring zygomatic arches was on the line leading toward *Platygonus vetus*. (Gazin, 1938; Woodburne, 1968.)

† Cope's Peccary, *Platygonus bicalcaratus* Cope, 1892 (*P. texanus* Gidley, 1903)

This species was described from Blanco on the basis of a tooth fragment which Cope believed was the posterior part of M_3; however, Gazin (1938) thought that the fragment was actually P_3 or P_4, and it has little diagnostic value. Additional material from Blanco was provisionally referred to the species by Gidley (1903), who thought that a second species, *Platygonus texanus*, was also present in the same beds. Material from other faunas (Cita Canyon, Red Light, Keefe Canyon, Rexroad, Deer Park, Santa Fe River IB, Inglis IA) has been referred to *Platygonus bicalcaratus*, and it seems likely that only one large peccary inhabited the Great Plains and Florida during the Blancan and early Irvingtonian. *Platygonus* has also been found at Sand Draw, Broadwater, White Rock, Borchers, and Donnelly Ranch, but the material is not diagnostic to species.

Platygonus bicalcaratus was larger than *Platygonus pearcei*, and the molars have high cross crests completely divided by cross valleys. Woodburne (1968) believes that *Platygonus bicalcaratus* was probably a member of the lineage leading toward *Platygonus compressus*. (Gazin, 1938; Gidley, 1903; Woodburne, 1968.)

† Leidy's Peccary, *Platygonus vetus* Leidy, 1882 (*P. cumberlandensis* Gidley, 1920; *P. intermedius* Gidley, 1920)

During the late Irvingtonian–early Rancholabrean, a population of large peccaries inhabited the plains and open forests from the Appalachians and Florida westward. Three species have been described, but it now appears that only one widespread, variable species is represented. Two "species" were described from Cumberland Cave, where a minimum of 22 skulls and numerous postcranial elements were recovered. Gidley (1920) said that the animals differed from *Platygonus vetus*, a species based on fragmentary jaw material found at Mifflin, Pennsylvania, in their slightly smaller size and less squarish molars, characters that are quite variable in peccaries. The species has also been found at Coleman IIA, Fossil Lake(?), Gilliland, Gordon, Renick (West Virginia), and Rushville. *Platygonus* sp. has been recorded from Cudahy, Holloman, Java, Mullen I, Rock Creek, Vallecito Creek, and Wellsch Valley. Whether these Irvingtonian peccaries are *Platygonus vetus* or one of the Blancan species is presently unknown.

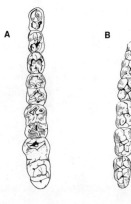

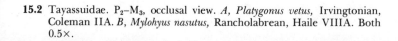

15.2 Tayassuidae. P_2–M_3, occlusal view. *A*, *Platygonus vetus*, Irvingtonian, Coleman IIA. *B*, *Mylohyus nasutus*, Rancholabrean, Haile VIIIA. Both 0.5×.

Platygonus vetus is one of the largest peccaries known. Viewed from above, the massive, elongate skull shows a narrow snout, widely flaring zygomatic arches, and a backward projection of the inion. The most distinctive feature of the skull is the pronounced expansion of the zygoma downward and forward from the orbit (fig. 15.3). In males, the zygoma is two or more times deeper than the diameter of the orbit. Although the development of the zygoma varies with age and sex, it is greater in *Platygonus vetus* than in any other species of peccary. The lower third incisor has been lost, and the molars are lophodont (see fig. 15.1). The limb bones are long and massive, but the foot bones are relatively short proportionately.

The ancestry of *Platygonus vetus* can probably be traced back to the Blancan species, *Platygonus pearcei*. Leidy's peccary died out by the end of the Illinoian and was replaced by *Platygonus compressus*. (Gidley, 1920; Gidley and Gazin, 1938; Guilday, personal communication, 1975.)

†Flat-headed Peccary, *Platygonus compressus* Le Conte, 1848 (*P. leptorhinus* Williston, 1894)

One of the first North American fossil animals to be discovered was the flat-headed peccary (fig. 15.1). A skull was found in a Kentucky cave in 1804 or 1805, and Caspar Wistar identified it as a peccary in 1806, but it wasn't until 1853 that Leidy provided a description of the specimen. Meanwhile, in 1848, LeConte named some fragmentary

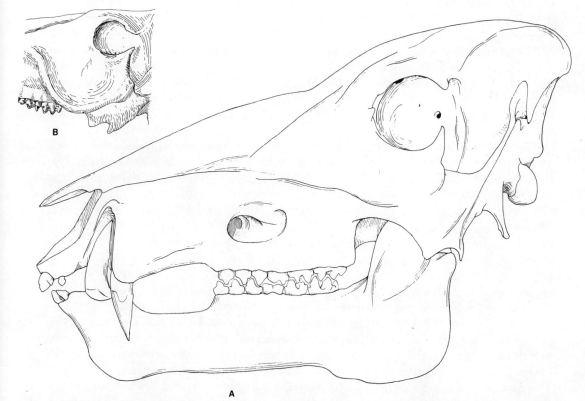

15.3 Tayassuidae. *A*, *Platygonus compressus*, Rancholabrean, Ann Arbor, Washtenaw Co., Michigan, skull and mandible. *B*, *Platygonus vetus*, Irvingtonian, Cumberland Cave, zygomatic arch. Note the great development of the zygomatic arch in *P. vetus*. *B*, after Gidley and Gazin (1938). *A*, 0.5×; *B*, 0.25×.

material found at Galena, Illinois, *Platygonus compressus*. This widespread late Pleistocene peccary was relatively common over much of its range, an area extending from New York to California and from the edge of the Wisconsinan ice sheet to Mexico. In the Appalachian region, its remains make up about 90 percent of the large mammal fauna in some caves. Formerly, *Platygonus compressus* was regarded as an indicator of interglacial conditions or southern faunas, but it is now known that the species had wide environmental tolerances.

Unlike finds of *Mylohyus nasutus*, which usually consist of single individuals, those of *Platygonus compressus* often consist of several individuals, or even small herds. Several of the finds (Hickman, Ky.; Goodland, Kans.; Denver, Colo.; Belding, Mich.; Columbus, Ohio) consist of specimens found close together and apparently represent herds that perished in storms (Finch, Whitmore, and Sims, 1972). Cave deposits have yielded large numbers of peccaries, and it is believed the animals sought shelter in caves. At Welsh Cave, a minimum of 31 individuals were recovered; at Bat Cave, 98 individuals (based on 77 lower left canines and 21 deciduous upper left canines); and at Zoo Cave, 81 individuals (based on 41 right upper canines and 40 deciduous right upper canines) (Guilday, Hamilton, and McCrady, 1971; Hawksley, Reynolds, and Foley, 1973; Hood and Hawksley, 1975). Good samples have also been obtained from Laubach, Cherokee, and Peccary caves, and the species is known from approximately 70 Sangamonian and Wisconsinan sites.

Platygonus compressus was about the size of a European wild boar (*Sus scrofa*) and stood about 760 mm at the shoulder. Apparently, there was no clinal variation in size, since the large samples from Bat and Welsh caves show all gradations. The species differed from *Platygonus vetus* in possessing a shorter, deeper snout with a shorter postcanine diastema and less-flaring zygomatic arches (see fig. 15.3). Generally, *Platygonus compressus* was considerably smaller than *Platygonus vetus*, although a few specimens from Laubach Cave, a late Wisconsinan deposit, approach the latter species in size. *Platygonus compressus* had longer legs, a larger nasal cavity, and higher, sharper transverse cusps on the teeth than living peccaries. Reduction of the side toes is complete, and the species did not have dew claws.

The dentition is more specialized than that of the extant species. I^2 is reduced to a small peg, I_3 is lost, and the incisors and canines have a more gracile appearance. The hypsodont cheek teeth were adapted to chew coarse vegetation, and the dentition suggests browsing habits.

A sagittal section of the skull (see Guilday, Hamilton, and McCrady, 1971) shows a relatively small brain cavity and a greatly enlarged olfactory area. The orbit has migrated caudally and is situated higher and more laterally in the cranium than in *Tayassu*. The auditory bullae are small. These adaptations, plus the increased length of the limbs, indicate that *Platygonus compressus* inhabited open country, where there was an increased need for keen eyesight and fleetness of foot. The extensive turbinals in the nasal cavity filtered the dry, dusty air that often prevailed on the plains.

The flat-headed peccary lived in small, loosely organized herds made up of individuals of both sexes and all ages. There was little sexual dimorphism, although the relative development of the canines, canine buttresses, and skull crests was greater in males. These peccaries were well adapted for both defensive and offensive fighting. The razor-sharp lower canine was the animals' main weapon, and the cranium shows several modifications designed to protect the eye and facial musculature from severe injury.

The ancestry of *Platygonus compressus* can be traced back to the *Platygonus bicalcaratus* line, and the late Pleistocene species shows close affinities to the living *Tayassu tajacu*, the collared peccary. By the end of the Pleistocene, *Platygonus compressus* was extinct; the latest date is 11,900 ± 750 years B.P. (Ray, Denny, and Rubin, 1970), obtained on a specimen from Mosherville, Pennsylvania. *Platygonus* remains have not been found at any North American archeological site, and there is no evidence that man played any part in the species' extinction. Loss of habitat due to climatic and vegetational changes, coupled with direct competition and predation (probably from the black bear), reduced the population to the point where the low reproductive rate could no longer keep pace with environmental attrition, and the species became extinct. No remains of the collared peccary (*Tayassu tajacu*) have been found in association with *Platygonus*, and the former species apparently did not enter the Southwest until after *Platygonus* was extinct. In the Appalachian and Ozark regions, its niche has apparently been filled by *Ursus americanus* and the introduced *Sus scrofa*. (Eshelman, Evenson, and Hibbard, 1972; Finch, Whitmore, and Sims, 1972; Guilday, Hamilton, and McCrady, 1971; Hawksley, Reynolds, and Foley, 1973; Hood and Hawksley, 1975; Lewis, 1970; Ray, Denny, and Rubin, 1970; Simpson, 1949; Slaughter, 1966a; Williston, 1894.)

Family Camelidae—Camels and Llamas

The camel family underwent most of its evolution in North America, and its fossil history can be traced back to the early Tertiary. By the late Pliocene, camels had reached Asia, eastern Europe, and eastern Africa, and the closely related llamas were widespread in South America. Both camels and llamas were common in the North American Pleistocene but became extinct at the end of the epoch. Today, the family survives in the wild state in South America and the Gobi Desert of Mongolia, and domesticated camels are found from northern Africa through Central Asia.

Camelids are hornless artiodactyls that have a long, slender neck and long limbs with a much reduced ulna and fibula. The distinctive cannon bones diverge distally, and no trace of lateral digits remains. The feet are characteristic—digitigrade, with broad, spreading toes that lie nearly flat on the ground and terminate in a small nail beneath which is a thick supporting pad of connective tissue rather than a hoof. Most of the camelids were grazers and traveled in herds.

The subfamily Camelinae is divided into four tribes: the ancestral Protolabidini, the Camelini (*Procamelus, Titanotylopus, Blancocamelus, Paracamelus, Camelus*), the Camelopini (*Megatylopus, Camelops*), and the Lamini (*Pliauchenia, Hemiauchenia, Palaeolama, Lama, Vicugna*). Although the Camelopini and the Lamini are more closely related to each other than to the Camelini, the three tribes have had separate histories since the Barstovian.

Six genera of camelids inhabited North America during the Pleistocene. Of these, the camels (*Titanotylopus, Blancocamelus*) and camelopines (*Megatylopus, Camelops*) were found only in the western half of the continent, whereas the llamas (*Hemiauchenia, Palaeolama*) ranged from coast to coast. (Romer, 1966; Walker, 1975; Webb, 1965.)

†Nebraska Camel, *Titanotylopus nebraskensis* Barbour and Schultz, 1934

The giant-camelid genus *Titanotylopus* is characterized by long, massive limbs, a relatively small braincase, a convex interorbital region, and well-developed P3 in both jaws. Height was about 3.5 m. Long neural spines on the thoracic vertebrae indicate a large hump. The genus differed from *Megatylopus* and *Camelops* in the lack of lacrimal vacuities and the small I^3, large upper canine, and broad anterior lobe on P_4. It had a rostrum shorter than that of *Camelops*. *Titanotylopus* shows many resemblances to the *Procamelus–Paracamelus–Camelus* line, and a Hemphillian species, *Titanotylopus merriami* Frick, from the Mt. Eden fauna in California, seems to bridge the gap, both temporally and morphologically, between the New World and Old World camelini. The stratigraphic range of *Titanotylopus* is Hemphillian to Irvingtonian.

The genotypic species, *Titanotylopus nebraskensis,* is known only from the type locality in Webster County, about 13 km northwest of Red Cloud in south-central Nebraska. The locality is believed to be Kansan in age, and McGrew (1944) thought that *Titanotylopus nebraskensis* occurred in younger deposits than *Titanotylopus spatulus*. It differs from the latter species in the absence of P_1 and the significantly shorter postcanine diastema. (Breyer, 1976; McGrew, 1944; Webb, 1965.)

† Spatulate-toothed Camel, *Titanotylopus spatulus* (Cope), 1893 (*Pliauchenia spatula* Cope; *Gigantocamelus fricki* Barbour & Schultz, 1939)

Originally described from the Blanco fauna, this species has also been found at Keefe Canyon, Broadwater, Sand Draw, Sandahl, and Holloman. At various times, this material has been referred to *Megatylopus* or *Gigantocamelus;* the latter taxon is now considered to be congeneric with *Titanotylopus*.

At Keefe Canyon, parts of at least 21 individuals were recovered. Pronounced sexual dimorphism is evident, with males, characterized by larger size, a more robust skull, and larger teeth (especially canines), outnumbering females two to one. The species gets its name from the tips of the lower incisors, which are broad in young animals and then become spatulate. The second phalanges are broader than those of *Camelops,* which perhaps indicates better development of the pad.

Titanotylopus sp. has been reported from Grand View, Red Light, Hudspeth, Donnelly Ranch, White Rock, Mullen II, Sandahl, and Vallecito Creek. (Breyer, 1976; Hibbard and Riggs, 1949; Webb, 1965.)

† Meade's Camel, *Blancocamelus meadei* Dalquest, 1975

This large, extremely long-limbed camel, genus *Blancocamelus* (monotypic), is known only from Blanco. Compared to *Titanotylopus, Blancocamelus* has limb bones as long or longer, more slender, and less transversely flattened. The limb bones are similar to those of *Hemiauchenia blancoensis* in proportion but are much longer. As yet, no skulls or teeth have been found in association with skeletal parts. *Blancocamelus* may be a relict taxon of the Aepycamelinae (giraffe-camels), whose heyday was in the Miocene; additional material is needed before its relationships can be established. (Dalquest, 1975; Webb, personal communication, 1975.)

† Cochran's Camel, *Megatylopus cochrani* (Hibbard and Riggs), 1949
(*Pliauchenia cochrani*)

The large-camelid genus *Megatylopus* closely resembled *Pliauchenia* and was probably ancestral to *Camelops*. The main character that distinguishes *Megatylopus* from its probable descendant *Camelops* is the presence of a well-developed P_3. The stratigraphic record of this North American genus extends from the Hemphillian to the late Blancan.

The type locality of *Megatylopus cochrani* is Keefe Canyon, and the species is also known from White Bluffs. *Megatylopus* sp. is recorded from Hagerman and Cita Canyon; neither collection has been studied. *Megatylopus cochrani* was the most advanced

ORDER ARTIODACTYLA

species of *Megatylopus* and shows an overall resemblance to *Camelops*. It has a vestigial P_1 (perhaps a sex-linked character), no P_2, and a large P_3. The species was about the size of *Camelops sulcatus*. (Hibbard and Riggs, 1949; Webb, 1965.)

†Kansas Camel, *Camelops kansanus* Leidy, 1854

Animals of the genus *Camelops* were large llamalike camelids that inhabited the western half of North America from late Blancan to late Wisconsinan/early Holocene times (see fig. 15.4). The genus is characterized by a long cranium, flattened dorsally with no

15.4 Camelidae. Distribution of *Camelops*: ▲ = *C. hesternus*; △ = *C. minidokae*; ■ = *C. sulcatus*; □ = *C. kansanus*; ● = *C. huerfanensis*; ◓ = *C. traviswhitei*; ○ = *Camelops* sp.

interorbital depression, a large lacrimal vacuity, M^3 with a posteriorly elongated metastyle, no P_3, and long, robust limbs. It differs from *Hemiauchenia* in having a deeper, more robust premaxilla, a broader, more rugose tip of the rostrum, and a large upper incisor lying opposite the anterior border of the incisive foramen. It differs from *Titanotylopus* in having a large upper incisor and a smaller canine lying far posterior to it. *Camelops* probably resembled the living dromedary (*Camelus dromedarius* Linnaeus) and, like it, had a single, middorsal hump. Six species are currently recognized, but the genus is in need of revision. *Camelops* has been reported from several Blancan sites (Cita Canyon, Red Light, Deer Park, Borchers, Sand Draw, Broadwater, Hagerman), but the material has not been identified as to species.

The poorly known genotypic species, *Camelops kansanus* was described from a premaxillary–maxillary fragment found in glacial drift in Kansas Territory. Some workers (Savage, 1951) thought that the type specimen lacked diagnostic characters, but Webb (1965) has shown that it can be distinguished from *Hemiauchenia*. Specimens found at Cragin Quarry, Butler Spring, Angus, Mullen II, Hay Springs, Rushville, Gordon, and Afton have been referred to this taxon. (Savage, 1951; Webb, 1965.)

† Traviswhite's Camel, *Camelops traviswhitei* Mooser and Dalquest, 1975

First described from Cedazo, this large camelopine has also been tentatively identified at Blanco. It resembles the Hemphillian species *Megatylopus matthewi* Webb. *Camelops traviswhitei* differs from the other species of *Camelops* in having sharply V-shaped lakes on both the upper and lower molars; the lakes, lacking cementum, have moderately thick enamel on the labial sides and extremely thin enamel on the lingual sides. Other species of *Camelops* have bent-oval-shaped lakes that are filled with cementum and surrounded by thick enamel. *Camelops traviswhitei* was as large as the largest known specimens of *Camelops hesternus*. (Mooser and Dalquest, 1975a.)

† Furrow-toothed Camel, *Camelops sulcatus* (Cope), 1893 (*Holmeniscus sulcatus*)

An Irvingtonian species, *Camelops sulcatus* was originally described from Rock Creek, and a large sample has since been found at Slaton. The small size of the teeth and limb bones distinguish it from other species of *Camelops*. Deep, broadly rounded grooves are present on the inner faces of M_1 and M_2. (Dalquest, 1967; Savage, 1951.)

† Huerfano Camel, *Camelops huerfanensis* (Cragin), 1892 (*Auchenia huerfanensis*)

First described from Huerfano County, Colorado, this late Rancholabrean species has been identified at Moore Pit, Coppell, Hill-Shuler, American Falls, and Denver. It was about the same size and had the same proportions as *Camelops hesternus* but differed from it in minor cranial details, namely, the location of the postpalatine foramen (opposite M^1 in *C. huerfanensis*, opposite P^3 or P^4 in *C. hesternus*) and the configuration of the lower border of the mandible from the symphysis to a point below M_3 (nearly straight in *C. huerfanensis*, strongly convex in *C. hesternus*). The two species are closely related and may be conspecific. (M. L. Hopkins, 1955; Savage, 1951.)

† Minidoka Camel, *Camelops minidokae* Hay, 1927

Camelid material found in a gravel bed at Minidoka, Minidoka County, Idaho, was described as a new species of *Camelops*, and specimens have been recorded from American Falls, Irvington, and Medicine Hat (Kansan). Measurements of this species are consistently smaller than those of *Camelops hesternus*. A prominent internal median groove bordered anteriorly by a stylid is present on M_1 and M_2. This condition is approached on some specimens of *Camelops hesternus*. Savage (1951) has found no morphological characters that would rule out *Camelops minidokae* from the ancestry of at least the Rancho La Brea sample of *Camelops hesternus*. (Savage, 1951.)

† Yesterday's Camel, *Camelops hesternus* (Leidy), 1873 (*Auchenia hesterna*)

Large herds of *Camelops hesternus*, the best known and most widely distributed camelopine, roamed across western North America during Rancholabrean times. Originally described from Livermore, Alameda County, California, the species has been reported from some 30 local faunas plus many isolated sites in Alberta, Arizona, California, Idaho, Mexico (Aguascalientes and State of Mexico), Nevada, Oregon, Saskatchewan, Texas, Utah, Wyoming, and Yukon Territory, and many Rancholabrean records listed as *Camelops* sp. are probably referable to this species.

Camelops hesternus has a relatively long, slender skull with a long rostrum, a deep facial region dominated by a large masseteric fossa and well-developed lacrimal vacuities, posteriorly placed orbits (above M^3), a large brain cavity, and a long basicranial region (see fig. 15.5). The basifacial axis is flexed downward from the basicranial axis, though not to the extent seen in *Lama*. This flexion is a progressive character, found in the Camelopini and the Lamini but not in the Camelini. The large, deeply split, mobile upper lips provided a prehensile grasping device for procuring food. Although primarily a grazer, *Camelops*, with its long neck and legs, was probably an occasional browser.

In *Camelops hesternus*, I^3 is larger than the canine, P^3 is narrow and bladelike, P^4 is

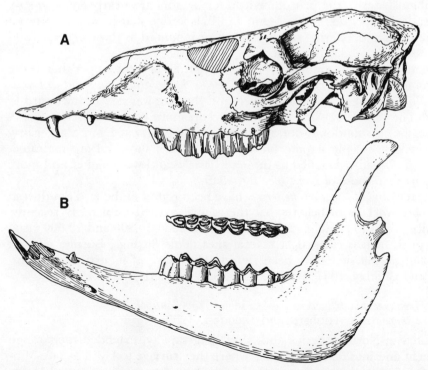

15.5 Camelidae. *Camelops hesternus*, Rancholabrean, Rancho La Brea. *A*, skull; *B*, left mandible, with occlusal view of lower dentition. After Webb (1965). 0.2×.

15.6 Camelidae, *Camelops hesternus*, Rancholabrean, Rancho La Brea, right metacarpus and phalanges, anterior view. After Webb (1965). 0.11×.

quadrate and submolariform, the procumbent lower incisors are nearly equal in size, the canine is laterally compressed, P_3 is absent, and P_4 is wedge-shaped with a posterior enamel fold. The styles and external ribs are not as pronounced as those of *Lama*, and the "llama buttresses" are weakly developed.

Yesterday's camel was a large animal, with limbs about 20 percent longer than those of the extant dromedary. Compared to that of *Camelus dromedarius*, the head of *Camelops hesternus* was longer and relatively narrower, and the face was flexed downward to a greater extent. The muzzle was longer, with thicker, more heavily muscled lips. A high shoulder region, a single mid-dorsal hump, probably situated more anteriorly than in the dromedary, steeply sloping hindquarters, and a short tail are indicated. The toes were not as broadly flattened as those of *Camelus* and were higher and more widely spread apart than those of *Lama* (see fig. 15.6).

Well-preserved remains of *Camelops hesternus* have been found in the arid Southwest, which has led some workers to believe that the animal existed until fairly recently. However, evidence points to a late glacial extinction, between 10,800 and 12,600 years ago. Association with man is claimed at several sites (Tule Springs, Burnet, Casper, Sandia, Clovis), but no kill sites have been found. The cause of its extinction is unknown. (Martin and Guilday, 1967; Romer, 1929; Savage, 1951; Webb, 1965.)

†Blanco Llama, *Hemiauchenia blancoensis* (Meade), 1945 (*Tanupolama blancoensis; T. seymourensis* Hibbard and Dalquest, 1962)

Llamas originated in North America and, spreading rapidly, extended their range across the continent and into South America, where they survive today. The late Miocene genus *Pliauchenia* was apparently ancestral to the later llamas. Characters that distinguish the llamas from the camels include a higher-domed skull, larger brain, re-

15.7 Camelidae. Distribution of *Hemiauchenia* (□ = *H. blancoensis*; ○ = *H. macrocephala*; △ = *Hemiauchenia* sp.) and *Palaeolama* (▲ = *P. mirifica*).

duced number of premolars, well-developed anterior stylids on the lower molars ("llama buttresses"), and absence of a dorsal hump.

During the Blancan and Pleistocene, two genera and three species of llamas inhabited North America (see fig. 15.7). The North American species became extinct at the end of the epoch, as did the South American species of *Palaeolama* and *Hemiauchenia*. *Lama* and *Vicugna* survived in South America and today are represented by *Lama glama*, the llama; *Lama pacos*, the alpaca; *Lama guanicoe*, the guanaco; and *Vicugna vicugna*, the vicuña. Of these, only the guanaco and the vicuña survive in the wild; the others are domesticated. The Pleistocene llamas were recently revised by Webb (1974c), who clarified the relationships between the North American and South American forms and showed that the name *Tanupolama* is a junior synonym of *Hemiauchenia*.

The genus *Hemiauchenia* originated in North America in the middle Pliocene and by the early Pleistocene had reached southeastern South America. *Hemiauchenia vera* (Matthew), a more primitive species than *Hemiauchenia blancoensis*, is known from several Hemphillian faunas in the United States and may be ancestral to the latter species.

Llamas probably ranged over much of North America during the Blancan, and the species *Hemiauchenia blancoensis* is known from Blanco, Cita Canyon, Rexroad, White Rock, Donnelly Ranch, Santa Fe River IB, Gilliland, and the basal fossiliferous layer of Medicine Hat. Material from Curtis Ranch and Vallecito Creek is probably referable to this species (Webb, 1974c). *Hemiauchenia seymourensis*, described from the Gilliland fauna, is now considered to be conspecific with *Hemiauchenia blancoeniss* (Breyer, 1977).

Like other species of *Hemiauchenia*, the Blanco llama had a large, two-rooted P_3, a simple, triangular P_4, weak "llama buttresses' on the lower molars, and gracile limbs with relatively long metapodials. The Blanco llama was larger than *Hemiauchenia macrocephala* and had more strongly developed styles on the upper teeth than the latter species. P_4 was narrow and bilobed. (Breyer, 1977; Hibbard and Dalquest, 1962; Webb, 1974c.)

†Large-headed Llama, *Hemiauchenia macrocephala* (Cope), 1893 (*Holmeniscus macrocephala; Camelus americanus* Wortman, 1898; *Tanupolama stevensi* Merriam and Stock, 1928; *T. hollomani* Hay and Cook, 1930)

The best-known Pleistocene llama is *Hemiauchenia macrocephala*, an animal whose stratigraphic range extended from the late Blancan (Haile XVA) to the late Rancholabrean. It remains have been found at numerous localities from Florida to California, and many specimens listed in the literature as *Tanupolama* sp. are probably referable to this species.

Hemiauchenia macrocephala was smaller than *Hemiauchenia blancoensis* but larger than the extant *Lama glama*. It had higher-crowned cheek teeth than *Hemiauchenia blancoensis*, and the molars had moderately developed styles and buttresses. P_4 is relatively wide and labially convex, with an elongate posterior fossetid; the anterolingual groove is strong, but the posterolabial groove is weak or absent.

As Stock (1928) noted, the metapodials of *Hemiauchenia* show pronounced elongation and slenderness. Webb (1974c) showed that the metapodials are longer and the epipodials shorter in relation to the whole limb in *Hemiauchenia* than in the other species of llamas. This indicates that *Hemiauchenia* had a long stride and was highly cursorial. It was a plains-dweller and probably fed primarily on grass.

Although *Hemiauchenia macrocephala* has a long geologic history, there do not seem to be any morphological difference between specimens from late Blancan, Irvingtonian, or Rancholabrean faunas. Its extinction at the end of the Wisconsinan is unexplained. (Stock, 1928; Webb, 1974c.)

†Stout-legged Llama, *Palaeolama mirifica* (Simpson), 1929 (*Tanupolama mirifica*)

The genus *Palaeolama* is characterized by a large, double-rooted P_3 and weak "llama buttresses" on the lower molars and is distinguished from *Hemiauchenia* by its complex P_4 (with multiple fossetids), low-crowned cheek teeth, and stocky limbs. P_4 is the most distinctive tooth, with folding more complex than that of any other camelid. The teeth have an angular appearance, and the enamel is often crenulated. The diet probably consisted of shoots and leaves of bushes and trees, as well as grass. Short, stocky metapodials and longer epipodials suggest that *Palaeolama* was adapted to live in rugged terrain. Blancan and Pleistocene occurrences are shown in figure 15.6.

The center of origin of *Palaeolama* and *Lama* appears to be the Andean region, and it is believed that the two genera had a common ancestry in a *Hemiauchenia*-like stock. *Palaeolama* rapidly expanded its range west to Peru and Ecuador and north to Central America and had reached the southern United States in both California and the Gulf Coast region by Irvingtonian times. Apparently, *Palaeolama* did not extend its range inland in North America; it became extinct in the late Wisconsinan.

The stratigraphic range of *Palaeolama mirifica* extends from the Irvingtonian (Coleman IIA, the richest known sample) to the late Wisconsinan of Florida (about 15 faunas), Texas (Ingleside), and California (Emery Borrow Pit). This llama was formerly considered to be robust species of *Tanupolama*, but Webb (1974c) showed that it is more closely allied to the South American species *Palaeolama weddelli* (P. Gervais) and *Palaeolama aequatorialis* Hoffstetter. (Langenwalter, personal communication, 1973; Simpson, 1929a; Webb, 1974c.)

FAMILY CERVIDAE—DEER

The history of the deer family, which has a good fossil record, begins in the early Miocene. Deer are ruminants, mostly browsing in habits, with brachyodont or moderately hypsodont cheek teeth; the molars usually have well-developed ribs, which are external on the upper molars and internal on the lower molars. Males carry antlers, which are shed and regenerated annually; females, except in *Rangifer,* do not have antlers. The feet, with few exceptions, are functionally tetradactyl.

Modern deer evolved in the Old World and entered North America in the Pliocene, replacing various aberrant deerlike animals of American origin. Several migrations occurred during the Pleistocene, populating the North American continent with stocks of Eurasian origin.

The Pleistocene deer of North America are grouped in two subfamilies. The Odocoileinae, or telemetacarpalian deer, are characterized by the great reduction of the metacarpals of the side toes (digits II and V), of which only the distal parts remain. This group comprises all the endemic deer of the Americas, as well as the moose and *Rangifer* groups and the Eurasian roe deer. In the subfamily Cervinae, the proximal as well as the distal parts of the side metacarpals are retained; this condition is termed plesiometacarpalian. Confined mainly to the Old World, this group is represented in the New World by *Cervus.* In addition, there are some Pleistocene forms, known mainly or only from antler and jaw remains, which cannot at present be definitely classified.

CERVIDAE *incertae sedis*

†False Elk, *Bretzia pseudalces* Fry and Gustafson, 1974

The sole known species of its genus, *Bretzia pseudalces* is known from antler and skull fragments found in the early Blancan deposits of White Bluffs. About the size of a mule deer, it had short-beamed antlers with a single well-developed tine and a large terminal palm that was diamond-shaped or semirectangular, oriented in the same plane as the long axis of the head, and had denticulate distal borders. The antlers distantly resembles *Alces* and *Dama* (fallow deer), but the details are very different, and *Bretzia* is probably not closely related to either genus (see fig. 15.8). Relationship to Miocene Old World forms like *Cervavitus* is a more likely alternative. *Bretzia* may have been an immigrant from Eurasia. It cannot at present be assigned to a subfamily. (Fry and Gustafson, 1974.)

Subfamily Odocoileinae

†Brachydont Deer, *Odocoileus brachyodontus* Oelrich, 1953

The genus *Odocoileus* comprises medium-sized deer with large, dichotomously forking antlers on a steeply rising beam, and a long, narrow face and slender limbs. The teeth are distinctly smaller than those of other common Pleistocene cervids of North America. The genus appeared at the beginning of the Blancan and thus forms a useful time-marker. It is by far the most common cervid genus in the Blancan and Pleistocene of North America.

Odocoileus brachyodontus is present in the Fox Canyon and Rexroad faunas of the early Blancan, whereas *Odocoileus* cf. *brachyodontus* has been reported from Blanco, here regarded as late Blancan. Fragmentary deer remains from other early Blancan sites, e.g., the Upper Etchegoin of the North Coalinga region in California, may belong to the same species. A possible ancestor is *Procoileus edensis* Frick from the Eden Pliocene in California.

As the name implies, the species differs from later *Odocoileus* in its lower-crowned and somewhat larger cheek teeth. It may be ancestral to one or both of the living species, *Odocoilus hemionus* and *Odocoileus virginianus,* the latter of which was present in the later Blancan. (Oelrich, 1953.)

White-tailed Deer, *Odocoileus virginianus* (Zimmermann), 1780 (*Dama virginiana; O. sheridanus* Frick, 1937; *O. cooki* Frick, 1937)

This species appeared in the later Blancan (Santa Fe River IB, Haile XVA) and may also be represented by remains of this age provisionally referred to *Odocoileus* sp. (Cita Canyon, Hudspeth, Red Light). It is not uncommon in the Irvingtonian of the central and eastern states (Coleman IIA, Conard Fissure, Cumberland, Hay Springs, Inglis IA, Port Kennedy, Punta Gorda, Rushville, Slaton, and possibly Curtis Ranch and Gilliland). In the Rancholabrean, the species was apparently very common and widespread. It has been recorded from more than 150 localities in Arkansas, Florida, Georgia, Illinois, Indiana, Iowa, Kansas, Kentucky, Louisiana, Maryland, Michigan, Mississippi, Missouri, Nebraska, Nevada, New Jersey, New Mexico, New York, North Carolina, Ohio, Oklahoma, Ontario, Pennsylvania, South Carolina, Tennessee, Texas, Virginia, West Virginia, and Wisconsin. These records are concentrated in the central and eastern parts of the continent. Numerous finds of *Odocoileus* sp. are known from the western part of North America, however, and some of these may well be *Odocoileus virginianus*. The species was (and is) also present in South America.

Odocoileus virginianus is slightly smaller than *Odocoileus hemionus* and has distinctly different antlers. The antlers have a forward-curving beam with a stout brow tine; the posterior prong is upright, and the anterior prong is directed forward and carries a number of mostly unbranched, vertical tines.

The species may probably be derived from the early Blancan *Odocoileus brachyodontus*. Although no definite evolutionary trends can be observed in the species, there is some

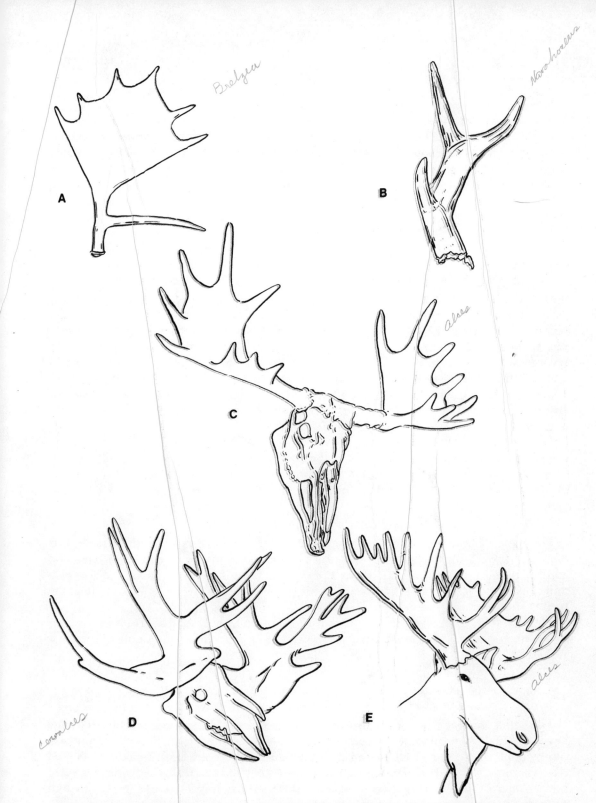

15.8 Cervidae. Antler shapes in cervids. *A*, *Bretzia pseudalces*, Blancan, White Bluffs. *B*, *Navahoceros fricki*, Rancholabrean, San Josecito Cave. *C*, *Alces latifrons*, Elster Glaciation, Europe. *D*, *Cervalces scotti*, Rancholabrean or early Holocene bog, New Jersey. *E*, *Alces alces*, Recent. *A*, after Fry and Gustafson (1974); *B*, after G. J. Miller (unpublished manuscript); *C*, after Kurtén (1972); *D–E*, after Guilday (1971b). Not to scale.

variation in size and morphology, temporally as well as locally, which may be on the subspecies level in some instances. A dwarfed form occurs in the Florida Keys.

The white-tailed deer is an inhabitant of woodlands, forest edges, and stream borders. Its present distribution extends from British Columbia in the Northwest and Quebec and the maritime provinces in the Northeast into South America but does not include Baja California or the arid regions of the Southwest. It is an important game animal; the numerous remains at Holocene occupation sites testify to its importance to Paleo-Indians. (Frick, 1937; Guilday, 1971a, b; Hay, 1923, 1924, 1927; Slaughter et al., 1962; Webb, 1974a.)

Mule Deer, *Odocoileus hemionus* (Rafinesque), 1817 (*Cervus hemionus; Cariacus ensifer* Cope, 1889; *Odocoileus cascensis* Frick, 1937)

Mule deer remains have been identified at about 15 sites in Arkansas, British Columbia, California, Idaho, Nevada, New Mexico, Texas, and Wyoming. Many additional records of *Odocoileus* sp. from the western region of North America probably represent this species; most of these sites are Rancholabrean, but there are also Irvingtonian records of *Odocoileus* in this area (Bautista, El Casco, and Irvington, all in California) that may be mule deer. The Pleistocene *Odocoileus* are in need of revision, however.

The species is somewhat larger and more heavily built than *Odocoileus virginianus*, and the antlers show repeated dichotomous branching into subequal prongs. It may have evolved from *Odocoileus brachyodontus*, early *Odocoileus virginianus*, or some hitherto unknown species.

Mule deer are found in a variety of habitats, ranging from woods to open plains and broken terrain. It is the most important game animal in the West today. (Frick, 1937; Harington, 1971c; Savage, 1951; Schultz and Howard, 1935.)

†Florida Marsh Deer, *Blastocerus extraneus* Simpson, 1928

The genus *Blastocerus*, which is related to and perhaps derived from *Odocoileus*, is now confined to South America, where it also occurred in the later Pleistocene. The South American marsh deer, *Blastocerus dichotomus* Illiger, reaches about the size of a mule deer and has stoutly built antlers with a bifurcating beam; it inhabits marshy forests from Guyana to northern Argentina.

The Floridian species, *Blastocerus extraneus*, is known only from Sabertooth Cave and may date from the Sangamonian. The sole find is a mandible; the dental characters resemble those of *Blastocerus dichotomus*. (Simpson, 1928.)

†Mountain Deer, *Navahoceros fricki* (Schultz and Howard), 1935 (*Rangifer fricki; Cervus lascrucensis* Frick, 1937; *Odocoileus halli* Alvarez, 1969)

The genus *Navahoceros*, with only one species, consists of large deer, in the size range between mule deer and wapiti, with stocky limbs, very short and heavy metapodials, simply built, three-tined antlers (see fig. 15.8), and molar teeth with weakly developed ribs (see figs. 15.9 and 15.10). It is probably closely related to the living Andean deer of South America, *Hippocamelus*, from which it differs in the presence of a brow tine and the more-flattened cross section of the antlers.

The species *Navahoceros fricki* is known from several Rancholabrean cave sites: Little Box Elder, Burnet (type locality), San Josecito, Cueva Las Cruces and Slaughter Canyon Cave, New Mexico, and additional caves, as well as the open-air site Tlapacoya, in Mexico. Its characters and the location of the fossil finds indicate adaptation to a climbing mode of life similar to that of the huemul, *Hippocamelus bisulcus* (Molina). Limb proportions of climbing bovids, e.g., ibex and chamois, are modified in the same way.

The North American mountain deer may be an immigrant from the south, derived from *Hippocamelus* or a common ancestor; G. J. Miller (personal communication) is in the process of reporting on an early *Navahoceros* from the Anza Borrego. The range of the Rancholabrean *Navahoceros fricki*, as far as is known, was confined to the Rocky Mountains and nearby areas. Archeological evidence from Burnet Cave suggests survival to about 11,500 years B.P.; the Little Box Elder occurrence may be almost as recent. (Alvarez, 1969; Hester, 1967; Kurtén, 1975; G. J. Miller, personal communication, 1979; Schultz and Howard, 1935.)

†Fugitive Deer, *Sangamona fugitiva* Hay, 1920 [*Cervus whitneyi* Allen, 1876 (*nomen oblitum*)]

Sangamona, like *Navahoceros*, had teeth and jaws intermediate in size between those of *Odocoileus hemionus* and *Cervus elaphus*; however, *Sangamona* probably approached *Cervus elaphus* in stature since it had unusually long limbs. This genus differs from most other deer in the absence or weak development of the ribs on the molars and the extremely slender limbs with stilt like metapodials (see figs. 15.9 and 15.10). The antlers are unknown. The resemblance in molar structure to *Navahoceros* suggests a phyletic relationship, and *Sangamona* is thus referred to the Odocoileinae. Only one species is known.

Sangamona fugitiva has been found at Alton, Baker Bluff, Brynjulfson, Cavetown, Frankstown, Peccary, Robinson, and Whitesburg (type locality). The same species is probably also represented by remains from Dubuque, Iowa, which were described as *Cervus whitneyi* by Allen (1876). All of the records are probably Wisconsinan in age. The species was thus a Rancholabrean inhabitant of the east-central United States.

The characters of the species indicate a strong cursorial adaptation, and the superficially bovidlike molars suggest that it was a grazing animal. Thus open ground can be assumed to have been the favored habitat of the fugitive deer. Although many finds come from caves, the species has also been found at open-air sites and, in one case (Alton), in a deposit of the loess type.

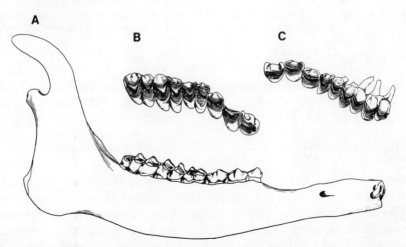

15.9 Cervidae. *A, B, Navahoceros fricki*, Rancholabrean, San Josecito Cave. *A*, right mandible, external view; *B*, right P^2–M^3, occlusal view. *C, Sangamona fugitiva*, Rancholabrean, Frankstown Cave, left P^2–M^3, occlusal view. 0.33×.

314   ORDER ARTIODACTYLA

15.10 Cervidae. *A–C*, tibia; *D–E*, radius. *A, Odocoileus* sp., early Holocene, Washtenaw Co., Michigan. *B, D, Navahoceros fricki*, Rancholabrean, San Josecito Cave. *C, E, Sangamona fugitiva*, Rancholabrean, Frankstown Cave. All 0.33×.

Extinction coincides with that of the megafauna in general. The terminal date is that for Brynjulfson, 9,440 ± 760 years B.P. (Allen, 1876; Hay, 1920; Kurtén, 1975, 1979; Parmalee and Oesch, 1972; Peterson, 1926.)

Caribou, *Rangifer tarandus* (Linnaeus), 1758 (*Cervus tarandus; Rangifer muscatinensis* Leidy, 1879)

The earliest record of *Rangifer* in North America is from the Irvingtonian Cape Deceit fauna, and the species has also been reported from Illinoian deposits near Fairbanks. During the Wisconsinan, caribou ranged from Alaska south to Tennessee (Guy Wilson, Baker Bluff, Beartown Cave), and remains have been found at approximately 35 sites in Alaska, Alberta, Idaho, Illinois, Iowa, Kentucky, Nebraska, New York, Ontario, Pennsylvania, South Dakota, Tennessee, Virginia, and Wisconsin. Banfield (1961) referred all the Pleistocene specimens to the living species, *Rangifer tarandus*, which today has an Arctic distribution and is found in rugged mountain country, forests, and hummocky lowlands.

Caribou are superbly adapted to life in the Far North. For protection against the cold, they have a thick, woolly undercoat protected by stiff, tubular guard hairs, and their broad, flat, deeply cleft hooves are adapted for travel across boggy ground and snow. Gregarious and nomadic, caribou, especially the barren-ground caribou (*Rangifer tarandus groenlandicus*), travel in large herds and frequently migrate long distances. They feed on grasses, sedges, mushrooms, and low shrubs; in winter, the lichen genus *Cladonia* is a staple food.

Caribou are the only member of the deer family in which both sexes are antlered. Although antler development varies with age, sex, physiological condition, available food, and mineral content of the soil, the antlers are usually long, sweeping beams with a well-developed brow tine extending over the face (the two sides are never mirror images of each other, however). Since antlers of females and subadults are smaller and weaker, they are less likely to be preserved; most of the fossil antlers are large and presumably belonged to males.

Caribou have been found at several archeological sites, and at Dutchess Quarry Cave, Orange County, New York, split limb bones were found associated with a Cumberland fluted point. A flesher made from a caribou tibia was found at Locality 14N, Old Crow River (Irving and Harington, 1973), and has been dated at $27,000 \pm {}^{3000}/_{2000}$ years B.P., which makes it one of the oldest artifacts found in the New World. Paleo-Indians hunted caribou in New York, Northwest Territories, Nova Scotia, and Ontario, but not to the same extent as their counterparts did in Eurasia, where many cultures were based on reindeer economy.

Rangifer probably originated in Beringia or in the mountains of northeastern Asia. The ancestry is unknown. The earliest European record is from Süssenborn, a middle Pleistocene deposit in Germany. Although caribou extended their range south during the Wisconsinan, they moved north as the ice sheets retreated and today are found across Alaska, Canada, and coastal Greenland, with relict populations in northern Idaho, the northern Great Lakes region, and Maine. The presence of caribou in a fauna south of its present range is a good indicator of colder climates during the last glaciation. (Anderson and White, 1975; Banfield, 1961; Guilday, Hamilton, and Parmalee, 1975; Irving and Harington, 1973; Kelsall, 1968; Walker, 1975.)

† Broad-fronted Moose, *Alces latifrons* (Johnston), 1874 (*Cervus latifrons;*
  Cervalces alaskensis Frick, 1937)

The mooselike deer (Old World elks), constituting the tribe Alcini, are represented in the Pleistocene of the Holarctic by three main groups, which are recognized here as the genera *Libralces, Alces,* and *Cervalces. Libralces gallicus,* of the late Villafranchian in Europe, is characterized by very long-beamed, small-palmed antlers and a "generalized" skull with nasals only moderately reduced. In modern *Alces*, the antlers are short-beamed with large palms, and the nasals are very much shortened in connection with the development of an overhanging, highly movable muzzle. The Nearctic Pleistocene *Cervalces* retains longer nasals than *Libralces* but has more complex antlers.

The position of *Alces latifrons* is not entirely clear. Azzaroli (1953) referred it to *Libralces,* but a skull from the middle Pleistocene in Germany, in the Göttingen collection, displays an *Alces*-like nasal structure. The nomenclatural problem is further compounded by the fact that the generic name *Praealces,* which was proposed by Portis in

1920, would take precedence over *Libralces* Azzaroli (1952) if *Libralces gallicus* and *Alces latifrons* are judged to be congeneric. The best course at present seems to be to refer the species provisionally to *Alces*.

Alces latifrons was the largest known deer, exceeding in size the living moose as well as the great *Cervalces*. It was common in the mid-Pleistocene faunas of northern Eurasia. It also occurs in deposits in Alaska and Yukon evidently dating from the Illinoian and Wisconsinan; an antler beam from Old Crow River Loc. 22 has been radiocarbon dated at $33,800 \pm 2,000$ years B.P. The Alaskan material includes the type of *Cervalces alaskensis*, with a beam length of 430 mm, which is outside the known size range for antlers of *Alces alces* and *Cervalces* but within that of 15 antlers of *Alces latifrons* from the middle Pleistocene of Europe (305–465 mm; mean, 364 ± 13 mm).

Thus it may be concluded that the species extended its range in Illinoian times, or even earlier, from northeastern Asia into the Alaska refugium, where a population became isolated during Sangamonian flooding of Beringia. If these forms retained longer nasal bones than *Libralces*, they could be ancestral to *Cervalces*.

As noted above, the beam length of the *Alces latifrons* antlers greatly exceeds that of *Alces alces*, and the palm, in addition, is somewhat more simply built, without the tendency to bipartition seen in the modern form (see fig. 15.8). The great width of the antlers suggests that these animals were less exclusively adapted to a forest and bog habitat than the living moose; however, the possible differentiation within what is here tentatively regarded as the single species *Alces latifrons* needs further study. (Azzaroli, 1952, 1953; Frick, 1937; Harington, personal communication, 1976; Kahlke, 1958a, 1960, 1969; Péwé and Hopkins, 1967; Portis, 1920; Sher, 1975.)

Moose, *Alces alces* (Linnaeus), 1758 (*Cervus alces; Alces runnymedensis* Hay, 1923)

The living moose differs from *Alces latifrons* in the shape of the antlers, which have a shorter beam and a relatively larger, bipartite palmation (see fig. 15.8). The shortening of the nasal bones is an important character but, as noted above, is also found in material referred to *Alces latifrons* and having the characteristic antlers of that species. The American moose, sometimes regarded as a distinct species, *Alces americana* (Clinton), is now thought to be conspecific with the European form.

Fossil remains of moose occur in Wisconsinan and possibly earlier deposits in the Alaska–Yukon refugium (Eschscholtz Bay, Fairbanks, Gold Run, Old Crow River), although with incomplete materials, distinction from *Alces latifrons* is difficult. In addition, the species has been recorded at some 15 sites in British Columbia, Illinois, Kentucky, Michigan, Minnesota, Ohio, South Carolina, and Washington, and possibly also Missouri, Oklahoma, and Pennsylvania; however, in these and some other cases, confusion with *Cervalces* is possible. The sites south of the ice are either of uncertain age or definitely late Wisconsinan to Holocene, and so it appears that the species was a late Wisconsinan immigrant in these areas. Several occurrences are well south of the present range.

The species probably evolved in Eurasia from an *Alces latifrons*–like ancestor. The gradual shortening of the antler beam is illustrated by the fact that Eemian interglacial European moose have beams markedly longer (an average of 246 mm for two antlers from the Weimar travertines) than those of living animals. When the route into North America south of the ice was opened in the late Wisconsinan, the species was able to populate this area.

The moose is the largest living cervid. The size forms a cline, with the largest individuals in areas on both sides of Beringia. Moose inhabit the coniferous forests of Holarctica, preferring wet and boggy ground, where their spreading hooves give good support. They browse on conifers as well as deciduous trees and also feed on aquatic plants, grasses, and sedges. (Harington, 1971c; Harington and Clulow, 1973; Hay, 1923, 1924, Kahlke, 1958b; R. L. Wilson, 1967.)

†Stag-Moose, *Cervalces scotti* (Lydekker), 1898 [*Alces scotti; Cervus americanus* Harlan, 1825 (preoccupied name); *Cervalces roosevelti* Hay, 1913; *C. borealis* Bensley, 1913]

The genus *Cervalces* comprises mooselike animals of large size that resemble *Libralces* in their retention of long nasal bones; the average size is only slightly less than that of *Alces latifrons,* and the limb proportions are similar to those of the latter species. The beam of the antlers is somewhat shorter than that of *Alces latifrons* and less deflected downward. The palmation is unique and highly complicated; the internal part is directed upward and bipartite, terminating in several tines, and the outer part curves backward and upward and may carry one or more tines (see fig. 15.8). The total antler span is 1,620 mm in one skeleton but may have been greater in other individuals. All known specimens are now referred to the species *Cervalces scotti,* but some material referred to *Cervalces* may in reality be *Alces latifrons.*

It is somewhat difficult to delimit the stratigraphic range of the species, especially since the pre-Wisconsinan material lacks the diagnostic antler palms. Sangamonian finds come from Fort Qu'Appelle, Toronto, and Bradford, Ontario; a specimen from Giltner, Nebraska, has been referred to the Kansan. Wisconsinan-age remains have been found in Arkansas, Illinois (five sites), Indiana, Iowa, Kansas, Kentucky, Michigan, Missouri (two sites), Nebraska, New Jersey (two good skeletons in postglacial bogs), Ohio, Oklahoma, Pennsylvania, and Virginia. Frgamentary finds, of limb bones, for instance, are hard to tell from *Alces.* Like the moose, *Cervalces* appears to have been a muskeg inhabitant, at least in the late Wisconsinan and early Holocene.

The length of the antler beam tended to diminish through time; in pre-Wisconsinan specimens, it averages 323 ± 7 mm (three specimens, range 310–330 mm) and in Wisconsinan specimens, 238 ± 17 mm (nine specimens, range 180–300 mm). The trend thus parallels that seen in the ancestry of *Alces alces.* It is highly probable that *Cervalces* arose near *Libralces,* and it is also likely that the Beringian form here referred to *Alces latifrons* represents a transitional stage, but the dichotomy between the *Cervalces* and *Alces alces* lineages must obviously go back to an ancestral form with long nasal bones.

No absolute terminal date for *Cervalces scotti* is available as yet, but the numerous postglacial occurrences probably date from the Wisconsinan retreat stages, between 15,000 and 10,000 years B.P.; it is questionable whether the species survived into the Holocene proper. Competition with immigrant true moose, which seems to have entered North America south of the ice during this interval, may have contributed to the extinction. The stag-moose evidently had a restricted geographic range. (Hay, 1913c, 1914; E. A. Hibbard, 1958; Khan, 1970; Schultz, Frankforter, and Toohey, 1951; Scott, 1885.)

Subfamily Cervinae

Wapiti, *Cervus elaphus* Linnaeus, 1758

The genus *Cervus* comprises the plesiometacarpalian deer and includes numerous species, living and extinct, in the Old World, where it has a long evolutionary history from the Villafranchian on. The size ranges from small to very large. Antlers have a short to moderately long beam, oriented obliquely upward, backward, and to the side; the number of tines varies from 3 to 10 or more. In cross section, the antlers are rounded, with some apical flattening, and the surface is rugosely grooved.

Although the wapiti is the only North American species definitely referable to *Cervus*, various records of other species of this genus may be found in the literature. In some cases, the material is too fragmentary to be definitely determined at present; examples are *Cervus fortis* Cope, 1878, from Loup Fork, Oregon, *Cervus lucasi* Hay, 1927, from Sand Hollow, Ada County, Idaho, and *Cervus aguangae* Frick, 1937, from Aguanga, southern California. Here may also be added *Alces shimeki* Hay, 1914, based on a mandible fragment from Harrison County, Iowa; the teeth seem much too small to permit reference to *Alces*. *Cervus? brevitrabalis* is discussed below.

The wapiti (American elk) is often regarded as a distinct species, *Cervus canadensis* Erxleben, 1777. In recent literature, however, its specific identity with the Eurasian form is usually recognized. The species is characterized by large size and antlers carrying brow and bez tines and usually at least three additional tines.

Wapiti is present in the Irvingtonian (Cape Deceit) and Illinoian (Fairbanks I) of Alaska. Irvingtonian records from south of the ice (Cumberland, Conard Fissure) need further substantiation. In Sangamonian times, the wapiti was a member of the boreal fauna, as indicated by its presence at Medicine Hat. The species is common in Wisconsinan and early Holocene deposits; it has been recorded at about a hundred localities in Alberta, Arkansas, Idaho, Illinois, Indiana, Iowa, Kansas, Kentucky, Maryland, Michigan, Minnesota, Missouri, New Jersey, New York, North Carolina, Oklahoma, Ontario, Pennsylvania, South Carolina, Tennessee, Vermont, Virginia, Washington, Wisconsin, Wyoming, and the Yukon. Records from California, Florida, and Georgia are doubtful. Most of the records are within or close to the historical range of the species.

In Eurasia, *Cervus elaphus* was widespread from mid-Pleistocene times on. An early form called *Cervus acoronatus* by Beninde (1937) is probably not a distinct species; it resembles Asian and American specimens more than living European representatives. The origin of the species is not known with certainty, but there are Villafranchian species which may be ancestral.

Wapiti, like most other deer, are mainly inhabitants of woodland and forests and feed on twigs, bark, herbs, and grasses. Their original, very wide range in North America has been broken up into relict patches by extirpation. Prized game animals, wapiti were also hunted in prehistoric times, but remains at human occupation sites are usually few and quite overshadowed by those of *Odocoileus*. (Beninde, 1937; Geist, 1971; Guilday, 1971b; Guthrie, 1966c; Hay, 1923, 1924, 1927; Péwé and Hopkins, 1967; Ray, 1957; Stalker and Churcher, 1970.)

† Cope's Deer, *Cervus? brevitrabalis* (Cope), 1889 (*Alces brevitrabalis*; *A. semipalmatus* Cope, 1889)

This species, of uncertain affinities, is only known from antler fragments in the Delight (Washtuckna Lake) fauna, probably of Irvingtonian or Rancholabrean age. The beam is somewhat flattened; distal portions are unknown. Certain similarities to wapiti suggest that the species be tentatively included in the genus *Cervus*. (Fry and Gustafson, 1974.)

Family Antilocapridae—Pronghorns

An endemic North American family, the Antilocapridae range stratigraphically from the middle Miocene to the Recent. Two subfamilies are recognized, the Merycodontinae and the Antilocaprinae. The Antilocaprinae evolved from the Merycodontinae, replaced this group in the early Pliocene, and became a widespread and diversified group until the late Pleistocene, when all but one of the species became extinct. Only *Antilocapra americana,* the pronghorn, survived.

The Antilocaprinae have molariform premolars, hypsodont cheek teeth, horns consisting of paired, superorbital, non-deciduous horn-cores with deciduous sheaths, and widely separated parietal ridges on the skull.

In their numbers and diversity, the late Cenozoic antilocaprids resemble the extant African antelope; about 12 genera inhabited the plains and deserts of North America. The Pleistocene antilocaprids are in need of revision. Several of the "species" were based on fragmentary jaws or horn-cores, with other material later being tentatively referred to the species. Only a few sites, e.g., Papago Springs and San Josecito, have yielded adequate samples showing age, sexual, individual, and geographic variation. As Stock (1930) pointed out, differences in size of the horn-cores may be due to age or sex; this was illustrated in the sample from San Josecito (see Furlong, 1943). We have made no attempt to determine the validity of the named species—that awaits the reviser of the group.

† Prentice's Pronghorn, *Ceratomeryx prenticei* Gazin, 1935

The monotypic genus *Ceratomeryx* is known only from Hagerman. This late Pliocene antilocaprid is characterized by a relatively large anterior prong over the orbit and a much smaller posterior prong. The anterior prong is tilted backward. Gazin (1935) thought that *Ceratomeryx prenticei* was ancestral to *Tetrameryx*. However, the discovery of *Texoceros* Frick in the Pliocene of Oklahoma (Frick, 1937) makes it more probable that *Ceratomeryx* was a somewhat separated phylogenetic type, possibly having a common ancestry with *Sphenophalos* Merriam, a poorly known Pliocene form. (Colbert and Chaffee, 1939; Frick, 1937; Gazin, 1935.)

† Matthew's Pronghorn, *Capromeryx furcifer* Matthew, 1902 (*C. minimus* Meade, 1942)

Four species of *Capromeryx,* a small, four-horned antilocaprid that ranged stratigraphically from the Blancan to the late Rancholabrean and geographically from California to Florida, have been identified in Pleistocene faunas. The genus is distinguished from other antilocaprids by its small size and the configuration of the horn-cores.

Capromeryx furcifer was described from a lower jaw found at Hay Springs; much later, Frick (1937) figured two horn-cores from the type locality. Since Hay Springs is a mixed fauna, there exists doubt concerning the age of the specimens. Antilocaprid material from Angus, Slaton, and Cragin Quarry has been referred to this species. Meade (1942) described *Capromeryx minimus* from Slaton, but Hibbard and Taylor (1960) noted that the differences between it and the holotype of *Capromeryx furcifer* are not as great as the individual variation seen in a Recent sample of *Antilocapra americana.*

Capromeryx furcifer was about two-thirds the size of *Antilocapra americana* and showed numerous resemblances to the latter species in tooth and jaw morphology. Descriptions of referred horn-cores (see Meade, 1942; G. E. Schultz, 1969) from Slaton and Cragin

Quarry show that the two prongs arise from a common, laterally constricted base. The anterior prong is triangular in cross section, with a deeply concave posterior face; the posterior prong is long and slender, and the forward inclination is not as pronounced as in *Capromeryx mexicana* from Tequixquiac, Mexico. The probable habitat of *Capromeryx furcifer* was open, grassy uplands. (Frick, 1937; Hibbard and Taylor, 1960; Meade, 1942; G. E. Schultz, 1969.)

†Skinner's Pronghorn, *Capromeryx arizonensis* Skinner, 1942

The stratigraphic range of this species is Blancan and Irvingtonian. *Capromeryx arizonensis* was described from an isolated locality, Dry Mountain, Graham County, Arizona, and specimens from Broadwater, Santa Fe River IB, and Inglis IA have been referred to this species.

The anterior prong is as large as the posterior one, and both are situated over the posterior part of the orbit on a relatively large core base. The horn-cores are larger than those of the type of *Capromeryx mexicana* or specimens referred to *Capromeryx furcifer*. Apparently, the horn-cores were nearly erect on the skull, and each was covered by a separate sheath. (Skinner, 1942.)

†Diminutive Pronghorn, *Capromeryx minor* Taylor, 1911 (*Breameryx minor* Furlong, 1946)

The diminutive pronghorn has been identified at Blackwater Draw, Ingleside, McKittrick, Rancho La Brea, and Schuilling. Although Furlong (1946) proposed the genus *Breameryx* for the tiny antilocaprid from Rancho La Brea, most workers believe that it is a species of *Capromeryx*.

Capromeryx minor has more hypsodont premolars and a shorter tooth row than the other species of *Capromeryx*, a P_3 with a single enamel indentation on the lingual side of the crown, a relatively broad skull, large orbits, and tubular auditory bullae. The jaw is shorter and heavier than that of the living pronghorn, and the animal probably had a shorter snout (see fig. 15.11). The paired horn-cores arise from a common pedestal;

15.11 Antilocapridae. Restoration of the Pleistocene antilocaprids *Capromeryx minor* and *Stockoceros onusrosagris*, along with the extant *Antilocapra americana*.

the anterior prong is short, spurlike, and triangular in cross section, and the posterior prong is long, slender, and circular in cross section.

Capromeryx minor stood about 560 mm at the shoulder and weighed about 10 kg; it had long, light limbs. The species is more common than *Antilocapra americana* at Rancho La Brea. A plains-dweller, it probably sought refuge from its many enemies by hiding in copses of trees and shrubs. (Chandler, 1916; Furlong, 1946; Hibbard and Taylor, 1960; W. E. Miller, 1971; Stock, 1965; Taylor, 1911.)

†Mexican Pronghorn, *Capromeryx mexicana* Furlong, 1925

Originally described from the late Pleistocene of Tequixquiac, State of Mexico, this species has also been identified at Cedazo (early Rancholabrean). The horn-cores arise from a common base and are somewhat convergent; deep external sulci are present on both horn-cores. This tiny antilocaprid was about the size of *Capromeryx minor*. (Furlong, 1925.)

†Shuler's Pronghorn, *Tetrameryx shuleri* Lull, 1921

The genus *Tetrameryx* Lull includes large four-horned antilocaprids with long posterior prongs and short anterior prongs of the horn-cores. Five species are currently recognized.

Tetrameryx shuleri was described from the Dallas sand pits, and material from Moore Pit, Slaton, and Hill-Shuler have been referred to it. The paired horn-cores are slender, straight, and rounded in cross section; coalesced at the base, they diverge strongly anteroposteriorly. The posterior prong is about twice as long as the anterior one, and the external sulcus spirals anteriorly on the posterior prong. The cheek teeth are similar to but slightly larger than those of *Antilocapra americana*. (Colbert and Chaffee, 1939; Mooser and Dalquest, 1975b.)

†Irvington Pronghorn, *Tetrameryx irvingtonensis* Stirton, 1939

Known only from the middle Irvingtonian fauna of Irvington, this four-horned antilocaprid was about the same size as *Antilocapra americana*. The posterior prong is longer and heavier than the anterior one and is strongly curved anteriorly; the sulcus is confined to the outer surface, though tending to spiral anteriorly. (Mooser and Dalquest, 1975b; Stirton, 1939.)

†Knox Pronghorn, *Tetrameryx knoxensis* Hibbard and Dalquest, 1960

A skull fragment with attached horn-cores was found at Gilliland and tentatively described as a new species of *Tetrameryx*. Both the anterior and posterior prongs are strongly flattened. The posterior prong is larger and longer than the anterior one and has a shallow sulcus along the midline of the external surface. The horn-cores have a greater diameter than those of *Stockoceros* or *Tetrameryx shuleri* and are more elliptical in cross section than those of *Tetrameryx irvingtonensis*. (Hibbard and Dalquest, 1960.)

†Mooser's Pronghorn, *Tetrameryx mooseri* Dalquest, 1974

This relatively large four-horned antilocaprid is known only from the early Rancholabrean Cedazo local fauna in central Mexico. It was about the size of *Antilocapra americana*. The anterior prong is short, straight, blunt-ended, and flat; the much longer posterior prong is slender and broadly oval in cross section, with a distinct external sulcus. A unique characteristic is the poor development of the supraorbital shelf compared to the condition in *Antilocapra, Stockoceros,* and species of *Tetrameryx* in which this character is known. (Dalquest, 1974.)

†Tacubaya Pronghorn, *Tetrameryx tacubayensis* Mooser and Dalquest, 1975

Known only from the Cedazo local fauna, this large four-horned species has a short, slender anterior prong that is strongly inclined anteriorly and a slender, straight posterior prong that is rounded in cross section and lacks an external sulcus. The horn-core base is narrowed and relatively elevated at the point of divergence of the tines. (Mooser and Dalquest, 1975b.)

†Hay's Pronghorn, *Hayoceros falkenbachi* Frick, 1937

Although given generic rank by most authors, *Hayoceros* was originally described as a subgenus of *Tetrameryx*. According to Frick (1937), this species was "six-pronged." The anterior prong is large, shaped like the male horn-core of *Antilocapra americana*, and, like the horn-core of that species, supposedly had a forked sheath. The posterior prong is tall and straight. The metapodials are approximately the same length but are heavier than those of the extant pronghorn. It is known only from Hay Springs. (Frick, 1937.)

†Conkling's Pronghorn, *Stockoceros conklingi* (Stock), 1930 (*Tetrameryx conklingi*)

Frick (1937) described *Stockoceros* as a subgenus of *Tetrameryx;* however, most authors have treated it as a separate genus. Two closely related species are recognized, both represented by large samples.

Stockoceros conklingi was first described from Shelter Cave, Dona Ana County, New Mexico, and later reported from Ventana, San Josecito, and Cedazo. More than 50 individuals were recovered at San Josecito, which makes this one of the best known species. It has forked horn-cores arising from a common base. Furlong (1943) pointed out the following differences between horn-cores of young and old animals: (1) in young animals, the horn-cores are erect; with increasing age, they tilt outward from the sagittal plane (the inclination being most pronounced in old animals); (2) in young animals, the anterior prong is somewhat heavier than the posterior one but in anteroposterior alignment with it; this position changes with frontal growth, until the anterior prong shows greater inclination than the posterior one; and (3) in young animals, the pedicel is relatively thin; with age, it thickens anteriorly but remains relatively thin and flat at the base of the rear prong.

Stockoceros conklingi was smaller than *Tetrameryx shuleri* and intermediate in size between *Antilocapra americana* and *Capromeryx minor*. The skull and mandible are shown in figure 15.12. The limb bones are slightly heavier and the cannon bones shorter and heavier than those of the living pronghorn. It was less fleet-footed and probably lived in rougher terrain than the extant species.

Stock (1930) tentatively referred this species to *Tetrameryx,* saying that its small size and different horn-cores perhaps separate it generically from *Tetrameryx shuleri* but that these differences might be due to age or sex. Frick (1937) transferred *Tetrameryx conklingi* and *Tetrameryx onusrosagris* to his new subgenus *Stockoceros*. Skinner (1942) thought that both species were distinct, but Furlong (1943) did not believe *Tetrameryx onusrosagris* to be specifically distinct from *Stockoceros conklingi*. Only when detailed statistical studies are done on the large samples of both species and comparisons are made with specimens of *Antilocapra americana* will the question be resolved. (Frick, 1937; Furlong, 1943; Skinner, 1942; Stock, 1930, 1932a.)

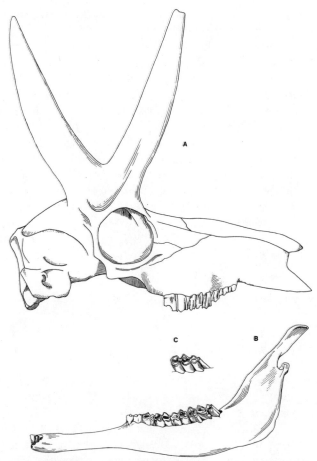

15.12 Antilocapridae. *Stockoceros conklingi*, Rancholabrean, San Josecito Cave. *A*, skull; *B*, mandible; *C*, occlusal view of M_3. *A, B*, ca. 0.33×; *C*, 0.66×.

†Quentin's Pronghorn, *Stockoceros onusrosagris* (Roosevelt and Burden), 1934
  (*Tetrameryx onusrosagris*)

One of the better known Pleistocene antilocaprids is *Stockoceros onusrosagris*. It was first described from Papago Springs, a site that has yielded more than 60 individuals, and is also known from Burnet and Mullen II. The material from the latter site, early Illinoian in age, was described as a distinct subspecies, *Stockoceros onusrosagris nebrascensis* Skinner, although this is questionable.

Quentin's pronghorn was slightly larger than *Stockoceros conklingi* and smaller than *Antilocapra americana*. The symmetrically forked horn-cores arise over each orbit from a short core base, and separate sheaths covered each prong, as indicated by the basal rings of nutritive foramina. Skinner (1942) believed that both sexes were horned since no hornless skulls were found. He mentioned the possibility that cave habitation might have been a seasonal phenomena and that perhaps only males were present in the Papago Springs sample.

In his comparison between *Stockoceros onusrosagris* and *Antilocapra americana*, Skinner (1942) noted the following dental and cranial differences: *Stockoceros* has smaller, more hypsodont cheek teeth, an M_3 that is more variable (having either three or four lobes), strongly separated double roots on the lower premolars, a more slender nose (the rostrum was not inflated as it is in *Antilocapra*), and auditory bullae that are swollen, smoothly rounded, and larger. (See restoration, figure 15.11.)

Stockoceros onusrosagris was probably as swift as the living pronghorn since the limb bones are equally slender and light, although the proportions are slightly different. For example, the tibia of *Stockoceros* is longer than that of *Antilocarpa* even though the latter is a larger animal. This suggests a greater bounding ability in *Stockoceros* (see fig. 15.13).

Skinner (1942) distinguished *Stockoceros onusrosagris* from *Stockoceros conklingi* by the larger size, larger, more divergent, and more outwardly flaring horn-cores, and more gracile skeleton of the former species. Webb (1973) believes that a relatively close relationship exists between the four-horned Pleistocene genera, *Stockoceros* and *Tetrameryx*, and the six-horned Pliocene genera, *Hexameryx* and *Hexobelomeryx*, because the pedicels in all four genera are short and the anterior and posterior prongs of *Hexobelomeryx* closely resemble those of *Stockoceros* in shape, position, and texture. (Colbert and Chaffee, 1939; Skinner, 1942; Webb, 1973.)

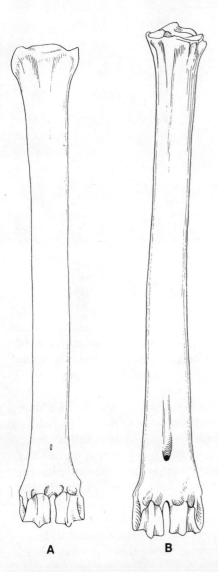

15.13 Antilocapridae. *Stockoceros conklingi*, Rancholabrean, San Josecito Cave. *A*, metacarpus; *B*, metatarsus. 0.33×.

Pronghorn, *Antilocapra americana* (Ord), 1815 (*Antilope americana; Neomeryx finni* Parks, 1925)

The genus *Antilocapra* includes the living pronghorn, *Antilocapra americana*, of western North America and its middle Pliocene ancestor, *Antilocapra garciae* Webb, which is known only from Florida.

The only surviving antilocaprid is the pronghorn, an animal that ranges across the western plains and deserts from southern Canada to north-central Mexico. The species has been recorded at a number of late Pleistocene sites, but some of these finds may represent extinct species. Several of the Pleistocene occurrences (Emery Borrow Pit, Maricopa, McKittrick, Rancho La Brea, Ben Franklin, Kyle, Brynjulfson) fall outside the present range of the species.

Antilocapra americana is one of the most advanced living artiodactyls with regard to dentition and limb structure (Vaughan, 1972). High-crowned cheek teeth enable it to browse on grit-covered shrubs and forbs. The fleetest American mammal, the pronghorn has been clocked at speeds of 65 km per hour. The legs are long and slender, and there is no trace of lateral digits. The tarsus distal to the calcaneum and astragalus consists of three bones: the fused navicular–cuboid, the fused ecto–middle cuneiform, and the internal cuneiform. Beneath the posterior part of each hoof is a pad of connective tissue which protects the foot. The front feet, with their larger hooves, carry most of the weight when the animal is running. (See figure 15.11.)

In *Antilocapra americana*, the anterior prong of the horn-core is absent, but the forked sheath and the anterior vascular groove that passes up the pedicel toward the anterior prong testify to its former presence. The posterior prong is a single, somewhat laterally compressed blade covered by a forked sheath composed of agglutinated hair. The sheath is shed annually. Both sexes have horns, but those of the female are small and lack the anterior prong. The horn-cores are situated over the orbits and show considerable individual variation in size, degree of divergence, and outward flare.

Adult pronghorns weigh between 36 and 60 kg and stand 810 to 1,040 mm at the shoulder; females are slightly smaller than males. The animals have large eyes, and the prominent orbits are situated high and far back on the skull. Pronghorns are gregarious and travel in small groups; in winter, herds numbering 50 to 100 are common. Although pronghorns are hunted for trophy, sport, and meat, their numbers are now increasing owing to successful conservation practices.

The ancestry of *Antilocapra americana* has been debated for years (see Barbour and Schultz, 1934; Chandler, 1916; Hesse, 1935; Stirton, 1938). *Proantilocapra platycornea* Barbour and Schultz from Cherry County, Nebraska, was thought to represent a late Tertiary (cf. Claredonian) stage leading toward *Antilocapra*. Compared to the extant pronghorn, this species has smaller, blunt-tipped, laterally compressed horn-cores, a similar occipital region, and somewhat shorter limb bones. Webb (1973) recently described *Antilocapra (Subantilocapra) garcia* from late Hemphillian deposits in Florida and believes that this species represents an early stage in the evolution of the pronghorn. It has a tall-based, laterally compressed horn-core with a weak anterior prong and a strong posterior prong. The bifurcation of the horn-core tip indicates close affinities with the living species. (Barbour and Schultz, 1934; Chandler, 1916; Hesse, 1935; Stirton, 1938; Vaughan, 1972; Walker, 1975; Webb, 1973.)

Family Bovidae—Bovids

The characteristic feature of this artiodactyl family is the presence, usually in both sexes, of true horns—an unbranched bony core covered by a horny sheath that is not shed. Females usually have smaller horns than males. The part of the skull behind the horns (the posthorn part) is long in the primitive condition but has been much shortened in some lineages. As in cervids, the shape of the horn functions as an identification signal; the horns may also be used in ramming bouts. Most bovids are grazers, with more or less hypsodont cheek teeth. The family comprises the majority of living artiodactyls and has an impressive history in the Tertiary of the Old World, beginning with antelopelike animals in the early Miocene. The variation in size and morphology is great. Several subfamilies are recognized, but only two entered North America. The center of bovid differentiation lay in Asia and Africa, but since the majority of the bovids were adapted to a warm climate, they tended to become confined to southern Asia and Africa; however, northern Asia and Beringia witnessed the emergence of various subarctic bovid groups. From Irvingtonian times on, the American part of Beringia was populated by bovids, and most of these stocks also succeeded in colonizing North America south of the ice. None, however, appears to have reached South America.

Of the two subfamilies that reached the Nearctic, the Caprinae comprises sheep, goats, muskoxen, and related extinct forms. The range in size is great. Several tribes are recognized, among which the Saigini (saiga), Rupricaprini (chamoislike bovids, including *Oreamnos*), Caprini (true sheep and goats), and Ovibovini (muskoxen and related forms) concern us here. The subfamily Bovinae (large, cattlelike bovids) is represented in North America by the genera *Bison* (bison, American "buffalo"), *Platycerabos* (flat-horned oxen), and *Bos* (true cattle, yaks).

The fossil history of many bovids, especially *Bison*, is excellently documented in North America. However, the infraspecific variability of many morphological characters, notably horn-core shape and size, has resulted in taxonomic confusion and the creation of many superfluous names.

Subfamily Caprinae

Saiga, *Saiga tatarica* (Linnaeus), 1758 (*Capra tatarica; Saiga ricei* Frick, 1937)

The monotypic genus *Saiga* is known in the New World only from deposits near Fairbanks, the Arctic coastal plain, and the Baillie Islands, Northwest Territories. This steppe-adapted Eurasian animal invaded North America in Illinoian times, surviving in this continent until the late Wisconsinan. Mountainous terrain, forests, glaciers, and deep snow were barriers to its southward advance. Frick described a new species, *Saiga ricei*, from Fairbanks in 1937, and the Siberian Pleistocene material was later referred to this extinct species. However, additional studies have shown no discernible differences between this material and the living species (Harington, personal communication, 1975). During the Pleistocene, the range of the saiga extended from England to the Northwest Territories. Simpson (1945a) considered the saiga to be related to the chiru (*Pantholops*) of the Pleistocene–Recent of Tibet and Ladakh and put them in the same tribe, the Saigini.

Standing 750 to 800 mm at the shoulder, the saiga resembles a small, lightly-built sheep with a large head and downward-pointing nostrils. The distinctive inflated and mobile muzzle, containing highly convoluted choanae and numerous mucous glands and hairs, is an adaptation for warming and moistening dry air. Only males possess horns; the horn-cores are characterized by their pronounced backward bend.

Today, saigas inhabit the dry, cold, flat plains of Siberia, where they travel in large herds and feed on the semidesert vegetation. Formerly, they were hunted excessively for their horns, which were prized in the Chinese pharmaceutical trade, but saigas have been strictly protected since 1920 and are now the most numerous hoofed animal in the USSR. (Bannikov et al., 1961; Frick, 1937; Harington, 1971c and personal communication, 1975; Péwé, 1975; Sher, 1968; Simpson, 1945a; Walker, 1975.)

Mountain Goat, *Oreamnos americanus* (Blainville), 1816 (*Rupicapra americana*)

Oreamnos is not a true goat, but rather a mountain antelope closely related to the European chamois, *Rupicapra;* it is placed in the tribe Rupicaprini. Common ancestry with a form similar to *Pachygazella grangeri* from the Lower Pliocene of China is indicated. Perhaps a closer relative of *Oreamnos* is *Neotragocerus*, from the Hemphillian of Nebraska and the Blancan of Glenn's Ferry Formation of Idaho. *Neotragocerus* is known only from two horn-cores, which are longer, straighter, and more laterally compressed than those of the living species. Harington (1971b) believes that the genus *Oreamnos* entered North America from Eurasia before the Wisconsinan glaciation. Two species are recognized.

Mountain goats are rare as fossils partly because their habitat is not conducive to fossil preservation. The earliest record is from Sangamonian interglacial deposits near Quesnel Forks, British Columbia. Wisconsinan occurrences include Bell, Horned Owl, Little Box Elder, Samwel, and Potter Creek caves. *Oreamnos* was reported from Washtuckna Lake (now known as the Delight fauna), but later studies have shown the specimen to be the root of a mastodon tooth (Fry and Gustafson, 1974). Mountain goats are found on rocky slopes above the timberline in the Rocky Mountain and Coast ranges from southern Alaska and eastern Yukon Territory south to western Montana, central Idaho, and northern Oregon. They were introduced into the Black Hills, South Dakota, in 1924 and are thriving there.

With their shaggy white coats, black, slightly curving horns, and black hooves, mountain goats are quite distinctive. Stocky and sure-footed, they are renowned for their climbing ability. These animals are both browsers and grazers and make seasonal migrations from high mountain slopes to sheltered valleys in search of food. (Fry and Gustafson, 1974; Harington, 1971b; Walker, 1975.)

†Harrington's Mountain Goat, *Oreamnos harringtoni* Stock, 1936

This small extinct mountain goat is known from four late Wisconsinan cave deposits: Rampart, San Josecito, Smith Creek (type locality), and Stanton. It was about two-thirds the size of *Oreamnos americanus,* and the horn-cores are slightly longer, with greater backward curvature, than those of the living species.

Oreamnos harringtoni probably represents a southern population of *Oreamnos americanus* that spread into the Southwest during a glacial period, became isolated in the higher, cooler areas, and survived as a relict population until the end of the Wisconsinan. Its habits were probably similar to those of the living mountain goat. At Smith Creek Cave, it is associated with cold-adapted species, including pika, noble marten, and marmot. (See fig. 15.14.) (Harington, 1971b; Stock, 1936a; R. W. Wilson, 1942.)

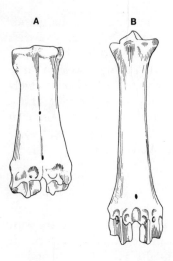

15.14 Bovidae. *Oreamnos harringtoni*, Rancholabrean, Smith Creek Cave. *A*, metacarpus; *B*, metatarsus. 0.33×.

?†Wild Goat, *Capra iowensis* (Palmer), 1956 (*Ibex iowensis*)

Goats (genus *Capra*) are similar to sheep (genus *Ovis*), but goats have a convex rather than a concave forehead, possess beards, lack scent glands on the feet, and males are odorous. Some workers (see Van Gelder, 1977) believe that *Capra* and *Ovis* are congeneric (*Capra* has page priority). Wild goats inhabit rugged terrain from Spain, North Africa, and the Mediterranean islands east to India and north to Siberia.

A partial skull with horn-cores of a goat was found in a sand bar of the Iowa River. It is similar to the Old World *Capra hircus*, but the age and significance of the specimen are uncertain at present. Although the degree of mineralization appears to be too high for the specimen to date from the early settlement of Iowa (ca. 1830), the possibility remains that it may represent a feral descendant of stock introduced elsewhere at an earlier date. The specimen is in a private collection. Tests on the skull, or future discoveries, will be necessary to establish the presence of *Capra* in the Pleistocene of North America. (Palmer, 1956; Reed and Palmer, 1964; Van Gelder, 1977; Walker, 1975.)

Dall Sheep, *Ovis dalli* Nelson, 1884

True sheep, genus *Ovis*, first appear in Villafranchian deposits in Europe and Asia. By the late Pleistocene, these large-horned, stocky animals had spread into mountainous regions of North Africa, Eurasia, and North America. Although still widely distributed, wild sheep have been reduced to relict populations in some areas. Their ancestry probably lies in the Rupricaprini, the goat antelopes.

The earliest records of wild sheep in North America are from Illinoian deposits near Fairbanks. Although not yet described in detail, the specimens have been referred to *Ovis nivicola*, the Siberian snow sheep, which is often considered to be conspecific with *Ovis dalli* (Guthrie, 1968a). Recent studies by Korobitsyna et al. (1974) show, however, that *Ovis nivicola* has 54 chromosomes, compared to 52 in *Ovis dalli* and *Ovis canadensis;* these authors believe that pachycerine sheep evolved in Beringia. The Bering Sea isolated populations in Siberia from those in Alaska in the Eemian–Sangamonian and again in the early Holocene, which resulted in speciation of the two populations, whose modern descendants are *Ovis nivicola* in eastern Siberia and *Ovis dalli* in the rugged mountains of Alaska, Yukon Territory, western Northwest Territories, and northern British Columbia. *Ovis dalli* has been identified in the Lost Chicken Creek fauna and is relatively common in late Pleistocene deposits near Dawson and Sixtymile River in Yukon Territory. A radiocarbon date of 23,000 ± 600 years B.P. on a horn-core of *Ovis dalli* from Bonanza Creek in the Dawson area indicates that Dall sheep inhabited eastern Beringia during the peak of the Wisconsinan glaciation (Harington, personal communication).

Ovis dalli differs from *Ovis canadensis* in its slightly smaller size and the configuration of the horns, which in males are narrow-based, wide-spreading, and amber-colored. Coloration is generally pure white, except in the subspecies *Ovis dalli stonei*, which varies from silvery gray to glossy black with white underparts. Like *Ovis canadensis*, Dall's sheep is a specialized grazer and feeds on grasses and herbs that are often dry and grit-covered. (Geist, 1971; Guthrie, 1968a; Harington, personal communication; Korobitsyna et al., 1974.)

Mountain or Bighorn Sheep, *Ovis canadensis* Shaw, 1804 (*O. catclawensis* Hibbard and Wright, 1956)

Remains of *Ovis canadensis* have been identified at Bell, Chimney Rock, Cochrane, Dry, Glendale, Gypsum, Horned Owl, Isleta, Jaguar, Kokoweef, Little Box Elder, Medicine Hat (Sangamonian), Parsnip River (British Columbia), Rampart, Red Willow, Stanton (Arizona), and several localities along the eastern margin of Lake Bonneville, Utah. Many of the late Wisconsinan specimens are characterized by large size, Hibbard and Wright (1956) described a new species, *Ovis catclawensis*, from Catclaw Cave, Mojave County, Arizona, on the basis of its larger size and wider premolars, and specimens from several other sites have been referred to it. However, Harris and Mundel (1974) believe that *Ovis catclawensis* is a temporal subspecies of *Ovis canadensis*. Recently, complete skeletons of large, long-legged sheep were found in Natural Trap Cave, and studies of this material may show that *Ovis catclawensis* is a distinct species (L. Martin, personal communication, November, 1976). In the Holocene, suitable sheep habitat decreased in area, and today *Ovis canadensis* has a discontinuous distribution in the mountains of the western half of North America, from southern British Columbia to northern Mexico and Baja California, in areas where stable climax grass and sedge communities predominate.

Compared to other wild sheep, *Ovis canadensis* has the largest and widest skull, the longest and thickest horn-cores, and a relatively wide rostrum. The horn-cores of adult rams are massive, broad-based, and not as wide-spreading as those of *Ovis dalli*. Horns are the species' chief weapon, and the skull, especially that of males, shows adaptations for withstanding the heavy concussions that result when two rams clash. The teeth are long, broad, and subhypsodont. Fully grown rams weigh about 102 kg; females are smaller and weigh about 72 kg.

Apparently, all North American fossil sheep belong to the subgenus *Pachyceros*. Recent studies (Korobitsyna et al., 1974) indicate that these sheep evolved their distinctive characteristics while isolated in the Beringian refugium, probably in Riss–Illinoian times, and migrated south into the western United States in the Sangamonian. Harris and Mundel (1974) believe that the specific characters of *Ovis canadensis* were set by the mid-Wisconsinan, that populations with relatively large males (the subspecies *Ovis canadensis catclawensis*) were widespread in the western half of the country in the late Wisconsinan, and that climatic deterioration at the end of the period resulted in selection for smaller animals. In recent years, competition with livestock, diseases introduced by domestic sheep, habitat destruction, disturbance by man, and indiscriminate hunting have reduced mountain sheep populations. (Geist, 1971; Harris and Mundel, 1974;

Hibbard and Wright, 1956; Korobitsyna et al, 1974; Martin, Gilbert, and Adams, 1977; Stock and Stokes, 1969; Stokes and Condie, 1961.)

†Shrub-Ox, *Euceratherium collinum* Furlong and Sinclair, 1904 (*Preptoceras sinclairi* Furlong, 1905; *Taurotragus americanus* Gidley, 1913; *Aftonius calvini* Hay, 1914; *Euceratherium bizzelli* Stovall, 1937)

The shrub-oxen (genus *Euceratherium*, monotypic) were among the first bovids to reach the New World; they were present well back in the Irvingtonian (Cumberland, Irvington, Vallecito, Wellsch Valley) and may have immigrated at the same time as the mammoths. Rancholabrean (mostly Wisconsinan), records are from Albiquiu (Rio Arriba Co., New Mexico), Alton, Burnet, Cox gravel pit (Harrison Co., Iowa), Hawver, Hydro (Caddo Co., Oklahoma), the Klamath River (Siskiyou Co., California), Kokoweef, McKittrick, Potter Creek, Samwel, Sullivan gravel pit (Grant Co., Kansas), Tequixquiac, and Turlin (Monona Co., Iowa).

Shrub-oxen were heavily built, large-horned bovids that probably attained about four-fifths of the stature of a modern bison and were somewhat larger than the living muskox. The horn-cores are characteristic; slightly flattened in cross section, they arise steeply near the posterior edge of the frontals, swing upward and slightly backward, then curve outward and forward and finally upward at the tips. In specimens assigned to *Preptoceras*, the horn-cores are shorter and less upright, somewhat more horizontally directed, and more wide-spreading. It has been suggested that these specimens are immature individuals of *Euceratherium*, and it seems unlikely that two so closely related species would occur together at so many sites (where found, *Preptoceras* is invariably associated with *Euceratherium*, except at McKittrick). The dentition is strongly hypsodont, with well-developed styles on the upper molars.

The shrub-oxen are probably related to the muskoxen (Harington, personal communication). They probably represent a primitive, high-horned stage in ovibovine evolution, which was succeeded, in more advanced muskoxen by a progressive deflection downward of the horncores—an evolutionary process reflected in the ontogeny of *Ovibos* calves (see fig. 15.15). Alternatively, the shrub-oxen have been referred to a tribe of their own, the Euceratheriini.

Euceratherium collinum was a large, specialized grazing bovid that probably inhabited lower hills, somewhat in the manner of sheep and goats, rather than high mountains (see fig. 15.16). The construction of the skull base and occipital condyles suggests that

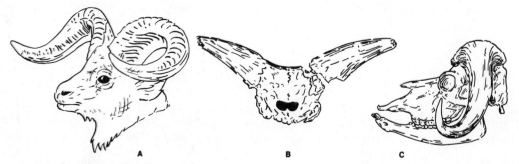

15.15 Bovidae. Ovibovine horn shapes, showing transition from primitive high-horned form to modern low-horned muskox. *A, Euceratherium collinum*, Irvingtonian to Rancholabrean, life reconstruction. *B, Soergelia* sp., posterior view of skull, based upon a European specimen, Irvingtonian in North America. *C, Ovibos moschatus*, Recent, skull. *A*, after Hibbard (1955b); *B*, after Kahlke (1969); *C*, after Gromova (1962). Not to scale.

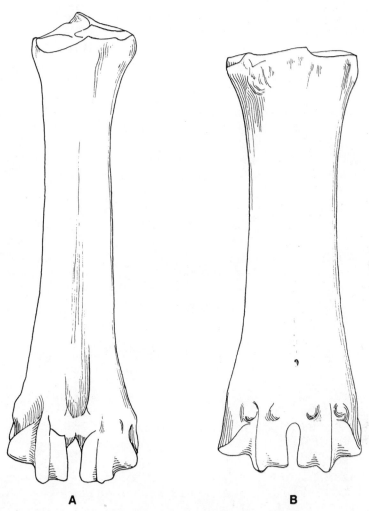

15.16 Bovidae. *Euceratherium collinum*, Rancholabrean, Samwel Cave. *A*, metatarsus; *B*, metacarpus. 0.33×.

the shrub-ox was a tremendous "butter," using its horns in head-on intraspecific encounters, as do modern rams. Association with artifacts at Burnet Cave indicates that the species survived to about 11,500 years B.P. (Frick, 1937; Furlong, 1905; Furlong and Sinclair, 1904; Gidley and Gazin, 1938; Harington, 1971c and personal communication, 1976; Opdyke et al., 1977; Savage, 1951; Schultz and Howard, 1935; Simpson, 1963; Sinclair, 1905; Sinclair and Furlong, 1905.)

†Soergel's Ox, *Soergelia mayfieldi* (Troxell), 1915 (*Preptoceras mayfieldi;*
  Soergelia elisabethae Schaub, 1951)

The genus *Soergelia*, a primitive muskox, was widespread in Eurasia in the middle Pleistocene, occurring from Europe to eastern Siberia, and has been recognized in a few Irvingtonian deposits in North America. *Soergelia* was slightly larger than *Euceratherium*, attaining the size of a steer, and was heavily built, with massive limbs. The horncores, which are set further apart than those of *Euceratherium*, and at a distance of some 3 cm from the posterior end of the frontals, are comparatively short and stout; they curve outward and downward and then somewhat forward (see fig. 15.15). The cheek teeth are very hypsodont and carry strong styles and stylids.

In the opinion of Harington (personal communication, 1976), *Soergelia* occupies an intermediate position between *Euceratherium* and the true muskoxen, being less high-horned than the former. Thus both genera could be regarded as primitive members of the Ovibovini.

Like *Euceratherium*, *Soergelia* probably arose in the Old World, and both genera seem to have entered North America at about the same time as *Mammuthus*, with *Euceratherium* perhaps a little earlier. They were thus the first bovids to reach the Nearctic.

There are only a few finds of *Soergelia* from North America. *Soergelia mayfieldi* was described from Rock Creek, early Irvingtonian, and *Soergelia* cf. *elisabethae* occurs at Old Crow River Loc. 11A (age uncertain, but it may be Illinoian or earlier). An additional find from Cloud County, Kansas, may be Kansan in age. Thus the stratigraphic range of *Soergelia* in North America may span a large part of the Irvingtonian.

Certain differences between the Old World *Soergelia elisabethae* and the Rock Creek form may indicate specific distinction; on the other hand, the Yukon specimen is very similar to the European ones. Eastern Siberian specimens may represent yet another taxon. However, the rank of these taxa is at present uncertain. (Eshelman, personal communication, 1976; Harington, personal communication, 1976; Schaub, 1951; Sher, 1975; Troxell, 1951b.)

† Woodland Muskox, *Symbos cavifrons* (Leidy), 1852 [*Ovibos cavifrons;*
   Scaphoceros tyrelli Osgood, 1905 (name preoccupied, changed to *Symbos* Osgood, 1905); *Bison appalachicolus* Rhoads, 1895; *Liops zuniensis* Gidley, 1906; *Bootherium sargenti* Gidley, 1908; *Symbos australis* Brown, 1908; *Bootherium nivicolens* Hay, 1915; *Symbos promptus* Hay, 1920; *S. convexifrons* Barbour, 1934; *Ovibos giganteus* Frick, 1937; *Bootherium brazosis* Hesse, 1942]

The monotypic genus *Symbos* ranged stratigraphically from the late Irvingtonian (Mullen II, Conard Fissure) to the end of the Wisconsinan and geographically from Alaska south to Mississippi and from the Pacific to the Atlantic coasts. Most of the records are from east-central United States and eastern Beringia. It is known from about 90 specimens found in Alaska, Alberta, Arkansas, British Columbia, California, Colorado, Idaho, Illinois, Indiana, Iowa, Kansas, Kentucky, Michigan, Missouri, Nebraska, New Mexico, North Dakota, Ohio, Oklahoma, Pennsylvania, Utah, Texas, Virginia, Washington, Wisconsin, Yukon Territory, and on the Continental Shelf, 64.3 km southeast of Atlantic City, New Jersey. *Symbos cavifrons* probably inhabited the plains and woodlands and was adapted to warmer conditions than *Ovibos moschatus*, the extant muskox.

Gidley (1908) described *Bootherium sargenti* from a partial skull found in a bog near Grand Rapids, Michigan. Several authors (Allen, 1913; Hibbard and Hinds, 1960; Semken, Miller, and Stevens, 1964) considered the specimen to be a female *Symbos cavifrons*, and the evidence for this view is quite compelling (Harington, personal communication, 1976): the similar basic confirmation of the horn-cores, the smaller size of the cranium, and the broad space between the horn-core bases parallels the differences between male and female *Ovibos moschatus;* thus the specimen resembles the expected appearance of a female *Symbos*. In addition, both genera apparently had similar geographic and habitat preferences, for they have been found not only in the same states and provinces but at many of the same sites. For these reasons, we believe that *Bootherium sargenti* is conspecific with *Symbos cavifrons*.

Symbos cavifrons was taller and had a more slender build than the extant muskox. Diagnostic of *Symbos* is the pitted, rough-basined surface (exostosis) between the horn-cores. The horn-cores are situated higher on the skull than those of *Ovibos* and curve outward, downward, and forward with a relatively high lateral flare. The horn sheaths were fused over the frontals (a specimen from Upper Cleary Creek, Alaska, with beautifully preserved hornsheaths shows this fusion). Compared to *Ovibos*, the skull of *Symbos* is longer and deeper (especially in the occipital region) and has a broad preorbital region, long nasals, and slightly protruding orbits. The large, shield-shaped basioccipital is divided by a deep median groove (see fig. 15.17). The cervical vertebrae are massive, with a constricted neural canal.

Virtually nothing is known about the ancestry of *Symbos*. Flerov (1967) thinks that the genus originated in the New World since it is unknown in the Old. However, Harington (1970, and personal communication, 1976) believes that *Symbos* had a Eurasian origin and entered North America in the late Irvingtonian. *Symbos* and *Ovibos* probably arose from the same basic stock, but *Symbos* shows a few features which are regarded as "less advanced," (e.g., higher-set horns) (Harington, personal communication, 1976). It is possible that *Symbos* evolved in broad intermontane valleys and later dispersed across the steppe grasslands and parklands, becoming widespread and relatively abundant by the late Wisconsinan. Pollen and sediments from a skull found in Kalamazoo County, Michigan, have been dated at $11,100 \pm 400$ years B.P. The cause of extinction is unknown. (J. A. Allen, 1913; Flerov, 1967; Gidley, 1908; Harington, 1970a, 1975, personal communication, 1976; Hibbard and Hinds, 1960; Jakway, 1961; Osgood, 1905; Semken, Miller, and Stevens, 1964.)

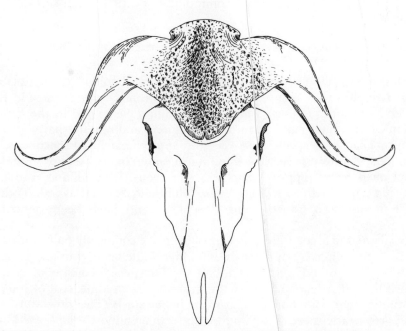

15.17 Bovidae. *Symbos cavifrons*, Rancholabrean, skull, composite drawing. Not to scale.

†Harlan's Muskox, *Bootherium bombifrons* (Harlan), 1825 (*Bos bombifrons*)

The status of the genus *Bootherium* is in doubt; it has often been considered to be congeneric with *Symbos*, but this is undemonstrated (Ray, Wills, and Palmquist, 1968).

The type of *Bootherium bombifrons* was collected at Big Bone Lick by the party sent out by Thomas Jefferson, and specimens from Frankstown Cave have been referred to this species. The horns are smaller and more slender than those of *Symbos*, rounded in cross section, and not broadened at the base so as to encroach on the upper surface of the skull. The posterior part of the skull (behind the horns) slopes abruptly downward in a manner similar to that seen in sheep and goats. The species was smaller than *Ovibos*. Allen (1913) thought that *Bootherium* and *Preptoceras*, although not closely related, were more similar to each other than to any other genus. Presuming that the type of *Bootherium bombifrons* does not represent an abnormal individual, Harington (personal communication) believes that the species is not closely related to *Symbos*. (Allen, 1913; Harington, personal communication, 1976; Ray, Wills, and Palmquist, 1968.)

†Staudinger's Muskox, *Praeovibos priscus* Staudinger, 1908

This poorly known genus has been found in Günz-Mindel–age deposits in England, France, Germany, Czechoslovakia, and Poland and has been identified in Illinoian-age deposits in Alaska (Cripple Creek Sump, Gold Hill, and Lower Cleary Creek, all near Fairbanks) and Yukon Territory (Old Crow River, Upper Porcupine River). *Praeovibos* may be an indicator of Illinoian-age deposits in eastern Beringia. Originally, it was thought to be ancestral to *Ovibos*, but *Ovibos* has since been found in deposits of the same age. It is probable that *Praeovibos*, *Ovibos*, and *Symbos* had a common ancestry in the Villafranchian. Larger, longer-legged, and more slender than the modern genus, *Praeovibos* was less specialized and probably represents a conservative holdover of the ancestral form. (Harington, personal communication, 1976; Kahlke, 1964; Kurtén, 1968; Péwé, 1975.)

Muskox, *Ovibos moschatus* (Zimmerman), 1780 (*Bos moschatus*; *Ovibos yukonensis* Gidley, 1908; *O. proximus* Bensley, 1923)

A surviving species, *Ovibos moschatus* first appears in Illinoian-age deposits in Alaska (Nome and Cripple Creek Sump); other pre-Wisconsinan records include Jinks Hollow, Henderson County, Illinois, and Morrill County, Nebraska. In the Wisconsinan, muskoxen inhabited tundra and arctic steppe regions, and remains have been found in Alaska (approximately 70 specimens), Canada (approximately 20 specimens, from Alberta, Manitoba, Northwest Territories, Ontario, and Yukon Territory), and the conterminous United States (approximately 12 specimens, from Illinois, Indiana, Iowa, Minnesota, Montana, Nebraska, New York, Ohio, and South Dakota). Today, muskoxen dwell in northern Canada and Greenland and have been reintroduced in Alaska.

Superbly adapted to arctic conditions, muskoxen are compactly built, with short legs, neck, and tail and a slight hump over the shoulders. The living muskox has a shoulder height of 132–138 cm and weighs 263–650 kg. The pelage consists of coarse, dark-brown guard hairs that reach nearly to the ground and an undercoat of fine, soft hair so dense that neither cold nor moisture can penetrate it. The downward-sweeping, closely set horns nearly meet at the midline to form an almost solid boss. The median groove separating the horn bases is distinctive. Both sexes are horned, but the horns of the female are smaller, more widely separated between the bases, and less rugose. The face is of moderate length, with protruding orbits and short nasal bones that do not reach the premaxillae. The dentition is less hypsodont than that of *Bison*, and the molars show a reduction of the accessory column on the inner side.

Muskoxen are gregarious, and herds totaling approximately 100 may gather in an area. When attacked, the herd takes up a defensive formation in which the calves are protected. This is effective against predators but not against man, who finds it easy to slaughter the whole herd. Muskoxen feed mainly on grasses, sedges, and willows.

The earliest known *Ovibos* remains are from Günz-age deposits at Süssenborn Obergünzberg, Germany (Kahlke, 1964). In Siberia, *Ovibos* has been identified in deposits of maximum-Riss age. Muskoxen became extinct in Eurasia about 3,000 years ago. The presence of *Ovibos* in deposits of possible pre-Wisconsinan age in Nebraska and Illinois suggests that muskoxen ranged south of the area covered by the Illinoian ice sheet. They probably retreated northward during the Sangamonian and perhaps became isolated in unglaciated refugia in the north and in a narrow tundra band south of the ice during the Wisconsinan. (Barbour, 1931; Harington, 1970b,c, personal communication, 1976; Kahlke, 1964; Kitts, 1953; Maher, 1966, Ray, Wills, and Palmquist, 1968; Tener, 1965; Walker, 1975.)

Subfamily Bovinae

†Steppe Bison, *Bison priscus* (Bojanus), 1827 (*Bos priscus; Bison crassicornis* Richardson, 1854; *B. alaskensis* Rhoads, 1897; *B. geisti* Skinner and Kaisen, 1947; *B. praeoccidentalis* Skinner and Kaisen, 1947)

The genus *Bison* is usually regarded as a diagnostic member of the Rancholabrean fauna in North America, although there is now evidence that may suggest a somewhat earlier appearance. Its origin lies in Eurasia, where early forms, transitional from the Villafranchian genus *Leptobos*, have been found. *Bison* differs from the related *Bos* in such characters as the shorter and broader forehead, more tubular orbits, parietals visible in dorsal aspect, and less reduced posthorn region of the skull (see fig. 15.18). However, some workers (see Van Gelder, 1977) believe that *Bison* and *Bos* are conspecific. The taxonomy of the genus is still obscure; the foundation for a modern treatment was laid by Skinner and Kaisen (1947), and much of the following tentative synthesis is based also on work by Michael Wilson (1969, 1974a, 1975; also Frison, 1974).

The steppe bison, a big and large-horned form, was widespread in Eurasia from the middle Pleistocene to the last glaciation, when it was portrayed by Ice Age man in pictures showing a very massively built animal with a black mane on top of the neck and behind the shoulders, giving a characteristic double-humped appearance. The earliest European records are pre-Cromerian and thus more than 0.7 m.y. old. Parts of Alaska and the Yukon were evidently populated, continuously or intermittently since Illinoian times and perhaps earlier, by bison conspecific with the Eurasian form. This population was probably the source of at least two distinct migrations into North America south of the ice. The earlier migration probably dates back to the Yarmouthian interglacial, though the existence of even earlier migrations is by no means improbable. These animals (such as *B. priscus alaskensis* from Gold Run Creek) apparently equaled in size such enormous early Eurasian forms as *Bison priscus longicornis* and thus could have given rise to the American giant bison group. Later Eurasian *Bison priscus* were smaller and may have given rise to the living species *Bison bonasus* and *Bison bison*. A late Sangamonian/early Wisconsinan migration thus probably gave rise to *Bison bison*.

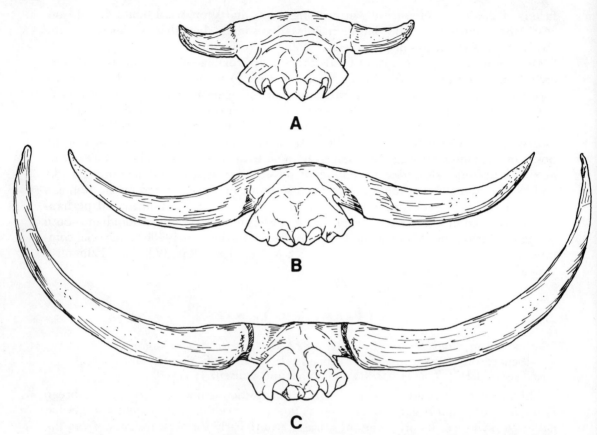

15.18 Bovidae. *Bison*, skulls with horn-cores, posterior view. *A, Bison bison*, Recent. *B, Bison priscus*, Rancholabrean, Fairbanks. *C, Bison latifrons*, Rancholabrean, near Hoxie, Kansas. After Skinner and Kaisen (1947). All ca. 0.10×.

Unfortunately, much of the bison material from Beringia is poorly dated. However, *priscus*-like *Bison* survived in the area through the late Wisconsinan. A desiccated and frozen mummy of a bison dated at 31,000 years B.P. has the skin and horn sheaths preserved; the latter have posteriad deflected tips as in Ice Age paintings of steppe bison in Europe. What may well be a terminal date for *Bison priscus crassicornis* is provided by the C-14 dates of 11,910 ± 180 and 12,460 ± 220 years B.P. from Locality 11(1) on Old Crow River. At this time, communication to North America south of the ice was being reestablished, and the local population may then have been swamped or replaced by *Bison bison* migrating from the south.

The morphological plasticity of the Alaskan bison is notable and may be explained as follows. During glaciations, the population was in contact with the Eurasian bison and affected by its evolutionary trends. During the Sangamonian interglacial, the population could well have been influenced by the American giant bison group if the specific separation was not complete. In addition, local adaptive trends undoubtedly occurred. (Flerov, 1975; Frison, 1974; Gromova, 1935; Guthrie, 1966a, 1972; Harington, personal communication, 1976; Harington and Clulow, 1973; Kurtén, 1968; Skinner and Kaisen, 1947; Stalker and Churcher, 1970; M. V. Wilson, 1969, 1974a, 1975.)

†Giant Bison, *Bison latifrons* (Harlan), 1825 (*Bos latifrons; Bison alleni* Marsh, 1877; *B. ferox* Marsh, 1877; *Bos crampianus* Cope, 1894; *B. arizonica* Blake, 1898; *Bison regius* Hay, 1913; *B. willistoni* Martin, 1924; *B. chaneyi* Cook, 1928; *B. angularis* Figgins, 1933; *B. rotundus* Figgins, 1933; *B. aguascalientes* Mooser and Dalquest, 1975)

Giant bison, *Bison latifrons*, are characterized by their very large size and extremely long horn-cores, which have a maximum span of 213 cm (compared to 66 cm in modern *B. bison*). The variation is considerable, sexual dimorphism is pronounced, and the picture is greatly complicated by the presence of shifting clines in space and time. Locally—in the Great Plains, for instance—this may take the form of a gradual reduction in size over time.

The giant bison presumably arose from early, large-horned *Bison priscus*. Whether the two should be regarded as distinct species is uncertain. The time of first appearance in North America south of the ice is also somewhat uncertain for although the species is definitely present in late Illinoian and Sangamonian times, finds from Medicine Hat and perhaps Nebraska may be earlier.

In the Sangamonian, the species was very widespread. It ranged to Mexico, Texas, Louisiana and Florida (Bradenton) in the south, California in the west, Idaho, Alberta, and Saskatchewan in the north, and South Carolina in the east. During the Wisconsinan, the giant bison evidently became rare and may finally have survived only as a relict in restricted areas. Mid-Wisconsinan records come from Texas (Moore Pit, Coppell), California (Newport Bay Mesa, San Pedro), and Alberta, and Rainbow Beach, Idaho, gives a terminal date of between 21,000 and 30,000 years B.P. (late Wisconsinan). A few specimens also occur in the Rancho La Brea fauna.

Bison latifrons thus survived into the late Wisconsinan. Extinction may possibly have been due to competition from *Bison bison* or, if the two forms were still able to interbreed, genetic swamping. VanderHoof (1942) has suggested that a possible reason for the adaptive inferiority of the giant bison might be the smaller size of the teeth relative to skull size (used as an index of body size), but Skinner and Kaisen (1947) have found that this ratio is dependent upon individual age and that the relationship is not found when large samples of various age classes are examined. Although skulls with attached horn-cores are distinctive, postcranial material is not, and great care must be taken in assigning skeletal parts to species. Recently, an associated skull and skeleton of *Bison latifrons* was found at American Falls (Soiset, personal communication, 1975). (Fuller and Bayrock, 1965; McDonald and Anderson, 1975; W. E. Miller, 1971; Schultz and Frankforter, 1946; Schultz and Hillerud, 1977; Schultz and Martin, 1970b; Skinner and Kaisen, 1947; Soiset, personal communication, 1975; Stalker and Churcher, 1970; VanderHoof, 1942; M. V. Wilson, 1969, 1974b, 1975.)

Bison, or American Buffalo, *Bison bison* (Linnaeus), 1758 (*Bos bison; Bison antiquus* Leidy, 1852; *B. californicus* Rhoads, 1897; *B. occidentalis* Lucas, 1898; *B. kansasensis* McClung, 1905; *B. pacificus* Hay, 1927; *B. figginsi* Hay and Cook, 1928; *B. taylori* Hay and Cook, 1928; *B. texanus* Hay and Cook, 1928; *B. oliverhayi* Figgins, 1933)

For a discussion of the subspecies of *Bison bison*, see Skinner and Kaisen (1947). The Eruopean wisent, *Bison bonasus* (Linnaeus), is sometimes considered conspecific with the American bison.

The *Bison bison* lineage is here considered to begin with a second bison immigration from Beringia in the Sangamonian interglacial and the partition of the bison population into a Beringian stock and a main continental population by the development of the Wisconsinan ice field. With the concomitant emergence of the Beringian land bridge, the Alaskan population was reunited with the Eurasian *Bison priscus* stock and remained so until bipartition of the ice sheet toward the end of the glaciation. It should be remembered, however, that speciation may well have been incomplete during this time. (Delimitation of "species" involves cutting through a continuum at the most convenient point in space and time.)

Early members of this lineage appear generally to have been large and are placed in the subspecies *Bison bison antiquus* (average tip to tip span of male horn-cores, 881 mm). Later evolution tended toward size reduction, and a north–south cline developed in the late Wisconsinan and Holocene, with the smaller *Bison bison occidentalis* (average male horn-core span, 747 mm) to the north and *Bison bison antiquus* to the south. The clines apparently showed some fluctuation, however, and bison in the *B. bison antiquus* size class were in existence in Wyoming (Casper) as late as 10,000 years B.P.; a herd of *B. bison occidentalis*–type bison in Alberta is only slightly younger, $9,670 \pm 60$ years B.P. (Harington, personal communication, 1976). The size reduction is also seen in Eurasian bison, e.g., in the European transition from *Bison priscus mediator* to *Bison bonasus* (Flerov, 1975).

Bison bison antiquus spread into the greater part of North America south of the Wisconsinan ice. There are records from Canada (Alberta, Manitoba), the United States (Arizona, California, Colorado, Florida, Idaho, Illinois, Indiana, Iowa, Kansas, Kentucky, Louisiana, Minnesota, Missouri, Nebraska, Nevada, New Mexico, Oklahoma, Oregon, Texas, Utah, Virginia, Washington, Wyoming), and Mexico. From Paleo-Indian times on, bison was hunted by man, and numerous kill sites, with remains of up to 200 individuals, have been discovered. Population analyses of such samples, in addition to revealing the age and sex structure of the herd, have disclosed clear-cut annual age groups, which indicates that entire herds were destroyed simultaneously.

With continued size decrease, the two present-day subspecies, *Bison bison bison* (the plains bison; average male horncore span, 581 mm) and *Bison bison athabascae* (the wood bison; average horn-core span, 665 mm) came into being. Intermediate stages are known, e.g., the Hawken bison (northeastern Wyoming), dated at $6,470 \pm 140$ years B.P. (this specimen has been referred to *Bison bison occidentalis*). Much of the story remains to be resolved, but material for a very detailed charting of the late Wisconsinan and Holocene evolution of the bison is available. (Butler, 1968; Flerov, 1975; Frison, 1974; Frison, Wilson, and Wilson, 1976; Guthrie, 1970; Harington, 1971c;, and personal communication, 1976; Hillerud, 1966; Robertson, 1974; Skinner and Kaisen, 1947; Wheat, 1972; M. V. Wilson, 1969, 1974a,b, 1975.)

† Flat-horned Ox, *Platycerabos dodsoni* (Barbour and Schultz), 1941 (*Parabos dodsoni*)

Oxen of the monotypic genus *Platycerabos* are large bovines. The skull, with its shortened posthorn region, resembles *Bos* rather than *Bison* in its general characters, but the shape and orientation of the horn-cores is different. The horn-cores are set very far apart, are slightly depressed near the base, curve outward and forward, are much flattened in cross section, and have a thick, rugose, burrlike ridge at the base. The flattening of the horn-cores resembles that of the water buffalo (*Bubalus*), but in that genus the horns curve backward. The skull is long, low, and broad, with a flat upper profile as in *Bos* and a remarkably broad, shallow occiput overhung by the broad occipital crest.

Platycerabos dodsoni is the sole known species, and the only find, a skull, was discovered in a ravine near Nehawka, Cass Co., Nebraska, where it had apparently eroded out of nearby Pleistocene clay deposits [the specimen was referred to the late Irvingtonian by Hibbard et al. (1965)]. The species probably arose from a Eurasian stock related to or ancestral to *Bos*. Harington (personal communication, 1976) points to similarities to the Asiatic genus *Urmiabos*, also known from a single, but unfortunately fragmentary, specimen from the late Miocene or Pliocene of Maragha in northern Iran. (Barbour and Schultz, 1941; Gromova, 1962; Harington, personal communication, 1976; Hibbard et al., 1965.)

Yak, *Bos grunniens* Linnaeus, 1766 (*Bos bunnelli* Frick, 1937)

The characters of the genus *Bos* have been referred to in the description of the preceding species; the genus may have arisen from *Leptobos*-like ancestors. The subgenus *Poëphagus*, comprising the yaks, differs from the nominate subgenus in its shorter nasals and less overhung occiput and in the shape of the horn-cores, which are less inclined forward and turn inward at the apex. It is known from the Pleistocene of Asia, and a number of finds from the Rancholabrean of Fairbanks show that it was present in Beringia.

The yak is an alpine animal, found in the wild state in the Tibetan highlands. In spite of its large size (shoulder height upwards of 2 m), it is an excellent climber and makes seasonal migrations between its summer pastures in the mountains and its wintering areas on the high plains. The wild form has been extirpated in many areas, but the domesticated yak ranges into Mongolia and China. Wild yaks are black; the smaller and shorter-horned domestic yaks have colors ranging from white to reddish and brown. (Frick, 1937; Gromova, 1962.)

CHAPTER SIXTEEN

ORDER SIRENIA

Sirenians, believed to be the animals that gave rise to the legend of mermaids, are represented by two families, the Dugongidae and the Trichechidae. The order first appears in early Eocene deposits in Hungary and spread rapidly; early sirenians have also been reported from southern Europe, the southeastern United States, Jamaica, and Java. A Tethyan origin is indicated, and sirenians are probably an offshoot from the same stock as proboscideans. Massive bone structure, a downturned snout, and reduced dentition are characteristic of these aquatic, herbivorous mammals. Florida is the only area in the world showing evidence of continuous sirenian habitation from the Eocene to the present.

FAMILY DUGONGIDAE—DUGONGS AND SEA COWS

Dugongids have a fairly good fossil record extending from the Eocene to the Recent. The family was in its heyday in the Miocene and has been on the decline since. The existing populations are nearing extinction in many areas. The living *Dugong* is restricted to warm marine waters of the Red Sea, coastal East Africa, and the shoals of the Indian Ocean and southwestern Pacific. The extinct *Hydrodamalis* was the only sirenian adapted to live in cold waters, and it is recorded from the Pliocene of North America and Japan and the Pleistocene of North America.

Dugongs differ externally from manatees in having a deeply notched tail fin, no nails on the flippers, fewer teeth, and (with the exception of *Hydrodamalis*) a more downturned snout.

†Steller's Sea Cow, *Hydrodamalis gigas* (Zimmermann), 1780 (*Rhytina*)

The genus *Hydrodamalis* comprises dugonglike animals that inhabited shallow waters of the North Pacific Ocean from the late Pliocene until the middle of the eighteenth century, when the relict population of *Hydrodamalis gigas* became extinct.

Steller's sea cow, the largest of the Recent sirenians and the only species adapted to live in cold waters, is known from two Pleistocene localities. A badly eroded skull fragment brought up in a trawl from Monterey Bay, California, has been radiocarbon dated at $18,940 \pm 1,110$ years B.P. (R. E. Jones, 1967). Recently, a partial skeleton was found in an interglacial beach deposit on Amchitka Island, Alaska; a uranium series date of $135,000 \pm 12,000$ years B.P. has been obtained on the bone (Gard, Lewis, and Whitmore, 1972). This specimen is larger than those found on Bering Island in historic times.

Hydrodamalis gigas was discovered in 1741 in the Komandorskiye Islands by the Bering Expedition. From 1743 to 1763, the animal was relentlessly slaughtered by meat-hungry Russian crews who hunted sea otters and fur seals for 8 or 9 months of the year. By 1768, Steller's sea cow was extinct.

Observations made on the living animal by the expedition's naturalist, Georg Steller, reveal that it was about 7.5 m long and weighed approximately 10 tons (the Bering Island population was smaller in size, probably averaging 5–6 tons; Domning, personal

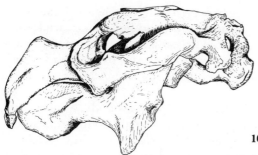

16.1 Dugongidae. *Hydrodamalis gigas*, Holocene (subfossil), Bering Sea, Alaska, skull. 0.10×.

communication, 1975.) A relatively small head, small, hooklike forelimbs lacking phalanges, no trace of hind limbs, very dense ribs, and a horizontally flattened tail characterized the animal (see fig. 16.1). It was covered by rough, barklike skin, beneath which was a thick layer of blubber. Like other sirenians, Steller's sea cow had horny plates covering the anterior part of the palate and mandible; however, it entirely lacked teeth. The anterior part of the rostrum was somewhat flattened and served as a grasping device to tear off the kelp and other marine algae on which it fed.

The ancestry of *Hydrodamalis gigas* can be traced back to the Miocene species *Metaxytherium jordani* Kellogg. The genus *Hydrodamalis* makes its appearance in late Pliocene deposits in California, with an un-named species that is intermediate between *Metaxytherium jordani* and *Hydrodamalis gigas*. These specimens are larger and more primitive in osteological characters than those of *H. gigas* from Bering Island. This evolutionary line is characterized by increased body size (absolute and also relative to head size) and loss of teeth and phalanges. *Hydrodamalis* had a wider range during the Pliocene and Pleistocene than in historic times, and its remains have been found throughout the North Pacific region. These large, easy-to-kill mammals may have provided food for the Paleo-Indians coming across Beringia (Domning, 1970). By the eighteenth century, Steller's sea cow existed only as a relict population in the suboptimum environment of the Bering Sea. (Domning, 1970 and personal communication, 1975; Gard, Lewis, and Whitmore, 1972; R. E. Jones, 1967; Walker, 1975; Whitmore, personal communication.)

Family Trichechidae—Manatees

Almost nothing is known about the evolution of manatees. The earliest records are from Miocene deposits in South America, and the animals are present in Florida by the late Blancan. Confined to warm tropical waters on both sides of the Atlantic, the range does not overlap with that of the dugong. Manatees are sensitive to temperature changes, and cold spells take many lives.

Manatee, *Trichechus manatus* Linnaeus, 1758

Trichechus is the only genus in the family and contains 3 extant species.

Remains of the manatee are abundant in Pleistocene deposits in Florida and have been found at Santa Fe River IB (late Blancan), Punta Gorda, Aucilla River, Chipola River, Rock Springs, Santa Fe River IIA, Seminole Field, Waccasassa River, and Withlacoochee River. The northernmost occurrence is from Fort Fisher, New Hanover Co., North Carolina, a deposit dated about 125,000 years (Funderberg, 1960.) The most common manatee fossils are the dense, banana-shaped ribs and the heart-shaped centra of vertebrae; occasionally, jaw fragments, isolated teeth, and limb bones are found. *Trichechus manatus* inhabits warm, shallow, often brackish coastal waters from the southeastern United States west to Texas from Vera Cruz southeast to northern South America, and in the West Indies; other species occur along the Atlantic coasts of northern South America and West Africa.

Manatees are characterized by a fusiform-shaped, nearly hairless body, a small head with a squarish snout and deeply cleft upper lip, and an evenly rounded, paddle-shaped tail. Unlike most other mammals, manatees have only six cervical vertebrae. The anterior appendages are modified into paddles; the posterior limbs are vestigial or absent. Manatees reach a maximum length of 4.5 m and weigh up to 600 kg. The cheek teeth are low crowned, with two simple transverse crests and a low cingulum at each end. Tooth replacement is by means of a horizontal forward movement similar to that occurring in elephants.

The Florida manatee probably came from South America, invading the Caribbean area in the late Pliocene. The arrival of *Trichechus manatus* in Florida coincided with the abrupt disappearance of the Dugongidae, probably as a result of direct competition. Today, manatee populations are greatly reduced because of persistent hunting and habitat destruction. (Funderberg, 1960; Reinhart, 1971; Walker, 1975.)

CHAPTER SEVENTEEN

ORDER PROBOSCIDEA

Proboscideans—the mastodonts, stegodonts, gomphotheres, and elephants—were a dominant group in the Tertiary and Pleistocene. The order was found in all continents except Australia and Antarctica, in habitats ranging from the Arctic tundra to tropical jungles. Only two species found in the Old World tropics, survive today.

The fossil history of this group is excellent and shows the gradual evolution of the different species, which are not sharply delineated. Proboscideans arose in Africa, probably in the late Eocene; common ancestry with sirenians and hyraxes is indicated. Three groups, the Moeritherioidea, Deinotherioidea, and Barytherioidea, were formerly considered to be proboscideans but are now excluded from the order; their exact taxonomic status is unknown. Proboscidean systematics has been characterized by over-splitting, which has obscured the true relationships of the taxa. Recent studies (Maglio, 1973) have clarified the picture. Two superfamilies and four families are now recognized (as compared to five superfamilies and seven families in Osborn, 1936.) The Gomphotherioidea were the basal stock and gave rise to the main line of proboscideans, the Gomphotheriidae, from which the Elephantidae arose, and the side branch Mammutoidea, from which the Mammutidae and Stegodontidae arose.

Graviportal in build, proboscideans have columnar legs with a long humerus and femur, short lower limb segments, and broad, five-toed, semiplantigrade feet. The large skull is highly pneumatic. From the bony nasal orifice located between the orbits projects the long, flexible trunk which is used in feeding and drinking, and as an organ of touch. The second upper incisors (and in some groups, the second lower incisors) are greatly enlarged to form tusks. The molars have either rounded cusps separated by valleys or ridgelike plates. Trends in proboscidean evolution include increased size, lengthening of limbs and trunk, shortening of the neck, elongation of the mandible (followed by a secondary shortening), and increased dental specializations.

Bones of these huge animals may have been the basis of the Greek and Roman legends about giants, and the study of proboscideans gave rise, much later, to the discipline of vertebrate paleontology. Mastodonts and gomphotheres were spectacular members of the North American Tertiary fauna; in the Irvingtonian, mammoths reached the New World, replaced most of the gomphotheres, and, together with the mastodonts, became the dominant herbivores until their extinction in early Holocene times.

FAMILY MAMMUTIDAE—MASTODONTS

The name mastodont, a term now restricted to members of the family Mammutidae, has been used indiscriminately for both mastodonts and gomphotheres. The North African Oligocene genus *Palaeomastodon* Andrews was probably ancestral to the Mammutidae. *Mammut,* the sole genus, ranged from the early Miocene to the early Pleistocene in Africa, Asia, and Europe and from the middle Miocene to the early Holocene in North America. A Eurasian branch probably gave rise to the Stegodontidae, a specialized group that evolved parallel to the true elephants but were not closely related to them. Throughout their long history, the mastodonts were a less prominent group than the gomphotheres.

Mastodont molars show a consistently simple pattern, with low cusps forming ridges without cement separated by open valleys. The upper tusks are well developed, and the lower tusks are small or absent. The mandible is short. *Mammut americanum* was a widespread, dominant member of American Blancan and Pleistocene faunas.

† American Mastodont, *Mammut americanum* (Kerr), 1791 (*Elephas americanus; Mammut* Blumenbach, 1799; *Mastodon giganteus* Cuvier, 1817; *Mammut progenium* Hay, 1914)

The genus *Mammut* is discussed above. The American mastodont (fig. 17.1) is one of the best-known Pleistocene mammals, and its remains have been found throughout the country. The earliest records are Blancan (Grand View, Hagerman, Santa Fe River IB, White Bluffs); Irvingtonian records include Cumberland, Delight, Inglis IA, Mullen II, and Port Kennedy. During the Rancholabrean, the American mastodont ranged from Alaska to Florida but was most numerous in the eastern forests. Dreimanis (1968) reports more than 600 late Wisconsinan occurrences in the glaciated and periglacial regions of eastern North America. These finds are clustered in two main areas, south of the Great Lakes and along the Atlantic Coast. Large numbers of mastodont molars (along with lesser numbers of mammoth teeth) have been recovered by fishermen from some 40 sites on the Continental Shelf off the northeastern coast. Specimens have been found as far as 300 km from the present shore line. Presumably, the animals lived on the conifer-covered shelf during times of glacially lowered sea levels during the last 25,000 years.

The preferred habitat was apparently open spruce woodlands and spruce forests, where they browsed on the sylvan vegetation. Mastodonts were not confined to spruce forests, however: they inhabited Florida, Texas, and the Great Plains, where they probably lived in valleys, lowlands, and swamps. Analysis of undigested plant remains found

17.1 Mammutidae and Elephantidae. The dominant proboscideans during the Pleistocene were the mastodonts and mammoths. Shown are *Mammut americanum* and *Mammuthus columbi*.

in the ribcages of some specimens has revealed twigs and cones of conifers, leaves, coarse grasses, swamp plants, and mosses.

Although Cuvier's descriptive term "mastodon" has been used frequently as a generic name, Blumenbach's name *Mammut* has priority and is the accepted scientific name. It is unfortunate that the two generic names *Mammut* and *Mammuthus* are so similar. The word *Mammut,* meaning "earth burrower," can be traced back to the Middle Ages, when eastern European farmers found the gigantic bones in their fields and believed that they belonged to monstrous burrowing beasts.

The name mastodont means "nipple tooth," and this aptly describes the large, low cusps arranged opposite one another to form low ridges separated by open valleys. Unlike the condition in gomphotheres, no intermediate conules or trefoil patterns are developed. During the life of the animal, six teeth, three deciduous premolars and three molars, are developed in each half of the jaw. The teeth increase progressively in size from the very small first premolar to the large third molar. Formed in the back part of the jaw, each tooth moved forward and downward, in the upper jaw, or forward and upward, in the lower jaw, to replace worn teeth. Three teeth are present in each half of the jaw in young animals; the first and second molars were in place during most of the animal's life, but only the third molar remained in old age.

Tusks were present in the upper jaws of both sexes, though those of males are larger and heavier. In addition, males had small, vestigial lower tusks, but these were usually lost by maturity. The upper tusks, extending 2 to 3 m from the maxilla, projected more or less horizontally from the skull, then curved outward and finally inward. Lacking enamel, the tusks are circular in cross section and show traces of annular growth rings. Mastodonts probably used their tusks to pry off and break branches into mouth-sized pieces. When both tusks are preserved, one is usually shorter than the other, which indicates that one tusk was used exclusively and that the animals were either right- or left-tusked.

The skull differs from that of a mammoth in its larger size, flattened brow, fewer sinuses, and absence of high occipital crests. The mandible is elongate, with a rounded angle. Like mammoths and modern elephants, mastodonts had a well-developed trunk. The head was carried horizontally.

A number of complete skeletons have been recovered. Of these, the "Warren mastodont" found near Newburgh, New York, in 1845 and described by J. C. Warren in 1852 is perhaps the best known. Mounted, it stands 2.7 at the shoulder. Mastodonts were more heavily built than mammoths, and their limb bones are noticeably shorter and thicker (see Olsen, 1972, for a detailed comparison of the skeletons of the two animals.) Compared to the extant Asiatic elephant, mastodonts had a deeper chest, broader pelvis, shorter legs, and a longer back. Height at the shoulder was between 2.7 and 3.0 m, and the total length of the body was about 4.5 m; females were smaller (a mounted skeleton at the University of Michigan measures 2.3 m at the shoulder.) Mastodonts were apparently covered with coarse, brownish hair about 2.75 cm long; they probably lacked the woolly undercoat characteristic of mammoths.

Unlike the mammoths, mastodonts did not undergo rapid evolution, and the single American species, *Mammut americanum*, was morphologically homogeneous in both time and space. The Villafranchian species *Mammut borsoni* (Hays) of Europe was closely related to *Mammut americanum*. The American mastodont probably survived until early Holocene times—most of the accepted terminal dates lie between 9,000 and 12,000 years B.P. Dates of 6,000–8,000 years B.P. are now discounted; the dated sample may have been contaminated by humic acids or other intrusive carbon. As yet, man and mastodont have not been found in clear association at any site. Increasing dryness, with resulting changes in vegetation, may have been a major factor in the extinction of the mastodont; relict populations would then have been subject to other environmental stresses, including hunting by man. (Dreimanis, 1968; Hay, 1914; Jepsen, 1960; Olsen, 1972; Osborn, 1936; Skeels, 1962; Warren, 1852; Whitmore et al., 1967.)

Family Gomphotheriidae—Gomphotheres

Of the four families of proboscideans, the Gomphotheriidae—the pig-toothed beasts—show the greatest diversity and include such specialized groups as the shovel-toothed amebelodonts, the beak-tusked rhynchotheres, and the short-jawed anancines (whose cranial specializations parallel those of the true elephants). The main lineage, the bunodont gomphotheres (called bunomastodonts by some authors), can be traced from *Phiomia* (from the Oligocene of North Africa) to *Gomphotherium* (a widespread Miocene genus), *Tetralophodon* (from the Old World and North America), and *Stegomastodon* (from the late Tertiary and early Pleistocene of North and South America). A group of primitive, poorly known African gomphotheres gave rise to the highly successful Elephantidae, a family which subsequently replaced most of the gomphotheres.

Although many diverse lines of gomphotheres developed, the family has remarkably uniform dentition. Compared to the simple molars of the mastodonts, gomphothere molars are complex, with additional rounded cusps and accessory conules that wear to a complicated trefoil pattern. Tusks were usually present in both jaws. Gomphotheres had a longer body and head and shorter limbs than the true elephants.

Much work needs to be done on the taxonomy and relationships of this widespread and diverse family of proboscideans.

†Precursor Rhynchothere, *Rhynchotherium praecursor* (Cope), 1893 (*Dibelodon praecursor; Tetrabelodon shepardii* Cope, 1893; *Rhynchotherium falconeri* Osborn, 1923; *Serridentinus praecursor* Osborn, 1935; *Serebelodon praecursor* Osborn, 1936)

Rhynchotheres, genus *Rhynchotherium,* are a poorly known, taxonomically oversplit group of gomphotheres that ranged from Africa to North America in the late Tertiary. Characteristic of rhynchotheres are the compressed, enamel-banded lower tusks, a short, sharply downturned symphysis, and a stout ramus. The tooth pattern is relatively simple; the teeth have thick enamel and lack secondary trefoils and cement.

Rhynchotherlum praecursor was described from Blanco, but the two specimens, a mandible with M_{2-3} and an isolated M_3, have been placed in separate genera and species by various workers. However, Dalquest (1975) noted only minor differences between the original specimens and so referred them to *Rhynchotherium praecursor* and synonymized the other names. Two other specimens from Blanco have been referred to this species. *Rhynchotherium* sp. has been reported from Deer Park, Santa Fe River IB, and an early Blancan site in Graham County, Arizona. Relationships between the various rhynchotheres and other gomphotheres are unclear, and much work remains to be done on the group. (Dalquest, 1975; Tobien, 1973.)

†Wonderful Stegomastodont, *Stegomastodon mirificus* (Leidy), 1858 (*Mastodon mirificus; M. successor* Cope, 1892; *Stegomastodon texanus* Osborn, 1924;

S. *arizonae* Gidley, 1926; *S. primitivus* Osborn, 1936; *S. rexroadensis* Woodburne, 1961; *Anancus bensonensis* Gidley, 1926)

The stratigraphic range of the stegomastodonts, a group of New World gomphotheres, extends from the Blancan to the early Irvingtonian in North America. The animals are distinguished by a shortened skull and jaw, large uniformly curved upper tusks lacking enamel (at least in the adult stage), no lower tusks, and molariform teeth wearing to an at least double trefoil pattern. Several species have been described, but since the differences among them are small (some descriptions were based on growth stages) and the range of variation in stegomastodonts is unknown, we believe that these "species" probably represent extremes in the cline of the complex-toothed stegomastodonts and that only one species, *Stegomastodon mirificus*, ranged across the western plains. A thorough revision of the group is needed.

The earliest-named stegomastodont was *Stegomastodon mirificus*, the type of which was picked up along the Loup Fork of the Platte River in Nebraska by the F. V. Hayden survey party. It has also been reported from Blanco, Cita Canyon, Deer Park, Rexroad, Seger, Broadwater, Sand Draw (largest sample), Grand View, Benson, Curtis Ranch, Gilliland, Holloman, and Mullen I.

Stegomastodon mirificus was shorter and stockier than the extant Asiatic elephant and stood about 2.4 m at the shoulder (specimens from Sand Draw and Curtis Ranch are

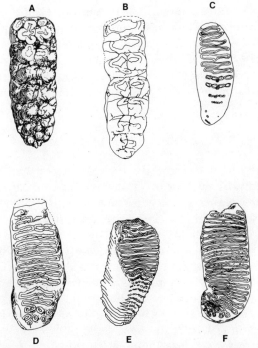

17.2 Proboscidean left upper molars. *A*, Gomphotheriidae, *Stegomastodon mirificus*, Blancan, Cita Canyon, M^3. *B*, Mammutidae, *Mammut americanum*, Rancholabrean, Fulton, Indiana, M^{2-3}. *C–F*, Elephantidae, M^3. *C*, *Mammuthus meridionalis*, early Irvingtonian, diagrammatic. *D*, *Mammuthus columbi*, late Irvingtonian to early Rancholabrean, Victoria, Victoria Co., Texas. *E*, *Mammuthus jeffersonii*, Rancholabrean, Ashland, Cass Co., Illinois. *F*, *Mammuthus primigenius*, Rancholabrean, Alaska (posterior end pathological). *A*, after Savage (1955); *B*, after Osborn (1936); *C*, after Maglio (1972); *D–F*, after Osborn (1942). All 0.12×.

slightly larger). It has a short skull, sharply downturned rostrum, and spoutlike symphysis. The last molars are the largest and most complex: M^3 has 5½ to 6½ lophs and M_3 has 7 lophids (see fig. 17.2). M^2 and M^3 often show extreme plication of the enamel (ptychodonty), a condition Osborn (1936) thought to be primitive.

Early Blancan specimens from Rexroad and Benson (called *S. rexroadensis* Woodburne and *Anancus bensonensis* Gidley, respectively) are slightly smaller than specimens from later deposits; the teeth show simple primitive trefoiling, and the enamel is not plicated. These early representatives resemble the South American genus *Haplomastodon* Hoffstetter. *Mammuthus* probably displaced *Stegomastodon* when it took over the latter's chief habitat, the grasslands; *Stegomastodon* became extinct in North America in the Irvingtonian. In South America, however, where neither *Mammuthus* nor *Mammut* penetrated, *Stegomastodon* survived until the late Pleistocene and was apparently associated with Early Man at Muaco, Venezuela. (Gidley, 1926; Martin and Guilday, 1967; Meade, 1945; Osborn, 1936; Savage, 1955; Webb, personal communication, 1977; Woodburne, 1961.)

† Cuvier's Gomphothere, *Cuvieronius* sp. Osborn, 1923

A side branch of the bunodont gomphothere line was *Cuvieronius*, a widespread Neotropical genus that ranged stratigraphically from the late Pliocene to the Rancholabrean in North and South America. A distinctive character is the spiral enamel band on the tusks. The molars have secondary trefoils, no cement, and little or no obliquity to the lophs. *Cuvieronius* has been identified at Benson, California Wash, and Tusker (Arizona), Crystal River Power Plant in Citrus County and Punta Gorda (Florida), near Las Cruces (Dona Ana Co., New Mexico), and Gilliland, Ingleside, and Sinton (Texas), but none of the material has been identified to species. Like *Mammut*, it was probably a browser. In tropical America, remains of *Cuvieronius* have been found in association with human artifacts. (Lundelius, 1972a; Martin and Guilday, 1967.)

Family Elephantidae—Elephants

The Elephantidae arose in Africa in the early Miocene. Rapid diversification of the African stock resulted in the establishment of three main lineages, of which the genera *Loxodonta*, *Elephas*, and *Mammuthus* were the most successful. These genera differ from each other in cranial and dental characteristics. They still inhabited Africa in the Pliocene, but by the end of the epoch, some members of the *Elephas* line had spread into Eurasia, where they became a dominant group. *Mammuthus* followed *Elephas*, entering Eurasia in the early Pleistocene and becoming an essentially Eurasian and Nearctic genus from middle Pleistocene times on. *Loxodonta* remained in Africa, where one species survives today. Classifications of the Elephantidae have been characterized by a proliferation of genera and species which has obscured relationships. We are following Maglio's excellent revision (1973) of the family.

The earliest representatives of the family were the Stegotetrabelodontinae, a widespread African group (early Miocene–early Pliocene) that was intermediate between the ancestral Gomphotheriidae and the more progressive subfamily Elephantinae. *Primelephas gomphotheriodes* Maglio, the earliest-known member of the Elephantinae, was probably ancestral to all later elephants.

Characteristic of the Elephantidae are large size, massive, columnar limbs, an elongated, flexible trunk, and well-developed upper tusks. The molars consist of a series of vertically oriented flattened plates, lying one behind another, which wear on edge to give a washboard of enamel ridges and of softer dentine and cementum hollows. Each plate is composed of a shell of enamel filled internally with dentine; cementum initially covers the crowns of the plates, fills the spaces between the plates, and covers the inner and outer surfaces to provide integrity to the molar teeth. The three permanent premolars were lost early in the history of the family, and the remaining three deciduous premolars became strongly molarized. During the life of an elephant, six cheek teeth develop in sequence in each half of the jaw. Each tooth is larger (with M3 the largest) and has a greater number of plates than the one preceding it. Formed plate by plate, the anterior part of the tooth is worn before the posterior part of the tooth is formed. Thus an upper molar erupts at an oblique angle from the alveolus, about 45° for M^2 or M^3. Lower molars erupt at shallower angles, with M_3 often erupting spirally from a distolingual position. When the anterior plates are worn down at the grinding surface, the root is resorbed, and the useless shell of worn dentine and enamel breaks off and is either swallowed or spit out. When the last plates of the last molar are worn down, the animal can no longer masticate its food and dies (at an age of about 70 years in modern elephants).

Phyletic trends in the molars of the Elephantidae include (a) nearly continuous increase in the number of plates on each tooth, (b) a decrease in the relative spacing of plates per unit length and a decrease in the mesiodistal diameter of the plates, which results in a greater lamellar frequency in later species, (c) a trend toward progressive thinning of the enamel forming the tooth plates (except in the dwarf species), and (d) an increase in the unworn crown height of the molar compared to its width. Elephant dentition is specialized for mesiodistal horizontal shearing, with food being cut between the enamel ridges. Other phyletic trends include reduction and loss of the lower tusks, shortening and deepening of the skull and jaw, and backward shift of the center of gravity of the skull and mandible. Elephants are gregarious and have a highly developed social organization.

Of the three lineages, the exclusively African (Middle Pliocene–Recent) *Loxodonta* is the most conservative in dental trends and rates of change. Reports of the *Loxodonta* stock in Europe were based on tooth characters; when cranial as well as dental morphology is considered, however, these specimens belong to the *Elephas* lineage (Maglio, 1973). The extant African elephant, *Loxodonta africana* (Blumenbach), found in parts of sub-Saharan Africa, is the largest living land mammal, with a shoulder height of 3–4 m and a weight of 5.0–7.5 tons.

Elephas, the most diverse genus in the family, originated in Africa in the early Pliocene. Two major lines developed. One survived in Africa and Europe until the late Pleistocene; the other migrated to southern Asia in the early Pleistocene, was able to compete successfully with the surviving gomphotheres and stegodonts, and gave rise to the living Asiatic elephant, *Elephas maximus* Linnaeus. Smaller than *Loxodonta africana,* the Asiatic elephant has a shoulder height of 2.5–3.0 m and weighs about 2.5 tons.

The third great lineage of the Elephantidae is represented by *Mammuthus*. With a stratigraphic range extending from the early Pliocene to the early Pleistocene in Africa, from the early Pleistocene to the early Holocene in Europe and Asia, and from the early Irvingtonian to the early Holocene in North America, *Mammuthus* had a wider geographic range than either *Loxodonta* or *Elephas*. Common ancestry with *Elephas* is indicated, and the teeth of the early species of both genera are similar. Mammoths have a very high, foreshortened skull, with dorsally expanded parietals, and the frontoparietal surface is concave vertically and convex transversely. Molar width decreased, the number of plates increased (from 8 to >30 on M3), the enamel became thinner, and crown height increased over evolutionary time (see fig. 17.2). The tusks are long, strongly curved, and spirally twisted.

In the middle Pleistocene, intense selection pressure resulting from increased competition with *Elephas* and vegetational changes due to a deteriorating climate caused evolutionary changes in the *Mammuthus* lineage that resulted in progressive adaptations in the teeth and skull.

Mammuthus meridionalis entered North America in the early Irvingtonian, and the genus is regarded as an index fossil for the Irvingtonian Land Mammal Age. In the Rancholabrean, a second migration across Beringia introduced *Mammuthus primigenius* into northern North America. Parallel evolutionary trends took place on both continents from ancestral *Mammuthus meridionalis,* and the descendant forms, *Mammuthus columbi* in North America and *Mammuthus armeniacus* (Falconer) in Europe, show many similarities.

Although diversity within *Mammuthus* is not great, many genera and species have been described. Osborn (1942), for example, recognized 16 specific taxa of mammoths in North America. Three of his "genera"— *Archidiskodon* (including the primitive mammoths *A. subplanifrons, A. meridionalis,* and *A. imperator*), *Parelephas* (the intermediate species *P. columbi* and *P. armeniacus*), and *Mammonteous* (the most progressive species, *M. primigenius*)—are names often seen in the literature. All mammoths are now included in the genus *Mammuthus.* We provisionally recognize four North American species or stages—*Mammuthus meridionalis, Mammuthus columbi, Mammuthus jeffersonii,* and *Mammuthus primigenius*—and note that these taxa represent a series of more or less successional populations. In fact, there is an almost continuous transition from the earliest known mammoth, *Mammuthus subplanifrons* (Osborn), of the early Pliocene of eastern and southern Africa, to the most advanced species, *Mammuthus primigenius,* of the last glaciation.

Identification of isolated and fragmentary mammoth teeth is difficult at best, and usually only the last molar is reliable for specific determination. A system using a combination of criteria (see Aguirre, 1969; Churcher, 1972; Whitmore et al., 1967) has been used with some success, although instances of overlapping measurements are frequent. (Aquirre, 1969; Churcher, 1972; Maglio, 1972, 1973; Osborn, 1942; Sikes, 1971; Whitmore et al., 1967.)

†Southern Mammoth, *Mammuthus meridionalis* (Nesti), 1825 (*Elephas meridionalis; Archidiskodon hayi* Barbour, 1915; *A. scotti* Barbour, 1925; *A. haroldcooki* Hay, 1928; *A. meridionalis nebrascensis* Osborn, 1932)

Mammoths arrived in North America sometime after the Oldavai event, probably about 1.5 m.y. ago (Lindsay, Johnson, and Opdyke, 1975), and are present in Irvingtonian faunas, including Crete (Saline Co., Nebr.), Angus, Cudahy, Holloman, Punta Gorda, Wellsch Valley, and Medicine Hat (Units D–E, Kansan). A specimen from Bruneau, Idaho, has been dated at 1.36 m.y. B.P. (Malde and Powers, 1962). These early mammoths have been described under several names, but it is now generally agreed that there was one species, *Mammuthus meridionalis.* This species inhabited the plains and steppe woodlands of Eurasia and North America during the early and middle Irvingtonian and was ancestral to the later mammoths, *Mammuthus columbi* and *Mammuthus armeniacus,* found in North America and Europe, respectively.

Mammuthus meridionalis had relatively broad molars with thick, widely spaced plates

(11–14 plates on M3), a lamellar frequency of 3.5–7.7, relatively thick enamel measuring 2.4–4.0 mm, and a crown height generally 10 to 60 percent greater than the width of the tooth. A conspicuous, lozenge-shaped median expansion (loxodont sinus) is evident on each ridge crest on worn molars. The massive tusks are straighter than those of later mammoths. Skull characteristics include dorsally expanded parietals and occipitals, a frontoparietal surface strongly concave anteroposteriorly, and large, slightly downturned external nares. The mandible has a well-developed anterior symphyseal process (the "chin"). Shoulder height was about 3.7 m. (Lindsay, Johnson, and Opdyke, 1975; Maglio, 1973; Malde and Powers, 1962.)

† Columbian Mammoth, *Mammuthus columbi* (Falconer), 1857 (*Elephas columbi;*
  E. imperator Leidy, 1858; *E. maibeni* Barbour, 1925; *Archidiskodon
  sonoriensis* Osborn, 1929)

The next stage in the mammoth lineage in North America is represented by a large proboscidean that inhabited North and Central America from the middle Irvingtonian to the middle Rancholabrean (fig. 17.1). It has been identified from faunas in Alaska, Alberta, Colorado, Florida, Idaho, Kansas, Mexico, Nebraska, Oklahoma, Texas, and Yukon Territory. Specimens from other states and provinces have been referred to this species, but many are much younger in age and should probably be referred to *Mammuthus jeffersonii*.

Much confusion exists over the taxonomy of this form. In 1857, Falconer named a new species, *Elephas columbi*, from a fragmentary tooth (M^3) found in Brunswick Canal, Darien, Georgia. A year later, Leidy described a fragmentary, poorly preserved M^3 from the Loup Fork, Niobrara River, Nebraska, as *Elephas imperator*. Unfortunately, the holotypes are inadequate for specific diagnosis. Osborn (1922) attempted to rectify the situation by designating neotypes for both species; the neotype of *Mammuthus imperator* compares favorably with the holotype, but the neotypes for *Mammuthus columbi* are close if not identical with that of *Mammuthus imperator*. Several authors (Falconer, 1863; Osborn, 1922; Maglio, 1973 and personal communication, 1972) have considered the holotypes of *Mammuthus columbi* and *Mammuthus imperator* to be indistinguishable. Both names have been used almost interchangeably for both the intermediate mammoth and the more progressive late Rancholabrean form. Osborn (1922) suggested calling the more advanced species *Mammuthus jeffersonii* and restricting *Mammuthus columbi* to the late middle Pleistocene mammoth (unfortunately, he did not adhere to this view in some of his later papers). We concur with this provisional solution and feel that, until extensive comparative studies are carried out, it is preferable to refer mammoth finds, especially isolated teeth and limb bones, to *Mammuthus* sp. As previously noted, species delineation in mammoths is not sharp. In *Mammuthus columbi*, for instance, late specimens grade morphologically into the more progressive late-Pleistocene species. A possible solution is to recognize substages of *Mammuthus columbi* based on the increasing number of plates on the molars.

Larger than *Mammuthus meridionalis*, the Columbian mammoth stood between 3.6 and 4.0 m at the shoulder; a specimen found in Lincoln County, Nebraska (named *Elephas maibeni* Barbour), had an estimated shoulder height of 3.97 m, which makes it one of the largest known American mammoths. The cranium is high and broad, with a deep, short rostrum. The massive mandible has a rounded symphysis, and the upper border of the coronoid process extends above the grinding surface of the molars. The Columbian mammoth has relatively short, broad molars. The molar plates are broad, with thick (2.5–3.2 mm) enamel; progressive stages show an increase in plate numbers, with 16–18 plates (5–8 plates/100 mm) on M3. The tusks are large, divergent, and greatly curved. (Barbour, 1925; Falconer, 1863; Maglio, 1973; Osborn, 1922, 1942.)

†Jefferson's Mammoth, *Mammuthus jeffersonii* (Osborn), 1922 (*Elephas jeffersonii; E. roosevelti* Hay, 1922; *E. washingtonii* Osborn, 1923; *E. eellsi* Hay, 1926; *E. exilis* Stock and Furlong, 1928; *E. floridanus* Osborn, 1929; *Parelephas progressus* Osborn, 1924)

The culmination of the *Mammuthus meridionalis–Mammuthus columbi* line was the progressive Wisconsinan species *Mammuthus jeffersonii*. It has been erroneously called *Mammuthus columbi* since most workers overlooked Osborn's (1942, p. 1084) name change: "The American elephant heretofore widely known as *Elephas columbi*, the Columbian Mammoth, will hereafter be known as *Elephas jeffersonii*, the Jefferson Mammoth." It has also been called *Mammuthus imperator*, and advanced stages have been frequently confused with the northern species, *Mammuthus primigenius*.

Remains of Jefferson's mammoth have been discovered at hundreds of sites in northern and western Canada, the western, central, and southern United States, and Mexico. At several sites, the species has been found in direct association with man, who successfully hunted it about 12,000 years ago. Apparently, this widespread, abundant species inhabited open prairies, where, like the other mammoths, it fed primarily on grasses. A relatively small (maximum shoulder height, 2.4–2.7 m) mammoth, *Elephas exilis* Stock and Furlong, has been found on Santa Rosa and San Miguel islands off the coast of southern California; its reduced stature is attributed to insular isolation, and it is now considered a subspecies (*M. jeffersonii exilis*).

Mammuthus jeffersonii stood between 3.2 and 3.4 m at the shoulder, had large lyrate, or incurved, tusks, and was probably covered with a thin coat of hair that in winter was supplemented by a fine, woolly undercoat. It has a long, broad cranium, a concave forehead, a prominent orbital region, and a shallow lower jaw with a slightly downturned beak.

Major evolutionary trends leading from *Mammuthus columbi* to *Mammuthus jeffersonii* are an increase in molar plates and in lamellar frequency (reflecting closer packing of the plates) with a concomitant thinning of the enamel. This is clearly shown in the M3: early representatives have 20–24 plates, a lamellar frequency of 5–7, and an enamel thickness of 2.0–2.3 mm, and later and more progressive representatives have 24–30 plates, a lamellar frequency of 7–9, and an enamel thickness of 1.5–2.0 mm. These extreme dental specializations approach the conditions in *Mammuthus primigenius*, and isolated teeth of the two species are often difficult to distinguish.

Paleo-Indians hunted Jefferson's mammoth, and a number of kill sites [Dent, Colo.; Union Pacific, Wyo.; Miami, Tex.; Domebo, Okla.; Blackwater Draw (Gray Sand Wedge), N. Mex.; Naco, Leikem, and Escapule, Ariz.] and hunting camps (Clovis and Mockingbird Gap, N. Mex.; Lehner Ranch and Murray Springs, Ariz.) have been discovered (Haynes, 1970) (see fig. 17.3). All of these sites belong to the Llano Complex, which is characterized by the distinctive fluted Clovis projectile points. Radiocarbon dates from these sites range from $11,310 \pm 240$ years B.P. (Clovis) to $11,160 \pm 500$ years B.P. (Domebo); no Clovis points are known after 10,000 years B.P. A mammoth bone shaft wrench found at Murray Springs is similar to batons found in Moldavia, USSR (Haynes and Hemmings, 1968). The extremely successful Clovis mammoth hunters rapidly extended their range over most of the continent, but although Clovis points have been found in every U.S. mainland state and many Canadian provinces, in none

ORDER PROBOSCIDEA

17.3 Elephantidae and Hominidae. Paleo-Indians attacking *Mammuthus jeffersonii*, Rancholabrean, Union Pacific Mammoth Kill Site, Wyoming.

of the eastern sites is there an association between the points and extinct animals. The change from Clovis to Folsom cultures coincides with the decline and the disappearance of the mammoths and the concomitant switch to bison hunting on the High Plains.

Mammuthus jeffersonii became extinct about 11,000 years ago. Although increased hunting pressure was undoubtedly a factor, overspecialization, climatic change, and a change in the composition of grasses (J. W. Wilson, personal communication) were probably equally responsible for the extinction of the last members of the long lineage of *Mammuthus*. (Harington, Tipper, and Mott, 1974; Haynes, 1970; Haynes and Hemmings, 1968; Maglio, 1973; Osborn, 1922, 1942; Skeels, 1962; C. Stock, 1935; J. W. Wilson, personal communication, 1974.)

† Woolly Mammoth, *Mammuthus primigenius* (Blumenbach), 1803 (*Elephas primigenius; E. jacksoni* Mather, 1838; *E. americanus* DeKay, 1842)

The most advanced stage in mammoth evolution is represented by *Mammuthus primigenius*, an inhabitant of tundra and northern taiga regions of Eurasia and northern North America. This species came across Beringia in the early Rancholabrean (Fairbanks I) and was a prominent member of the Beringian fauna (Eschscholtz Bay, Fairbanks II, Lost Chicken Creek, Old Crow River, Porcupine River, Gold Run Creek, Hunker Creek, and many isolated occurrences) and spread south as far as the periglacial regions of southern Canada (Medicine Hat, Woodbridge, Ontario) and the northeastern United States. Reported occurrences of the woolly mammoth south of this area should probably be referred to *Mammuthus jeffersonii,* a similar progressive late-Pleistocene species.

The external appearance of *Mammuthus primigenius* is well known from countless cave paintings and engravings in Eurasia and several well-preserved carcasses found in Siberia and Alaska. Smallest of the mammoths, the species had a shoulder height of about 2.8 m. It had a high-domed head, a humped, sloping back, a relatively short tail, a rather short trunk with two fingers at the tip, and small ears to reduce heat loss. The body was covered with long, probably black hair [Oakley (1975) believes that the reported reddish-brown hair of mammoths is the result of degradation of the black pigment, melanin] with a thick undercoat; beneath this was an insulating layer of fat some 80 mm thick. The large, greatly curved tusks often show wear facets on the ventral surface, which indicates that they were used in scraping snow and ice off the vegetation.

Mammuthus primigenius had the most complex dentition of any elephant. Molar characteris include an increased number of extremely thin, closely appressed plates (20–27 on M3, 9–13/100 mm), a lamellar frequency of 7–12, no median enamel loops, very thin (1–2 mm), finely ribbed enamel, a crown height 50 to 150 percent greater than the tooth width, and a crown heavily covered with cement. The woolly mammoth was a macrograzer and fed on siliceous plants. Analysis of stomach contents of preserved carcasses in Siberia has revealed a preponderance of grasses and tundra legumes; in winter, food probably included leaves, twigs, and bark. The molar adaptations are correlated with the development of the tundra–steppe environment, with its coarser vegetation. Guthrie and Matthews (1971) noted convergent evolution in the lemming, *Dicrostonyx,* a micrograzer; in the middle Pleistocene, both of these Holarctic genera added plates or triangles and acquired enamel specializations that enabled them to deal with the coarse vegetation on which they fed.

The skull has a very high, pointed, expanded parietal region and a rounded occiput. The tusk sheaths are parallel and closely spaced. The long neural spines on the thoracic vertebrae may account for the prominent dorsal hump shown on the cave paintings. For a detailed study of the skeleton, see Olsen (1972).

The woolly mammoth was of great importance to many Ice Age tribes as a game animal but, unfortunately, could not cope with a changing environment and increased predation by man. After the woolly mammoth had become extinct, the value of fossil ivory was discovered, and the tusks were collected, traded, and carved. This became a major industry in Siberia, where more than 25,000 woolly mammoths have been recovered. Radiocarbon dates obtained on flesh, hide, and hair of woolly mammoths from the Fairbanks area range from $15,380 \pm 300$ to $32,700 \pm 980$ years B.P. (Péwé, 1975). The species probably became extinct about 11,000 years ago. (Guthrie and Matthews, 1971; Kurtén, 1972; Maglio, 1973; Oakley, 1975; Olsen, 1972; Osborn, 1942; Péwé, 1975; Tikhomirov, 1958.)

CHAPTER EIGHTEEN

Order Primates

Prosimians, monkeys, apes, and man make up the order Primates. Although primates of the prosimian grade were common in North America in the early Tertiary, and presumably gave rise to the living Neotropical monkeys, the order had no Pleistocene representatives in the area under consideration before the late Rancholabrean, when human beings entered North America.

The human family, Hominidae, is an offshoot of the catarrhine primates (Old World monkeys and apes).

Family Hominidae—Man

The family has a long, though still imperfectly known, history in the Old World and is currently thought to have arisen in Africa or Asia in the Miocene. It is related to, and originated from, the ape family (the Pongidae). Hominid evolutionary trends include adoption of erect stance, increase of brain capacity, and reduction of incisor and, especially, canine size, with a corresponding reduction of the face. Important behavioral features include language, tool-making, and a highly developed social organization. Details of the process are much debated, but there can be no doubt that mankind evolved in the Old World and entered North America at a comparatively late date.

The earliest-known hominid may be the African–Eurasian genus *Ramapithecus* (late Miocene), of which little is known but facial fragments and teeth. Undoubtedly of hominid status is *Australopithecus* of the Pliocene and early Pleistocene, known mainly from Africa; these creatures were bipedal and small-brained and possessed an essentially human dentition. Early forms tentatively referred to *Homo* appeared in the late Pliocene of Africa, ca. 3 m.y. B.P., and were widespread around 1 m.y. B.P., when the species *Homo erectus* (Dubois) was present in Eurasia and Africa. The postcranial skeleton of *Homo erectus* is similar to that of modern man, but the smaller brain, heavy eyebrow ridges, protruding face, and other cranial and dental characters are distinctive. This species probably gave rise to *Homo sapiens*.

Modern Man, *Homo sapiens* Linnaeus, 1758

Modern man originated in the Old World, probably from the species *Homo erectus*. Morphological characters include a large brain, steep forehead without pronounced brow ridges, rounded skull, small face and teeth, and presence of canine fossa and a well-developed chin. Whether Neanderthal man and related forms, with a number of specialized or archaic characters, should be included in the same species is still debated. Depending on classification, earliest records of modern man may be dated 0.1–0.2 m.y. B.P. [the youngest records of *Homo erectus* (e.g. Choukoutien, China) are at most slightly older].

Man almost certainly entered North America by way of Beringia. The time of entrance in the New World, however, is much disputed, and there are two schools of thought. The earliest widely documented traces of human beings pertain to the Paleo-Indian culture group. As examples may be noted Wilson Butte, 15,000 ± 800 and

14,500 ± 250 years B.P.; Fort Rock Cave, Oregon, 13,200 ± 720 years B.P.; Jaguar Cave, 11,580 ± 250 years B.P; Clovis, 11,310 ± 240 years B.P.; Ventana Cave 11,290 ± 1,000 years B.P. Lehner Ranch, 11,260 ± 360 years B.P.; and Murray Springs, 11,230 ± 340 years B.P. Several Mexican and Central and South American sites are in the same time range. This may indicate that the Paleo-Indians entered the Nearctic at the same time as *Ursus arctos* and *Alces alces,* from a Beringian bridgehead. Artifacts from the Yukon–Alaska area are in some cases of greater antiquity. Included are worked bones of mammoth and caribou from Old Crow River Loc. 14N, dated at 25,000 to 29,000 years B.P., and worked bone of bison and horse from Trail Creek Cave, Alaska, dating up to 15,750 years B.P.

The picture is complicated, however, by finds of artifacts or human skeletal remains of apparently much greater age within or south of the glaciated areas. The skull of an approximately 18-month-old child was found at Taber, Alberta, in deposits antedating the Wisconsinan maximum and estimated to be between 30,000 and 40,000 years old, and possibly much older. At Medicine Hat (Unit K), a Sangamonian deposit with a rich fauna has yielded large numbers of worked stones regarded as artifacts, but this interpretation has been disputed. Well south of the glaciated area, some Californian sites have produced material which may also antedate the Paleo-Indians. Racemization dates for human skulls and bones from Del Mar and Los Angeles River are 46,000 ± 3,000 and 26,000 years B.P., respectively; this dating method is perhaps not as reliable as some others, but the last-mentioned date is supported by a radiocarbon date of 23,600+ years B.P. A human skull from Laguna Beach has a radiocarbon age of 17,150 ± 1,470 years, and human bones and artifacts from Yuba have been dated at 21,500 ± 2,000 years B.P. In spite of the large standard errors of many dates, the geological setting of some finds is such that an early date is strongly suggested. On the somewhat weaker basis of archeological typology, some artifact assemblages have also been assigned to pre-Paleo-Indian cultures, notably, the "core tools" of Calico, California.

Also in the pre–Paleo-Indian range fall some sites in Latin America: Tlapacoya and Valsequillo, Mexico, 23,150 ± 950 and 21,000 ± 850 years B.P., respectively; the Ayacucho and Paccacaisa complexes of Pikimachay Cave, Peru, with dates of 14,150 ± 180, and 16,050 ± 1,200, and 19,600 ± 3,000 years B.P., and older; as well as other sites. A date of more than 28,000 years B.P. for "Otavalo Man" from a cave in Ecuador (Davies, 1973) has turned out to be incorrect; the skull is Recent (Brothwell and Burleigh, 1977).

At face value, the evidence suggests a pre–Paleo-Indian immigration, preceding the Wisconsinan glacial maximum, which populated the western part of the Americas. If such was the case, the population seems to have remained sparse throughout, with a low cultural profile; it may have been swamped by the vigorously spreading Paleo-Indian invaders of the late Wisconsinan, with their marked reliance on big game. This concept has met with serious criticism, however, and the issue is still undecided. The literature is voluminous and widely scattered. For recent discussion and references, see Bada and Helfman, 1975; Brothwell and Burleigh, 1977; Bryan, 1969, 1973; Davies, 1973; Frison, 1975; Haynes, 1967, 1973, 1974, Kennedy, 1975; Lynch, 1974; Macneish, 1971, 1976; P. S. Martin, 1967, 1973; Stalker, 1977; and Stalker and Churcher, 1970.)

CHAPTER NINETEEN

EXTINCTION

Extinction, the end of a phyletic line without replacement, has occurred throughout the history of life on earth and is the ultimate destiny of every species. The extinction of dinosaurs at the end of the Mesozoic and the "sudden" disappearance of the megafauna at the end of the Wisconsinan are the best-known examples of widespread extinctions. Hundreds of causes have been suggested to explain extinction (Van Valen, 1969, lists 86 reasons for the late Pleistocene alone) yet, for the most part, extinction remains an enigma.

The dramatic extinction of many large mammals at the end of the Wisconsinan has received wide publicity, and many people wrongly assume that extinction took place only at the end of the Pleistocene and affected only the megafauna. Tables 19.1 and 19.2 show that extinction occurred throughout the Blancan and Pleistocene and affected mammals of all sizes.

By grouping species by order (see table 19.3), it can be seen that extinction rates for insectivores, bats, carnivores, rodents, and rabbits were highest in the Blancan, whereas those for edentates, artiodactyls, perissodactyls, and proboscideans peaked in the Wisconsinan. In the Blancan, 72.2 percent of the mammalian fauna became extinct; in the Irvingtonian, 41 percent; and in the Wisconsinan, 26.4 percent (for the whole Rancholabrean, the figure is 32.2 percent). The average extinction percentage for the dif-

TABLE 19.1
BLANCAN/PLEISTOCENE EXTINCTIONS OF MAMMAL SPECIES ACCORDING TO STRATIGRAPHIC RANGE

	Number of Species
Blancan extinctions	
1. Very Early Blancan	12
2. Early Blancan	49
3. Middle Blancan	27
4. Late Blancan	49
Blancan, exact age unknown	1
Total	138
Irvingtonian extinctions	
Early Irvingtonian	39
Middle Irvingtonian	14
Late Irvingtonian	33
Irvingtonian, exact age unknown	3
Total	89
Rancholabrean extinctions	
Late Illinoian	7
Sangamonian	21
Wisconsinan	77
Rancholabrean, exact age unknown	3
Total	108
Extinctions during historic times	3
Surviving species (with Pleistocene/Holocene records)	229

TABLE 19.2
EXTINCTIONS OF MAMMAL SPECIES BY SIZE

Size*	Blancan	Irvingtonian	Rancholabrean	Total	Surviving Species
Small	97	55	29	181	166
Medium	31	25	33	89	50
Large	5	12	35	52	16
Very large	1	2	6	9	1

*Small, 1–907 g; medium, 908 g–181 kg; large, 182 kg–1.9 tons; very large, ≥2.0 tons.

ferent orders is about 35 but ranges from 0 to 100. Extinction rates are higher for small mammals in the Blancan and for large mammals in the Wisconsinan.

Genera, rather than species, have generally been used in extinction tables (see P. S. Martin, 1967). This may introduce bias against more diverse genera, and some genera listed as extinct in North America survived on other continents. Species of *Tapirus*, *Hydrochoerus*, and *Tremarctos* became extinct in North America, for example, yet three species of *Tapirus*, two species of *Hydrochoerus*, and one species of *Tremarctos* are alive and well in South America today. *Bos grunniens* and *Saiga tatarica* disappeared from Alaska at the end of the Pleistocene, yet these two species continue to thrive in central Asia. Since extinction acts on species, not genera or higher categories, we believe that using species gives a more accurate picture. Of course, not everyone agrees which species (or genera) are valid, and for groups in need of revision, such as horses, where a number of poorly defined species are currently recognized, using species may be misleading.

Table 19.4 lists the Blancan/Pleistocene mammals by family and shows the number of species that became extinct from Blancan through historic times and the number of surviving North American species with a stratigraphic range extending back into the Pleistocene. Species now absent in North America but surviving elsewhere (*Cuon alpinus*, *Saiga tatarica*, *Bos grunniens*) are not included, nor are introduced or domestic species lacking a fossil record (*Equus caballus*, *Sus scrofa*). Extinct subspecies of surviving species (*Panthera leo atrox*, *Panthera onca augusta*, *Mustela eversmanni beringiae*) are included. Three species, *Mustela macrodon*, *Monachus tropicalis*, and *Hydrodamalis gigas*, have become extinct within historic times.

Some families that are present in the North American Pleistocene vanished from the continent but survived on other continents. These include the Hyaenidae, Hydrochoeridae, Tapiridae, Camelidae, Dugongidae, and Elephantidae. Other families, including the Glyptodontidae, Megalonychidae, Megatheriidae, Mylodontidae, Mammutidae, and Gomphotheriidae, were globally extinct by the end of the Pleistocene.

Another way of looking at extinction is to compare rates of extinction between the native and immigrant fauna. Table 19.5 shows the extinction rates for autochthonous and allochthonous species. The latter are species of Eurasian or Beringian origin (e.g., species of *Rangifer*, *Alces*, *Cervus*, and *Ovis*. Only the "megafauna" and the Carnivora are treated. The megafauna shows a striking difference between the extinction rates of autochthones (very high) and allochthones (moderately high). An analogous but less marked difference is seen among the carnivores. It is evident that the endemic American species were more prone to extinction at the end of the Pleistocene than Beringian immigrants. Much work remains to be done on this topic.

Having looked at how extinction affected orders, families, and species, we turn to the question of the causes of extinction.

TABLE 19.3
BLANCAN, IRVINGTONIAN, AND WISCONSINAN EXTINCTIONS, WITH SPECIES GROUPED BY ORDER

Order	Blancan			Irvingtonian			Wisconsinan		
	No. of species	No. becoming extinct	Percentage	No. of species	No. becoming extinct	Percentage	No. of species	No. becoming extinct	Percentage
Marsupialia	1	0	0	1	0	0	1	0	0
Insectivora	18	14	72.2	17	7	41.1	20	0	0
Chiroptera	2	1	50.0	8	2	25.0	30	5	10.6
Edentata	5	1	20.0	10	5	50.0	8	8	100
Carnivora	36	21	58.3	41	15	36.5	47	13	27.6
Rodentia	91	77	83.5	90	46	50.1	113	10	8.8
Lagomorpha	13	12	92.3	7	0	0	16	1	6.2
Perissodactyla	6	4	66.6	11	1	9.0	13	13	100
Artiodactyla	14	7	50.0	28	11	39.2	34	23	67.6
Proboscidea	4	1	25.0	3	2	66.6	3	3	100
Sirenia	1	0	0	1	0	0	2	0	0
Primates	0	0	0	0	0	0	1	0	0
Totals	191	138	72.2	217	89	41	288	76	26.4

TABLE 19.4
EXTINCTIONS OF MAMMAL SPECIES BY FAMILY

Family	Blancan	Irvingtonian	Rancholabrean	Historic times	Surviving No. Amer. spp.
Didelphidae	—	—	—	—	1
Soricidae	12	7	2	—	16
Talpidae	2	—	—	—	4
Mormoopidae	—	—	—	—	1
Phyllostomatidae	—	—	1	—	1
Vespertilionidae	1	2	4	—	22
Molossidae	—	—	1	—	3
Dasypodidae	—	1	2	—	1
Glyptodontidae	1	1	2	—	0
Megalonychidae	—	2	1	—	0
Megatheriidae	—	—	2	—	0
Mylodontidae	—	1	1	—	0
Mustelidae	12	5	3	1	14
Canidae	2	5	3	—	9
Procyonidae	3	—	1	—	2
Ursidae	1	—	3	—	3
Felidae	3	4	5	—	7
Hyaenidae	—	1	—	—	0
Odobenidae	—	—	—	—	1
Otariidae	—	—	—	—	3
Phocidae	—	—	—	1	7
Aplodontidae	—	—	—	—	1
Sciuridae	15	4	1	—	28
Geomyidae	9	4	3	—	7
Heteromyidae	6	3	—	—	15
Castoridae	5	1	2	—	1
Cricetidae	39	31	14	—	49
Zapodidae	2	1	—	—	3
Erethizontidae	1	2	—	—	1
Hydrochoeridae	—	—	2	—	0
Ochotonidae	—	—	1	—	2
Leporidae	12	—	1	—	13
Equidae	4	—	13	—	0
Tapiridae	—	1	3	—	0
Tayassuidae	2	1	3	—	0
Camelidae	2	4	7	—	0
Cervidae	2	1	5	—	5
Antilocapridae	1	4	8	—	1
Bovidae	—	1	9	—	5
Dugongidae	—	—	—	1	0
Trichechidae	—	—	—	—	1
Mammutidae	—	—	1	—	0
Gomphotheriidae	1	1	1	—	0
Elephantidae	—	1	3	—	0
Hominidae	—	—	—	—	1

Blancan and early Pleistocene extinctions are poorly known; faunas in many cases are widely separated and dating is inexact. Late Pliocene extinctions, however, are comparable in number to late Pleistocene extinctions (see Webb, 1969). Table 19.1 shows that 138 species became extinct in the Blancan, as compared to 108 species in the Rancholabrean.

TABLE 19.5
WISCONSINAN/HOLOCENE EXTINCTIONS
ACCORDING TO ORIGIN

	Number of species	Number becoming extinct	Percentage
Megafauna			
Autochthonous	49	45	91.2
Allochthonous	13	6	46.1
Carnivora			
Autochthonous	35	11	32.9
Allochthonous	10	2	20.0

A major cause of Blancan extinctions was increasing aridity. The forest savanna on the Great Plains, for example, was replaced by vast grasslands. Forest dwellers and browsers died out, surviving grazers developed strongly hypsodont teeth, rodents underwent an evolutionary explosion, and small carnivores multiplied. Along the Gulf Coast, climatic deterioration was not so severe, and this area was a refuge for many South American immigrants.

Many typical Pliocene mammals disappeared at the end of the Blancan, some leaving descendant species, others, representing the end of their lineage, vanishing forever. Competition was probably intense between the native fauna and invading species better adapted to the deteriorating climatic conditions.

Proposed causes of late Pleistocene extinctions fall into two categories, climate and man. Both theories have their devout adherents, to whom the opposing view seems too illogical to consider. Proponents of both views have oversimplified the conditions to make their point, and this has confused the issue. The climate and physiography of North America were not (and are not) uniform, and different species do not have the same ecological requirements; thus to assume that a single factor affected all species is misleading. Unfortunately, a comprehensive analysis of late Pleistocene extinctions is still lacking.

Considering the wide range of ecotones and habitats inhabited by late Pleistocene species, it is hard to see how any one climatic factor could have been the main cause of extinction. Local climatic conditions, however, did cause changes in vegetation and reduction or loss of habitat. In central Alaska, for example, grasslands decreased, and forest and muskeg increased. In the Great Lakes region, increasing aridity caused pine and hardwood forests to increase at the expense of spruce forests. These changes affected many species—some were able to migrate, others either couldn't or had no place to go as the habitat they were adapted to disappeared. Local weather phenomena such as blizzards, droughts, dust storms, and floods also played havoc with local populations. Slaughter (1967) has noted that extinctions on the southern Great Plains coincide with adjustments in range made by surviving species; perhaps the species that became extinct were not able to make these adjustments. J. W. Wilson (personal communication, 1974) believes that a change in the composition of grasses, with an increase in those having C_4, rather than C_3, metabolism, probably occurred at the close of the Pleistocene because of less equable climates with hotter, drier summers and colder winters. The C_4 grasses are less nutritious, with less nitrogen and more cellulose, lignin, and silica, and tend to inhibit reproduction of the animals that feed on them. Such changes in the composition of grasses would not be reflected in pollen diagrams that just record "grasses."

Factors such as competition and overspecialization continued to affect populations in the late Pleistocene. For example, there was competition between *Castoroides ohioensis* and *Castor canadensis*, *Martes nobilis* and *Martes americana*, *Arctodus simus* and *Ursus arctos*, *Platygonus compressus* and *Ursus americanus*, and *Cervalces scotti* and *Alces alces*. Competition may have occurred between *Bison latifrons* and *Bison bison*, but it is also possible that the larger form graded into the smaller species, for chronoclinal studies on Holocene bison (M. V. Wilson, 1975) show a continuous size decrease. Glyptodonts and the Wisconsinan mammoths (*Mammuthus jeffersonii* and *M. primigenius*), among others, were overspecialized, unable to adapt to changing conditions. Other factors such as disease, parasitism, low reproduction rates, and the appearance of new predators may have also affected local populations.

As Guilday (1967a) has noted, these factors, though perhaps not in themselves causing extinction, weakened populations. Some species were affected more than others. Large animals were affected more than small ones, and females and juveniles were probably affected more than adult males. The net result, however, was to make local populations vulnerable to a new factor in the environment—man, who by the late Pleistocene was a skilled hunter.

Paul S. Martin is the foremost proponent of "prehistoric overkill," the theory which cites decimation of the late Pleistocene megafauna by man as the cause of Pleistocene extinctions. He believes that only the phenomenon of overkill can explain the global extinctions occurring at the close of the Pleistocene (Martin, 1967). He notes that, in North America, extinction occurred without replacement by evolution or immigration (many ecological niches were never refilled) and that this unbalanced extinction had not occurred anywhere before the advent of man. Man, he argues, was the only new element and, in a relatively short time after his arrival in the New World, wiped out many genera of herbivores, which precipitated the loss of many carnivores. Using radiocarbon dates, Martin has postulated that this extinction culminated about 11,000 years B.P.

In a later paper, Mosimann and Martin (1975) have used a mathematical model to show that overkill was possible. The model proposes that highly skilled hunters entered North America in a single migration about 12,000 years ago, moved rapidly southward (16 km a year, or from central Alberta to Patagonia in 1,000 years), and wiped out the native megafauna by about 10,000 years B.P. The authors assume (a) that the prey biomass was large and healthy (comparable to that of the African game parks), (b) that the human population increased rapidly (at a growth rate 3.5 percent a year, the population would double in 20 years), and (c) that the population moved through an area too rapidly to leave behind any traces. The model projects that 300,000 hunters could have wiped out 100 million large animals in 300 years. The authors ignore the effects of natural predation, disease, weather, and hardships of moving through different terrains.

Richard MacNeish (1976) has refuted much of Mosimann and Martin's (1975) hypothesis. He points out that more than 50 excavated sites from Alaska to Patagonia containing human artifacts and extinct animal bones have yielded radiocarbon dates earlier than 12,000 years B.P. He divides Paleo-Indian prehistory into four stages based on the development of stone-tool assemblages that perhaps extend back $70,000 \pm 30,000$ years B.P. The early stages represent relatively unskilled hunters who had little effect on the fauna. Only in the last stage (13,000–8,500 years B.P.) was man a factor. MacNeish points out that the artifacts of the last stage (Stage IV) are so varied, widespread, and seemingly indigenous to the New World that it is hard to see how they could be used to support a theory of a single late migration into North America. None of the tool assemblages seems to be related to slightly earlier complexes in northeastern Asia or to indicate that the Stage IV complex came from there. Mac-Neish notes that there is no evidence for a 3.4 or 2.4 percent annual growth rate (a rate of 0.1 percent may be too large) or for a rapid increase in population about 12,000 years ago followed by a population crash (after the extinction of the megafauna).

Of all the extinct species, man apparently hunted only two, mammoth and bison, to any extent. Although remains of extinct mammals have been found at the same level as artifacts and hearths at several sites (Jaguar Cave, Little Box Elder Cave, Casper, Gypsum Cave, Burnet Cave, Lubbock Lake, etc.), few of these specimens show evidence of having been used by man. But as Hester (1967) has observed, it is often impossible to determine if the animals present at a site were killed by man or some other agency.

Antti Järvi (1978) has pointed out that, judging from coprolites, early and middle Paleolithic Europeans appear to have been obligate predators and that predators always live in dynamic equilibrium with their prey. Overkill is possible only in the case of an omnivorous population that can turn to plant foods when game resources are depleted. Conversely, according to Järvi's hypothesis, the occurrence of overkill should be taken as a sign of omnivory in the human population concerned. Presence of seed fragments in early Holocene coprolites from Tehuacán (Bryant and Williams-Dean, 1975) indicates that early Americans were facultative plant feeders.

Radiocarbon dates have been obtained from a number of late Wisconsinan sites (see table 19.6 and Appendix 1). However, most sites, and the majority of species, have not been critically dated.

A number of objections have been raised regarding the uncritical acceptance of radiocarbon dates. In some cases, material is contaminated by humic acids or other intrusive carbons, which results in much younger dates; in other cases, there is ambiguous stratigraphic association between the sample and the extinct fauna. In addition, dates on bone tend to be too young; new techniques are remedying this problem, however.

Radiocarbon dates can be used to determine the approximate date of the last appearance of a species (see table 19.6). Of course, this method must be used with caution since dates are not available for most species, and a species may have survived longer in other areas. Most of the terminal dates fall between 9,400 and 12,700 years B.P. Martin (1967) has questioned the validity of dates younger than 10,000 years, but a number of new dates in the 9,000-year range appear to be valid. The unpublished C-14 dates of 7,000–8,000 years B.P. at Devil's Den (Martin and Webb, 1974) should be regarded as questionable until more is known about the material dated and its stratigraphic association. It is quite likely, however, that Florida was an early Holocene refugium and that several species survived longer there.

Extinction did not occur uniformly across the continent. Local conditions affected local populations. No one cause can account for it; rather, a mosaic of adverse conditions prevailed. We believe that changes in vegetation, sudden storms, droughts, loss of habitat, interspecific competition, low reproduction rates, and overspecialization, to name a few factors, reduced or weakened populations, making them vulnerable to environmental pressures, including man, the hunter, who probably delivered the *coup de grâce* to some of the megafauna between 12,000 and 9,000 years ago.

Many questions about extinction remain unanswered. For example, Why did species closely related to extinct North American species survive in South America, Eurasia, and Africa (e.g., *Tremarctos ornatus, Catagonus wagneri, Lama glama, Pathera leo,*

TABLE 19.6
LAST APPEARANCES OF SPECIES BECOMING EXTINCT DURING THE WISCONSINAN

Species	Date B.P.	Fauna	Reference
Myotis rectidentis	13,970 ± 310– 28,340 ± 1,710	Laubach	Valastro, Davis, and Varela, 1977
Myotis magnamolaris	13,970 ± 310– 28,340 ± 1,710	Laubach	Valastro, Davis, and Varela, 1977
Holmesina septentrionalis	9,880 ± 270	Hornsby Springs	Webb, 1974a
Dasypus bellus	7,200 ± 300	Miller's Cave	Patton, 1963
D. bellus	9,440 ± 760	Brynjulfson**	Parmalee and Oesch, 1972
Glyptotherium floridanum	13,970 ± 310– 28,340 ± 1,710	Laubach	Valastro, Davis, and Varela, 1977
Megalonyx jeffersonii	7,000–8,000	Devil's Den*	Martin and Webb, 1974
M. jeffersonii	9,440 ± 250	Evansville	Hester, 1967
M. jeffersonii	9,440 ± 760	Brynjulfson**	Parmalee and Oesch, 1972
Eremotherium rusconii	10,000	Watkin's Quarry	Voorhies, 1971
Nothrotheriops shastensis	9,900 ± 400	Rampart	P. S. Martin, 1967
Glossotherium harlani	9,880 ± 270	Hornsby Springs	Webb, 1974a
Martes nobilis	10,370 ± 350	Jaguar	Kurtén and Anderson, 1972
Brachyprotoma obtusata	9,440 ± 760	Brynjulfson**	Parmalee and Oesch, 1972
Canis dirus	7,000–8,000	Devil's Den*	Martin and Webb, 1974
C. dirus	9,440 ± 760	Brynjulfson**	Parmalee and Oesch, 1972
C. dirus	9,880 ± 270	Hornsby Springs	Webb, 1974a
Tremarctos floridanus	7,000–8,000	Devil's Den*	Martin and Webb, 1974
Arctodus simus	12,770 ± 900	Natural Trap	Martin, Gilbert, and Adams, 1977
Smilodon fatalis	9,410 ± 155	First American Bank Site	Guilday, 1977
Panthera leo atrox	10,370 ± 350	Jaguar	Kurtén and Anderson, 1972
P. leo atrox	10,370 ± 160	Lost Chicken Cr.	Harington, 1978
Panthera onca augusta	7,000–8,000	Devil's Den*	Martin and Webb, 1974
Acinonyx trumani	12,770 ± 900	Natural Trap	Martin, Gilbert, and Adams, 1977
Castoroides ohioensis	9,550 ± 375	Ben Franklin	Slaughter and Hoover, 1963
Synaptomys australis	7,000–8,000	Devil's Den*	Martin and Webb, 1974
Mylohyus nasutus	9,550 ± 375	Ben Franklin	Slaughter and Hoover, 1963
M. nasutus	11,300 ± 1,000	New Paris No. 4	Guilday, Martin, and McCrady, 1964
Platygonus compressus	7,000–8,000	Devil's Den*	Martin and Webb, 1974
P. compressus	11,900 ± 750	Mosherville	Ray, Denny, and Rubin, 1970
Camelops huerfanensis	37,000+	Moore Pit	Slaughter, 1966b

Acinonyx jubatus, and *Equus hemionus*? Why did some large herbivores survive (*Cervus elaphus, Alces alces, Antilocapra americana, Ovis canadensis, Oreamnos americanus, Ovibos moschatus, Bison bison*)? Was there a difference in the time of extinction of browsers and grazers? What were the environmental tolerances of the extinct species? We don't know. The answers are not all in.

More data are needed before any of the extinction hypotheses can be confirmed or refuted. More interdisciplinary studies, involving paleontology, zoology, ecology, botany, biogeography, geology, and anthropology, are needed. More sites and species need to be critically dated. Poorly known groups need to be revised. Since we are living

TABLE 19.6 *(Continued)*

Species	Date B.P.	Fauna	Reference
Camelops hesternus	9,940 ± 160– 11,680 ± 160	Smith Creek	Valastro, Davis, and Varela, 1977
C. hesternus	10,370 ± 350	Jaguar	Kurtén and Anderson, 1972
C. hesternus	10,780 ± 135	Lindenmeier	P. S. Martin, 1967
Hemiauchenia macrocephala	11,690 ± 250	Gypsum	P. S. Martin, 1967
Palaeolama mirifica	9,880 ± 270	Hornsby Springs	Webb, 1974a
Navahoceros fricki	11,500	Burnet	Hester, 1967
Sangamona fugitiva	9,440 ± 760**	Brynjulfson	Parmalee and Oesch, 1972
Cervalces scotti	10,230 ± 150	Carter	Mills and Guilday, 1972
Capromeryx minor	11,170 ± 360	Blackwater Draw	Slaughter, 1964
Tetrameryx shuleri	37,000+	Moore Pit	Slaughter, 1966b
Stockoceros onusrosagris	11,500	Burnet	Hester, 1967
Oreamnos harringtoni	10,050	Rampart	Hester, 1967
Euceratherium collinum	11,500	Burnet	Hester, 1967
Symbos cavifrons	10,370 ± 160	Lost Chicken Cr.	Harington, 1978
Bootherium bombifrons	17,200 ± 600	Big Bone Lick	Ives et al., 1967
Bison priscus	10,370 ± 160	Lost Chicken Cr.	Harington, 1978
Bison latifrons	21,000–30,000	Rainbow Beach	McDonald and Anderson, 1975
Equus occidentalis	11,300 ± 1,200	Ventana	P. S. Martin, 1967
Equus complicatus	17,200 ± 600	Big Bone Lick	Ives et al. 1967
Equus scotti	11,170 ± 360	Blackwater Draw	Slaughter, 1964
Equus conversidens	10,370 ± 170	Jaguar	Kurtén and Anderson, 1972
Equus lambei	10,370 ± 160	Lost Chicken Cr.	Harington, 1978
Equus excelsus	11,170 ± 360	Clovis	P. S. Martin, 1967
Tapirus copei	37,000+	Moore Pit	Slaughter, 1966b
Tapirus veroensis	9,880 ± 270	Hornsby Springs	Webb, 1974a
Mammut americanum	7,000–8,000	Devil's Den*	Martin and Webb, 1974
M. americanum	9,880 ± 270	Hornsby Springs	Webb, 1974a
M. americanum	10,690 ± 360	La Mirada	W. E. Miller, 1971
Mammuthus jeffersoni	11,160 ± 500	Domebo	Leonhardy, 1966
M. jeffersoni	11,260 ± 360	Lehner Ranch	Haynes, 1970
M. jeffersoni	11,230 ± 340	Murray Springs	Haynes, 1970
M. jeffersoni	11,280 ± 350	Union Pacific	Haynes, 1970
Mammuthus jeffersoni exilis	11,800 ± 800	Santa Rosa Is.	Berger and Libby, 1966
Mammuthus primigenius	10,370 ± 160	Lost Chicken Cr.	Harington, 1978
M. primigenius	15,380 ± 300	Fairbanks	Péwé, 1975

*Dated at 7,000–8,000 years B.P.—unpublished C-14 data (Martin and Webb, 1974).
**New C-14 dates are older (Coleman and LiLiu, 1975).

in an interglacial, it behooves us to learn more about past ice ages for another one may be on its way, and knowledge of Pleistocene environments, faunas, and species may be a key to mankind's future.

(Bryant and Williams-Dean, 1975; Guilday, 1967a; Hester, 1967; Järvi, 1978; Mac-Neish, 1976; P. S. Martin, 1967; Martin and Webb, 1974; Mosimann and Martin, 1975; Slaughter, 1967; Van Valen, 1969; Webb, 1969; J. W. Wilson, personal communication, 1976; M. V. Wilson, 1975.)

REFERENCES

Adams, D. B. 1979. The Cheetah: Native American. *Science* 205:1155–58.
Adams, R. McC. 1953. The Kimmswick bone bed. *Mo. Archaeol.* 15:40–56.
Aguirre, E. 1969. Evolutionary history of the elephant. *Science* 164:1366–76.
Akersten, W. A. 1972. Red Light local fauna (Blancan) of the Love Formation, southeastern Hudspeth County, Texas. *Bull. Texas Mem. Mus.* 20:1–53.
———. 1973. Evolution of Geomyine rodents with rooted cheek teeth. Ph.D. dissertation, Univ. Mich., Ann Arbor.
Alekseev, M. N., et al. 1973. Scheme of correlation. *Coll. Papers Internat. Colloq. Neogene-Quaternary* 4:160 (insert).
Alexander, H. L., Jr. 1963. The Levi site: A paleo-Indian campsite in central Texas. *Amer. Antiq.* 28:510–28.
Allen, G. M. 1932. A Pleistocene bat from Florida. *J. Mammal.* 13:256–59.
Allen, J. A. 1876. Description of some remains of an extinct species of wolf and an extinct species of deer from the lead region of the Upper Mississippi. *Amer. J. Sci., Ser. 3,* 11:47–51.
———. 1913. Ontogenetic and other variations in muskoxen with a systematic review of the muskox group, recent and extinct. *Mem. Amer. Mus. Nat. Hist. NS* 1(4):103–226.
Allison, I. S. 1966. Fossil Lake, Oregon: Its geology and fossil faunas. *Oreg. State Monogr. Stud. Geol.* 9:1–48.
Alt, D., and H. K. Brooks. 1964. Age of the Florida marine terraces. *J. Geol.* 73:406–411.
Alvarez, T. 1969. Restos fósiles de mamíferos de Tlapacoya, Estado de México (Pleistoceneo–Recienta). *Misc. Publ. Univ. Kans. Mus. Nat. Hist.* 51:93–112.
American Geological Institute. 1976. *Dictionary of Geological Terms.* Garden City, N.Y.: Doubleday.
Anderson, E. 1968. Fauna of the Little Box Elder Cave, Converse County, Wyoming: The Carnivora. *Univ. Colo. Stud. Ser. Earth Sci.* 6:1–59.
———. 1970. Quaternary evolution of the genus *Martes* (Carnivora, Mustelidae). *Acta Zool. Fennica* 130:1–132.
———. 1973. Ferret from the Pleistocene of central Alaska. *J. Mammal.* 54(3):778–79.
———. 1974. A survey of the late Pleistocene and Holocene mammal fauna of Wyoming. In M. V. Wilson (ed.), *Applied geology and archaeology: The Holocene history of Wyoming; Geol. Surv. Wyo. Rept. Inv.* 10:78–87.
———. 1977. Pleistocene Mustelidae (Mammalia, Carnivora) from Fairbanks, Alaska. *Bull. Mus. Comp. Zool.* 148(1):1–21.
Anderson, E., and White, J. A. 1975. Caribou (Mammalia, Cervidae) in the Wisconsinan of southern Idaho. *Tebiwa* 17(2):59–65.
Arata, A. A., and Hutchison, J. H. 1964. The raccoon (*Procyon*) in the Pleistocene of North America. *Tulane Stud. Geol.* 2:21–27.
Armstrong, D. M., and Jones, J. K., Jr. 1972. *Sorex merriami. Mammal. Species* 2:1–2.
———. 1972. *Notiosorex crawfordi. Mammal. Species* 17:1–5.
Armstrong, R. L., Leeman, W. P., and Malde, H. E. 1975. K-Ar dating, Quaternary and Neogene volcanic rocks of the Snake River Plain, Idaho. *Amer. J. Sci.* 275:225–51.
Auffenberg, W. 1957. A note on an unusually complete specimen of *Dasypus bellus* (Simpson) from Florida. *Quart. J. Fla. Acad. Sci.* 20(4):233–37.
———. 1958. Fossil turtles of the genus *Terrapene* in Florida. *Bull. Fla. State Mus.* 3:53–92.
———. 1963. The fossil snakes of Florida. *Tulane Stud. Zool.* 10:131–216.
———. 1967. Further notes on fossil box turtles of Florida. *Copeia* 1967:319–25.
Ayer, M. Y. 1937. The archaeological and faunal material from Williams Cave, Guadelupe Mountains, Texas. *Proc. Acad. Nat. Sci. Phila.* 88:599–618.

Azzaroli, A. 1952. L'alce di Sénèze. *Palaeontogr. Ital.* 47:133–41.
——. 1953. The deer of Weybourn Crag and Forest Bed of Norfolk. *Bull. Brit. Mus. (Nat. Hist.) Geol.* 2:1–96.
Bada, J. L., and Helfman, P. M. 1975. Amino acid racemization dating of fossil bone. *World Archaeol.* 7:160–73.
Bader, R. S. 1957. Two Pleistocene mammalian faunas from Alachua County, Florida. *Bull. Fla. State Mus. Biol. Ser.* 2(5):53–75.
Bandy, O. L. 1972. The Plio–Pleistocene boundary, Europe and California, and the paleomagnetic scale. *Coll. Papers Internat. Colloq. Neogene-Quaternary*, 1:15–62.
Bandy, O. L., and Wilcoxon, J. A. 1970. The Plio–Pleistocene boundary, Italy and California. *Bull. Geol. Soc. Amer.* 81:2939–48.
Banfield, A. W. F. 1961. A revision of the reindeer and caribou, genus *Rangifer*. *Bull. Nat. Mus. Canada Biol. Ser.* 66:vi–177.
——. 1974. *The Mammals of Canada.* Toronto: Univ. Toronto Press. 438 pp.
Bannikov, A. G., et al. 1961. *Biology of the Saiga.* (Israel Program for Scientific Translation, 1967. 252 pp.)
Barbour, E. H. 1916. A giant Nebraska bear, *Dinarctotherium merriami*. *Bull. Nebr. Geol. Surv.* 4:349–53.
——. 1925. Skeletal parts of the Columbian mammoth, *Elephas maibeni* sp. nov. *Bull. Nebr. State Mus.* 1(10):95–118.
——. 1931. The musk-oxen of Nebraska. *Bull. Nebr. State Mus.* 1(25):211–33.
Barbour, E. H., and C. B. Schultz. 1934. A new antilocaprid and a new cervid from the late Tertiary of Nebraska. *Amer. Mus. Novitates* 734:1–4.
—— and ——. 1937. An early Pleistocene fauna from Nebraska. *Amer. Mus. Novitates* 942:1–10.
—— and ——. 1941. A new fossil bovid from Nebraska with notes of a new bison quarry in Texas. *Nebr. State Mus. Bull.* 2(7):63–68.
Barbour, R. W., and Davis, W. H. 1969. *Bats of America.* Lexington, Ky.: Univ. Press Ky. 268 pp.
Barnes, L. G., and E. D. Mitchell. 1975. Late Cenozoic northeast Pacific Phocidae. *Rapp. Proces-Verbaux Reunions Conseil Perm. Intern. Exploration Mer* 168:34–42.
Barton, J. B. 1975. A preliminary report of the American Falls Dam local fauna, Idaho. *Proc. Utah Acad. Sci. Arts Letters* 52(1):76. Abst.
Beaumont, G. de. 1967. Observations sur les Herpestinae (Viverridae, Carnivora) de l'Oligocène supérieur avec quelques remarques des Hyaenidae du Néogène. *Arch. Sci. Genève* 20:79–108.
Beninde, J. 1937. Über die Edelhirschformen von Mosbach, Mauer und Steinham a.d. Murr. *Paläontol. Zeitschr.* 19:79–116.
Bensley, B. A. 1913. A *Cervalces* antler from Toronto Interglacial. *Toronto Univ. Stud. Geol., Ser. 2*, 8:1–3.
Berger, R., and Libby, W. F. 1966. UCLA Radiocarbon dates V. *Radiocarbon* 8:467–97.
——. 1968. UCLA Radiocarbon dates VI. *Radiocarbon* 10(2):402–16.
Bjork, P. R. 1970. The Carnivora of the Hagerman local fauna (Late Pliocene) of southwestern Idaho. *Trans. Amer. Phil. Soc. NS* 60(7):1–54.
——. 1974. Additional carnivores from the Rexroad Formation (Upper Pliocene) of southwestern Kansas. *Trans. Kans. Acad. Sci.* 76(1):24–38.
Black, C. C. 1963. A review of the North American Tertiary Sciuridae. *Bull. Mus. Comp. Zool.* 130:109–248.
Black, C. C., ed. 1974. History and prehistory of the Lubbock Lake site. *J. West Tex. Mus. Assoc.* 15:1–160.
Blake, W. P. 1898. *Bison latifrons* and *B. arizonica*. *Amer. Geol.* 22:247–48.
Bojanus, L. H. 1827. De uro nostrate eiusque sceleto commentatio. *Nova Acta Leopoldina* 13(2):413–78.
Bonifay, M.-F. 1971. Carnivores quaternaires du Sud-Est de la France. *Mem. Mus. Nat. Hist. Nat. NS* C21:43–377.
Bovard, J. F. 1907. Notes on Quaternary Felidae from California. *Univ. Calif. Publ. Bull. Dept. Geol.* 5:155–70.

Brantley, A. G. 1971. Paleoenvironmental significance of bone orientation in Watkin's Quarry (late Pleistocene), Glynn County, Georgia. *Bull. Ga. Acad. Sci.* 29(2):128.
Breyer, J. 1976. *Titanotylopus* (= *Gigantocamelus*) from the Great Plains Cenozoic. *J. Paleontol.* 50(5):783–88.
———. 1977. Intra- and interspecific variation in the lower jaw of *Hemiauchenia. J. Paleontol.* 51(3):527–35.
Brodkorb, P. 1957. New passerine birds from the Pleistocene of Reddick, Florida. *J. Paleontol.* 31:129–38.
———. 1958. Fossil birds from Idaho. *Wilson Bull.* 70:237–42.
———. 1959. The Pleistocene avifauna of Arredondo, Florida. *Bull. Fla. State Mus.* 4:269–91.
———. 1963. A giant flightless bird from the Pleistocene of Florida. *Auk* 80:11–115.
Brothwell, D., and Burleigh, R. 1977. On sinking Otavalo Man. *J. Archaeol. Sci.* 1977:291–94.
Brown, B. 1903. A new genus of ground sloth from the Pleistocene of Nebraska. *Bull. Amer. Mus. Nat. Hist.* 19:569–83.
———. 1908. The Conard Fissure, a Pleistocene bone deposit in northern Arkansas: with description of two new genera and twenty new species of mammals. *Mem. Amer. Mus. Nat. Hist.* 9(4):157–208.
———. 1912. *Brachyostracon*, a new genus of glyptodonts from Mexico. *Bull. Amer. Mus. Nat. Hist.* 31:167–77.
Bryan, A. L. 1969. Early man in America and the late Pleistocene chronology of western Canada and Alaska. *Current Anthropol.* 10:339–65.
———. 1973. Paleoenvironments and cultural diversity in late Pleistocene South America. *Quatern. Res.* 3(2):237–56.
Bryant, V. M., Jr., and Williams-Dean, G. 1975. The coprolites of man. *Sci. Amer.* 232(1):100–9.
Burt, W. H., and Grossenheider, R. P. 1976. *A Field Guide to the Mammals,* 3rd ed. Boston: Houghton Mifflin Co. 289 pp.
Butler, B. R. 1968. An introduction to archaeological investigations in the Pioneer Basin locality of eastern Idaho. *Tebiwa* 11:1–30.
Butler, B. R., Gildersleeve, H., and Sommers, J. 1971. The Wasden Site bison: Sources of morphological variation. In A. H. Stryd and R. A. Smith (eds.), *Aboriginal Man and Environment on the Plateau of Northwest America.* Calgary: Archaeol. Assoc. Univ. Calgary. Pp. 126–52.
Buwalda, J. P. 1914. Pleistocene beds at Manix in the eastern Mohave Desert region. *Univ. Calif. Publ. Bull. Dept. Geol.* 7(24):443–64.
Cahn, A. R. 1922. *Chlamytherium septentrionale,* a fossil edentate new to the fauna of Texas. *J. Mammal.* 3(1):22–24.
Carr, W. J., and Trimble, D. E. 1963. Geology of the American Falls Quadrangle. *Bull. U.S. Geol. Surv.* 1121G:1–43.
Chamberlain, T. C. 1895. The classification of American glacial deposits. *J. Geol.* 3:270–77.
Chandler, A. C. 1916. Notes on *Capromeryx* material from the Pleistocene of Rancho La Brea. *Univ. Calif. Publ. Bull. Dept. Geol.* 9(10):111–20.
Chantell, C. J. 1970. Upper Pliocene frogs from Idaho. *Copeia* 1970:654–64.
Choate, J. R., and Hall, E. R. 1967. Two new species of bats, genus *Myotis,* from a Pleistocene deposit in Texas. *Amer. Midl. Nat.* 78(2):531–34.
Churcher, C. S. 1959a. Fossil *Canis* from the tar pits of La Brea, Peru. *Science* 130(3375):564–65.
———. 1959b. The specific status of the New World red fox. *J. Mammal.* 40(4):513–20.
———. 1966. The affinities of *Dinobastis serus* Cope, 1893. *Quaternaria* 8:263–75.
———. 1968. Pleistocene ungulates from the Bow River gravels at Cochrane, Alberta. *Canad. J. Earth Sci.* 5(6):1467–88.
———. 1972. Imperial mammoth and Mexican half-ass from near Bindloss, Alberta. *Canad. J. Earth Sci.* 9(11):1562–67.
———. 1975. Additional evidence of Pleistocene ungulates from the Bow River gravels at Cochrane, Alberta. *Canad. J. Earth Sci.* 12(1):68–76.
Churcher, C. S., and Morgan, A. V. 1976. A grizzly bear from the Middle Wisconsin of Woodbridge, Ontario. *Canad. J. Earth Sci.* 13:341–47.

Churcher, C. S., and Stalker, A. MacS. 1970. A late postglacial horse from Pashley, Alberta. *Canad. J. Earth Sci.* 7(3):1020–26.

Clark, T. W., Hoffmann, R. S., and Nadler, C. F. 1971. *Cynomys leucurus. Mammal. Species* 7:1–4.

Clutton-Brock, J., Corbet, G. B., and Hills, M. 1976. A review of the family Canidae, with a classification by numerical methods. *Bull. Brit. Mus. (Nat. Hist.), Zool.* 29:119–99.

Cockerell, T. D. A. 1909. A fossil ground sloth in Colorado. *Univ. Colo. Stud.* 6(4):309–12.

Colbert, E. H., and Chaffee, R. G. 1939. A study of *Tetrameryx* and associated fossils from Papago Springs Cave, Sonoita, Arizona. *Amer. Mus. Novitates* 1034:1–21.

Coleman, D. C., and LiLiu, C. 1975. Illinois State Geological Survey Radiocarbon Dates VI. *Radiocarbon* 17(2):160–73.

Compton, L. V. 1937. Shrews from the Pleistocene of the Rancho La Brea asphalt. *Univ. Calif. Publ. Bull. Dept. Geol. Sci.* 24(5):85–90.

Cook, H. J. 1928. A new fossil bison from Texas. *Proc. Colo. Mus. Nat. Hist.* 8(3):34–37.

Cooke, H. B. S. 1973. Pleistocene chronology: Long or short?. *Quatern. Res.* 3(2):206–20.

Cope, E. D. 1871. Preliminary report on the Vertebrata discovered in the Port Kennedy Bone Cave. *Proc. Amer. Phil. Soc.* 12(1871–72):73–108.

———. 1878. Descriptions of new Vertebrata from the Upper Tertiary formations of the West. *Proc. Amer. Phil Soc.* 17:219–31.

———. 1879. The cave bear of California. *Amer. Nat.* 13:791.

———. 1880. On the extinct cats of North America. *Amer. Nat.* 14(12):833–58.

———. 1884. The extinct Mammalia of the Valley of Mexico. *Proc. Amer. Phil. Soc.* 22:1–21.

———. 1889. The vertebrate fauna of the *Equus* beds. *Amer. Nat.* 23:160–65.

———. 1892a. A hyaena and other Carnivora from Texas. *Proc. Acad. Nat. Sci. Phila.* 44:326–27.

———. 1892b. A contribution to the vertebrate paleontology of Texas. *Proc. Amer. Phil. Soc.* 30:123–31.

———. 1893. A preliminary report on the vertebrate paleontology of the Llano Estacado. *Fourth Ann. Rept. Tex. Geol. Surv.*, 136 pp.

———. 1894. Extinct Bovidae, Canidae and Felidae from the Pleistocene of the plains. *J. Acad. Nat. Sci. Phila.* 9:453–59.

———. 1895. The fossil vertebrates from the fissure at Port Kennedy, Pa. *Proc. Acad. Nat. Sci. Phila.* 1895:446–51.

———. 1896. New and little known Mammalia from the Port Kennedy bone deposit. *Proc. Acad. Nat. Sci. Phila.* 1896:378–94.

———. 1899. Vertebrate remains from the Port Kennedy bone deposit. *J. Acad. Nat. Sci. Phila.*, 11:193–267.

Corner, R. G. 1977. A late Pleistocene–Holocene vertebrate fauna from Red Willow County, Nebraska. *Trans. Nebr. Acad. Sci.* 4:77–93.

Corner, R. G., and Myers, T. P. 1976. Paleontological and archaeological remains from gravel pits in Red Willow County, Nebraska. *Proc. Nebr. Acad. Sci.*, 86th Ann. Meeting Abst., p. 45.

Cox, A., Doell, R. R., and Dalrymple, G. B. 1963. Geomagnetic polarity epochs and Pleistocene geochronology. *Nature* 198:1049–51.

Crook, W. W., and Harris, R. K. 1957. Hearths and artifacts of early man near Lewisville, Texas, and associated faunal material. *Bull. Tex. Archaeol. Paleontol. Soc.* 28:7–97.

Cuffey, R. J., Johnson, G. H., and Rasmussen, D. L. 1964. A microtine rodent and associated gastropods from the upper Pleistocene of southwestern Indiana. *J. Paleontol.* 38:1109–11.

Curry, R. P. 1966. Glaciations about 3,000,000 years ago in the Sierra Nevada. *Science* 154:770–71.

Cushing, J. E., Jr. 1945. Quaternary rodents and lagomorphs of San Josecito Cave, Nuevo Leon, Mexico. *J. Mammal.* 26:182–85.

Dalquest, W. W. 1962. The Good Creek Formation, Pleistocene of Texas, and its fauna. *J. Paleontol.* 36(3):568–82.

———. 1964. A new Pleistocene local fauna from Motley County, Texas. *Trans. Kans. Acad. Sci.* 67(3):499–505.

———. 1965. New Pleistocene formation and local fauna from Hardeman County, Texas. *J. Paleontol.* 39(1):63–79.
———. 1967. Mammals of the Pleistocene Slaton local fauna of Texas. *Southw. Nat.* 12(1):1–30.
———. 1969. Pliocene carnivores of the Coffee Ranch. *Bull. Texas Mem. Mus.* 15:1–43.
———. 1972. On the Upper Pliocene skunk, *Spilogale rexroadi* Hibbard. *Trans. Kans. Acad. Sci.* 74(2):234–36.
———. 1974. A new species of four-horned antilocaprid from Mexico. *J. Mammal.* 55(1):96–101.
———. 1975. Vertebrate fossils from the Blanco local fauna of Texas. *Occ. Papers Mus. Tex. Tech. Univ.* 30:1–52.
———. 1977. Mammals of the Holloman local fauna, Pleistocene of Oklahoma. *Southw. Nat.* 22(2):255–68.
———. 1978. Phylogeny of American horses of Blancan and Pleistocene age. *Ann. Zool. Fennici* 15:191–99.
Dalquest, W. W., and Donovan, T. J. 1973. A new three-toed horse (*Nannipus*) from the late Pliocene of Scurry County, Texas. *J. Paleontol.* 47(1):34–35.
Dalquest, W. W., and Hughes, J. T. 1965. The Pleistocene horse, *Equus conversidens. Amer. Midl. Nat.* 74(2):408–17.
Dalquest, W. W., Roth, E., and Judd, F. 1969. The mammal fauna of Schulze Cave, Edwards County, Texas. *Bull. Fla. State Mus.* 13(4):206–76.
Dalrymple, G. B. 1972. Potassium-argon dating of geomagnetic reversals and North American glaciations. In W. W. Bishop and J. A. Miller (eds.), *Calibration of Hominoid Evolution, Recent Advances in Isotopic and Other Dating Methods Applicable to the Origin of Man.* New York: Wenner-Gren Foundation for Anthropol. Res. Pp. 107–34.
Davies, D. M. 1973. Fossil man in Ecuador. *Spectrum* 106(1973):5.
Davis, L. C. 1969. The biostratigraphy of Peccary Cave, Newton County, Arkansas. *Proc. Ark. Acad. Sci.* 23:192–96.
Davis, W. B. 1939. *The Recent Mammals of Idaho.* Caldwell, Idaho: Caxton Printers. 400 pp.
Dawkins, W. B. 1888. On *Ailurus anglicus*, a new carnivore from Red Crag. *Quart. J. Geol. Soc. London* 44:228–31.
De Geer, G. 1940. Geochronologica Suecica principles. *K. Svenska Vet. Akad. Handl. Stockholm* (3)18(6):1–360.
Del Campana, D. 1913. I cani pliocenici di Toscana. *Paleontogr. Ital.* 19:189–254.
Denton, G. H., and Armstrong, R. L. 1969. Miocene–Pliocene glaciations in southern Alaska. *Amer. J. Sci.* 267:1121–42.
Dice, L. R. 1917. Systematic position of several American Tertiary lagomorphs. *Univ. Calif. Publ. Bull. Dept. Geol.* 10:179–83.
———. 1925. Rodents and lagomorphs of the Rancho La Brea deposits. *Carnegie Inst. Washington Publ.* 349:119–30.
Dietrich, W. O. 1959. *Hemionus* Pallas im Pleistozän von Berlin. *Vert. Palasiat.* 3:13–22.
Dolan, E. M., and Allen, G. T. 1961. An investigation of the Darby and Hornsby Springs sites, Alachua County, Florida. *Spec. Pub. Fla. Geol. Surv.* 7:1–124.
Domning, D. P. 1969. A list, bibliography, and index of the fossil vertebrates of Louisiana and Mississippi. *Trans. Gulf Coast Assoc. Geol. Soc.* 19:385–422.
———. 1970. Sirenian evolution in the North Pacific and the origin of Steller's sea cow. *Proc. 7th Ann. Conf. Biol. Sonar Diving Mammals Stanford Res. Inst.* 970:217–20.
Dort, W., Jr. 1975. Archaeo-geology of Jaguar Cave, Upper Birch Creek Valley, Idaho. *Tebiwa* 17(2):33–57.
Downey, J. S. 1962. Leporidae of the Tusker local fauna from southeastern Arizona. *J. Paleontol.* 36:1112–15.
———. 1968. Late Pliocene lagomorphs of the San Pedro Valley, Arizona. *U.S. Geol. Surv. Prof. Paper* 600D:169–73.
Downs, T. 1954. Pleistocene birds from the Jones fauna of Kansas. *Condor* 56:207–21.
Downs, T., and White, J. A. 1968. A vertebrate faunal succession in superposed sediments from

late Pliocene to middle Pleistocene in California. *Proc. 23rd Internat. Geol. Cong. Prague* 10:41–47.

Downs, T., et al. 1959. Quaternary animals from Schuiling Cave in the Mojave Desert, California. *Contrib. Sci. Los Angeles Co. Mus.* 29:1–21.

Dreimanis, A. 1968. Extinction of mastodons in eastern North America: Testing a new climatic environmental hypothesis. *Ohio J. Sci.* 68(6):257–72.

Eames, A. J. 1930. Report on ground sloth coprolite from Dona Ana County, New Mexico. *Amer. J. Sci.* 20(119):353–56.

Eddleman, C. D., and Akersten, W. A. 1966. Margay from the post-Wisconsin of southeastern Texas. *Tex. J. Sci.* 18(4):378–85.

Einsohn, S. D. 1971. The stratigraphy and fauna of a Pleistocene outcrop in Doniphan County, northeastern Kansas. M.S. thesis. Univ. Kans., Lawrence.

Eldredge, N., and Gould, S. J. 1972. Punctuated equilibria: An alternative to phyletic gradualism. In T. J. M. Schopf (ed.), *Models in Paleobiology*. San Francisco: Freeman, Cooper & Co. Pp. 82–115.

Elftman, H. O. 1931. Pleistocene mammals of Fossil Lake, Oregon. *Amer. Mus. Novitates* 481:1–21.

Erdbrink, D. P. 1953. *A Review of Fossil and Recent Bears of the Old World with Remarks on Their Phylogeny Based upon Their Dentition*, 2 vols. Deventer, Neth.: Jan de Lange. 597 pp.

Erickson, B. R. 1962. A description of *Castoroides ohioensis* from Minnesota. *Proc. Minn. Acad. Sci.*, 30:6–13.

Eshelman, R. E. 1975. Geology and paleontology of the early Pleistocene (late Blancan) White Rock fauna from north-central Kansas. In *Studies on Cenozoic Paleontology and Stratigraphy, Claude W. Hibbard Memorial Vol.*, 4, *Univ. Mich. Papers Paleontol.*, 13:1–60.

Eshelman, R. E., Evenson, E. B., and Hibbard, C. W. 1972. The peccary, *Platygonus compressus* LeConte from beneath Wisconsinan till, Washtenaw County, Michigan. *Mich. Academician* 5(2):243–56.

Etheridge, R. 1958. Pleistocene lizards of the Cragin Quarry fauna of Meade County, Kansas. *Copeia* 1958:94–101.

Evans, G. L. 1961. The Freisenhahn Cave. *Bull. Tex. Mem. Mus.* 2:1–22.

Evans, G. L., and Meade, G. E. 1945. Quaternary of the Texas High Plains. *Univ. Tex. Publ.* 4401:485–507.

Evernden, J. F., et al. 1964. Potassium-argon dates and the Cenozoic mammalian chronology of North America. *Amer. J. Sci.* 262:145–98.

Ewer, R. F. 1973. *The Carnivores*. Ithaca, N.Y.: Cornell Univ. Press. 494 pp.

Fahlbusch, V. 1976. Report on the International Symposium on mammalian stratigraphy of the European Tertiary. *Newsl. Stratigr.* 5:160–67.

Falconer, H. 1863. On the American fossil elephant of the regions bordering the Gulf of Mexico (*Elephas columbi* Falc.), with general observations on the living and extinct species. *Nat. Hist. Rev.* 3:43–114.

Feduccia, J. A. 1967. *Ciconia maltha* and *Grus americana* from the upper Pliocene of Idaho. *Wilson Bull.* 79:316–18.

Figgins, J. D. 1933. The bison of the western area of the Mississippi basin. *Proc. Colo. Mus. Nat. Hist.* 12(4):16–33.

Finch, W. C., Whitmore, F. C., Jr., and Sims, J. D. 1972. Stratigraphy, morphology, and paleoecology of a fossil peccary herd from western Kentucky. *U.S. Geol. Surv. Prof. Paper* 790:1–25.

Findley, J. S. 1953. Pleistocene Soricidae from San Josecito Cave, Nuevo Leon, Mexico. *Univ. Kans. Publ Mus. Nat. Hist.* 5(36):633–39.

Fleischer, R. L., Price, P. B., and Walker, R. M. 1969. Quaternary dating by the fission track technique. In D. Brothwell and E. Higgs (eds.), *Science in Archaeology*. London: Thames & Hudson. Pp. 58–61.

Flerov, C. C. 1967. On the origin of the mammalian fauna of Canada. In D. M. Hopkins (ed.), *The Bering Land Bridge*. Stanford, Calif.: Stanford Univ. Press. Pp. 271–80.

REFERENCES

———. 1975. Die Bison-Reste aus den Travertinen von Weimar-Ehringsdorf. *Abhandl. Zent. Geol. Inst. Berlin* 23:171–99.

Flint, R. F. 1971. *Glacial and Quaternary Geology.* New York: John Wiley & Sons. 829 pp.

Fortsch, D. E. 1972. A late Pleistocene vertebrate fauna from the northern Mojave Desert of California. M.S. thesis, Univ. So. Calif., Los Angeles.

Frazier, M. F. 1977. New records of *Neofiber leonardi* (Rodentia, Cricetidae) and the paleoecology of the genus. *J. Mammal.* 58(3):368–73.

Freudenberg, W. 1910. Die Säugetierfauna des Pliocäns und Postpliocäns von Mexiko. *Geol. Paleontol. Abhandl. NS* 9:195–231.

Frick, C. 1921. Extinct vertebrate faunas of the badlands of Bautista Creek and San Timoteo Canyon, southern California. *Univ. Calif. Publ. Bull. Dept. Geol.* 12:277–424.

———. 1937. Horned ruminants of North America. *Bull. Amer. Mus. Nat. Hist.* 69:1–699.

Friedman, I., Smith, R. L., and Clark, D. 1969. Obsidian dating. In D. Brothwell and E. Higgs (eds.), *Science in Archaeology.* London: Thames & Hudson. Pp. 62–75.

Frison, G. C., ed. 1974. *The Casper Site: A Hell Gap Bison Kill on the High Plains.* New York: Academic Press. 266 pp.

———. 1975. Man's interaction with Holocene environments on the Plains. *Quatern. Res.* 5:289–300.

Frison, G. C., Wilson, M., and Wilson, D. J. 1976. Fossil bison and artifacts from an early altithermal period arroyo trap in Wyoming. *Amer. Antiq.* 41(1):28–57.

Frison, G. C., et al. 1978. Paleo-Indian procurement of *Camelops* on the northwestern plains. *Quatern. Res.* 10(3):385–400.

Fry, W. E., and Gustafson, E. P. 1974. Cervids from the Pliocene and Pleistocene of central Washington. *J. Paleontol.* 48(2):375–86.

Frye, J. C., Leonard, A. B., and Hibbard, C. W. 1943. Westward extension of the Kansas "*Equus* beds." *J. Geol.* 51:33–47.

Fuller, W. A., and Bayrock, L. A. 1965. Late Pleistocene mammals from central Alberta, Canada. In R. E. Folinsbee and D. M. Ross (eds.), *Vertebrate Paleontology in Alberta.* Edmonton: Univ. Alberta Press. Pp. 53–63.

Funderberg, J. B. 1960. Fossil manatee from North Carolina. *J. Mammal.* 41(4):521.

Furlong, E. L. 1905. *Preptoceras,* a new ungulate from the Samwel Cave, California. *Univ. Calif. Publ. Bull. Dept. Geol.* 4(8):163–69.

———. 1906. The exploration of Samwel Cave. *Amer. J. Sci.* 22:235–47.

———. 1925. Notes on the occurrence of mammalian remains in the Pleistocene of Mexico with the description of a new species, *Capromeryx mexicana. Univ. Calif. Publ. Bull. Dept. Geol.* 15(5):137–52.

———. 1943. The Pleistocene antelope *Stockoceros conklingi* from San Josecito Cave, Mexico. *Publ. Carnegie Inst. Washington* 551:1–8.

———. 1946. Generic identification of the Pleistocene antelope from Rancho La Brea. *Publ. Carnegie Inst. Washington* 551:135–40.

Furlong, E. L., and Sinclair, W. J. 1904. Preliminary description of *Euceratherium collinum. Univ. Calif. Publ. Amer. Archaeol. Ethnol.* 2:18.

Gabunia, L. K. 1972. On the Neogene–Quaternary boundary in Europe (as based on the data of mammalian fauna). *Coll. Papers Internat. Colloq. Neogene-Quaternary* 2:34–46.

Galbreath, E. C. 1938. Post-glacial fossil vertebrates from east-central Illinois. *Field Mus. Nat. Hist. Geol. Ser.* 6:303–13.

———. 1964. A dire wolf skeleton and Powder Mill Creek Cave, Missouri. *Trans. Ill. State Acad. Sci.,* 57(4):224–42.

Gard, L. M., Jr., Lewis, G. E., and Whitmore, F. C., Jr. 1972. Steller's sea cow in Pleistocene Interglacial beach deposits on Amchitka, Aleutians. *Bull. Geol. Soc. Amer.* 83:867–70.

Gardner, A. L. 1973. The systematics of the Genus *Didelphis* (Marsupialia, Didelphidae) in North and Middle America. *Spec. Publ. Mus. Tex. Tech. Univ.* 4:1–81.

Gazin, C. L. 1933a. A new shrew from the Upper Pliocene of Idaho. *J. Mammal.* 14:142–44.

Gazin, C. L. 1933b. New felids from the Upper Pliocene of Idaho. *J. Mammal.* 14:251–56.
———. 1934a. Fossil hares from the late Pliocene of southern Idaho. *Proc. U.S. Nat. Mus.* 83:111–21.
———. 1934b. Upper Pliocene mustelids from the Snake River Basin of Idaho. *J. Mammal.* 15(2):137–49.
———. 1935. A new antilocaprid from the Upper Pliocene of Idaho. *J. Paleontol.* 9:390–93.
———. 1936. A study of the fossil horse remains from the Upper Pliocene of Idaho. *Proc. U.S. Nat. Mus.* 83(2985):281–320.
———. 1938. Fossil peccary remains from the Upper Pliocene of Idaho. *J. Washington Acad. Sci.* 28(2):41–49.
———. 1942. The late Cenozoic vertebrate faunas from the San Pedro Valley, Ariz. *Proc. U.S. Nat. Mus.* 92(3155):475–518.
———. 1950. Annotated list of fossil Mammalia associated with human remains at Melbourne, Florida. *J. Washington Acad. Sci.* 40(12):397–404.
———. 1956. Exploration for the remains of giant ground sloths in Panama. *Smithsonian Inst. Ann. Rept.* 4272:341–54.
Geist, V. 1971. *Mountain sheep: A Study in Behavior and Evolution.* Chicago and London: Univ. Chicago Press. 383 pp.
Getz, L. 1960. Middle Pleistocene carnivores from southwestern Kansas. *J. Mammal.* 41:361–65.
Gidley, J. W. 1900. A new species of Pleistocene horse from the Staked Plains of Texas. *Bull. Amer. Mus. Nat. Hist.* 13:111–16.
———. 1901. Tooth characters and revision of the North American species of the genus *Equus*. *Bull. Amer. Mus. Nat. Hist.* 14:91–142.
———. 1903. On two species of *Platygonus* from the Pliocene of Texas. *Bull. Amer. Mus. Nat. Hist.* 19(14):477–81.
———. 1906. A fossil raccoon from a California Pleistocene deposit. *Proc. U.S. Nat. Mus.* 29:553–54.
———. 1908. Descriptions of two new species of Pleistocene ruminants of the genera *Ovibos* and *Bootherium*, with notes on the latter genus. *Proc. U.S. Nat. Mus.* 34(1627):681–84.
———. 1913a. Preliminary report on a recently discovered Pleistocene cave deposit near Cumberland, Maryland. *Proc. U.S. Nat. Mus.* 46:93–102.
———. 1913b. An extinct American eland. *Smithsonian Misc. Coll.* 60(27):1–3.
———. 1920. Pleistocene peccaries from the Cumberland Cave deposit. *Proc. U.S. Nat. Mus.* 57:651–78.
———. 1922a. Field explorations in the San Pedro Valley and Sulphur Springs Valley of southern Arizona. *Explorations and field-work of the Smithsonian Institution in 1921*, pp. 25–30.
———. 1922b. Preliminary report on fossil vertebrates of the San Pedro Valley, Arizona, with descriptions of new species of rodents and lagomorphs. *U.S. Geol. Surv. Prof. Paper* 131E:119–31.
———. 1926. Fossil Proboscidea and Edentata of the San Pedro Valley, Arizona. *U.S. Geol. Surv. Prof. Paper* 140B:83–95.
———. 1928. A new species of bear from the Pleistocene of Florida. *J. Washington Acad. Sci.* 18:430–33.
Gidley, J. W., and Gazin, C. L. 1933. New Mammalia in the Pleistocene fauna from Cumberland Cave. *J. Mammal.* 14:343–57.
———. 1938. The Pleistocene vertebrate fauna from Cumberland Cave, Maryland. *Bull. U.S. Nat. Mus.* 171:1–99.
Giles, E. 1960. Multivariate analysis of Pleistocene and Recent coyotes (*Canis latrans*) from California. *Univ. Calif. Publ. Geol. Sci.* 36(8):369–90.
Gillette, D. D. 1973. A review of North American glyptodonts (Edentata, Mammalia): Osteology, systematics, and paleontology. Ph.D. dissertation, Southern Methodist Univ., Dallas, Tex.
———. 1976. A new species of small cat from the late Quaternary of southeastern United States. *J. Mammal.* 57(4):664–76.

Goldman, E. A. 1941. A Pleistocene otter from Iowa. *Papers Mammalogy Field Mus. Nat. Hist. Zool. Ser.* 27:229–31.
Goodrich, C. 1940. Mollusks of a Kansas Pleistocene deposit. *Nautilus* 53:77–79.
Graham, R. W. 1972. Biostratigraphy and paleoecological significance of the Conard Fissure local fauna with emphasis on the genus *Blarina*. M.Sc. thesis, Univ. Iowa, Iowa City.
Graham, R. W., and Semken, H. A. 1976. Paleoecological significance of the short-tailed shrew (*Blarina*), with a systematic description of *Blarina ozarkensis*. *J. Mammal.* 57(3):433–49.
Green, M. 1963. Some late Pleistocene rodents from South Dakota. *J. Paleontol.* 37(3):688–90.
Gromova, V. I. 1935. Ueber neue Funde von *Bison priscus longicornis* mihi. *Trav. Inst. Paleozool. Acad. Sci. URSS*, 4:137–47. (Russian with German summary.)
———. 1949. Histoire des chevaux (genre *Equus*) de l'Ancien Monde. *Trav. Inst. Paleontol. Acad. Sci. URSS.* 7:1–373. (French translation by Piedresson de St. Auboin. Ann. Cent. Etud. Docum. Paleont. 13, Paris, 1955.)
Gromova, V. I., ed. 1962. *Fundamentals of Paleontology*, vol. 13. (Israel Program for Scientific Translation, 1968.)
Gruhn, R. 1961. The archaeology of Wilson Butte Cave, south-central Idaho. *Occ. Papers Idaho State College Mus.* 6:1–198.
Guilday, J. E. 1958. The prehistoric distribution of the opossum. *J. Mammal.* 39(1):39–43.
———. 1961. Prehistoric record of *Scalopus* from western Pennsylvania. *J. Mammal.* 42(1):117–18.
———. 1962a. Notes on Pleistocene vertebrates from Wythe County, Virginia. *Ann. Carnegie Mus.* 36:77–86.
———. 1962b. The Pleistocene local fauna of the Natural Chimneys, Augusta County, Virginia. *Ann. Carnegie Mus.* 36:87–122.
———. 1963. Pleistocene zoogeography of the lemming, *Dicrostonyx*. *Evolution* 17(2):194–97.
———. 1967a. Differential extinction during late-Pleistocene and Recent times. In P. S. Martin and H. E. Wright (eds.), *Pleistocene Extinctions: The Search for a Cause*. New Haven and London: Yale Univ. Press. Pp. 121–40.
———. 1967b. Notes on the Pleistocene big brown bat. *Ann. Carnegie Mus.* 39(7):105–14.
———. 1967c. The climatic significance of the Hosterman's Pit local fauna, Centre County, Pennsylvania. *Amer. Antiq.* 32(2):231–32.
———. 1968a. Grizzly bears from eastern North America. *Amer. Midl. Nat.* 79(1):247–50.
———. 1968b. Pleistocene zoogeography of the lemming *Dicrostonyx*. *Univ. Colo. Stud. Ser. Earth Sci.* 6:61–71.
———. 1969. Small mammal remains from the Wasden Site (Owl Cave), Bonneville County, Idaho. *Tebiwa* 12(1):47–57.
———. 1971a. Biological and archaeological analysis of bones from a 17th century Indian village (46Pu31), Putnam County, West Virginia. *Rept. Archaeol. Inv. No. 4, W. Va. Geol. Econ. Surv.*, 64 pp.
———. 1971b. The Pleistocene history of the Appalachian mammal fauna. *Res. Div. Mongr. Va. Polytechnic Inst. State Univ.* 4:233–62.
———. 1977. Sabertooth cat, *Smilodon floridanus* (Leidy), and associated fauna from a Tennessee cave (40Dv40), the First American Bank site. *J. Tenn. Acad. Sci.* 52(3):84–94.
Guilday, J. E., and Adam, E. K. 1967. Small mammal remains from Jaguar Cave, Lemhi County, Idaho. *Tebiwa* 10(1):26–36.
Guilday, J. E., and Bender, M. S. 1960. Late Pleistocene records of the yellow-cheeked vole, *Microtus xanthognathus* (Leach). *Ann. Carnegie Mus.* 35:315–30.
Guilday, J. E., and Doutt, J. K. 1961. The collared lemming (*Dicrostonyx*) from the Pennsylvania Pleistocene. *Proc. Biol. Soc. Washington* 74:249–50.
Guilday, J. E., and Hamilton, H. W. 1973. The late Pleistocene small mammals of Eagle Cave, Pendleton County, West Virginia. *Ann. Carnegie Mus.* 44(5):45–58.
Guilday, J. E., and Handley, C. O., Jr. 1967. A new *Peromyscus* (Rodentia, Cricetidae) from the Pleistocene of Maryland. *Ann. Carnegie Mus.* 39(6):91–103.

Guilday, J. E., and Irving, D. C. 1967. Extinct Florida spectacled bear *Tremarctos floridanus* (Gidley) from central Tennessee. *Bull. Nat. Speleol. Soc.* 29(4):149–62.
Guilday, J. E., and McCrady, A. D. 1966. Armadillo remains from Tennessee and West Virginia caves. *Bull. Nat. Speleol. Soc.* 28(4):183–84.
Guilday, J. E., and McGinnis, H. 1972. Jaguar (*Panthera onca*) from Big Bone Cave, Tennessee and east central North America. *Bull. Nat. Speleol. Soc.* 34(1):1–14.
Guilday, J. E., and Parmalee, P. W. 1972. Quaternary Periglacial records of voles of the genus *Phenacomys* Merriam (Cricetidae; Rodentia). *Quatern. Res.* 2(2):170–75.
Guilday, J. E., Hamilton, H. W., and Adam, E. K. 1967. Animal remains from Horned Owl Cave, Albany County, Wyoming. *Contrib. Geol.* 6(2):97–99.
Guilday, J. E., Hamilton, H. W., and McCrady, A. D. 1966. The bone breccia of Bootlegger Sink, York County, Pennsylvania. *Ann. Carnegie Mus.* 38(8):145–63.
———, ———, and ———. 1969. The Pleistocene vertebrate fauna of Robinson Cave, Overton County, Tennessee. *Palaeovertebrata* 2:25–75.
———, ———, and ———. 1971. The Welsh Cave peccaries (*Platygonus*) and associated fauna, Kentucky Pleistocene. *Ann. Carnegie Mus.* 43(9):249–320.
Guilday, J. E., Hamilton, H. W., and Parmalee, P. W. 1975. Caribou (*Rangifer tarandus* L.) from the Pleistocene of Tennessee. *J. Tenn. Acad. Sci.* 50(3):109–12.
Guilday, J. E., Martin, P. S., and McCrady, A. D. 1964. New Paris No. 4: A Pleistocene cave deposit in Bedford County, Pennsylvania. *Bull. Nat. Speleol. Soc.* 26(4):121–94.
Guilday, J. E., Parmalee, P. W., and Hamilton, H. W. 1977. The Clark's Cave bone deposit and the late Pleistocene paleoecology of the central Appalachian Mountains of Virginia. *Bull. Carnegie Mus. Nat. Hist.* 2:1–87.
Guilday, J. E., et al. 1978. The Baker Bluff cave deposit, Tennessee, and the late Pleistocene faunal gradient. *Bull. Carnegie Mus. Nat. Hist.* 11:1–67.
Gunter, H. 1931. The mastodon from Wakulla Spring, Wakulla County, Florida. In *Florida Woods and Waters*. Dept. Game and Fresh Water Fish, Tallahassee, Fla., pp. 14–16.
Gustafson, E. P. 1978. The vertebrate faunas of the Pliocene Ringold Formation, south-central Washington. *Bull. Mus. Nat. Hist. Univ. Oreg., Eugene* 23:1–62.
Gut, H. J. 1959. A Pleistocene vampire bat from Florida. *J. Mammal.* 40(4):534–38.
Gut, H. J., and Ray, C. E. 1963. The Pleistocene vertebrate fauna of Reddick, Florida. *Quart. J. Fla. Acad. Sci.* 26(4):315–28.
Guthrie, R. D. 1966a. Bison horn cores—character choice and systematics. *J. Paleontol.* 40:738–40.
———. 1966b. Pelage of fossil bison—a new osteological index. *J. Mammal.* 47(4):725–27.
———. 1966c. The extinct wapiti of Alaska and Yukon Territory. *Canad. J. Zool.* 44(1):45–47.
———. 1968a. Paleoecology of the large mammal community in interior Alaska during the late Pleistocene. *Amer. Midl. Nat.* 79(2):346–63.
———. 1968b. Paleoecology of a late Pleistocene small mammal community from interior Alaska. *Arctic* 21(4):223–44.
———. 1970. Bison evolution and zoogeography in North America during the Pleistocene. *Quart. Rev. Biol.* 45(1):1–15.
———. 1972. Re-creating a vanished world. *Nat. Geog. 141(3):294*–301.
———. 1973. Mummified pika (*Ochotona*) carcass and dung pellets from Pleistocene deposits in interior Alaska. *J. Mammal.* 54(4):970–71.
Guthrie, R. D., and Matthews, J. V. 1971. The Cape Deceit Fauna—Early Pleistocene mammalian assemblage from the Alaskan Arctic. *Quatern. Res.* 1(4):474–510.
Hager, M. W. 1972. A late Wisconsin–Recent vertebrate fauna from the Chimney Rock Animal Trap, Larimer County, Colorado. *Contrib. Geol.* 11(2):63–71.
———. 1975. Late Pliocene and Pleistocene history of the Donnelly Ranch vertebrate site, southeastern Colorado. *Contrib. Geol.*, Spec. Paper No. 2, 62 pp.
Hall, E. R. 1926. A new marten from the Pleistocene cave deposits of California. *J. Mammal.* 7(2):127–30.

———. 1930. A bassarisk and a new mustelid from the later Tertiary of California. *J. Mammal.* 11(1):23–26.

———. 1936. Mustelid mammals from the Pleistocene of North America with systematic notes on some recent members of the genera *Mustela, Taxidea,* and *Mephitis. Carnegie Inst. Washington Publ.* 473:41–119.

———. 1944. A new genus of American Pliocene badger with remarks on the relationships of the badgers of the Northern Hemisphere. *Carnegie Inst. Washington Publ.* 551:9–23.

———. 1951a. American weasels. *Univ. Kans. Publ. Mus. Nat. Hist.* 4:1–466.

———. 1951b. A synopsis of the North American lagomorphs. *Univ. Kans. Publ. Mus. Nat. Hist.* 5:119–202.

Hall, E. R., and Kelson, K. 1959. *The Mammals of North America.* New York: The Ronald Press. 2 vols., 1,083 pp.

Hall, E. T. 1969. Dating pottery by thermoluminescence. In D. Brothwell and E. Higgs (eds.), *Science in Archaeology.* London: Thames & Hudson. Pp. 106–8.

Hallberg, G. R., Semken, H. A., and Carson, L. C. 1974. Quaternary records of *Microtus xanthognathus* (Leach), the yellow-cheeked vole, from northwestern Arkansas and southwestern Iowa. *J. Mammal.* 55(3):640–45.

Hamon, J. H. 1964. Osteology and paleontology of the passerine birds of the Reddick, Florida Pleistocene. *Bull. Fla. Geol. Surv.* 44:1–210.

Handley, C. O., Jr. 1955. A new Pleistocene bat *Corynorhinus* from Mexico. *J. Washington Acad. Sci.* 45(2):48–49.

———. 1956. Bones of mammals from West Virginia caves. *Amer. Midl. Nat.* 56(1):250–56.

Haq, B. U., Berggren, W. A., and Van Couvering, J. A. 1977. Corrected age of the Pliocene/Pleistocene boundary. *Nature* 269(5628):483–88.

Harington, C. R. 1969. Pleistocene remains of the lion-like cat (*Panthera atrox*) from the Yukon Territory and northern Alaska. *Canad. J. Earth Sci.* 6(5):1277–88.

———. 1970a. Ice age mammal research in the Yukon Territory and Alaska. In R. A. Smith and J. W. Smith (eds.), *Early Man and Environments in Northwest North America.* Calgary, Alberta: Univ. Calgary Archaeol. Assoc. Pp. 35–51.

———. 1970b. A postglacial muskox (*Ovibos moschatus*) from Grandview, Manitoba, and comments on the zoogeography of *Ovibos. Nat. Mus. Canada Publ. Paleontol.* 2:1–13.

———. 1970c. A Pleistocene muskox (*Ovibos moschatus*) from gravels of Illinoian age near Nome, Alaska. *Canad. J. Earth Sci.* 7(5):1326–31.

———. 1971a. A Pleistocene lion-like cat (*Panthera atrox*) from Alberta. *Canad. J. Earth Sci.* 8(1):170–74.

———. 1971b. A Pleistocene mountain goat from British Columbia with comments on the dispersal history of *Oreamnos. Canad. J. Earth Sci.* 8(9):1081–93.

———. 1971c. Ice Age mammals in Canada. *Arctic Circ.* 22(2):66–88.

———. 1975. Pleistocene muskoxen (*Symbos*) from Alberta and British Columbia. *Canad. J. Earth Sci.* 12(6):903–19.

———. 1977. Marine mammals in the Champlain Sea and the Great Lakes. *Ann. N.Y. Acad. Sci.* 288:508–37.

———. 1978. Quaternary vertebrate faunas of Canada and Alaska and their suggested chronological sequence. *Syllogeus* 15:1–105.

Harington, C.R., and Clulow, F. V. 1973. Pleistocene mammals from Gold Run Creek, Yukon Territory. *Canad. J. Earth Sci.* 10(5):697–759.

Harington, C. R., and Sergeant, D. E. 1972. Pleistocene ringed seal skeleton from Champlain Sea deposits near Hull, Quebec—a reidentification. *Canad. J. Earth Sci.* 9(8):1039–51.

Harington, C. R., and Shackleton, D. M. 1978. A tooth of *Mammuthus primigenius* from Chestermere Lake near Calgary, Alberta, and the distribution of mammoths in southwestern Canada. *Canad. J. Sci.* 15(8):1272–83.

Harington, C.R., Tipper, H.W., and Mott, R.J. 1974. Mammoth from Babine Lake, British Columbia. *Canad. J. Earth Sci.* 11(2):285–303.

Harlan, R. 1825. *Fauna Americana*. Philadelphia.

Harlow, R. F. 1964. Osteometric data for the Florida black bear. *Quart. J. Fla. Acad. Sci.* 25:257–74.

Harris, A. H. 1970. The Dry Cave mammalian fauna and late Pluvial conditions in southeastern New Mexico. *Tex. J. Sci.* 22(1):3–27.

Harris, A. H., and Findley, J. S. 1964. Pleistocene–Recent fauna of the Isleta caves, Bernalillo County, New Mexico. *Amer. J. Sci.* 262:114–20.

Harris, A. H., and Mundel, P. 1974. Size reduction in bighorn sheep (*Ovis canadensis*) at the close of the Pleistocene. *J. Mammal.* 55(3):678–80.

Haury, E. W. 1950. *The Stratigraphy and Archaeology of Ventana Cave, Arizona*. Albuquerque: Univ. N. Mex. Press. 599 pp.

Haury, E. W., Saylor, E. B., and Wasley, W. W. 1959. The Lehner mammoth site, southeastern Arizona. *Amer. Antiq.* 25(1):2–30.

Hausman, L. A. 1929. The "ovate bodies" of the hair of *Nothrotherium shastense*. *Amer. J. Sci.* 18(5):331–36.

Hawksley, O. 1965. Short-faced bear (*Arctodus*) fossils from Ozark caves. *Bull. Nat. Speleol. Soc.* 27(3):77–92.

Hawksley, O., and McGowan, J. 1963. The dire wolf in Missouri. *Mo. Speleol.* 5:63–72.

Hawksley, O., Reynolds, J. E., and Foley, R. L. 1973. Pleistocene vertebrate fauna of Bat Cave, Pulaski County, Missouri. *Bull. Nat. Speleol. Soc.* 35(3):61–87.

Hay, O. P. 1899. A census of the fossil Vertebrata of North America. *Science, ser. 2*, 10:681–84.

———. 1913a. Notes on some fossil horses, with descriptions of four new species. *Proc. U.S. Nat. Mus.* 44:569–94.

———. 1913b. The extinct bisons of North America; with description of one new species, *Bison regius*. *Proc. U.S. Nat. Mus.* 46:166–200.

———. 1913c. Description of two new species of ruminants from the Pleistocene of Iowa. *Proc. Biol. Soc. Washington* 26:5–8.

———. 1914. The Pleistocene mammals of Iowa. *Ann. Rept. Iowa Geol. Surv.* 23:1–662.

———. 1915. Contributions to the knowledge of the mammals of the Pleistocene of North America. *Proc. U.S. Nat. Mus.* 48:515–75.

———. 1917a. Description of a new species of extinct horse, *Equus lambei*, from the Pleistocene of Yukon Territory. *Proc. U.S. Nat. Mus.* 53:435–43.

———. 1917b. On a collection of fossil vertebrates made by Dr. F. W. Cragin in the *Equus* beds of Kansas. *Kans. Univ. Sci. Bull.* 10(4):39–51.

———. 1917c. Vertebrata, mostly from stratum no. 3, Vero, Florida, together with descriptions of new species. *Fla. Geol. Surv. 9th Ann. Rept.* pp. 43–68.

———. 1919. Descriptions of some mammalian and fish remains from Florida of probably Pleistocene age. *Proc. U.S. Nat. Mus.* 56:103–12.

———. 1920. Descriptions of some Pleistocene vertebrates found in the United States. *Proc. U.S. Nat. Mus.* 58:83–146.

———. 1921. Descriptions of Pleistocene Vertebrata, types or specimens of which are preserved in the United States National Museum. *Proc. U.S. Nat. Mus.* 59:617–38.

———. 1923. The Pleistocene of North America and its vertebrated animals from the states East of the Mississippi River and from the Canadian Provinces East of Longitude 95°. *Carnegie Inst. Washington Publ.* 322:1–499.

———. 1924. The Pleistocene of the Middle Region of North America and its vertebrated animals. *Carnegie Inst. Washington Publ.* 322A:1–385.

———. 1926. A collection of Pleistocene vertebrates from southwestern Texas. *Proc. U.S. Nat. Mus.* 68(24):1–18.

———. 1927. The Pleistocene of the Western Region of North America and its vertebrated animals. *Carnegie Inst. Washington Publ.* 322B:1–346.

Hay, O. P., and Cook, H. J. 1928. Preliminary descriptions of fossil mammals recently discovered in Oklahoma, Texas, and New Mexico. *Proc. Colo. Mus. Nat. Hist.* 8(2):33.

REFERENCES

―― and ――. 1930. Fossil vertebrates collected near or in association with human artifacts at localities near Colorado, Texas, Frederick, Oklahoma, and Folsom, New Mexico. *Proc. Colo. Mus. Nat. Hist.* 9(2):2–40.

Haynes, C. V., Jr. 1967. Carbon-14 dates and early man in the New World. In P. S. Martin and H. E. Wright, Jr. (eds.), *Pleistocene Extinctions: The Search for a Cause.* New Haven and London: Yale Univ. Press. Pp. 267–86.

――. 1969. The Murray Springs Clovis site. *Plains Anthropol.* 14(46): 298–99.

――. 1970. Geochronology of Man–Mammoth sites and their bearing on the origin of the Llano Complex. In W. Dort, Jr., and J. K. Jones, Jr. (eds.), *Pleistocene and Recent Environments of the Central Great Plains; Dept. Geol. Univ. Kans. Spec. Publ.* 3:77–92.

――. 1973. The Calico site: Artifacts or geofacts? *Science* 181:305–10.

――. 1974. Paleoenvironments and cultural diversity in late Pleistocene South America: A reply to A. L. Bryan. *Quatern. Res.* 4(3):378–82.

Haynes, C. V., Jr., and Hemmings, E. T. 1968. Mammoth-bone shaft wrench from Murray Springs, Arizona. *Science* 159:186–87.

Hays, J. D., Imbrie, J., and Shackleton, N. J. 1976. Variations in the earth's orbit: Pacemaker of the Ice Ages. *Science* 194:1121–32.

Hemmer, H. 1971. Zur Charakterisierung und straigraphischen Bedeutung von *Panthera gombaszoegensis* (Kretzoi, 1938). *Neues Jahrb. Geol. Palaeontol. Monatsch.* 1971:701–11.

――. 1973. Neue Befunde zur Verbreitung und Evolution der pliozänpleistozänen Gattung *Dinofelis* (Mammalia, Carnivora, Felidae). *Neues Jahrb. Geol. Palaeontol. Monatsch.* 1973:157–69.

――. 1974. Zur Artgeschichte des Löwen *Panthera (Panthera) leo* (Linnaeus, 1758). *Veröff. Zool. Staatssamml. München* 17:167–280.

Hershkovitz, P. 1966. Mice, land bridges and Latin American faunal interchange. In R. L. Wenzel and V. J. Tipton (eds.), *Ectoparasites of Panama.* Chicago: Field Museum Nat. Hist. Pp. 725–51.

Hesse, C. J. 1935. New evidence on the ancestry of *Antilocapra americana. J. Mammal.* 16(4):307–15.

Hester, J. J. 1967. The agency of man in animal extinctions. In P. S. Martin and H. E. Wright (eds.), *Pleistocene Extinctions: The Search for a Cause.* New Haven and London: Yale Univ. Press. Pp. 169–92.

Hibbard, C. W. 1937. A new *Pitymys* from the Pleistocene of Kansas. *J. Mammal.* 18(2):235.

――. 1938. An Upper Pliocene fauna from Meade County, Kansas. *Trans. Kans. Acad. Sci.* 40:239–65.

――. 1939a. Four new rabbits from the Upper Pliocene of Kansas. *Amer. Midl. Nat.* 21:506–13.

――. 1939b. A new *Synaptomys* from the Pleistocene. *Univ. Kans. Sci. Bull.* 26(8):367–71.

――. 1939c. Notes on some mammals from the Pleistocene of Kansas. *Trans. Kans. Acad. Sci.* 42:463–70.

――. 1939d. *Nekrolagus*, a new name for *Pediolagus* Hibbard, not Marelli. *Amer. Midl. Nat.* 21: Table of contents.

――. 1940a. The occurrence of *Cervalces scotti* Lydekker in Kansas. *Trans. Kans. Acad. Sci.* 43:411–15.

――. 1940b. A new Pleistocene fauna from Meade County, Kansas. *Trans. Kans. Acad. Sci.* 43:417–25.

――. 1941a. Mammals of the Rexroad fauna from the Upper Pliocene of southwestern Kansas. *Trans. Kans. Acad. Sci.* 44:265–313.

――. 1941b. New mammals from the Rexroad fauna, Upper Pliocene of Kansas. *Amer. Midl. Nat.* 26(2):337–68.

――. 1941c. Paleoecology and correlation of the Rexroad fauna from the Upper Pliocene of southwestern Kansas, as indicated by the mammals. *Univ. Kans. Sci. Bull.* 27:79–104.

――. 1941d. The Borchers fauna, a new Pleistocene interglacial fauna from Meade County, Kansas. *Bull. Kans. Geol. Surv.* 38:197–220.

――. 1942. Pleistocene mammals from Kansas. *Bull. Kans. Geol. Surv.* 41:261–69.

Hibbard, C. W. 1943a. The Rezabek fauna, a new Pleistocene fauna from Lincoln County, Kansas. *Univ. Kans. Sci. Bull.* 29(2):235–47.
———. 1943b. *Etadonomys*, a new Pleistocene heteromyid rodent, and notes on other Kansas mammals. *Trans. Kans. Acad. Sci.* 46:185–91.
———. 1944. Stratigraphy and vertebrate paleontology of Pleistocene deposits of southwestern Kansas. *Bull. Geol. Soc. Amer.* 55:718–44.
———. 1949a. Pleistocene stratigraphy and paleontology of Meade County, Kansas. *Contrib. Mus. Paleontol. Univ. Mich.* 7(4):63–90.
———. 1949b. Pliocene Saw Rock Canyon fauna in Kansas. *Contrib. Mus. Paleontol. Univ. Mich.* 7:91–105.
———. 1949c. Techniques of collecting microvertebrate fossils. *Contrib. Mus. Paleontol. Univ. Mich.* 8:7–19.
———. 1950. Mammals of the Rexroad Formation from Fox Canyon, Kansas. *Contrib. Mus. Paleontol. Univ. Mich.* 8(6):113–92.
———. 1951a. Vertebrate fossils from the Pleistocene Stump Arroyo Member, Meade County, Kansas. *Contrib. Mus. Paleontol. Univ. Mich.* 9(7):227–45.
———. 1951b. A new jumping mouse from the Upper Pliocene of Kansas. *J. Mammal.* 32:351–52.
———. 1952a. A contribution to the Rexroad fauna. *Trans. Kans. Acad. Sci.* 55(2):196–208.
———. 1952b. A new *Bassariscus* from the Upper Pliocene of Kansas. *J. Mammal.* 33(3):379–81.
———. 1953a. The insectivores of the Rexroad fauna, Upper Pliocene of Kansas. *J. Paleontol.* 27(1):21–32.
———. 1953b. The Saw Rock Canyon fauna and its stratigraphic significance. *Papers Mich. Acad. Sci. Arts Letters* 38:387–411.
———. 1953c. *Equus (Asinus) calobatus* Troxell and associated vertebrates from the Pleistocene of Kansas. *Trans. Kans. Acad. Sci.* 56(1):111–26.
———. 1954a. Second contribution of the Rexroad fauna. *Trans. Kans. Acad. Sci.* 57(2):221–37.
———. 1954b. A new Pliocene vertebrate fauna from Oklahoma. *Papers Mich. Acad. Sci. Arts Letters* 39:339–59.
———. 1955a. Notes on the microtine rodents from Port Kennedy Cave deposit. *Proc. Acad. Nat. Sci. Phila.* 107:87–97.
———. 1955b. Pleistocene vertebrates from the Upper Becerra (Becerra Superior) Formation, Valley of Tequixquiac, Mexico, with notes on other Pleistocene forms. *Contrib. Mus. Paleontol. Univ. Mich.* 12:47–96.
———. 1955c. The Jinglebob interglacial (Sangamon?) fauna from Kansas and its climatic significance. *Contrib. Mus. Paleontol. Univ. Mich.* 12(10):179–228.
———. 1956a. *Microtus pennsylvanicus* (Ord) from the Hay Springs local fauna of Nebraska. *J. Paleontol.* 30(5):1263–66.
———. 1956b. Vertebrate fossils from the Meade Formation of southwestern Kansas. *Papers Mich. Acad. Sci. Arts Letters* 41:145–203.
———. 1957. Two new Cenozoic microtine rodents. *J. Mammal.* 38(1):39–44.
———. 1958a. New stratigraphic names for early Pleistocene deposits in southwestern Kansas. *Amer. J. Sci.* 256:54–59.
———. 1958b. Summary of North American Pleistocene mammalian local faunas. *Papers Mich. Acad. Sci. Arts Letters* 43:3–32.
———. 1959. Late Cenozoic microtine rodents from Wyoming and Idaho. *Papers Mich. Acad. Sci. Arts Letters* 44:3–40.
———. 1962. Two new rodents from the early Pleistocene of Idaho. *J. Mammal.* 43(4):482–85.
———. 1963a. A late Illinoian fauna from Kansas and its climatic significance. *Papers Mich. Acad. Sci. Arts Letters* 68:187–221.
———. 1963b. The origin of the P_3 pattern of *Sylvilagus, Caprolagus, Oryctolagus,* and *Lepus. J. Mammal.* 44(1):1–15.
———. 1964. A contribution to the Saw Rock local fauna of Kansas. *Papers Mich. Acad. Sci. Arts Letters* 49:115–27.

––––. 1967. New rodents from the late Cenozoic of Kansas. *Papers Mich. Acad. Sci. Arts Letters* 52:115–31.

––––. 1969. The rabbits (*Hypolagus* and *Pratilepus*) from the Upper Pliocene, Hagerman local fauna of Idaho. *Mich. Academician* 1:81–97.

––––. 1970a. A new microtine rodent from the Upper Pliocene of Kansas. *Contrib. Mus. Paleontol. Univ. Mich.* 23:99–103.

––––. 1970b. Pleistocene mammalian local faunas from the Great Plains and central lowland provinces of the United States. In W. Dort, Jr., and J. K. Jones, Jr. (eds.), *Pleistocene and Recent Environments of the Central Great Plains; Dept. Geol. Univ. Kans. Spec. Publ.* 3:395–433.

Hibbard, C. W., and Bjork, P. 1971. The insectivores of the Hagerman local fauna, Upper Pliocene of Idaho. *Contrib. Mus. Paleontol. Univ. Mich.* 23(9):171–80.

Hibbard, C. W., and Dalquest, W. W. 1960. A new antilocaprid from the Pleistocene of Knox County, Texas. *J. Mammal.* 41(1):20–23.

–––– and ––––. 1962. Artiodactyls from the Seymour Formation of Knox County, Texas. *Papers Mich. Acad. Sci. Arts Letters* 67:83–99.

–––– and ––––. 1966. Fossils from the Seymour Formation of Knox and Baylor counties, Texas, and their bearing on the late Kansan climate of that region. *Contrib. Mus. Paleontol. Univ. Mich.* 21(1):1–66.

–––– and ––––. 1973. *Proneofiber*, a new genus of vole (Cricetidae, Rodentia) from the Pleistocene Seymour Formation of Texas, and its evolutionary and stratigraphic significance. *Quatern. Res.* 3(2):269–74.

Hibbard, C. W., and Hinds, F. J. 1960. A radiocarbon date for a woodland musk ox in Michigan. *Papers Mich. Acad. Sci. Arts Letters* 45:103–8.

Hibbard, C. W., and Riggs, E. S. 1949. Upper Pliocene vertebrates from Keefe Canyon, Meade County, Kansas. *Bull. Geol. Soc. Amer.* 60:829–60.

Hibbard, C. W., and Rinker, G. C. 1943. A new mouse (*Microtus ochrogaster taylori*) from Meade County, Kansas. *Univ. Kans. Sci. Bull.* 29:255–68.

Hibbard, C. W., and Taylor, D. W. 1960. Two late Pleistocene faunas from southwestern Kansas. *Contrib. Mus. Paleontol. Univ. Mich.* 16(1):1–223.

Hibbard, C. W., and Wright, B. A. 1956. A new Pleistocene bighorn sheep from Arizona. *J. Mammal.* 37(1):105–7.

Hibbard, C. W., and Zakrzewski, R. J. 1967. Phyletic trends in the late Cenozoic microtine *Ophiomys* gen. nov. from Idaho. *Contrib. Mus. Paleontol. Univ. Mich.* 21:255–71.

–––– and ––––. 1972. A new species of microtine from the late Pliocene of Kansas. *J. Mammal.* 53(4):834–39.

Hibbard, C. W., et al. 1965. Quaternary mammals of North America. In H. E. Wright and D. G. Frye (eds.), *The Quaternary of the United States*. Princeton, N.J.: Princeton Univ. Press. Pp. 509–25.

Hibbard, C. W. et al. 1978. Mammals from the Kanopolis local fauna, Pleistocene (Yarmouth) of Ellsworth County, Kansas. *Contrib. Mus. Paleontol. Univ. Mich.* 25(2):11–44.

Hibbard, E. A. 1958. Occurrence of the extinct moose *Cervalces* in the Pleistocene of Michigan. *Papers Mich. Acad. Sci. Arts Letters* 43:33–37.

Hill, J. E. 1942. *Citellus parryi* from the Pleistocene of Alaska. *Bull. Geol. Soc. Amer.* 53:1842.

Hillerud, J. W. 1966. The Duffield site and its fossil *Bison*, Alberta, Canada. M.S. thesis, Univ. Nebr., Lincoln.

Hirschfeld, S. E., and Webb, S. D. 1968. Plio–Pleistocene megalonychid sloths of North America. *Bull. Fla. State Mus. Biol. Sci.* 12(5):213–96.

Ho, T. Y., Marcus, L., and Berger, R. 1969. Radiocarbon dating of petroleum-impregnated bone from tar pits at Rancho La Brea, California. *Science* 164(3883):1051–52.

Hoffstetter, R. 1952. Les mammiferes pleistocenes de la Republique de l'Equateur. *Mem. Soc. Geol. France* 66:1–391.

Holland, W. J. 1908. A preliminary account of the Pleistocene fauna discovered in a cave opened at Frankstown, Pennsylvania in April and May, 1907. *Ann. Carnegie Mus.* 4(11):228–33.

Holman, J. A. 1959a. Amphibians and reptiles from the Pleistocene (Illinoian) of Williston, Florida. *Copeia* 1959:96–102.

———. 1959b. Birds and mammals from the Pleistocene of Williston, Florida. *Bull. Fla. State Mus.* 5:1–25.

———. 1968. Upper Pliocene snakes from Idaho. *Copeia*, 1968:152–58.

Hood, C. H., and Hawksley, O. 1975. A Pleistocene fauna from Zoo Cave, Taney County, Missouri. *Mo. Speleol.* 15(1):1–42.

Hopkins, D. M., ed. 1967. *The Bering Land Bridge.* Stanford, Calif.: Stanford Univ. Press. 495 pp.

Hopkins, M. L. 1955. Skull of a fossil camelid from American Falls lake bed area of Idaho. *J. Mammal.* 36:278–82.

Hopkins, M. L., Bonnichsen, R., and Fortsch, D. 1969. The stratigraphic position and faunal associates of *Bison (Gigantobison) latifrons* in southeastern Idaho, a progress report. *Tebiwa* 12(1):1–7.

Howard, H. 1952. The prehistoric avifauna of Smith Creek Cave, Nevada, with a description of a new giant raptor. *Bull. So. Calif. Acad. Sci.*, 51(2):50–54.

———. 1962. A comparison of avian assemblages from individual pits at Rancho La Brea, California. *Contrib. Sci. Los Angeles Co. Mus.* 58:1–24.

Hutchison, J. H. 1967. A Pleistocene vampire bat (*Desmodus stocki*) from Potter Creek Cave, Shasta County, California. *Contrib. Mus. Paleontol. Univ. Calif.* 3:1–6.

———. 1968. Fossil Talpidae (Insectivora, Mammalia) from the later Tertiary of Oregon. *Bull. Mus. Nat. Hist. Univ. Oreg., Eugene* 11:1–117.

Huxley, J. S., ed. 1940. *The New Systematics.* Oxford: Clarendon Press. 583 pp.

Ikeya, M. 1977. Electron spin resonance dating and fission track detection of Petralona stalagmite. *Anthropos* (Athens) 4:152–68.

Irving, W. N., and Harington, C. R. 1973. Upper Pleistocene radiocarbon-dated artefacts from the northern Yukon. *Science* 179(4071):335–40.

Ives, P. C., et al. 1967. United States Geological Survey radiocarbon dates IX. *Radiocarbon* 9:505–29.

Izett, G. A., Wilcox, R. W., and Borchardt, G. A. 1972. Correlation of a volcanic ash near Mount Blanco, Texas with the Guaje pumice bed of the Jemex Mountains, New Mexico. *Quatern. Res.* 2(4):554–78.

Izett, G. A., et al. 1970. The Bishop ash bed: A Pleistocene marker bed in the western United States. *Quatern. Res.* 1(1):121–32.

Jackson, C. G., and Kaye, J. M. 1975. Giant tortoises in the late Pleistocene of Mississippi. *Herpetologica* 31:421.

Jakway, G. E. 1958. Pleistocene Lagomorpha and Rodentia from the San Josecito Cave, Nuevo Leon, Mexico. *Trans. Kans. Acad. Sci.* 61:313–27.

———. 1961. *Symbos convexifrons* an invalid species. *J. Mammal.* 42(1):114–15.

———. 1962. The Pleistocene faunal assemblages of the Middle Loup River Terrace Fills of Nebraska. Ph.D. dissertation, Univ. Nebr., Lincoln.

James, G. T. 1957. An edentate from the Pleistocene of Texas. *J. Paleontol.* 31(4):796–808.

Jammot, D. 1972. Relationships between the new species of *Sorex scottensis* and the fossil shrews *Sorex cinereus* Kerr. *Mammalia* 36:449–58.

Järvi, A. 1979. Overkill and plant food. *Cur. Anthropol.* 19:451–52.

Jefferson, T. 1799. A memoir on the discovery of certain bones of a quadruped of the clawed kind in the western parts of Virginia. *Trans. Amer. Phil. Soc.* 4(30):246–60.

Jegla, T. C., and Hall, J. S. 1962. A Pleistocene deposit of the free-tailed bat in Mammoth Cave, Kentucky. *J. Mammal.* 43(4):477–81.

Jelinek, A. J. 1957. Pleistocene faunas and early man. *Papers Mich. Acad. Sci. Arts Letters* 42:225–37.

Jepsen, G. L. 1960. A New Jersey mastodon. *Bull. N.J. State Mus.* 6:1–20.

Johnson, D. L. 1978. The origin of Island mammoths and the Quaternary Land Bridge history of the Northern Channel Islands, California. *Quatern. Res.* 10(2):204–25.

Johnson, N., Opdyke, N. D., and Lindsay, E. 1975. Magnetic polarity stratigraphy of Pliocene–Pleistocene terrestrial deposits and vertebrate fauna, San Pedro Valley, Arizona. *Bull. Geol. Soc. Amer.* 86:5–11.

Johnston, C. S. 1935. An extension in the range of fossil peccaries. *Amer. Midl. Nat.* 16:117–19.

——. 1938. Preliminary report on the vertebrate type locality of Cita Canyon and the description of an ancestral coyote. *Amer. J. Sci.* 35(5):383–90.

——. 1939. Preliminary report on the late Middle Pliocene Axtel Locality, and the description of a new member of the genus *Osteoborus*. *Amer. J. Sci.* 237:895–98.

Johnston, C. S., and Savage, D. E. 1955. A survey of various late Cenozoic vertebrate faunas of the Panhandle of Texas, I: Introduction, description of localities, preliminary faunal lists. *Univ. Calif. Publ. Bull. Dept. Geol. Sci.* 31(2):27–49.

Johnston, R. 1874. Notice of a new species of deer from the Forest Bed. *Ann. Mag. Nat. Hist.* 4:13.

Jones, J. K., Jr. 1958. Pleistocene bats from San Josecito Cave, Nuevo Leon, Mexico. *Univ. Kans. Publ. Mus. Nat. Hist.* 9:389–96.

——. 1964. Distribution and taxonomy of mammals in Nebraska. *Univ. Kans. Publ. Mus. Nat. Hist.* 16:1–356.

Jones, J. K., Jr., Carter, D. C., and Genoways, H. H. 1973. Revised checklist of North American mammals north of Mexico. *Occ. Papers Mus. Tex. Tech Univ.* 12:1–14.

Jones, R. E. 1967. A *Hydrodamalis* skull fragment from Monterey Bay, California. *J. Mammal.* 48:143–44.

Kahlke, H. D. 1958a. Die Cervidenreste aus den altpleistozänen Tonen von Voigtstedt bei Sangerhausen, I: Die Schädel, Geweihe und Gehörne. *Abhandl. Deutsch. Akad. Wiss. Berlin, Kl. Chem. Geol. Biol.* 1958(9):1–51.

——. 1958b. Die jungpleistozänen Säugetierfaunen aus dem Travertingebiet von Taubach-Weimar-Eringsdorf. *Alt-Thüringen* 3:97–130.

——. 1960. Die Cervidenreste aus den altpleistozänen Sanden von Mosbach (Biebrich-Wiesbaden), I: Die Geweihe, Gehörne und Gebisse. *Abhandl. Deutsch. Akad. Wiss. Berlin, Kl. Chem. Geol. Biol.* 1959(7):1–75.

——. 1964. Early Middle Pleistocene (Mindel/Elster) *Praeovibos* and *Ovibos*. *Comment. Biol. Soc. Sci. Fennica* 26(5):1–15.

——. 1969. Die Cerviden-Reste aus den Kiesen von Süssenborn bei Weimar. *Paläontol. Abhandl. A* 3:547–610.

——. 1972. Upper Pliocene and lower Pleistocene mammalian associations of eastern and southeastern Asia and the Plio-Pleistocene boundary. *Coll. Papers Internat. Colloq. Neogene-Quaternary* 1:109–19.

Kapp, R. O. 1965. Illinoian and Sangamon vegetation in southwestern Kansas and adjacent Oklahoma. *Contrib. Mus. Paleontol. Univ. Mich.* 19(14):167–255.

Kaspar, T. C., and McClure, W. L. 1976. The Taylor Bayou local fauna (Pleistocene) near Houston, Texas. *Southw. Nat.* 21(1):9–16.

Kaye, C. A. 1962. Early post-glacial beavers in southeastern New England. *Science* 138(3543):906–7.

Kaye, J. M. 1974. Pleistocene sediment and vertebrate fossil associations in the Mississippi Black Belt: A genetic approach. Ph.D. dissertation, La. State Univ., Baton Rouge.

Kellogg, L. 1910. Rodent fauna of the late Tertiary beds at Virginia Valley and Thousand Creek, Nevada. *Univ. Calif. Publ. Bull. Dept. Geol.* 5:421–37.

——. 1911. A fossil beaver from the Kettleman Hills, California. *Univ. Calif. Publ. Bull. Dept. Geol.* 6(17):401–2.

——. 1912. Pleistocene rodents of California. *Univ. Calif. Publ. Bull. Dept. Geol.* 7(8):151–68.

Kellogg, R. 1927. Fossil pinnipeds from California. *Carnegie Inst. Washington Publ.* 346:25–37.

Kelsall, J. P. 1968. The caribou. *Canad. Wildlife Ser. Monogr.* 3:1–340.

Kennedy, G. E. 1975. Early man in the New World. *Nature* 255:274–75.

Kennerly, T. E., Jr. 1956. Comparisons between fossil and recent species of the genus *Perognathus*. *Tex. J. Sci.* 8(1):74–86.

Khan, E. 1970. Biostratigraphy and paleontology of a Sangamon deposit at Fort Qu'Appelle, Saskatchewan. *Nat. Mus. Canada Publ. Paleontol.* 5:1–82.

Kilmer, F. H. 1972. A new species of sea otter from the later Pleistocene of northwestern California. *Bull. So. Calif. Acad. Sci.* 71(3):150–57.

Kinsey, P. E. 1974. A new species of *Mylohyus* peccary from the Florida Early Pleistocene. In S. D. Webb (ed.), *Pleistocene Mammals of Florida*. Gainesville, Fla.: Univ. Fla. Presses. Pp. 158–69.

Kitts, D. B. 1953. A Pleistocene musk ox from New York and the distribution of musk oxen. *Amer. Mus. Novitates* 1607:1–8.

Klein, J. 1971. Fossil mammals from Inglis IA, late Blancan of Citrus County, Florida. M.S. thesis, Univ. Fla., Gainesville.

Klingener, D. 1963. Dental evolution of *Zapus*. *J. Mammal.* 44(2):248–60.

Kolb, K. K., Nelson, M. E., and Zakrzewski, R. J. 1975. The Duck Creek molluscan fauna (Illinoian) from Ellis County, Kansas. *Trans. Kans. Acad. Sci.* 78:63–74.

Kontrimavichus, B. L. ed. 1976. *Beringia in Cenozoic*. Vladivostok: Acad. Sci. USSR. 594 pp.

Koopman, K. F. 1971. The systematic and historical status of the Florida *Eumops* (Chiroptera, Molossidae). *Amer. Mus. Novitates* 2478:1–6.

Kormos, T. 1914. Drei neue Raubtiere aus den Präglazialschichten des Somlyóhegy bei Püspökfürdö. *Mitteil. Jahrb. Ungar. Geol. Reichsanst. Budapest* 22:225–47.

Korobitsyna, K. V., et al. 1974. Chromosomes of the Siberian snow sheep, *Ovis nivicola*, and implications concerning the origin of Amphiberingian wild sheep (subgenus *Pachyceros*). *Quatern. Res.* 4(3):235–45.

Kowalski, K. 1977. Fossil lemmings (Mammalia, Rodentia) from the Pliocene and early Pleistocene of Poland. *Acta Zool. Cracoviensia* 22:297–317.

Kraglievich, J. L. 1952. Un canido del Eocuartario de Mar del Plata y sus relaciones con otras formas brasileñas y norte americanas. *Rev. Mus. Munic. Cienc. Nat. Mar del Plata* 1:53–70.

Kretzoi, M. 1962. Fauna und Faunenhorizont von Csarnóta. *Ann. Rept. Hung. Geol. Inst.* 1959:344–95.

Kukla, G. J. 1970. Correlations between loesses and deep-sea sediments. *Geol. Fören. Stockholm Förh.* 92:148–80.

Kurtén, B. 1952. The Chinese *Hipparion* fauna: A quantitative survey with comments on the ecology of the machairodonts and hyaenids and the taxonomy of the gazelles. *Comment. Biol. Soc. Sci. Fennica* 13(4):1–82.

———. 1960. A skull of the grizzly bear (*Ursus arctos* L.) from Pit 10, Rancho La Brea. *Contrib. Sci. Los Angeles Co. Mus.* 39:1–7.

———. 1963a. Fossil bears from Texas. *Pearce Sellards Ser. Tex. Mem. Mus.* 1:1–15.

———. 1963b. Notes on some Pleistocene mammal migrations from the Palearctic to the Nearctic. *Eiszeitalter Gegenwart* 14:96–103.

———. 1964. The evolution of the polar bear, *Ursus maritimus* Phipps. *Acta Zool. Fennica* 108:1–26.

———. 1965. The Pleistocene Felidae of Florida. *Bull. Fla. State Mus.* 9(6):215–73.

———. 1966. Pleistocene bears of North America, I: Genus *Tremarctos*, spectacled bears. *Acta Zool. Fennica* 115:1–120.

———. 1967a. Pleistocene bears of North America, II: Genus *Arctodus*, short-faced bears. *Acta Zool. Fennica* 117:1–60.

———. 1967b. Präriewolf und Säbelzahntiger aus dem Pleistozän des Valsequillo. Mexiko. *Quartär* 18:173–178.

———. 1968. *Pleistocene Mammals of Europe*. London: Weidenfeld and Nicolson. 317 pp.

———. 1970. The Neogene wolverine *Plesiogulo* and the origin of *Gulo* (Carnivora, Mustelidae). *Acta Zool. Fennica* 131:1–20.

———. 1972. *The Ice Age*. New York: G. P. Putnam's Sons. 179 pp.

———. 1973a. Geographic variation in size in the puma (*Felis concolor*). *Comment. Biol. Soc. Sci. Fennica* 63:1–8.

———. 1973b. Pleistocene jaguars in North America. *Comment. Biol. Soc. Sci. Fennica* 62:1–23.
———. 1973c. The genus *Dinofelis* (Carnivora, Mammalia) in the Blancan of North America. *Pearce Sellards Ser. Tex. Mem. Mus.* 19:1–7.
———. 1973d. Transberingian relationships of *Ursus arctos* Linne (brown and grizzly bears). *Comment. Biol. Soc. Sci. Fennica* 65:1–10.
———. 1974. A history of coyote-like dogs (Canidae, Mammalia). *Acta Zool. Fennica* 140:1–38.
———. 1975. A new Pleistocene genus of American mountain deer. *J. Mammal.* 56(2):507–8.
———. 1976. Fossil puma (Mammalia: Felidae) in North America. *Ned. J. Zool.* 26(4):502–34.
———. 1979. The stilt-legged deer *Sangamona* of the North American Pleistocene. *Boreas* 8:313–21.
Kurtén, B., and Anderson, E. 1972. The sediments and fauna of Jaguar Cave, II: The fauna. *Tebiwa* 15(1):21–45.
——— and ———. 1974. Association of *Ursus arctos* and *Arctodus simus* (Mammalia: Ursidae) in the late Pleistocene of Wyoming. *Breviora* 426:1–6.
Kurtén, B., and Crusafont, M. 1977. Villafranchian carnivores (Mammalia) from La Puebla de Valverde (Tereul, Spain). *Comment. Biol. Soc. Sci. Fennica* 85:1–39.
Kurtén, B., and Rausch, R. 1959. Biometric comparisons between North American and European mammals. *Acta Arctica* 11:1–45.
Lambe, L. M. 1911. On *Arctotherium* from the Pleistocene of Yukon. *Ottawa Nat.* 25:21–26.
Lammers, G. E. 1970. The late Cenozoic Benson and Curtis Ranch faunas of the San Pedro Valley, Cochise County, Arizona. Ph.D. dissertation, Univ. Ariz., Tucson.
Lance, J. F. 1966. Zoogeographic significance of capybara in Arizona. *Spec. Paper Geol. Soc. Amer.* 87:313.
Langenwalter, P. E. 1975. The fossil vertebrates of the Los Angeles–Long Beach harbors region. In D. F. Soule and M. Oguri (eds.), *Marine Studies of San Pedro Bay, California,* Part 9: *Paleontology.* Allan Hancock Fountain Harbors Environmental Projects and Office of Sea Grant Programs, Univ. So. Calif., pp. 36–54.
Langguth, A. 1969. Die südamerikanischen Canidae unter besonderer Berücksichtigung des Mähnenwolfes *Chrysocyon brachyurus* Illiger. *Zeitschr. Wiss. Zool.* 179:1–188.
Laudermilk, J. D. 1938. Plants in the dung of *Nothrotherium* from Rampart and Muav caves, Arizona. *Carnegie Inst. Washington Publ.* 487:271–81.
Laudermilk, J. D., and Munz, P. A. 1934. Plants in the dung of *Nothrotherium* from Gypsum Cave, Nevada. *Carnegie Inst. Washington Publ.* 453:29–37.
Lawrence, B. 1960. Fossil *Tadarida* from New Mexico. *J. Mammal.* 41(3):320–22.
———. 1967. Early domestic dogs. *Säugetierkunde* 32(1):44–59.
Lawrence, B., and Bossert, W. H. 1969. The cranial evidence for hybridization in New England *Canis. Breviora* 330:1–13.
Leffler, S. B. 1964. Fossil mammals from the Elk River Formation, Cape Blanco, Oregon. *Jour. Mammal.* 45(1):53–61.
Leidy, J. 1852. Remarks on two crania of extinct species of ox. *Proc. Acad. Nat. Sci. Phila.* 6:71.
———. 1853a. (*Description of Ursus amplidens.*) *Proc. Acad. Nat. Sci. Phila.* 6:303.
———. 1853b. Description of an extinct species of American lion: *Felis atrox. Trans. Amer. Phil. Soc. NS* 10:319–21.
———. 1854. Remarks on *Sus americanus* or *Harlanus americanus* and on other extinct mammals. *Proc. Acad. Nat. Sci. Phila.* 7:89–90.
———. 1858a. Notice of fossil Mammalia from valley of Niobrara River. *Proc. Acad. Nat. Sci. Phila.* 1858:11.
———. 1858b. Notice of remains of extinct Vertebrata, from the valley of The Niobrara River, collected during the exploring expedition of 1857, in Nebraska, under the command of Lieut. G. K. Warren, U. S. Top. Eng., by Dr. F. V. Hayden. *Proc. Acad. Nat. Sci. Phila.* 1858:20–29.
———. 1860. Description of vertebrate fossil. *Holme's Post-Pliocene Fossils of South Carolina,* pp. 99–122.
———. 1865. (Bones and teeth of horses from California and Oregon.) *Proc. Acad. Nat. Sci. Phila.* 1865:94.

Leidy, J. 1868a. Notice of some vertebrate remains from Hardin County, Texas. *Proc. Acad. Nat. Sci. Phila.* 1868:174–76.
———. 1868b. Notice of some remains of horses. *Proc. Acad. Nat. Sci. Phila.* 1868:195.
———. 1869. The extinct mammalian fauna of Dakota and Nebraska, including an account of some allied forms from other localities, together with a synopsis of the mammalian remains of North America. *J. Acad. Nat. Sci. Phila.* 7(2):1–472.
———. 1872. Remarks on some extinct vertebrates. *Proc. Acad. Nat. Sci. Phila.* 1872:37–38.
———. 1873. Remarks on extinct mammals from California. *Proc. Acad. Nat. Sci. Phila.* 1873:259–60.
———. 1889. The sabre-toothed tiger of Florida. *Proc. Acad. Nat. Sci. Phila.* 1889:29–31.
Leonard, A. B. 1950. A Yarmouthian molluscan fauna in the midcontinent region of the United States. *Univ. Kans. Paleontol. Contrib. Mollusca* 3:1–48.
Leonhardy, F. C., ed. 1966. Domebo: A Paleo-Indian mammoth kill in the prairie-plains. *Contrib. Mus. Great Plains* 1:1–53.
Lewis, G. E. 1970. New discoveries of Pleistocene bisons and peccaries in Colorado. *U.S. Geol. Surv. Prof. Paper* 700B:137–140.
Libby, W. F. 1955. *Radiocarbon Dating,* 2nd ed. Chicago: Univ. Chicago Press. 124 pp.
Ligon, J. D. 1965. A Pleistocene avifauna from Haile, Florida. *Bull. Fla. State Mus.* 10:127–58.
Lillegraven, J. A. 1967. *Bison crassicornis* and the ground sloth *Megalonyx jeffersoni* in the Kansas Pleistocene. *Trans. Kans. Acad. Sci.* 69:294–300.
Lindsay, E. H., and Tessman, N. T. 1974. Cenozoic vertebrate localities and faunas in Arizona. *J. Arizona Acad. Sci.* 9:3–24.
Lindsay, E. H., Johnson, N. M., and Opdyke, N. D. 1975. Preliminary correlation of North American land mammal ages and geomagnetic chronology. In *Studies on Cenozoic Paleontology and Stratigraphy, Claude W. Hibbard Memorial Vol. 3, Univ. Mich. Papers Paleontol.* 12:111–19.
Lipps, L., and Ray, C. E. 1967. The Pleistocene fossiliferous deposit at Ladds, Bartow County, Georgia. *Bull. Ga. Acad. Sci.* 25(3):113–19.
Long, A., and Martin, P. S. 1974. Death of American ground sloths. *Science* 186:638–40.
Long, C. A. 1964. Taxonomic status of the Pleistocene badger, *Taxidea marylandica. Amer. Midl. Nat.* 72(1):176–80.
———. 1972. Taxonomic revision of the North American badger, *Taxidea taxus. J. Mammal.* 53(4):725–59.
Loomis, F. B. 1924. Artifacts associated with the remains of a Columbian elephant at Melbourne, Florida. *Amer. J. Sci.* 8:503–8.
Lull, R. S. 1929. A remarkable ground sloth. *Mem. Peabody Mus. Yale Univ.* 3(2):1–21.
———. 1930. The ground sloth *Nothrotherium. Amer. J. Sci.* 20:344–56.
Lundelius, E. L., Jr. 1960. *Mylohyus nasutus,* long-nosed peccary from the Texas Pleistocene. *Bull. Tex. Mem. Mus.* 1:1–40.
———. 1963. Non-human skeletal material (in Centipede and Damp caves: Excavation in Val Verde County, Texas, 1958). *Tex. Archaeol. Soc.* 33:127–29.
———. 1967. Late Pleistocene and Holocene faunal history of central Texas. In P. S. Martin and H. E. Wright (eds.), *Pleistocene Extinctions: The Search for a Cause.* New Haven: Yale Univ. Press. Pp. 287–319.
———. 1972a. Fossil vertebrates from the late Pleistocene Ingleside fauna, San Patricio County, Texas. *Rept. Inv. Bur. Econ. Geol. Univ. Tex.* 77:1–74.
———. 1972b. Vertebrate remains from the Gray Sand. In *Blackwater Locality No. 1, a stratified early man site in eastern New Mexico; Publ. Fort Burgwin Res. Center* 8:148–63.
Lundelius, E. L., Jr., and Slaughter, B. H. 1976. Notes on American Pleistocene tapirs. In C. S. Churcher (ed.), *Athlon: Essays on Paleontology in Honour of Loris Shano Russell.* Royal Ontario Mus. Life Sci. Misc. Publ., pp. 226–243.
Lundelius, E. L., Jr., and Stevens, M. S. 1970. *Equus francisci* Hay, a small stiltlegged horse, middle Pleistocene of Texas. *J. Paleontol.* 44(1):148–53.
Lydekker, R. 1898. *The Deer of All Lands: A History of the Family Cervidae Living and Extinct.* London. 318 pp.

Lyell, C. 1834. *Principles of Geology,* 3rd ed. London, 4 vols.
Lyon, G. M. 1938. *Megalonyx milleri,* a new Pleistocene ground sloth from southern California. *Trans. San Diego Nat. Hist. Soc.* 9(6):15–30.
Lynch, T. F. 1974. The antiquity of man in South America. *Quatern. Res.* 4:356–77.
McClung, C. E. 1904. The fossil bison of Kansas. *Trans. Kans. Acad. Sci.* 9:157–59.
McCoy, J. J. 1963. The fossil avifauna of Ichetucknee River, Florida. *Auk* 80:335–51.
McCrady, E., Kirby-Smith, H. T., and Templeton, H. 1951. New finds of Pleistocene jaguar skeletons from Tennessee caves. *Proc. U.S. Nat. Mus.* 101:497–511.
McDonald, H. G. 1977. Description of the osteology of the extinct gravigrade edentate, *Megalonyx,* with observations on its ontogeny, phylogeny and functional anatomy. M.A. thesis, Univ. Fla., Gainesville.
McDonald, H. G., and Anderson, E. 1975. A late Pleistocene fauna from southeastern Idaho. *Tebiwa* 18(1):19–37.
Macdonald, J. R. 1956. A Blancan mammalian fauna from Wichman, Nevada. *J. Paleontol.* 30:213–16.
———. 1966. Dig that horseheaven. *Quart. Los Angeles Co. Mus.* 5:12–14.
———. 1967. The Maricopa Brea. *Mus. Alliance Quart. Los Angeles Co. Mus. Nat. Hist.* 6(2):21–24.
McGrew, P. O. 1944. An early Pleistocene (Blancan) fauna from Nebraska. *Field Mus. Nat. Hist. Geol. Ser.* 9(2):33–66.
———. 1948. The Blancan faunas, their age and correlation. *Bull. Geol. Soc. Amer.* 59(6):549–52.
McKenna, M. C. 1975. Toward a phylogenetic classification of the Mammalia. In W. P. Luckett and F. S. Szalay (eds.), *Phylogeny of the Primates.* New York: Plenum Press. pp. 21–46.
McMullen, T. L. 1975. Shrews from the late Pleistocene of central Kansas, with the description of a new species of *Sorex. J. Mammal.* 56(2):316–20.
MacNeish, R. S. 1971. Early man in the Andes. *Sci. Amer.* 224(4):36–46.
———. 1976. Early man in the New World. *Amer. Sci.* 64(3):316–27.
Macpherson, A. H. 1965. The origin of diversity in mammals of the Canadian Arctic tundra. *Syst. Zool.* 14:153–73.
Maglio, V. J. 1972. Evolution of mastication in the Elephantidae. *Evolution* 26(4):638–58.
———. 1973. Origin and evolution of the Elephantidae. *Trans. Amer. Phil. Soc.* NS 63(3):1–149.
Maher, W. J. 1966. Muskox bone of possible Wisconsin age from Banks Island, Northwest Territories. *Arctic* 21:260–66.
Malde, H. E., and Powers, H. A. 1962. Upper Cenozoic stratigraphy of western Snake River Plain, Idaho. *Bull. Geol. Soc. Amer.* 73:1197–1220.
Marcus, L. F. 1960. A census of the abundant large Pleistocene mammals from Rancho La Brea. *Contrib. Sci. Los Angeles Co. Mus.* 38:1–11.
Marsh, O. C. 1877 New vertebrate fossils. *Amer. J. Sci., ser. 3,* 14:242.
Martin, H. T. 1924. A new bison from the Pleistocene of Kansas with notice of a new locality for *Bison occidentalis. Kans. Univ. Sci. Bull.* 15:273–78.
Martin, L. D. 1972. The microtine rodents of the Mullen assemblage from the Pleistocene of north central Nebraska. *Bull. Univ. Nebr. State Mus.* 9(5):173–182.
Martin, L. D., Gilbert, B. M., and Adams, D. B. 1977. A cheetah-like cat in the North American Pleistocene. *Science* 195(4282):981–82.
Martin, P. S. 1967. Prehistoric overkill. In P. S. Martin and H. E. Wright (eds.), *Pleistocene Extinctions: The Search for a Cause.* New Haven and London: Yale Univ. Press. Pp. 75–120.
———. 1973. The discovery of America. *Science* 179:969–74.
———. 1975. Sloth droppings. *Nat. Hist.* 84(7):74–81.
Martin, P. S., and Guilday, J. E. 1967. A bestiary for Pleistocene biologists. In P. S. Martin and H. E. Wright (eds.), *Pleistocene Extinctions: The Search for a Cause.* New Haven and London: Yale Univ. Press. Pp. 1–62.
Martin, P. S., Sabels, B. E., and Shutler, D. 1961. Rampart Cave coprolite and ecology of the Shasta ground sloth. *Amer. J. Sci.* 259(2):102–27.
Martin, R. A. 1967. A comparison of two mandibular dimensions in *Peromyscus* with regard to identification of Pleistocene *Peromyscus* from Florida. *Tulane Stud. Zool.* 14(2):75–79.

Martin, R. A. 1968a. Late Pleistocene distribution of *Microtus pennsylvanicus*. *J. Mammal.* 49(2):265-71.
———. 1968b. Further study of the Friesenhahn Cave *Peromyscus*. *Southw. Nat.* 13:253-66.
———. 1969. Taxonomy of the giant Pleistocene beaver *Castoroides* from Florida. *J. Paleontol.* 43(4):1033-41.
———. 1970. Line and grade in the extinct *medius* species group of *Sigmodon*. *Science* 167:1504-6.
———. 1973a. Description of a new genus of weasel from the Pleistocene of South Dakota. *J. Mammal.* 54(4):924-29.
———. 1973b. The Java local fauna, Pleistocene of South Dakota: A preliminary report. *Bull. N. J. Acad. Sci.* 18(2):48-56.
———. 1974a. Fossil mammals from the Coleman IIA fauna, Sumter County. In S. D. Webb (ed.), *Pleistocene Mammals of Florida*. Gainesville, Fla.: Univ. Fla. Presses. Pp. 35-99.
———. 1974b. Fossil vertebrates from the Haile XIVA fauna, Alachua County. In S. D. Webb (ed.), *Pleistocene Mammals of Florida*. Gainesville, Fla.: Univ. Fla. Presses. Pp. 100-113.
———. 1975. *Allophaiomys* Kormos from the Pleistocene of North America. *Studies on Cenozoic Paleontology and Stratigraphy, Claude W. Hibbard Memorial Vol. 3, Univ. Mich. Papers Paleontol.* 12:97-100.
Martin, R. A., and Harksen, J. C. 1974. The Delmont local fauna, Blancan of South Dakota. *Bull. N. J. Acad. Sci.* 19(1):11-17.
Martin, R. A., and Webb, S. D. 1974. Late Pleistocene mammals from the Devil's Den fauna, Levy County. In S. D. Webb (ed.), *Pleistocene Mammals of Florida*. Gainesville, Fla.: Univ. Fla. Presses. Pp. 114-145.
Matthew, W. D. 1902. List of the Pleistocene fauna from Hay Springs, Nebraska. *Bull. Amer. Mus. Nat. Hist.* 16(24):317-22.
———. 1918. Contributions to the Snake Creek Fauna with notes on the Pleistocene of western Nebraska. *Bull. Amer. Mus. Nat. Hist.* 38(7):183-229.
———. 1924a. A new link in the ancestry of the horse. *Amer. Mus. Novitates* 131:1-2.
———. 1924b. Third contribution to the Snake Creek fauna. *Bull. Amer. Mus. Nat. Hist.* 50:69-210.
———. 1939. *Climate and Evolution*, 2nd ed. *Spec. Publ. New York Acad. Sci.* 1:1-223.
Matthew, W. D., and Stirton, R. A. 1930. Osteology and affinities of *Borophagus*. *Univ. Calif. Publ. Geol. Sci.* 10:171-216.
Matthews, J. V., Jr. 1975. Insects and plant macrofossils from two Quaternary exposures in the Old Crow-Porcupine region, Yukon Territory, Canada. *Arctic Alpine Res.* 7(3):249-59.
Mawby, J. E. 1965. Machairodonts from the late Cenozoic of the Panhandle of Texas. *J. Mammal.* 46(4):573-87.
———. 1967. Fossil vertebrates of the Tule Springs site, Nevada. *Papers Nev. State Mus. Anthropol.* 13:105-28.
Mayr, E. 1963. *Animal Species and Evolution*. Cambridge Mass.: Belknap Press of Harvard Univ.
Mead, R. A. 1968a. Reproduction in western forms of the spotted skunk (genus *Spilogale*). *J. Mammal.* 49(3):373-90.
———. 1968b. Reproduction in eastern forms of the spotted skunk (genus *Spilogale*). *J. Zool.* 156:119-36.
Meade, G. E. 1942. A new species of *Capromeryx* from the Pleistocene of west Texas. *Bull. Tex. Archaeol. Paleontol. Soc.* no. 14.
———. 1945. The Blanco fauna. *Univ. Tex. Publ.* 4401:509-56.
———. 1953. An early Pleistocene vertebrate fauna from Frederick, Oklahoma. *J. Geol.* 61(5):452-60.
———. 1961. The saber-toothed cat, *Dinobastis serus*. *Bull. Tex. Mem. Mus.* 2(2):24-60.
Mech, L. D. 1966. The wolves of Isle Royale. *U.S. Dept. Int. Nat. Park Serv. Fauna Ser.* 7, 210 pp.
———. 1970. *The wolf: The Ecology and Behavior of an Endangered Species*. Garden City, N.Y.: Natural History Press. 384 pp.
Mehl, M. G. 1962. Missouri's Ice Age animals. *Educ. Ser. Mo. Geol. Surv. Water Res.* 1:1-104.
Mehringer, P. J., Jr. 1965. Late Pleistocene vegetation in the Mohave Desert of southern Nevada. *J. Ariz. Acad. Sci.* 3(3):172-88.

———. 1967. The environment of extinction of the late Pleistocene megafauna in the arid southwestern United States. In P. S. Martin and H. E. Wright (eds.), *Pleistocene Extinctions. The Search for a Cause.* New Haven and London: Yale Univ. Press. Pp. 247–66.

Mehringer, P. J., Jr., and Haynes, C. V. 1965. The pollen evidence for the environment of early man and extinct mammals at the Lehner Mammoth Site, southeastern Arizona. *Amer. Antiq.* 31(1):17–23.

Mehringer, P. J., Jr., King, J. E., and Lindsay, E. H. 1970. A record of Wisconsin-age vegetation and fauna from the Ozarks of western Missouri. In W. Dort, Jr., and J. K. Jones, Jr. (eds.), *Pleistocene and Recent Environments of the Central Great Plains; Dept. Geol. Univ. Kans. Spec. Publ.* 3:173–83.

Melton, W. G., Jr. 1964. *Glyptodon fredericensis* (Meade) from the Seymour Formation of Knox County, Texas. *Papers Mich. Acad. Sci. Arts Letters* 49:129–46.

Merriam, J. C. 1903. The Pliocene and Quaternary Canidae of the Great Valley of California. *Univ. Calif. Publ. Bull. Dept. Geol. Sci.* 3:277–90.

———. 1905. A new sabre-tooth from California. *Univ. Calif. Publ. Bull. Dept. Geol.* 4:171–74.

———. 1910. New Mammalia from Rancho La Brea. *Univ. Calif. Publ. Bull. Dept. Geol.* 5(25):391–95.

———. 1911. Note on a gigantic bear from the Pleistocene of Rancho La Brea. *Univ. Calif. Publ. Bull. Dept. Geol.* 6:163–66.

———. 1912. The fauna of Rancho La Brea, 2: Canidae. *Mem. Univ. Calif.* 1(2):217–72.

———. 1913. Tapir remains from late Cenozic beds of the Pacific Coast region. *Univ. Calif. Publ. Bull. Dept. Geol.* 7(9):169–75.

———. 1916. Relationship of *Equus* to *Pliohippus* suggested by characters of a new species from the Pliocene of California. *Univ. Calif. Publ. Bull. Dept. Geol.* 9:525–34.

———. 1917. Relationships of Pliocene mammalian faunas from the Pacific Coast and Great Basin provinces of North America. *Univ. Calif. Publ. Bull. Dept. Geol.* 10:421–43.

———. 1918a. New Mammalia from the Idaho Formation. *Univ. Calif. Publ. Bull. Dept. Geol.* 10:523–30.

———. 1918b. New puma-like cat from Rancho La Brea. *Univ. Calif. Publ. Bull. Dept. Geol.* 10:535–37.

Merriam, J. C., and Stock, C. 1925. Relationships and structures of the short-faced bear, *Arctotherium*, from the Pleistocene of California. *Carnegie Inst. Washington Publ.* 347(1):1–35.

——— and ———. 1932. The Felidae of Rancho La Brea. *Carnegie Inst. Washington Publ.* 422:1–232.

Meyer, K. J. 1974. Pollenanalytische Untersuchungen und Jahresschichten-zählungen an der holstein-zeitlichen Kieselgur von Hetedorf. *Geol. Jahrb.* A21:87–105.

Miller, G. J. 1968. On the age distribution of *Smilodon californicus* Bovard from Rancho La Brea. *Contrib. Sci. Los Angeles Co. Mus.* 131:1–17.

———. 1969a. A new hypothesis to explain the method of food ingestion used by *Smilodon californicus* Bovard. *Tebiwa* 12(1):9–19.

———. 1969b. A study of cuts, grooves, and other marks on recent and fossil bone, 1: Animal tooth marks. *Tebiwa* 12(1):20–26.

Miller, G. S. 1899. A new fossil bear from Ohio. *Proc. Biol. Soc. Washington* 13:53–56.

Miller, L. H. 1943. The Pleistocene birds of San Josecito Cavern, Mexico. *Univ. Calif. Publ. Zool.* 47:143–68.

Miller, W. E. 1971. Pleistocene vertebrates of the Los Angeles Basin and vicinity (exclusive of Rancho La Brea). *Los Angeles Co. Mus. Nat. Hist. Sci. Bull.* 10:1–124.

———. 1973. Pleistocene mammal fauna from Utah. (Abst.). *Geol. Soc. Amer. Cordilleran Sect., Proc. 69th Ann. Meeting, Portland, Oreg.*, p. 81.

———. 1976. Late Pleistocene vertebrates of the Silver Creek local fauna from north central Utah. *Great Basin Nat.* 36(4):387–424.

Mills, R. S. 1975. A ground sloth, *Megalonyx*, from a Pleistocene site in Darke County, Ohio. *Ohio J. Sci.* 75(3):147–55.

Mills, R. S., and Guilday, J. E. 1972. First record of *Cervalces scotti* from the Pleistocene of Ohio. *Amer. Midl. Nat.* 88(1):255.

Milstead, W. M. 1956. Fossil turtles of Friesenhahn Cave, Texas, with the description of a new species of *Testudo. Copeia* 1956:162–71.

Mitchell, E. D. 1966. Northeastern Pacific Pleistocene sea otters. *J. Canada Fish Res. Bd.* 23:1897–1911.

Moore, J. 1890. Concerning a skeleton of the great fossil beaver, *Castoroides ohioensis. Cincinnati Soc. Nat. Hist.* 13:138–69.

Mooser, O. 1959. La fauna "Cedozo" del Pleistoceno en Aguascalientes. *Anal. Inst. Biol. Mexico* 29:409–52.

Mooser, O., and Dalquest, W. W. 1975a. A new species of camel (genus *Camelops*) from the Pleistocene of Aguascalientes, Mexico. *Southw. Nat.* 19(4):341–45.

—— and ——. 1975b. Pleistocene mammals from Aguascalientes, central Mexico. *J. Mammal.* 56(4):781–820.

Mosimann, J. E., and Martin, P. S. 1975. Simulating overkill by Paleoindians. *Amer. Sci.* 63:304–13.

Müller, H. 1974. Pollenanlytische Untersuchungen und Jahresschichtenzählungen an der eemzeitlichen Kieselgur von Bispingen/Luhe. *Geol. Jahrb.* A21:149–69.

Musil, R. 1970. Domestication of the dog already in the Magdalenian? *Anthropologie* 8:87–88.

Neff, N. A. 1975. Fishes of the Kanopolis local fauna (Pleistocene) of Ellsworth County, Kansas. *Studies on Cenozoic Paleontology* and *Stratigraphy, Claude W. Hibbard Memorial Vol. 3, Univ. Mich. Papers Paleontol.* 12:39–48.

Neill, W. T. 1957. Historical biogeography of present day Florida. *Fla. Mus. Biol. Sci. Bull.* 2:175–220.

Nelson, R. S., and Semken, H. A. 1970. Paleoecological and stratigraphic significance of the muskrat in Pleistocene deposits. *Bull. Geol. Soc. Amer.* 81:3733–38.

Nowak, R. M. 1973. North American Quaternary *Canis*. Ph.D. dissertation, Univ. Kans., Lawrence.

Oakley, K. P. 1975. *Decorative and Symbolic Uses of Vertebrate Fossils*. Oxford: Oxford Univ. Press. 60 pp.

Oelrich, T. M. 1953. Additional mammals from the Rexroad fauna. *J. Mammal.* 34(3):373–78.

Oesch, R. D. 1969. Fossil Felidae and Machairodontidae from two Missouri caves. *J. Mammal.* 50(2):367–68.

Ognev, S. I. 1947. *Mammals of the USSR and Adjacent Countries*, vol. 5. (Israel Program for Scientific Translations, Jerusalem, 1963.)

Olsen, S. J. 1956. A new species of *Osteoborus* from the Bone Valley Formation of Florida. *Fla. Geol. Surv. Spec. Publ.* 2:1–5.

——. 1958. The bog lemming from the Pleistocene of Florida. *J. Mammal.* 39(4):537–40.

——. 1960. Additional remains of Florida's Pleistocene vampire. *J. Mammal.* 41(4):458–62.

——. 1972. Osteology for the archaeologist, 3: The American mastodon and the woolly mammoth. *Papers Peabody Mus. Archaeol. Ethnol. Harvard Univ.* 56(3):1–47.

Olson, E. C. 1940. A late Pleistocene fauna from Herculaneum, Missouri. *J. Geol.* 48:32–57.

Opdyke, N. D. 1972. Paleomagnetism of deep-sea cores *Rev. Geophys.* 10:213–50.

Opdyke, N. D., et al., 1977. The paleomagnetism and magnetic polarity stratigraphy of the mammal-bearing section of Anza Borrego State Park, California. *Quatern. Res.* 7(3):316–29.

Orr, P. C. 1956. Radiocarbon dates from Santa Rosa Island, I. *Bull. Santa Barbara Mus. Nat. Hist. Dept. Anthropol.* 2:1–10.

——. 1969. *Felis trumani*, A new radiocarbon dated cat skull from Crypt Cave, Nevada. *Bull. Santa Barbara Mus. Nat. Hist. Dept. Geol.* 2:1–8.

Osborn, H. F. 1903. *Glyptotherium texanum*, a new glyptodont from the lower Pleistocene of Texas. *Bull. Amer. Mus. Nat. Hist.* 19:491–94.

——. 1922. Species of American Pleistocene mammoths, *Elephas jeffersonii*, new species. *Amer. Mus. Novitates* 41:1–16.

——. 1936. *Proboscidea*, Vol. I. New York: American Museum Press. 802 pp.

——. 1942. *Proboscidea*, Vol. II. New York: American Museum Press. Pp. 805–1675.

Osgood, W. H. 1905. *Symbos,* a substitute for *Scaphoceros. Proc. Biol. Soc. Washington* 18:223-24.
Owen, R. 1869. On fossil teeth of equines from Central and South America. *Proc. Roy. Soc. London* 17:267-68.
———. 1870. On fossil remains of equines from Central and South America referable to *Equus conversidens* Ow., *Equus tau* Ow., and *Equus arcidens* Ow. *Phil. Trans. Roy. Soc. London* 159:559-73.
Packard, R. 1960. Speciation and evolution of the pigmy mice, genus *Baiomys. Univ. Kans. Publ. Mus. Nat. Hist.* 9:579-670.
Palmer, H. A. 1956. *Ibex iowensis,* first evidence of fossil goat in North America. *Proc. Iowa Acad. Sci.* 63:450-52.
Parmalee, P. W. 1967. A Recent cave bone deposit in southwestern Illinois. *Bull. Nat. Speleol. Soc.* 29(4):119-47.
Parmalee, P. W., and Oesch, R. D. 1972. Pleistocene and Recent faunas from the Brynjulfson Caves, Missouri. *Ill. State Mus. Rept. Inv.* No. 25, 52 pp.
Parmalee, P. W., Bogan, A. E., and Guilday, J. E. 1976. First records of the giant beaver (*Castoroides ohioensis*) from eastern Tennessee. *J. Tenn. Acad. Sci.* 51(3):87-88.
Parmalee, P. W., Oesch, R. D., and Guilday, J. E. 1969. Pleistocene and Recent vertebrate faunas from Crankshaft Cave, Missouri. *Ill. State Mus. Rept. Inv. No.* 14, 37 pp.
Pascual, R., et al. 1965. Las edades del Cenozoico mamalifero de la Argentina, con especial atenciòn a aquellas del Terrirorio Bonaerense. *An. Com. Invest. Cienc. Buenos Aires* 6:165-93.
Patterson, B., and Pascual, R. 1972. The fossil mammal fauna of South America. In A. Keast, F. C. Erk, and B. Glass (eds.), *Evolution, Mammals and Southern Continents.* Albany: State Univ. N.Y. Press, pp. 247-309.
Patton, T. H. 1963. Fossil vertebrates from Miller's Cave, Llano County, Texas. *Bull. Tex. Mem. Mus.* 7:1-41.
———. 1965. A new genus of fossil microtine from Texas. *J. Mammal.* 46(3):466-71.
Paula Couto, C. de. 1954. Sobre um gliptodonte do Urugquai um tatu fossil do Brasil. *Serv. Geol. Min. Brasil Notas Prelim. Estud.* 80:1-10.
———. 1967. Pleistocene edentates of the West Indies. *Amer. Mus. Novitates* 2304:1-55.
———. 1971. On two small Pleistocene ground sloths. *An. Acad. Brasil. Cienc. Suppl.* 43:499-513.
Paulson, G. R. 1961. The mammals of the Cudahy Fauna. *Papers Mich. Acad. Sci. Arts Letters* 46:127-53.
Peterson, O. A. 1926. The fossils of the Frankstown Cave, Blair County, Pennsylvania. *Ann. Carnegie Mus.* 16:249-315.
Peterson, R. L. 1966. *The Mammals of Eastern Canada.* Toronto: Oxford Univ. Press. 465 pp.
Péwé, T. L. 1966. *Permafrost and its effect on life in the North.* Corvallis, Oreg.: Oregon State Univ. Press. 40 pp.
———. 1975. Quaternary geology of Alaska. *U.S. Geol. Surv. Prof. Paper* 835:1-145.
Péwé, T. L., and Hopkins, D. M. 1967. Mammal remains of Pre-Wisconsin age in Alaska. In D. M. Hopkins (ed.), *The Bering Land Bridge.* Stanford, Calif.: Stanford Univ. Press. Pp. 266-70.
Pizzimenti, J. J., and Hoffmann, R. S. 1973. *Cynomys gunnisoni. Mammal. Species* 25:1-4.
Portis, A. 1920. Elenco delle specie di cervicorni fossili in Roma e attorno a Roma. *Boll. Soc. Geol. Ital.* 39:132-39.
Powell L. H. 1948. The giant beaver, *Castoroides* in Minnesota. *Sci. Bull.* 2, Sci. Mus. St. Paul Inst., 30 pp.
Prentiss, D. W. 1903. Description of an extinct mink from the shellheaps of the Maine coast. *Proc. U.S. Nat. Mus.* 26:887-88.
Quackenbush, L. S. 1909. Notes on Alaskan mammoth expeditions of 1907-1908. *Bull. Amer. Mus. Nat. Hist.* 26:87-103.
Quinn, J. H. 1957. Pleistocene Equidae of Texas. *Univ. Tex. Bur. Econ. Geol. Rep. Inv.* 33:1-51.
Rasmussen, D. L. 1974. New Quaternary mammal localities in the Upper Clark Fork River valley, western Montana. *Northw. Geol.* 3:62-70.
Rausch, R. L. 1953. On the status of some Arctic mammals. *Arctic* 6(2):91-148.

Rausch, R. L. 1964. The specific status of the narrow-skulled vole (subgenus *Stenocranius* Kashchenko) in North America. *Zeitschr. Saugetierk.* 29:343–58.

Rausch, R. L., and Rausch, V. R. 1972. Observations on chromosomes of *Dicrostonyx torquatus stevensoni* Nelson and chromosomal diversity in varying lemmings. *Zeitschr. Säugetierk.* 37:372–84.

Ray, C. E. 1957. A list, bibliography and index of the fossil vertebrates of Florida. *Spec. Publ. Fla. Geol. Surv.* 3:1–175.

———. 1958. Additions to the Pleistocene mammalian fauna from Melbourne, Florida. *Bull. Mus. Comp. Zool.* 119:421–49.

———. 1961. The monk seal in Florida. *J. Mammal.* 42(1):113.

———. 1964a. *Tapirus copei* in the Pleistocene of Florida. *Quart. J. Fla. Acad. Sci.* 27(1)59–66.

———.1964b. The jaguarundi in the Quaternary of Florida. *J. Mammal.* 45:330–32.

———. 1965. A new chipmunk *Tamias aristus*, from the Pleistocene of Georgia. *J. Paleontol.* 39(5):1016–22.

———. 1966a. The identity of *Bison appalachicolus*. *Notulae Naturae Acad. Nat. Sci. Phila.* 384:1–5.

———. 1966b. The status of *Bootherium brazosis*. *Pearce-Sellards Ser. Tex. Mem. Mus.* 5:3–7.

———. 1967. Pleistocene mammals from Ladds, Bartow County, Georgia. *Bull Ga. Acad. Sci.* 25(3):120–50.

———. 1971. Polar bear and mammoth on the Pribilof Islands. *Arctic* 24:9–18.

Ray, C. E., and Lipps, L. 1968. Additional notes on the Pleistocene mammals from Ladds, Georgia. *Bull. Ga. Acad. Sci.* 26:63.

——— and ———. 1970. Southerly distribution of porcupine in eastern United States during late Quaternary time. *Bull. Ga. Acad. Sci.* 28(2):24.

Ray, C. E., Cooper, B. N., and Benninghoff, W. S. 1967. Fossil mammals and pollen in a late Pleistocene deposit at Saltville, Virginia. *J. Paleontol.* 41(3):608–22.

Ray, C. E., Denny, C. S., and Rubin, M. 1970. A peccary, *Platygonus compressus* LeConte, from drift of Wisconsinan age in northern Pennsylvania. *Amer. J. Sci.* 268:78–94.

Ray, C. E., Olsen, S. J., and Gut, H. J. 1963. Three mammals new to the Pleistocene fauna of Florida, and a reconsideration of five earlier records. *J. Mammal.* 44(3):373–95.

Ray, C. E., Wills, D. L., and Palmquist, J. C. 1968. Fossil musk oxen of Illinois. *Trans. Ill. State Acad. Sci.* 61(3):282–92.

Ray, C. E., et al. 1968. Fossil vertebrates from the marine Pleistocene of southeastern Virginia. *Smithsonian Misc. Coll.* 153(3):1–25.

Reed, C. A., and Palmer, H. A. 1964. A late Quaternary goat (*Capra*) in North America? *Zeitschr. Säugetierk.* 29:372–78.

Reed, K. M. 1962. Two new species of fossil talpid insectivores. *Breviora* 168:1–7.

Reichstein, H. 1957. Schädelvariabilität europaischer Mauswiesel (*Mustela nivalis*) und Hermeline (*Mustela erminea*) in Beziehung zu Verbreitung und Geschlecht. *Zeitschr. Säugetierk.* 22:151–82.

Reinhart, R. H. 1971. Fossil Sirenia of Florida. *Plaster Jacket* no. 15. Fla. State Mus., Gainesville.

Rensberger, J. M. 1975. *Haplomys* and its bearing on the origin of the aplodontid rodents. *J. Mammal.* 56(1):1–14.

Repenning, C. A. 1962. The giant ground squirrel, *Paenemarmota*. *J. Paleontol.* 36:540–56.

———. 1967a. Palearctic–Nearctic mammalian dispersal in the late Cenozoic. In D. M. Hopkins (ed.), *The Bering Land Bridge*. Stanford, Calif.: Stanford Univ. Press. Pp. 288–302.

———. 1967b. Subfamilies and genera of the Soricidae. *U.S. Geol. Surv. Prof. Paper* 565:1–74.

———. 1975. Otarioid evolution. *Rapp. Proces-Verbaux Reunions Conseil Perm. Intern. Exploration Mer* 169:27–33.

———. 1976. *Enhydra* and *Enhydriodon* from the Pacific Coast of North America. *J. Res. U.S. Geol. Surv.* 4(3):305–15.

Repenning, C. A., and Tedford, R. H. 1977. Otarioid seals of the Neogene. *U.S. Geol. Surv. Prof. Paper* 992:1–93.

Repenning, C. A., Hopkins, D. M., and Rubin, M. 1964. Tundra rodents in a late Pleistocene fauna from the Tofty Placer District, central Alaska. *Arctic* 17(3):177–97.

Repenning, C. A., Peterson, R. S., and Hubbs, C. L. 1971. Contributions to the systematics of the southern fur seals, with particular reference to the Juan Fernandez and Guadalupe species. In W. H. Burt (ed.), *Antarctic Pinnipedia Antarctic Res. Ser.* 18:1–34.

Repenning, C. A., Ray, C. E., and Grigorescu, D. 1979. "Pinniped Biogeography. In Jane Gray and A. J. Boucot (eds.), *Historical Biogeography, Plate Tectonics and the Changing Environment.* 37th Ann. Biol. Colloq. Oreg. State Univ. Press, Corvallis. Pp. 357–69.

Rhoads, S. N. 1897. Notes on living and extinct species of North American Bovidae. *Proc. Acad. Nat. Sci. Phila.* 49:483–502.

Richardson, J. 1854. *The Zoology of the Voyage of H. M. S. Herald: Vertebrates Including Fossil Mammals.* London.

Ridenour, J. 1969. Depositional environments of the late Pleistocene American Falls Formation, southeastern Idaho. M.S thesis, Idaho State Univ., Pocatello.

Robertson, J. S., Jr. 1974. Fossil *Bison* of Florida. In S. D. Webb (ed.), *Pleistocene Mammals of Florida.* Gainesville, Fla.: Univ. Fla. Presses, Pp. 214–46.

———. 1976. Latest Pliocene mammals from Haile XVA, Alachua County, Florida. *Bull. Fla. State Mus. Biol. Sci.* 20(3):111–86.

Romer, A. S. 1929. A fresh skull of an extinct American camel. *J. Geol.*, 37(3):261–67.

———. 1966. *Vertebrate Paleontology*, 3rd. ed. Chicago: Univ. Chicago Press, 468 pp.

Roth, E. L. 1972. Late Pleistocene mammals from Klein Cave, Kerr County, Texas. *Tex. J. Sci.* 24(1):75–84.

Russell, R. D., and VanderHoof, V. L. 1931. A vertebrate fauna from a new Pliocene formation in northern California. *Univ. Calif. Publ. Bull. Dept. Geol.* 20:11–21.

Russell, R. J. 1960. Pleistocene pocket gophers from San Josecito Cave, Nuevo Leon, Mexico. *Univ. Kans. Publ. Mus. Nat. Hist.* 9(21):539–48.

———. 1968. Evolution and classification of the pocket gophers of the subfamily Geomyinae. *Univ. Kans. Publ. Mus. Nat. Hist.* 16:473–579.

Sadek-Kooros, H. 1966. Jaguar Cave: An early man site in the Beaverhead Mountains of Idaho. Ph. D. dissertation, Harvard University, Cambridge, Mass.

Saunders, J. J. 1977. Late Pleistocene vertebrates of the western Ozark Highland, Missouri. *Rept. Inv. Ill. State Mus.* 33:1–118.

Savage, D. E. 1951. Late Cenozoic vertebrates of the San Francisco Bay region. *Univ. Calif. Publ. Bull. Dept. Geol. Sci.* 28(10):215–314.

———. 1955. A survey of various late Cenozoic vertebrate faunas of the Panhandle of Texas, II: Proboscidea. *Univ. Calif. Publ. Bull. Dept. Geol. Sci.* 31(3):51–74.

———. 1960. A survey of various late Cenozoic vertebrate faunas of the Panhandle of Texas, III: Felidae. *Univ. Calif. Publ. Bull. Dept. Geol. Sci.* 36(6):317–44.

Schaub, S. 1941. Ein neues Hyaenidengenus von der Montagne de Perrier. *Eclogae Geol. Helv.* 34:279–86.

———. 1951. *Soergelia* n. gen., ein Caprine aus dem thüringischen Altpleistocaen. *Eclogae Geol. Helv.* 44:375–81.

Scheffer, V. B. 1958. *Seals, Sea Lions and Walruses.* Stanford, Calif.: Stanford Univ. Press. 179 pp.

Schneider, D. 1973. The adaptable raccoon. *Nat. Hist.* 82(7):64–71.

Schneider, K. B., and Faro, J. B. 1975. Effects of sea ice on sea otters (*Enhydra lutris*). *J. Mammal.* 56(1):91–101.

Schultz, C. B. 1934. The Pleistocene mammals of Nebraska. *Bull. Nebr. State Mus.* 1(41):357–93.

———. 1972. Holocene Interglacial migrations of mammals and other vertebrates. *Quatern. Res.* 2(3):337–40.

Schultz, C. B., and Frankforter, W. D. 1946. The geologic history of the bison of the Great Plains. *Bull. Univ. Nebr. State Mus.* 3(1):1–9.

Schultz, C. B., and Hillerud, J. M. 1977. The antiquity of *Bison latifrons* in the Great Plains of North America. *Trans. Nebr. Acad. Sci.* 4:103–16.

Schultz, C. B., and Howard, E. B. 1935. The fauna of Burnet Cave, Guadalupe Mountains, New Mexico. *Proc. Acad. Nat. Sci. Phila.* 87:273–98.

Schultz, C. B., and Martin, L. D. 1970a. Machairodont cats from the early Pleistocene Broadwater and Lisco local faunas. *Bull. Univ. Nebr. State Mus.* 9(2):33–38.
—— and ——. 1970b. Quaternary mammalian sequence in the central Great Plains. In W. Dort, Jr., and J. K. Jones, Jr. (eds.), *Pleistocene and Recent Environments of the Central Great Plains; Spec. Publ. Dept. Geol. Univ. Kans.* 3:341–53.
—— and ——. 1972. Two lynx-like cats from the Pliocene and Pleistocene. *Bull. Univ. Nebr. State Mus.* 9(7):196–203.
Schultz, C. B., and Stout, T. M. 1948. Pleistocene mammals and terraces in the Great Plains. *Bull. Geol. Soc. Amer.* 59(6):553–88.
—— and ——. 1961. Field Conference on the Tertiary and Pleistocene of Western Nebraska. *Spec. Publ. Univ. Nebr. State Mus.* 2:1–54.
Schultz, C. B., and Tanner, L. G. 1957. Medial Pleistocene fossil vertebrate localities in Nebraska. *Bull. Univ. Neb. State Mus.* 4:59–81.
Schultz, C. B., Frankforter, W. D., and Toohey, L. 1951. A graphic resume of the Pleistocene of Nebraska (with notes on the fossil mammalian remains). *Bull. Univ. Nebr. State Mus.* 3(6):1–41.
Schultz, C. B., Tanner, L. G., and Martin, L. D. 1972. Phyletic trends in certain lineages of Quaternary mammals. *Bull. Univ. Nebr. State Mus.* 9(6):183–95.
Schultz, C. B., et. al. 1963. Paleontologic investigations at Big Bone Lick State Park, Kentucky: A preliminary report. *Science* 142:1167–69.
Schultz, C. B., et al. 1967. Big Bone Lick, Kentucky. Mus. Notes Univ. Nebr. State Mus. No. 33.
Schultz, G. E. 1965. Pleistocene vertebrates from the Butler Spring local fauna, Meade County, Kansas. *Papers Mich. Acad. Sci. Arts Letters* 50:235–65.
——. 1967. Four superimposed late Pleistocene vertebrate faunas from southwest Kansas. In P. S. Martin and H. E. Wright (eds.), *Pleistocene Extinctions: The Search for a Cause.* New Haven and London: Yale Univ. Press. Pp. 321–36.
——. 1969. Geology and paleontology of a late Pleistocene Basin in southwest Kansas. *Spec. Paper Geol. Soc. Amer.* 105:1–85.
——. 1977. Blancan and post-Blancan faunas in the Texas Panhandle. In G. E. Schultz (ed.), *Guidebook Field Conference on Late Cenozoic Biostratigraphy of the Texas Panhandle and Adjacent Oklahoma, August 1–6, 1977.* Spec. Publ. 1, Kilgore Res. Center, Dept. Geol. Anthropol., pp. 105–45.
Schultz, J. R. 1937. A late Cenozoic vertebrate fauna from the Coso Mountains, Inyo County, California. *Publ. Carnegie Inst. Washington* 487:77–109.
——. 1938. A late Quaternary mammal fauna from the tar seeps of McKittrick, California. *Publ. Carnegie Inst. Washington* 487(4): 111–215.
Schütt, G. 1973. Revision der *Cuon* und *Xenocyon*-Funde (Canidae, Mammalia) aus den altpleistozänen Mosbacher Sanden (Wiesbaden, Hessen). *Mainz. Naturwiss. Arch.* 12:49–77.
Scott. W. B. 1885. *Cervalces americanus*, a fossil moose or elk from the Quaternary of New Jersey. *Proc. Acad. Nat. Sci. Phila.* 1885:174–202.
Sellards, E. H. 1915. *Chlamytherium septentrionalis*, an edentate from the Pleistocene of Florida. *Amer. J. Sci. Ser. 4*, 40:139–45.
——. 1916a. Human remains and associated fossils from the Pleistocene of Florida. *Fla. Geol. Surv. 8th Ann. Rept.*, pp. 121–60.
——. 1916b. On the discovery of fossil human remains in Florida in association with extinct vertebrates. *Amer. J. Sci. Ser. 4*, 42:1–18.
——. 1918. The skull of a Pleistocene tapir including description of a new species and a note on the associated fauna and flora. *Fla. State Geol. Surv. 10th and 11th Ann. Repts.*, pp. 57–70.
——. 1919. The geology and mineral resources of Bexar County. *Univ. Tex. Bull.* 1932:1–202.
——. 1932. Geologic relations of deposits reported to contain artifacts at Frederick, Oklahoma. *Bull. Geol. Soc. Amer.* 43:783–96.
Semken, H. A., Jr. 1966. Stratigraphy and paleontology of the McPherson *Equus* beds (Sandahl local fauna), McPherson County, Kansas. *Contrib. Mus. Paleontol. Univ. Mich.* 20(6):121–78.

———. 1969. Paleoecological implications of micromammals from Peccary Cave, Newton County, Arkansas. Geol. Soc. Amer. Absts. South-Cent. Sect., 1969, part 2, p. 27.
Semken, H. A., Jr., and Griggs, C. D. 1965. The long-nosed peccary, *Mylohyus nasutus*, from McPherson County, Kansas. *Papers Mich. Acad. Sci. Arts, Letters,* 50:267–74.
Semken, H. A., Jr., Miller, B. B., and Stevens, J. B. 1964. Late Wisconsin woodland musk oxen in association with pollen and invertebrates from Michigan. *J. Paleontol.* 38(5):823–35.
Sher, A. V. 1968. Fossil saiga in northeastern Siberia and Alaska. *Int. Geol. Rev.* 10(11):1247–60.
———. 1975. Die nördlichste Variante der "Mindel"-Fauna in Eurasien und der Ursprung der subarktischen Mammalier. *Quartärpaläontologie* 1:235–42.
Shotwell, J. A. 1956. Hemphillian mammalian assemblage from northeastern Oregon. *Bull. Geol. Soc. Amer.* 67:717–38.
———. 1970. Pliocene mammals of southeast Oregon and adjacent Idaho. *Bull. Univ. Oreg. Mus. Nat. Hist.* 17:1–103.
Shutler, R. 1968. Tule Springs: Its implications to early man studies in North America. *Contrib. Anthropol. East. N.M. Univ.* 1(4):19–26.
Sikes, S. K. 1971. *The Natural History of the African Elephant.* New York: American Elsevier Publ. Co. 397 pp.
Simpson, G. G. 1928. Pleistocene mammals from a cave in Citrus County, Florida. *Amer. Mus. Novitates* 328:1–16.
———. 1929a. Pleistocene mammalian fauna of the Seminole Field, Pinellas County, Florida. *Bull. Amer. Mus. Nat. Hist.* 56(8):561–99.
———. 1929b. The extinct land mammals of Florida. *Fla. State Geol. Surv. 20th Ann. Rept.,* pp. 229–80.
———. 1930a. Additions to the Pleistocene of Florida. *Amer. Mus. Novitates* 406:1–14.
———. 1930b. *Holmesina septentrionalis,* extinct giant armadillo of Florida. *Amer. Mus. Novitates* 442:1–10.
———. 1941a. The function of saber-like canines in carnivorous mammals. *Amer. Mus. Novitates* 1130:1–12.
———. 1941b. Discovery of jaguar bones and footprints in a cave in Tennessee. *Amer. Mus. Novitates* 1131:1–12.
———. 1941c. Large Pleistocene felines of North America. *Amer. Mus. Novitates* 1136:1–27.
———. 1945a. Principles of classification and a classification of mammals. *Bull. Amer. Mus. Nat. Hist.* 85:1–350.
———. 1945b. Notes on Pleistocene and Recent tapirs. *Bull. Amer. Mus. Nat. Hist.* 86(2):37–81.
———. 1946. Bones in the brewery. *Nat. Hist.* 55(6):252–59.
———. 1947. Holarctic mammalian faunas and continental relationships during the Cenozoic. *Bull. Geol. Soc. Amer.* 58:613–87.
———. 1949. A fossil deposit in a cave in St. Louis. *Amer. Mus. Novitates* 1408:1–46.
———. 1961. *Horses.* Natural History Library Ed. Garden City, N.Y.: Doubleday. 323 pp.
———. 1963. A new record of *Euceratherium* or *Preptoceras* (extinct Bovidae) in New Mexico. *J. Mammal.* 44(4):383–84.
Sinclair, W. J. 1903. A preliminary account of the exploration of Potter Creek Cave, Shasta County, California. *Science* 17:708–12.
———. 1905. New Mammalia from the Quaternary caves of California. *Univ. Calif. Publ. Bull. Dept. Geol.* 4(7):145–61.
Sinclair, W. J., and Furlong, E. L. 1904. *Euceratherium,* a new ungulate from the Quaternary caves of California. *Univ. Calif. Publ. Bull. Dept. Geol.* 3(20):411–18.
Skeels, M. A. 1962. The mammoths and mastodons of Michigan. *Papers Mich. Acad. Sci. Arts Letters* 47:101–33.
Skinner, M. F. 1942. The fauna of Papago Springs Cave, Arizona, and a study of *Stockoceros. Bull. Amer. Mus. Nat. Hist.* 80(6):143–220.
Skinner, M. F., and Hibbard, C. W. 1972. Early Pleistocene and pre-glacial and glacial rocks and faunas of north-central Nebraska. *Bull. Amer. Mus. Nat. Hist.* 148(1):1–148.

Skinner, M. F., and Kaisen, O. C. 1947. The fossil *Bison* of Alaska and preliminary revision of the genus. *Bull. Amer. Mus. Nat. Hist.* 89(3):123–256.
Slaughter, B. H. 1960. A new species of *Smilodon* from a late Pleistocene alluvial terrace deposit of the Trinity River. *J. Paleontol.*, 34(3):486–92.
———. 1961a. A new coyote in the late Pleistocene of Texas. *J. Mammal.* 42:503–9.
———. 1961b. The significance of *Dasypus bellus* Simpson in Pleistocene local faunas. *Tex. J. Sci.* 13(3):311–15.
———. 1964. An ecological interpretation of the Brown Sand Wedge local fauna, Blackwater Draw, New Mexico, and a hypothesis concerning late Pleistocene extinction. *Paleoecol. Llano Estacado* 2:1–37.
———. 1966a. *Platygonus compressus* and associated fauna from the Laubach Cave of Texas. *Amer. Midl. Nat.* 75(2):475–94.
———. 1966b. The Moore Pit local fauna; Pleistocene of Texas. *J. Paleontol.* 40(1):78–91.
———. 1967. Animal ranges as a clue to late Pleistocene extinctions. In P. S. Martin and H. E. Wright (eds.), *Pleistocene Extinctions: The Search for a Cause.* New Haven and London: Yale Univ. Press. Pp. 155–67.
Slaughter, B. H., and Hoover, R. 1963. Sulphur River Formation and the Pleistocene mammals of the Ben Franklin local fauna. *J. Grad. Res. Center SMU* 31(3):132–48.
Slaughter, B. H., and McClure, W. L. 1965. The Sims Bayou local fauna: Pleistocene of Houston, Texas. *Tex. J. Sci.* 17(4):404–17.
Slaughter, B. H., and Ritchie, R. 1963. Pleistocene mammals of the Clear Creek local fauna, Denton County, Texas. *J. Grad. Res. Center SMU* 31(3):117–31.
Slaughter, B. H., et al. 1962. The Hill-Shuler local faunas of the Upper Trinity River, Dallas and Denton Counties, Texas. *Rept. Inv. Univ. Tex. Bur. Econ. Geol.* 48:1–75.
Sokolov, I. I. 1973. Trends of evolution and the classification of the subfamily Lutrinae (Mustelidae, Fissipedia). *Bull. Moscow Prir. Biol.* 78(6):45–52.
Stalker, A. MacS. 1969. Quaternary stratigraphy in southern Alberta, II: Sections near Medicine Hat. *Geol. Surv. Canada Paper* 69-26, 28 pp.
———. 1971. Quaternary studies in the southwestern prairies. *Geol. Surv. Canada Paper* 71-1, part A, pp. 180–81.
———. 1977. Indications of Wisconsin and earlier man from the southwest Canadian prairies. *Ann. N.Y. Acad. Sci.* 288:119–36.
Stalker, A. MacS., and Churcher, C. S. 1970. Deposits near Medicine Hat, Alberta, Canada. (Display chart with marginal notes.) Geol. Surv. Canada.
——— and ———. 1972. Glacial stratigraphy of the southwestern Canadian prairies; the Laurentide Record. *Proc. 24th Int. Geol. Cong.* 1972, Sect. 12, pp. 110–19.
Starrett, A. 1956. Pleistocene mammals of the Berends fauna of Oklahoma. *J. Paleontol.* 30(5):1187–92.
Stephens, J. J. 1959. A new Pliocene cat from Kansas. *Papers Mich. Acad. Sci. Arts Letters* 44:41–46.
———. 1960. Stratigraphy and paleontology of a late Pleistocene basin, Harper County, Oklahoma. *Bull. Geol. Soc. Amer.* 71:1675–1702.
Stevens, M. S. 1965. A new species of *Urocyon* from the Upper Pliocene of Kansas. *J. Mammal.* 46(2):265–69.
Stirton, R. A. 1931. A new genus of the family Vespertilionidae from the San Pedro Pliocene of Arizona. *Univ. Calif. Publ. Bull. Dept. Geol. Sci.* 20(4):27–30.
———. 1936. Succession of North American continental Pliocene mammalian faunas. *Amer. J. Sci.* 232:161–206.
———. 1938. Notes on some late Tertiary and Pleistocene antilocaprids. *J. Mammal.* 19(3):366–70.
———. 1939. Cenozoic mammal remains from the San Francisco Bay region. *Univ. Calif. Publ. Bull. Dept. Geol. Sci.* 24(13):339–410.
———. 1940. Phylogeny of North American Equidae. *Univ. Calif. Publ. Bull. Dept. Geol. Sci.* 25(4):165–98.
———. 1965. Cranial morphology of *Castoroides*. In *Dr. D. N. Wadia Commemorative Vol. Mining Metalurg. Inst. India*, pp. 273–85.

Stirton, R. A., and Christian, W. G. 1940. A member of the Hyaenidae from the Upper Pliocene of Texas. *J. Mammal.* 21:445–48.

Stirton, R. A., and VanderHoof, V. L. 1933. *Osteoborus,* a new genus of dogs, and its relations to *Borophagus* Cope. *Univ. Calif. Publ. Bull. Geol. Sci.* 23:175–82.

Stock, A. D., and Stokes, W. L. 1969. A re-evaluation of Pleistocene bighorn sheep from the Great Basin and their relationship to living members of the genus *Ovis. J. Mammal.* 50(4):805–7.

Stock, C. 1914a. Skull and dentition of the mylodont sloths of Rancho La Brea. *Univ. Calif. Publ. Bull. Dept. Geol.* 8(18):319–34.

——. 1914b. The systematic position of the mylodont sloths from Rancho La Brea. *Science NS* 39:761–63.

——. 1917a. Further observations on the skull structure of mylodont sloths from Rancho La Brea. *Univ. Calif. Publ. Bull. Dept. Geol.* 10(11):165–78.

——. 1917b. Recent studies of the skull and dentition of *Nothrotherium* from Rancho La Brea. *Univ. Calif. Publ. Bull. Dept. Geol.* 10(10):137–64.

——. 1918. The Pleistocene fauna of Hawver Cave. *Univ. Calif. Publ. Bull. Dept. Geol.* 10(24):461–515.

——. 1920. A mounted skeleton of *Mylodon harlani. Univ. Calif. Publ. Bull. Dept. Geol.* 12(6):425–30.

——. 1925. Cenozoic gravigrade edentates of western North America with special reference to the Pleistocene Megalonychidae and Mylodontidae of Rancho La Brea. *Carnegie Inst. Washington Publ.* 331:1–206.

——. 1928. *Tanupolama,* a new genus of llama from the Pleistocene of California. *Carnegie Inst. Washington Publ.* 393:29–37.

——. 1929. A census of the Pleistocene mammals of Rancho La Brea based on the collections of the Los Angeles Museum. *J. Mammal.* 10(4):281–89.

——. 1930. Quaternary antelope remains from a second cave deposit in the Organ Mountains, New Mexico. *Los Angeles Mus. Paleontol. Publ.* 2:1–18.

——. 1931. Problems of antiquity presented in Gypsum Cave, Nevada. *Sci. Monthly* 32:22–32.

——. 1932a. A further study of the Quaternary antelopes of Shelter Cave, New Mexico. *Los Angeles Mus. Paleontol. Publ.* 3:1–45.

——. 1932b. *Hyaenognathus* from the late Pliocene of the Coso Mountains, California. *J. Mammal.* 13:263–66.

——. 1935. Exiled elephants of the Channel Islands, California. *Sci. Monogr.* 41:205–14.

——. 1936a. A new mountain goat from the Quaternary of Smith Creek Cave, Nevada. *Bull. So. Calif. Acad. Sci.* 35(3):149–53.

——. 1936b. Sloth tracks in the Carson prison. *Westways,* July 1936, 28(7):26–27.

——. 1937. Tar babies. *Westways* 29(9):26–27.

——. 1942. A ground sloth in Alaska. *Science* 95:552–53.

——. 1943. Foxes and elephants of the Channel Islands. *Los Angeles Co. Mus. Quart.* 3:6–9.

——. 1950. Bears from the Pleistocene Cave of San Josecito, Nuevo Leon, Mexico. *J. Washington Acad. Sci.* 40(10):317–21.

——. 1965. Rancho La Brea, a record of Pleistocene life in California., 6th ed. Los Angeles Co. Mus. Sci. Ser. Paleontol, no. 11, 81 pp.

Stock, C., and Bode, F. D. 1936. The occurrence of flints and extinct animals in pluvial deposits near Clovis, New Mexico, III: Geology and vertebrate paleontology of the late Quaternary near Clovis, New Mexico. *Proc. Acad. Nat. Sci. Phila.* 88:219–41.

Stock, C., and Furlong, E. L. 1928. The Pleistocene elephants of Santa Rosa Island, California. *Science* 68(1754):140–41.

Stock, C., and Richards, H. G. 1949. A *Megalonyx* tooth from the Northwest Territories, Canada. *Science* 110(2870):709.

Stokes, W. L., and Condie, K. C. 1961. Pleistocene bighorn sheep from the Great Basin. *J. Paleontol.* 35:598–609.

Storer, J. E. 1975. Pleistocene prairie dog in south-central Alberta. *Blue Jay* 33(4):247.

Stovall, J. W. 1937. *Euceratherium bizelli*, a new ungulate from Oklahoma. *Jour. Paleontol.* 11(5):450–55.

———. 1940. *Megalonyx hogani* a new species of ground sloth from Gould, Oklahoma. *Amer. J. Sci.* 238:140–46.

Stovall, J. W., and Johnston, C. S. 1935. Two fossil grizzly bears from the Pleistocene of Oklahoma. *J. Geol.* 43(2):208–13.

Stovall, J. W., and McAnulty, W. N. 1950. The vertebrate fauna and geological age of Trinity River Terraces in Henderson County, Texas. *Amer. Midl. Nat.* 44(1):211–50.

Strain, W. S. 1966. Blancan mammalian fauna and Pleistocene formation of Hudspeth County, Texas. *Bull. Tex. Mem. Mus.* 10:1–55.

Szabo, B. J., Stalker, A. MacS., and Churcher, C. S. 1973. Uranium-Series ages of some Quaternary deposits near Medicine Hat, Alberta, Canada. *Canad. J. Earth Sci.* 10(9):1464–69.

Taylor, D. W. 1954. A new Pleistocene fauna and new species of fossil snails from the High Plains. *Occ. Papers Mus. Zool. Univ. Mich.* 557:1—16.

———. 1966. Summary of North American Blancan nonmarine mollusks. *Malacologia* 4:1–172.

Taylor, W. P. 1911. A new antelope from the Pleistocene of Rancho La Brea. *Univ. Calif. Publ. Bull. Dept. Geol.* 6(10):191–97.

Tedford, R. H. 1970. Principles and practices of mammalian geochronology in North America. *Proc. No. Amer. Paleontol. Conv.* Sept. 1969, vol F, pp. 666–703.

Tedford, R. H., and Gustafson, E. P. 1977. First North American record of the extinct panda *Parailurus*. *Nature* 265:621–23.

Tener, J. S. 1965. Muskoxen in Canada, a biological and taxonomic review. *Canad. Wildlife Serv. Monogr.* 2:1–166.

Thaeler, C. S., and Nevo, E. 1973. Cytotaxonomy and evolution of the *Thomomys talpoides* complex of pocket gophers from the southern Rocky Mountains. *Abst.* ICSEB, Boulder, Colo., 1 p.

Thenius, E. 1954. Zur Abstammung der Rotwölfe. *Österr. Zool. Zeitschr.* 5:377–87.

Tikhomirov, B. A. 1958. Natural conditions and vegetation in the mammoth epoch in northern Siberia. *Problems in the North* 1:168–88.

Tobien, H. 1957. *Cuon* Hodg. und *Gulo* Frisch (Carnivora, Mammalia) aus den altpleistozäner Sanden von Mosbach bei Wiesbaden. *Acta Zool. Cracoviensia* 2:433–51.

———. 1970. Biostratigraphy of the mammalian faunas at the Pliocene-Pleistocene boundary in middle and western Europe. *Palaeogeog. Palaeoclim. Palaeoecol.* 8:77–93.

———. 1973. On the evolution of mastodonts (Proboscidea, Mammalia). *Notizbl. hess. Landesamtes Bodenforsch.* 101:202–76.

Torre, D. 1967. I cani villafranchiani della Toscana. *Palaeontogr. Ital.* 63:113–38.

Trapp, G. R. 1972. Some anatomical and behavioral adaptations of ringtails, *Bassariscus astutus*. *J. Mammal.* 53(3):549–57.

Troxell, E. L. 1915a. The vertebrate fossils of Rock Creek, Texas. *Amer. J. Sci.*, ser. 4, 39:613–38.

———. 1915b. A fossil ruminant from Rock Creek, Texas, *Preptoceras mayfieldi* sp. nov. *Amer. J. Sci.*, ser. 4, 40:479–82.

Turekian, K. K., and Bada, J. L. 1972. The dating of fossil bones. In W. W. Bishop and J. A. Miller (eds.), *Calibration of Hominid Evolution*. Edinburgh: Scottish Academic Press. Pp. 171–86.

Valastro, S., Jr., Davis, E. M., and Varela, A. G. 1977. University of Texas at Austin Radiocarbon dates XI. *Radiocarbon* 17(2):160–73.

VanderHoof, V. L. 1933. Additions to the fauna of the Tehama upper Pliocene of northern California. *Amer. J. Sci.*, ser. 5, 25:382–84.

———. 1936. Notes on the type of *Borophagus diversidens* Cope. *J. Mammal.* 17:415–16.

———. 1942. A skull of *Bison latifrons* from the Pleistocene of northern California. *Univ. Calif. Publ. Bull. Dept. Geol. Sci.* 27:1–24.

Van der Meulen, A. J. 1972. Middle Pleistocene smaller mammals from the Monte Peglia (Orvieto, Italy) with special reference to the phylogeny of *Microtus* (Arvicolidae, Rodentia). *Quaternaria* 17:1–144.

———. 1978. *Microtus* and *Pitymys* (Arvicolidae) from Cumberland Cave, Maryland, with a comparison of some New and Old World species. *Ann. Carnegie Mus.* 47(6):101–45.
Van Gelder, R. G. 1959. A taxonomic revision of the spotted skunk (genus *Spilogale*). *Bull. Amer. Mus. Nat. Hist.* 117(5):233–92.
———. 1968. The genus *Conepatus* (Mammalia, Mustelidae): Variation within a population. *Amer. Mus. Novitates* 2322:1–37.
———. 1977. Mammalian hybrids and generic limits. *Amer. Mus. Novitates* 2635:1–25.
Vangengeim, E. A., and Zazhigin, V. S. 1972. Mammalian fauna of Siberia and the Neogene-Quaternary boundary. *Coll. Papers Internat. Colloq. Neogene-Quaternary* 2:278–87.
Van Valen, L. 1967. New Paleocene insectivores and insectivore classification. *Bull. Amer. Mus. Nat. Hist.* 135:221–84.
———. 1969. Late Pleistocene extinctions. *Proc. No. Amer. Paleontol. Conv.*, part E, pp. 469–85.
Van Zyll de Jong, C. G. 1972. A systematic review of the Nearctic and Neotropical river otters (genus *Lutra*, Mustelidae, Carnivora). *Life Sci. Contrib. Royal Ontario Mus.* 80:1–104.
Vaughan, T. A. 1972. *Mammalogy*. Philadelphia: W. B. Saunders Co. 463 pp.
Vaughan, T. A., and Bateman, G. C. 1970. Functional morphology of the forelimb of mormoopid bats. *J. Mammal.* 51:217–35.
Vereshchagin, N. K. 1971. The cave lion and its history in the Holarctic and on the territory of the USSR. *Trudy Zool. Inst. Akad. Nauk. SSSR* 49:123–97. (In Russian.)
Voorhies, M. R. 1971. The Watkins Quarry: A new late Pleistocene mammal locality in Glynn County, Georgia. *Bull. Ga. Acad. Sci.* 29(2):128.
———. 1973. Vertebrate fossils of coastal Georgia: A field geologist's guide. In R. W. Frey (ed.), *The Neogene of the Georgia Coast*. Georgia Geol. Soc., pp. 81–102.
———. 1974a. Pleistocene vertebrates with boreal affinities in the Georgia Piedmont. *Quatern. Res.* 4(1):85–93.
———. 1974b. The Pliocene horse *Nannipus minor* in Georgia: Geologic implications. *Tulane Stud. Geol. Paleontol.* 11:109–14.
———. 1975. A new genus and species of fossil kangaroo rat and its burrow. *J. Mammal.* 56(1):160–76.
Walker, D. 1974. A Pleistocene gyrfalcon. *Auk* 91(4):820–21.
Walker, E. P. 1975. *Mammals of the World*, 3rd ed. Baltimore, Md.: John Hopkins Press. 2 vols., 1,500 pp.
Warren, J. C. 1852. *"Mastodon giganteus" of North America*. Boston: John Wilson & Son. 219 pp.
Warter, J. K. 1976. Late Pleistocene plant communities—evidence from the Rancho La Brea tar pits. *Spec. Publ. Calif. Native Plant Soc.* 2:32–39.
Waters, J. H., and Mack, C. W. 1962. Second find of sea mink in southeastern Massachusetts. *J. Mammal.* 43:429–30.
Waters, J. H. and Ray, C. E. 1961. Former range of the sea mink. *J. Mammal.* 42(3):380–83.
Webb, S. D. 1965. The osteology of *Camelops*. *Los Angeles Co. Mus. Sci. Bull.* 1:1–54.
———. 1969. Extinction–origination equilibria in late Cenozoic land mammals of North America. *Evolution* 23:688–702.
———. 1973. Pliocene pronghorns of Florida. *J. Mammal.* 54(1):203–21.
———. 1974a. Chronology of Florida Pleistocene mammals. In S. D. Webb (ed.), *Pleistocene Mammals of Florida*. Gainesville, Fla.: Univ. Fla. Presses. Pp. 5–31.
———. 1974b. The status of *Smilodon* in the Florida Pleistocene. In S. D. Webb (ed.), *Pleistocene mammals of Florida*. Gainesville, Fla.: Univ. Fla. Presses. Pp. 149–53.
———. 1974c. Pleistocene llamas of Florida, with a brief review of the Lamini. In S. D. Webb (ed.), *Pleistocene Mammals of Florida*. Gainesville, Fla.: Univ. Fla. Presses. Pp. 170–213.
———. 1976. Mammalian faunal dynamics of the great American interchange. *Paleobiology* 2:220–34.
Weigel, R. D. 1962. Fossil vertebrates of Vero, Florida. *Spec. Publ. Fla. Geol. Surv.* 10:1–59.
Wells, P. V. 1976. Macrofossil analysis of woodrat (*Neotoma*) middens as a key to the Quaternary vegetational history of arid America. *Quatern, Res.* 6(2):223–48.

Wells, P. V., and Jorgensen, C. D. 1964. Pleistocene wood rat middens and climatic change in Mohave Desert: A record of juniper woodlands. *Science* 143(3611):1171–74.

Wetzel, R. M. 1977. The Chacoan peccary *Catagonus wagneri* (Rusconi). *Bull. Carnegie Mus. Nat. Hist.* 3:1–36.

Wetzel, R. M., et al. 1975. *Catagonus*, an "extinct" peccary, alive in Paraguay. *Science* 189(4200):379–81.

Wheat, J. B. 1972. The Olsen-Chubbuck site, a Paleo-Indian bison kill. *Mem. Soc. Amer. Archaeol.* 26:1–180.

Whitaker, J. O., Jr. 1974. *Cryptotis parva*. *Mammal. Species* 43:1–8.

White, J. A. 1966. A new *Peromyscus* from the late Pleistocene of Anacapa Island with notes on variation in *Peromyscus nesodytes* Wilson. *Contrib. Sci. Los Angeles Co. Mus.* 96:1–8.

———. 1968. A new porcupine from the middle Pleistocene of the Anza-Borrego Desert of California. *Contrib. Sci. Los Angeles Co. Mus.* 136:1–15.

———. 1969. Late Cenozoic bats (subfamily Nyctophylinae) from the Anza-Borrego Desert of California. *Univ. Kans. Mus. Nat. Hist. Misc. Publ.* 51:275–82.

———. 1970. Late Cenozoic porcupines (Mammalia, Erethizontidae) of North America. *Amer. Mus. Novitates* 2421:1–15.

White, J. A., and Downs, T. 1961. A new *Geomys* from the Vallecito Creek Pleistocene of California. *Contrib. Sci. Los Angeles Co. Mus.* 42:1–34.

Whitmore, F. C., Jr., and Foster, H. L. 1967. *Panthera atrox* (Mammalia, Felidae) from central Alaska. *J. Paleontol.* 41(1):247–51.

Whitmore, F. C., Jr., et al. 1967. Elephant teeth from the Atlantic Continental Shelf. *Science* 156(3781):1477–81.

Williston, S. W. 1894. A new Dicotyline mammal from the Kansas Pliocene. *Science* 23:164.

Willoughby, D. P. 1948. A statistical study of the metapodials of *Equus occidentalis* Leidy. *Bull. So. Calif. Acad. Sci.* 47(3):84–94.

Wilson, M. V. 1969. Problems in the speciation of American fossil bison. In R. G. Forbis et al. (eds.), *Post-Pleistocene Man and His Environment on the Northern Plains*. Calgary: Student's Press, Univ. Calgary. Pp. 178–99.

———. 1972. Review of "Biostratigraphy and paleontology of a Sangamon deposit at Fort Qu'Appelle, Saskatchewan," by E. Khan (1970). *Contrib. Geol.* 11(2):87–91.

———. 1974a. History of the bison in Wyoming, with particular reference to early Holocene forms. In M. Wilson (ed.), *Applied Geology and Archaeology: The Holocene History of Wyoming; Geol. Surv. Wyoming Rept. Inv.* 10:91–99.

———. 1974b. The Casper local fauna and its fossil bison. In G. C. Frison (ed.), *The Casper Site, a Hell Gap Bison Kill on the High Plains*. New York: Academic Press. Pp. 125–71.

———. 1975. Holocene fossil bison from Wyoming and adjacent areas. M.A. thesis, Univ. Wyo., Laramie.

Wilson, R. L. 1967. The Pleistocene vertebrates of Michigan. *Papers Mich. Acad. Sci. Arts letters* 52:197–234.

Wilson, R. W. 1932. *Cosomys*, a new genus of vole from the Pliocene of California. *J. Mammal.* 13:150–54.

———. 1933a. A rodent fauna from later Cenozoic beds of southwestern Idaho. *Carnegie Inst. Washington Publ.* 440:119–35.

———. 1933b. Pleistocene mammalian fauna from the Carpinteria Asphalt. *Carnegie Inst. Washington Publ.* 440:59–76.

———. 1935. A new species of porcupine from the later Cenozoic of Idaho. *J. Mammal.* 16(3):220–21.

———. 1936. A new Pleistocene deer-mouse from Santa Rosa Island, California. *J. Mammal.* 17:408–10.

———. 1942. Preliminary study of the fauna of Rampart Cave, Arizona. *Carnegie Inst. Washington Publ.* 530:169–85.

Woldstedt, P. 1969. *Quartär: Handbuch der straitigraphischen Geologie.* Stuttgart: Enke. Vol. 2, pp. 1–263.

Wood, A. E. 1935. Evolution and relationships of the heteromyid rodents with new forms from the Tertiary of western North America. *Ann. Carnegie Mus.* 24:73–262.

Wood, H. E., II, et al. 1941. Nomenclature and correlation of the North American continental Tertiary. *Bull. Geol. Soc. Amer.* 52:1–48.

Woodard, G. D., and Marcus, L. F. 1973. Rancho La Brea fossil deposits: A re-evaluation from stratigraphic and geological evidence. *J. Paleontol.* 47(1):54–69.

Woodburne M. O. 1961. Upper Pliocene geology and vertebrate paleontology of part of the Meade Basin, Kansas. *Papers Mich. Acad. Sci. Arts Letters* 46:61–101.

———. 1968. The cranial myology and osteology of *Dicotyles tajacu*, the collared peccary, and its bearing on classification. *Mem. So. Calif. Acad. Sci.* 7:1–48.

Woods, C. A. 1973. *Erethizon dorsatum*. *Mammal. Species* 29:1–6.

Wright, T., and Lundelius, E. L., Jr. 1963. Post-Pleistocene raccoons from central Texas and their zoogeographic significance. *Pearce-Sellards Ser. Tex. Mem. Mus.* 2:2–21.

Young, S. P. 1958. *The Bobcat of North America.* Washington: Wildlife Inst. North America. 193 pp.

Zakrzewski, R. J. 1967a. The primitive vole, *Ogmodontomys*, from the late Cenozoic of Kansas and Nebraska. *Papers Mich. Acad. Sci. Arts Letters* 52:133–50.

———. 1967b. The systematic position of *Canimartes* from the upper Pliocene of Idaho. *J. Mammal.* 48:293–97.

———. 1969. The rodents from the Hagerman local fauna, upper Pliocene of Idaho. *Contrib. Mus. Paleontol. Univ. Mich.* 23(1):1–36.

———. 1972. Fossil microtines from late Cenozoic deposits in the Anza-Borrego Desert, California, with the description of a new subgenus of *Synaptomys*. *Contrib. Sci. Los Angeles Co. Mus.* 221:1–12.

———. 1974. Fossil Ondatrini from western North America. *J. Mammal.* 55(2):284–92.

———. 1975a. Pleistocene stratigraphy and paleontology in western Kansas: The state of the art, 1974. *Studies on Cenozoic Paleontology and Stratigraphy, Claude W. Hibbard Memorial Vol. 3, Univ. Mich. Papers Paleontol.* 12:121–128.

———. 1975b. The late Pleistocene arvicoline rodent *Atopomys*. *Ann. Carnegie Mus.* 45(12):255–61.

Zakrzewski, R. J., and Maxfield, J. 1971. Occurrence of *Clethrionomys* in the late Pleistocene of Kansas. *J. Mammal.* 52(3):620–21.

Zeimens, G., and Walker, D. N. 1974. Bell Cave, Wyoming: Preliminary archaeological and paleontological investigations. In M. Wilson (ed.), *Applied Geology and Archaeology: The Holocene History of Wyoming; Geol. Surv. Wyo. Rept. Inv.* 10:88–90.

Zeuner, F. E. 1952. *Dating the past: An Introduction to Geochronology.* London: Methuen. 495 pp.

APPENDICES

APPENDIX 1
WISCONSINAN RADIOCARBON DATES

Fauna	Lab.	Date (years B.P.)	Material	References
Aden Crater	Y-1163b	11,080 ± 200	*Nothrotheriops* tissue	P. S. Martin, 1967
Ben Franklin	SM-532	9,340 ± 1,000	Charcoal	Slaughter and Hoover, 1963
Big Bone Lick	W-1617	17,200 ± 600	Plant-bearing silt from *Bootherium* skull	Ives et al., 1967
Blackwater Draw No. 1	L-304C	11,170 ± 360	Carbonized plant remains	Haynes, 1967
Boney Spring	M-2211	13,700 ± 600	Plant debris from *Mammut* tusk	Mehringer, King, and Lindsay, 1970
Boney Spring	I-3922	16,580 ± 220	Plant debris from *Mammut* tusk	Mehringer, King, and Lindsay, 1970
Brynjulfson	ISGS D-70	9,440 ± 760	Bone	Parmalee and Oesch, 1972
Brynjulfson	ISGS 166D	21,150 ± 430	Bone	Coleman and LiLiu, 1975
Brynjulfson	ISGS 204A	27,000+	Collagen fraction from *Platygonus*	Coleman and LiLiu, 1975
Brynjulfson	ISGS 204B	34,600 ± 2,100	Collagen fraction from *Platygonus*	Coleman and LiLiu, 1975
Casper	—	9,830 ± 350	Charcoal	Frison, 1974
Casper	—	10,060 ± 170	Bone	Frison, 1974
Chimney Rock	—	11,980 ± 180	Bone collagen	Hager, 1972
China Lake	—	18,600 ± 4,500	*Mammuthus* ivory	Fortsch, 1972
Clear Creek	—	28,840 ± 4,770	Mollusk shells	Slaughter and Hoover, 1963
Clovis	A-481	11,170 ± 360	Silty clay around *Mammuthus* skull	P. S. Martin, 1967
Cochrane	GSC-612	10,760 ± 160	—	Churcher, 1975
Cochrane	GSC-613	11,370 ± 170	—	Churcher, 1975
Dam	—	26,500 ± 3,500	Bone	Barton, unpublished manuscript
Domebo	SI-172	11,220 ± 500	*Mammuthus* bone	Leonhardy, 1966
Dry Cave	I-3365	14,470 ± 250	*Neotoma* dung	Harris, 1970
Dry Cave	TX-1774	29,290 ± 1,060	—	Valastro, Davis, and Varela, 1977
Dry Cave	TX-1775	25,160 ± 1,730	Waterlain silts	Valastro, Davis, and Varela, 1977
Dry Cave	TX-1773	33,590 ± 1,550	—	Valastro, Davis, and Varela, 1977
Fairbanks	SI-1631	11,735 ± 130	*Bison* horn sheath	Péwé, 1975
Fairbanks	SI-290	12,460 ± 320	*Bison* horn sheath	Péwé, 1975
Fairbanks	SI-453	15,380 ± 300	*Mammuthus* flesh	Péwé, 1975
Fairbanks	SI-839	21,065 ± 1,365	*Bison* horn sheath	Péwé, 1975
Fairbanks	SI-456	22,680 ± 300	*Panthera leo* tendon	Péwé, 1975
Fairbanks	SI-850	25,090 ± 1,070	*Symbos* horn sheath	Péwé, 1975

Site	Lab No.	Date	Material	Reference
Fairbanks	SI-1721	31,400 ± 2,040	*Bison* hide and hair	Péwé, 1975
Fairbanks	SI-840	39,000+	*Bison* horn sheath	Péwé, 1975
Guy Wilson	I-4163	19,700 ± 600	Bone collagen	Guilday, 1971
Gypsum Cave	LJ-452	11,690 ± 250	*Nothrotheriops* dung	P. S. Martin, 1967
Hosterman's Pit	M-1291	9,340 ± 1,000	Charcoal	Guilday, 1971
Howard Ranch	—	16,775 ± 565	Mollusk shells	Dalquest, 1965
Jaguar Cave	Geochron	10,370 ± 350	Hearth	Sadek-Kooros, 1966
Jaguar Cave	Geochron	11,580 ± 250	Hearth	Sadek-Kooros, 1966
La Mirada	—	10,690 ± 360	Wood	W. E. Miller, 1971
Laubach Cave (Laubach 1)	TX-1137	15,850 ± 500	—	Valastro, Davis, and Varela, 1977
Laubach Cave (Laubach 2)	TX-1138	13,970 ± 310	—	Valastro, Davis, and Varela, 1977
Laubach Cave (Laubach 3)	TX-1139	23,230 ± 490	Brown silt	Valastro, Davis, and Varela, 1977
Laubach Cave (Laubach 3)	TX-1419	28,340 ± 1,710	Red-brown silt	Valastro, Davis, and Varela, 1977
Lehner Ranch	A-378	10,940 ± 100	—	P. S. Martin, 1967
Lehner Ranch	K-544	11,170 ± 140	—	P. S. Martin, 1967
Lehner Ranch	M-811	11,290 ± 500	—	P. S. Martin, 1967
Levi Shelter	O-1106	10,000 ± 175	*Equus* bone	Haynes, 1967
Lost Chicken Creek	I-8582	10,370 ± 160	*Bison*	Harington, 1978
Lost Chicken Creek	SI-355	26,760 ± 300	*Equus* bone	Péwé, 1975
Lubbock Lake	C-558	9,883 ± 350	Bone	Black, 1974
Lubbock Lake	LL-308	12,650 ± 350	Clam shells	Black, 1974
McKittrick	UCLA 728	38,000 ± 2,500	Plant stems	Berger and Libby, 1966
Medicine Hat (Mitchell Bluff)	GSC-1370	25,000 ± 800	Wood	Stalker, 1977
Medicine Hat (Evil-smelling Bluff)	GSC-1044	38,000+	Wood	Stalker, 1977
Natural Trap Cave	Y-727	12,770 ± 900	*Equus* bone	L. Martin, Gilbert, and Adams, 1977
New Paris No. 4	A-744	11,300 ± 1,000	Charcoal	Guilday, Martin, and McCrady, 1964
Olsen-Chubbuck	GX-145	10,150 ± 500	Collagen (*Bison* hooves)	Wheat, 1972
Powder Mill Creek	L473A	13,170 ± 600	*Canis dirus* bone	Galbreath, 1964
Rampart Cave		9,900 ± 400	*Nothrotheriops* dung (surface)	P. S. Martin, 1967
Rampart Cave	L473C	11,900 ± 500	*Nothrotheriops* dung (45.7 cm)	P. S. Martin, 1967
Rampart Cave	L473D	38,300	*Nothrotheriops* dung (137 cm)	P. S. Martin, 1967
Rainbow Beach	WSU-1423	21,500 ± 700	Bone collagen	McDonald and Anderson, 1975
Rainbow Beach	WSU-1424	31,300 ± 2,300	Bone collagen	McDonald and Anderson, 1975

APPENDIX 1 (Continued)

Fauna	Lab.	Date (years B.P.)	Material	References
Rancho La Brea (Pit 3)	UCLA 1292B	12,650 ± 160	*Smilodon* bone	Berger and Libby, 1968
Rancho La Brea (Pit 9)	UCLA 773D	13,300 ± 160	Wood	Berger and Libby, 1966
Rancho La Brea (Pit 13)	UCLA 1292I	15,300 ± 200	*Smilodon* bone	Berger and Libby, 1968
Rancho La Brea (Pit 3)	UCLA 1292A	21,400 ± 560	*Smilodon* bone	Berger and Libby, 1968
Rancho La Brea (Pit 60)	UCLA 1292H	23,700 ± 600	*Smilodon* bone	Berger and Libby, 1968
Rancho La Brea	UCLA 737A	23,300 ± 510	Cypress wood	Berger and Libby, 1966
Rancho La Brea	UCLA 737B	32,350 ± 1,400	*Quercus* leaves	Berger and Libby, 1966
Rancho La Brea	UCLA 773C	37,000 ± 2,600	Wood	Berger and Libby, 1966
Rancho La Brea (Pit 12)	UCLA 773E	40,000+	Wood	Berger and Libby, 1966
Robert	—	11,100 ± 390	Mollusk shells	G. E. Schultz, 1967
Saltville	SI-461	13,460 ± 420	*Mammut* tusk	Guilday, 1971
Santa Rosa Is.	UCLA 748	11,300 ± 160	Black muck and charcoal	Berger and Libby, 1966
Santa Rosa Is.	UCLA 746	27,000 ± 800	Charcoal	Berger and Libby, 1966
Santa Rosa Is.	UCLA 749	37,000+	Rich charcoal	Berger and Libby, 1966
Scharbauer	L-304C	13,400 ± 1,200	*Equus* bone	Haynes, 1967
Schulze Cave	I-2741A	9,310 ± 310	*Odocoileus* bone	Dalquest, Roth, and Judd, 1969
Schulze Cave	SM-807	9,680 ± 700	*Ursus arctos* bone	Dalquest, Roth, and Judd, 1969
Smith Creek Cave	TX-1420	9,940 ± 160	Hearth 9 (Test Pit 6)	Valastro, Davis, and Varela, 1977
Smith Creek Cave	TX-1421	11,680 ± 160	Hearth 9 (Test Pit 5)	Valastro, Davis, and Varela, 1977
Smith Creek Cave	TX-1637	11,140 ± 200	Hearth 9 (Test Pit 4)	Valastro, Davis, and Varela, 1977
Smith Creek Cave	TX-1638	10,330 ± 190	Hearth 12 (Test Pit 6)	Valastro, Davis, and Varela, 1977
Taber Child Site	GSC-1233	49,000+	Wood	Stalker, 1977
Trollinger	I-3537	25,650 ± 700	Organic mud	Mehringer, King, and Lindsay, 1970
Trollinger	I-3599	32,200 ± 1,900–1,600	Sediments in *Mammut* tusk	Mehringer, King, and Lindsay, 1970
Ventana Cave	A-203	11,300 ± 1,200	Charcoal	P. S. Martin, 1967
Welsh Cave	I-2982	12,950 ± 550	*Platygonus* bone collagen	Guilday, Hamilton, and McCrady, 1971
Wilson Butte	M-1409	14,500 ± 500	—	Haynes, 1967
Wilson Butte	M-1410	15,000 ± 800	Bone	Haynes, 1967

APPENDIX 2
STRATIGRAPHIC RANGES[a,b]

	H	BL 1	BL 2	BL 3	BL 4	IRV E	IRV M	IRV L	RLB I	RLB S	RLB W	R
MARSUPIALIA												
Didelphidae												
Didelphis virginiana						——————————————————						
INSECTIVORA												
Soricidae												
†*Sorex rexroadensis*		——————										
†*S. taylori*		————————————————										
†*S. hagermanensis*		————										
†*S. powersi*		————										
†*S. meltoni*		————										
†*S. leahyi*				————								
†*S. sandersi*				————								
†*S. cudahyensis*							———					
S. cinereus							—————————————————					
†*S. scottensis*									———			
†*S. kansasensis*									————			
S. longirostris									————————————			
S. vagrans							————————————————					
S. ornatus							————————————————					
†*S. lacustris*							———					
†*S. megapalustris*							————					
S. palustris							————————————————					
S. fumeus							————————————————					
S. arcticus							————————————————					
S. dispar										————————		
S. trowbridgii										————————		
S. merriami										————————		
S. saussurei										————————		
†*Microsorex pratensis*							———					
†*M. minutus*							———					
M. hoyi							————————————————					
†*Planisorex dixonensis*			———									
†*Paracryptotis rex*		—										
†*P. gidleyi*			———									
†*Cryptotis meadensis*		———										
†*C. adamsi*			———									
C. parva						————————————————						
C. mexicana										————————		
†*Blarina ozarkensis*							———					
B. brevicauda					—————————————————							
†*Notiosorex jacksoni*		————————————————										
N. crawfordi										————————		
Talpidae												
Condylura cristata										————————		
Parascalops breweri										————————		
Scapanus latimanus										————————		
†*Scalopus rexroadi*		———										
†*S. blancoensis*												
S. aquaticus										————————		

[a] Dagger (†) indicates extinct species.
[b] Except Pinnipedia, for which stratigraphic range is unknown.
[c] Abbreviations: *H*, Hemphillian; *BL*, Blancan (divisions: 1, very early; 2, early; 3, middle; 4, late); *IRV*, Irvingtonian (*E*, early; *M*, middle; *L*, late); *RLB*, Rancholabrean (*I*, Illinoian; *S*, Sangamonian; *W*, Wisconsinan); *R*, Recent (Holocene).

	H	BL				IRV			RLB			
		1	2	3	4	E	M	L	I	S	W	R

CHIROPTERA
Mormoopidae
  Mormoops megaphylla
Phyllostomatidae
  Leptonycteris nivalis
  †*Desmodus stocki*
Vespertilionidae
  Myotis thysanodes
  M. evotis
  M. keenii
  M. leibi
  M. californicus
  M. sodalis
  M. grisescens
  †*M. rectidentis*
  †*M. magnamolaris*
  M. velifer
  M. lucifugus
  M. austroriparius
  M. volans
  Lasionycteris noctivagans
  Pipistrellus subflavus
  P. hesperus
  Eptesicus fuscus
  †*Histiotus stocki*
  Lasiurus borealis
  †*L. fossilis*
  L. cinereus
  †*L. golliheri*
  L. intermedius
  Nycticeius humeralis
  †*Plecotus tetralophodon*
  †*P. alleganiensis*
  P. rafinesquii
  P. townsendii
  Antrozous pallidus
  †*Anzanycteris anzensis*
Molossidae
  †*Tadarida constantinei*
  T. brasiliensis
  Eumops perotis
  E. glaucinus

EDENTATA
Dasypodidae
  †*Kraglievichia floridanus*
  †*Holmesina septentrionalis*
  †*Dasypus bellus*
  D. novemcinctus
Glyptodontidae
  †*Glyptotherium texanum*
  †*G. arizonae*
  †*G. floridanum*
  †*G. mexicanum*
Megalonychidae
  †*Megalonyx leptostomus*
  †*M. wheatleyi*
  †*M. jeffersonii*

APPENDICES

	H	BL				IRV			RLB			R
		1	2	3	4	E	M	L	I	S	W	

Megatheriidae
  †*Eremotherium rusconii*
  †*Nothrotheriops shastensis*
Mylodontidae
  †*Glossotherium chapadmalense*
  †*G. harlani*

CARNIVORA
Mustelidae
  †*Martes diluviana*
  M. pennanti
  †*M. nobilis*
  M. americana
  †*Mustela rexroadensis*
  M. frenata
  M. erminea
  M. rixosa
  †*M. meltoni*
  M. vison
  †*M. macrodon*
  †*M. eversmanni beringiae*
  M. nigripes
  †*Tisisthenes parvus*
  †*Gulo schlosseri*
  G. gulo
  †*Ferinestrix vorax*
  †*Trigonictis idahoensis*
  †*T. cookii*
  †*Sminthosinus bowleri*
  †*Canimartes cumminsi*
  Taxidea taxus
  †*Satherium piscinarium*
  Lutra canadensis
  †*Enhydriodon* n.sp.
  †*Enhydra macrodonta*
  E. lutris
  †*Buisnictis breviramus*
  †*B. burrowsi*
  †*Spilogale rexroadi*
  S. putorius
  †*Brachyprotoma obtusata*
  †*Osmotherium spelaeum*
  Mephitis mephitis
  Conepatus leuconotus
Canidae
  †*Borophagus direptor*
  †*B. diversidens*
  †*Canis lepophagus*
  C. latrans
  †*C. priscolatrans*
  C. rufus
  †*C. armbrusteri*
  C. lupus
  †*C. cedazoensis*
  C. familiaris
  †*C. dirus*
  Cuon alpinus

	H	BL				IRV			RLB			
		1	2	3	4	E	M	L	I	S	W	R
†*Protocyon texanus*						—						
†*Urocyon progressus*			——									
U. cinereoargenteus				——	– – – – – – – – – –							
U. littoralis												
Vulpes velox										——	——	
V. vulpes									——————————			
Alopex lagopus									——	——————		
Procyonidae												
†*Procyon rexroadensis*			——									
P. lotor									————————————			
†*Bassariscus casei*			——									
†*B. sonoitensis*			——									
B. astutus										——		
†*Parailurus anglicus*			——									
Ursidae												
†*Tremarctos floridanus*			– – – – – – – – – – – – – – –									
†*Arctodus pristinus*			——————									
†*A. simus*						————————————						
†*Ursus abstrusus*			——									
U. americanus						————————————————————						
U. arctos										——		——
U. maritimus												——
Felidae												
†*Megantereom hesperus*			——									
†*Smilodon gracilis*			————									
†*S. fatalis*						—————						
†*Ischyrosmilus ischyrus*			———————————									
†*Homotherium serum*			——————————————									
†*Dinofelis paleoonca*			————									
†*Panthera leo atrox*										——————		
P. onca										——————		——
†*Acinonyx studeri*			——									
†*A. trumani*									— – – – – – – – –			
Felis concolor									————————————			
†*F. lacustris*						—— ——						
F. pardalis									————————			
†*F. amnicola*						———————						
F. yagouaroundi												——
F. wiedii						– – – – – – – –						
†*Lynx issiodorensis*			——									
L. canadensis									————————————			
L. rufus				————————————————								
Hyaenidae												
†*Chasmaporthetes ossifragus*		—————										
Rodentia												
Aplodontidae												
Aplodontia rufa										——		
Sciuridae												
†*Paenemarmota barbouri*		——————————										
Marmota monax									———			——
†*M. arizonae*			——									
M. flaviventris												——
†*Spermophilus howelli*		——										
†*S. rexroadensis*		——————										
†*S. bensoni*			———									

APPENDICES

	H	BL				IRV			RLB			R
		1	2	3	4	E	M	L	I	S	W	
†*S. meltoni*				—								
†*S. johnsoni*				—								
†*S. boothi*				—								
†*S. magheei*				—								
†*S. finlayensis*				—								
†*S. cragini*				—								
†*S. tuitus*				—								
†*S. meadensis*				—								
†*S. cochisei*						—						
S. richardsonii												—
S. townsendii											—	—
S. parryii											—	—
S. columbianus											—	—
S. armatus										—	—	—
S. tridecemlineatus									—	—	—	—
S. spilosoma									—	—	—	—
S. mexicanus									—	—	—	—
S. beecheyi										—	—	—
S. variegatus									—	—	—	—
S. franklinii											—	—
S. lateralis										—	—	—
Ammospermophilus leucurus												—
†*Cynomys hibbardi*			—									
†*C. vetus*			—									
†*C. niobrarius*								—				
C. ludovicianus								—	—	—	—	—
C. leucurus								—	—	—	—	—
C. gunnisoni								—	—	—	—	—
†*Tamias aristus*										—		
T. striatus												—
Eutamias minimus									—	—	—	—
Sciurus niger									—	—	—	—
S. carolinensis												—
S. arizonensis									—	—	—	—
S. alleni									—	—	—	—
Tamiasciurus hudsonicus												—
T. douglasii										—	—	—
†*Cryptopterus webbi*					—							
Glaucomys volans												—
G. sabrinus										—	—	—
Geomyidae												
†*Thomomys gidleyi*			—									
†*T. potomacensis*								—				
†*T. orientalis*											—	
T. bottae										—	—	—
T. umbrinus											—	—
T. townsendii								—	—	—	—	—
T. talpoides								—	—	—	—	—
†*T. microdon*										—		
†*Pliogeomys parvus*			—									
†*Nerterogeomys smithi*		—										
†*N. minor*			—									
†*N. paenebursarius*				—								
†*N. persimilis*						—						
†*Geomys adamsi*		—										
†*G. jacobi*			—									

	H	BL				IRV			RLB			R
		1	2	3	4	E	M	L	I	S	W	R
†*G. quinni*				—								
†*G. garbanii*						—	—					
†*G. tobinensis*												
G. bursarius									—	—	—	—
G. pinetis									—	—	—	—
†*Cratogeomys bensoni*			—									
C. castanops												
†*Heterogeomys onerosus*									—	—		
Heteromyidae												
†*Prodipodomys centralis*			—	—	—							
†*P. idahoensis*				—	—							
†*Etadonomys tiheni*				—	—							
†*Dipodomys gidleyi*				- - -								
D. ordii									—	—	—	—
D. ingens										—	—	—
D. agilis										—	—	—
D. spectabilis										—	—	—
D. merriami									—	—	—	—
†*Perognathus rexroadensis*		—										
†*P. gidleyi*		—	—									
†*P. magnus*		—	—									
†*P. maldei*			—									
†*P. pearlettensis*												
P. californicus									—	—	—	—
P. flavus										—	—	—
P. inornatus										—	—	—
P. intermedius										—	—	—
P. parvus										—	—	—
P. hispidus									—	—	—	—
P. merriami										—	—	—
P. longimembris										—	—	—
P. apache										—	—	—
Liomys irroratus										—	—	—
Castoridae												
†*Dipoides rexroadensis*		—	—									
†*D. intermedius*		—	—									
†*Paradipoides stovalli*									—	—		
†*Procastoroides sweeti*		—	—	—								
†*P. idahoensis*				—	—							
†*Castoroides ohioensis*				—	—	—	—	—				
†*Castor californicus*				—	—	—	—					
†*C. accessor*				—	—	—	—	—				
C. canadensis				—	—	—	—	—	—	—	—	—
Cricetidae												
Oryzomys palustris									—	—	—	—
†*Reithrodontomys wetmorei*		—										
†*R. rexroadensis*		—										
†*R. pratincola*			—	—								
†*R. moorei*												
R. montanus									—	—	—	—
R. humulis									—	—	—	—
R. megalotis										—	—	—
R. fulvescens									—	—	—	—
†*Peromyscus kansasensis*		—	—									
†*P. baumgartneri*		—	—									
†*P. irvingtonensis*							—					

APPENDICES

	H	BL				IRV			RLB			
		1	2	3	4	E	M	L	I	S	W	R
†*P. cragini*				——	——	——	——					
P. crinitus										——	——	——
†*P. nesodytes*										——		
†*P. anyapahensis*										——		
†*P. imperfectus*										——		
P. eremicus										——	——	——
†*P. progressus*							——					
†*P. berendsensis*										——		
†*P. hagermanensis*			——									
†*P. cumberlandensis*								——				
†*P. cochrani*										——		
P. maniculatus										——	——	——
P. polionotus										——	——	——
P. leucopus										——	——	——
P. gossypinus										——	——	——
P. boylii										——	——	——
P. pectoralis										——	——	——
P. truei										——	——	——
P. difficilis										——	——	——
†*P. oklahomensis*										——		
P. floridanus									——			——
Ochrotomys nuttalli									——			——
†*Baiomys aquilonius*			——									
†*B. rexroadensis*		——										
†*B. minimus*		——										
†*B. kolbi*		——										
†*B. brachygnathus*						——						
B. taylori												——
†*Bensonomys arizonae*		——										
†*B. eliasi*		——										
†*B. meadensis*				——								
†*Symmetrodontomys simplicidens*		——										
†*Onychomys bensoni*		——	——									
†*O. gidleyi*			——									
†*O. fossilis*				——	——							
†*O. pedroensis*				——								
†*O. jinglebobensis*				——								
O. leucogaster							——	——				
O. torridus							——	——	——			——
†*Sigmodon medius*				——	——							
†*S. hudspethensis*					——							
†*S. curtisi*					——							
†*S. bakeri*									——			
S. hispidus										——	——	——
S. ochrognathus										——		——
†*Neotoma quadriplicata*			——									
†*N. taylori*			——									
†*N. spelaea*									——			
†*N. fossilis*		——										
N. floridana												——
N. micropus										——	——	——
N. albigula										——	——	——
N. lepida										——	——	——
N. mexicana										——	——	——
N. fuscipes										——		——
N. cinerea												——

	H	BL				IRV			RLB			
		1	2	3	4	E	M	L	I	S	W	R
†*Ogmodontomys poaphagus*		——————										
†*Cosomys primus*		————										
†*Ophiomys taylori*		——										
†*O. parvus*				————								
†*O. meadensis*				——								
†*O. magilli*				——								
†*O. fricki*				——								
†*Pliophenacomys finneyi*		————————										
†*P. primaevus*												
†*P. osborni*												
†*Pliomys deeringensis*							——					
†*Nebraskomys rexroadensis*		————										
†*N. mcgrewi*			——									
†*Atopomys texensis*							——					
†*A. salvelinus*								——				
†*Mimomys monahani*							——					
Clethrionomys gapperi								——	———————			
†*Pliolemmus antiquus*			————————									
Phenacomys intermedius							— — — — — — — —					
†*Microtus pliocaenicus*							——					
†*M. deceitensis*							——					
†*M. paroperarius*							————					
†*M. meadensis*							——					
M. pennsylvanicus									———————			
M. montanus										———		
M. californicus											———	
M. longicaudus										———		
M. mexicanus										———————		
M. chrotorrhinus									———————			
M. xanthognathus									———————			
M. miurus											———	
†*M. guildayi*							——					
†*M. llanensis*							——					
M. ochrogaster									———————			
†*M. cumberlandensis*							——					
†*M. aratai*							——					
†*M. hibbardi*										?		
†*M. mcnowni*									———			
M. pinetorum									———————			
Lagurus curtatus											———	
†*Proneofiber guildayi*						——						
†*Neofiber diluvianus*						——						
†*N. leonardi*							——					
N. alleni								——————————				
†*Pliopotamys minor*			——									
†*P. meadensis*				——								
†*Ondatra idahoensis*						——						
†*O. hiatidens*						——						
†*O. annectens*							——					
†*O. nebracensis*							——					
O. zibethicus								———————				
Lemmus sibiricus							——					
†*Predicrostonyx hopkinsi*							——					
Dicrostonyx torquatus								———————				
D. hudsonius											———	
†*Synaptomys rinkeri*			——									
†*S. bunkeri*										?		

APPENDICES

	H	BL				IRV			RLB			R
		1	2	3	4	E	M	L	I	S	W	
S. cooperi										—	—	—
†S. australis									——	——	——	
†S. vetus					—	—						
†S. landesi					—							
†S. anzaensis						—						
†S. meltoni					——	——						
†S. kansasensis					—	—						
S. borealis										—	—	—
Zapodidae												
†Zapus rinkeri			—									
†Z. sandersi				——	——	——						
Z. hudsonius										—	—	—
†Z. burti						—						
Z. princeps										—		—
Napaeozapus insignis										—	—	—
Erethizontidae												
†Coendou stirtoni						—						
†C. bathygnathus					—	—						
†C. cumberlandicus							—					
Erethizon dorsatum										—	—	—
Hydrochoeridae												
†Hydrochoerus holmesi					——	——	——					
†Neochoerus pinckneyi										—		
LAGOMORPHA												
Ochotonidae												
†Ochotona whartoni									– –	– –	– –	—
O. princeps									– –	– –	– –	—
O. collaris												—
Leporidae												
†Hypolagus limnetus		——	——									
†H. vetus		——	——									
†H. regalis			——	——	——							
†H. furlongi					—							
†H. arizonensis					—	—						
†H. browni					—	—						
†Notolagus lepusculus			—									
†Pratilepus vagus			—									
†P. kansasensis			—									
†Aluralagus bensonensis					—							
Brachylagus idahoensis										—		—
†Nekrolagus progressus		– –	– –									
Sylvilagus floridanus						——	——	——	——	——	——	—
S. nuttalli										—	—	—
S. transitionalis										—	—	—
S. audubonii										—	—	—
S. bachmani										—		
S. palustris										—	—	—
†S. leonensis										—		
S. aquaticus										—	—	—
†Lepus benjamini					—							
L. alleni												—
L. californicus					– –					—	—	—
L. townsendii										—	—	—
L. americanus										—	—	—
L. arcticus										—		—

APPENDICES

		BL				IRV			RLB			
	H	1	2	3	4	E	M	L	I	S	W	R

PERISSODACTYLA
Equidae
  †*Nannippus beckensis*
  †*N. phlegon*
  †*Equus simplicidens*
  †*E. parastylidens*
  †*E. cumminsii*
  †*E. calobatus*
  †*E. hemionus*
  †*E. tau*
  †*E. giganteus* group
  †*E. occidentalis*
  †*E. complicatus*
  †*E. fraternus*
  †*E. scotti*
  †*E. niobrarensis*
  †*E. conversidens*
  †*E. lambei*
  †*E. excelsus*
Tapiridae
  †*Tapirus merriami*
  †*T. californicus*
  †*T. copei*
  †*T. veroensis*

ARTIODACTYLA
Tayassuidae
  †*Mylohyus floridanus*
  †*M. nasutus*
  †*Platygonus pearcei*
  †*P. bicalcaratus*
  †*P. vetus*
  †*P. compressus*
Camelidae
  †*Titanotylopus nebraskensis*
  †*T. spatulus*
  †*Blancocamelus meadei*
  †*Megatylopus cochrani*
  †*Camelops kansanus*
  †*C. traviswhitei*
  †*C. sulcatus*
  †*C. huerfanensis*
  †*C. minidokae*
  †*C. hesternus*
  †*Hemiauchenia blancoensis*
  †*H. macrocephala*
  †*Palaeolama mirifica*
Cervidae
  †*Bretzia pseudalces*
  †*Odocoileus brachyodontus*
  O. virginianus
  O. hemionus
  †*Blastocerus extraneus*
  †*Navahoceros fricki*
  †*Sangamona fugitiva*
  Rangifer tarandus
  †*Alces latifrons*
  A. alces

	H	BL				IRV			RLB			R
		1	2	3	4	E	M	L	I	S	W	
†*Cervalces scotti*										——		
Cervus elaphus											——————	
†*C.? brevitrabalis*							?					
Antilocapridae												
†*Ceratomeryx prenticei*			——									
†*Capromeryx furcifer*										————		
†*C. arizonensis*			——————									
†*C. minor*										————		
†*C. mexicana*										————		
†*Tetrameryx shuleri*							——					
†*T. irvingtonensis*							————					
†*T. knoxensis*						——						
†*T. mooseri*												
†*T. tacubayensis*										——		
†*Hayoceros falkenbachi*										——		
†*Stockoceros conklingi*										————		
†*S. onusrosagris*										————		
Antilocapra americana											————	
Bovidae												
Saiga tatarica											——	
Oreamnos americanus											————	
†*O. harringtoni*											——	
?†*Capra iowensis*											?	?
Ovis dalli											——	
O. canadensis											————	
†*Eucertherium collinum*						——————————						
†*Soergelia mayfieldi*						————						
†*Symbos cavifrons*											————	
†*Bootherium bombifrons*											————	
†*Praeovibos priscus*										——		
Ovibos moschatus											————	
†*Bison priscus*										——		
†*B. latifrons*										————		
B. bison											————	
†*Platycerabos dodsoni*									——			
Bos grunniens												——
Sirenia												
Dugongidae												
†*Hydrodamalis gigas*											——	
Trichechidae												
Trichechus manatus		——									————————	
Proboscidea												
Mammutidae												
†*Mammut americanum*		————————————										
Gomphotheriidae												
†*Rhynchotherium praecursor*		————										
†*Stegomastodon mirificus*		————————										
†*Cuvieronius sp.*		- - - - - - - - - - - -										
Elephantidae												
†*Mammuthus meridionalis*						————						
†*M. columbi*										————		
†*M. jeffersonii*											——	
†*M. primigenius*											————	
Primates												
Hominidae												
Homo sapiens											——	

Index to Latin Names of Organisms

Page numbers for illustrations are in italics.

Abies concolor, 51
Acinonyx, 192–95
  jubatus, 192, 364
  pardinensis, 192, 193
  studeri, 20, 192–94, *193*
  trumani, 88, 194, 364
Adeloblarina, 111, 112
Aenocyon, 171
Aepycamelinae, 302
Aftonius calvini, 330
Agave, 142
Ailuraena johnstoni, 199
Ailurinae, 175, 177–78
Ailuropoda, 177
Ailurus, 175, 177
  fulgens, 177
Alces, 32, 41, 96, 310, 315–17, 318, 358
  alces, 39, 96, *311*, 316–17, 356, 362, 364
  americana, 316
  latifrons, *311*, 315–16, 317
  runnymedensis, 316
  semipalmatus, 318
  shimeki, 318
Alilepus, 278, 279
Allophaiomys, 34, 258–59, 262
Alopex, 174–75
  lagopus, 174–75
Aluralagus, 278, 279
  bensonensis, 277, 279
Amblonyx, 158
Amerhippus, 290–91
Ammospermophilus, 212, 217–18
  leucurus, 217–18
Amphicyon, 178
Anancus bensonensis, 347, 348
Anaptogonia, 266
Ancylopoda, 283
Anisonyx, 209
Antilocapra, 62, 322, 324, 325
  americana, 69, 319, *320*, 321, 322, 323, 325, 364
  garcia, 325
Antilocapridae, 319–25, 360
Antilocaprinae, 319
Antilope, 325
Antrozous, 125
  pallidus, 125
Anzanycteris, 125
  anzensis, 12, *123*, 125
Aonyx, 158
Aplodontia, 209–10
  rufa, 209–*210*
Aplodontidae, 209–10, 360
Archaeolaginae, 276, 277–78
Archidiskodon, 350
  haroldcooki, 350
  hayi, 350
  scotti, 350
  sonoriensis, 351

Arctocephalus, 203
Arctodus, 91, 178, 180–82
  pristinus, 78, 180, *181*
  simus, *41*, 52, 84, 88, 180–82, *181*, *182*, *183*, 184, *185*, 362, 364
Arctomys, 212, 214, 215, 217, 219
Arctotherium, 91, 180
  californicum, 180
  yukonense, 180
Argyrohyus, 91
Artemisia, 18, 28
Artiodactyla, 295–339, 359
Arvicola, 95, 247, 257, 260, 261, 262, 263, 270
  sigmodus, 262
  tetradelta, 261
Arvicolinae, 239
Asinus, 285, 287–88, 289
  aguascalientensis, 291
Astrohippus, 285, 287
Atopomys, 31, 256–57
  salvelinus, 35, *255*, 257
  texensis, 256–57
Atriplex, 142
Auchenia, 304, 305
Australopithecus, 355

Baiomys, 247–48
  aquilonius, 247, 248
  brachygnathus, 248
  kolbi, 248
  minimus, 248
  rexroadi, *241*, 247
  taylori, 247, 248
Barytherioidea, 343
Bassaris, 177
Bassariscus, 176–77
  astutus, 176, 177
  casei, *176*, 177
  sonoitensis, 48, 177
  sumichrasti, 177
Bauerus, 125
Bensonomys, 18, 248–49
  arizonae, 248, 249
  eliasi, *241*, 248
  meadensis, 248, 249
Bison, 3, 5, 22, 26, 32, 33, 35, 37, 39, 41, 45, 50, 57, 60, 62, 67, 71, 72, 73, 96, 203, 326, 335–38, *336*, 404, 405
  aguascalientes, 337
  alleni, 337
  angularis, 337
  appalachicolus, 76, 332
  bison, 335, *336*, 337–38, 362, 364
  b. antiquus, 56, 59, 66, 84, 87, 337, 338
  b. athabascae, 338
  b. bison, 338
  b. occidentalis, 337, 338
  bonasus, 335, 337, 338
  californicus, 337

  chaneyi, 337
  ferox, 337
  figginsi, 337
  geisti, 335
  kansasensis, 337
  latifrons, 39, 45, 56, 59, *61*, 62, 65, 66, 78, 80, 81, *336*, 337, 362, 365
  oliverhayi, 337
  pacificus, 337
  praeoccidentalis, 335
  priscus, 335–*336*, 337, 338, 365
  p. alaskensis, 335
  p. crassicornis, 335, 336
  p. longicornis, 335
  p. mediator, 338
  regius, 337
  rotundus, 337
  taylori, 337
  texanus, 337
  willistoni, 337
Blancocamelus, 301, 302
  meadei, 302
Blarina, 26, 103, 110, 111–13
  brevicauda, 108, 112–13
  b. brevicauda, 112
  b. carolinensis, 112
  b. kirtlandi, 112
  fossilis, 112
  ozarkensis, 111–12
  simplicidens, 112
Blarinini, 103, 109–13
Blastocerus, 58, 92, 312
  dichotomus, 312
  extraneus, 312
Bootherium, 86, 334, 404
  bombifrons, 66, 334, 365
  brazosis, 332
  nivicolens, 332
  sargenti, 332, 333
Boreostracon, 135
Borophagini, 165–67
Borophagus, 3, 9, 11, 12, 21, 22, 34, 165–67, 198
  direptor, 9, 165–66
  diversidens, 9, 19, *166*–67
Bos, 41, 326, 334, 335, 337, 338, 339
  arizonica, 337
  bunnelli, 339
  crampianus, 337
  grunniens, 339, 358
Bovidae, 326–39, 360
Bovinae, 326, 335
Brachylagus, 278, 279
  idahoensis, 279
Brachyostracon, 135
Brachyprotoma, 161, 163–64
  obtusata, *163*–64, 364
  pristina, 163
  spelaea, 163
Breameryx, 320

Bretzia, 21, 309–10
  pseudalces, 309–10, *311*
Brontornithidae, 13
Bubalus, 338
Buisnictis, 161–62
  breviramus, 161, 162
  burrowsi, 162
  meadensis, 161
  schoffi, 161
Burosor effossorius, 236

Callorhinus, 202–3
  ursinus, 202–3, 204
Callospermophilus, 212–13
Camelidae, 95, 295, 301–9, 358, 360
Camelinae, 301
Camelini, 301–2, 305
Camelopini, 301, 302–6
Camelops, 10, 25, 45, 51, 65, 71, 72, 73, 83, 87, 301, 302, *303*–306
  hesternus, *45*, *303*, 304, *305*–*6*, 365
  huerfanensis, *303*, 304, 364
  kansanus, 303–304
  minidokae, 51, *303*, 304–5
  sulcatus, *303*, 304
  traviswhitei, *303*, 304
Camelus, 301, 302, 306
  americanus, 308
  dromedarius, 304, 306
Canidae, 146, 165–75, 360
Canimartes, 155, 156
  cumminsii, 156
Caninae, 165, 167–75
Canis, 18, 167–72, 173, 174
  anderssoni, 167
  armbrusteri, 32, 169, *170*, 171
  arnensis, 167
  ayersi, 171
  caneloensis, 167
  cedazoensis, 170–71
  davisi, 167, 168, 174
  dirus, 52, 53, 56, 70, 72, 74, 80, 83, 85, *168*, 169, *170*, 171–72, 198, 364, 405
  edwardii, 168
  etruscus, 167, 168, 169, 170
  falconeri, 169
  familiaris, 62, 89, 171
  indianensis, 171
  irvingtonensis, 167
  latrans, 51, 74, 167–*168*, 169, *170*
  lepophagus, 20, 167, 168, 169
  lupus, 95, 167, 169–70, 171, 172
  milleri, 169
  mississippiensis, 171
  nehringi, 171
  orcutti, 167
  priscolatrans, 30, 168–69
  riviveronis, 167
  rufus, 167, 168, 169
Canoidea, 146–85
Capra, 326, 328
  hircus, 328
  iowensis, 328
Caprinae, 326–35
Caprini, 326
Capromeryx, 10, 18, 83, 84, 85, 319–21
  arizonensis, 310
  furcifer, 319–20
  mexicana, 320, 321
  minimus, 319
  minor, *320*–321, 322, 365
Cariacus ensifer, 312
Carnivora, 146–208, 358, 359, 361

Castor, 15, 74, 235, 236, 237–38, 266
  accessor, 238
  californicus, 237–38
  canadensis, 235, *237*, 238, 362
  fiber, 237
Castoridae, 235–38, 360
Castoroides, 30, 57, 74, 79, 235, 236–37, 238
  nebrascensis, 236
  ohioensis, *68*, 236–*237*, 238, 362, 364
Catagonus, 296
  wagneri, 296, 363
Caviomorpha, 209
Celtis, 14
Ceratomeryx, 319
  prenticei, 319
Ceratomorpha, 283
Cervalces, 63, 315, 316, 317
  alaskensis, 315, 316
  borealis, 317
  roosevelti, 317
  scotti, 66, *311*, 317, 362, 365
Cervavitus, 310
Cervidae, 309–18, 360
Cervinae, 309, 317–18
Cervus, 23, 309, 312, 314, 315, 316, 317–18, 358
  acoronatus, 318
  aguangae, 318
  americanus, 317
  brevitrabalis, 318
  canadensis, 318
  elaphus, 313, 317–18, 364
  fortis, 318
  lascrucensis, 312
  lucasi, 318
  whitneyi, 313
Chalinobus, 122
Chasmaporthetes, 13, 22, 27, 95, 167, 199
  borissiaki, 199
  lunensis, 199
  ossifragus, *199*, 211
Chelydra serpentina, 71
Chiroptera, 116–27, 359
Chlamytherium, 129
Citellus, 212, 213, 214
Cladonia, 315
Clethrionomys, 30, 60, 95, 257
  gapperi, 63, 64, 77, *255*, 257
  rutilus, 257
Coelodonta antiquitatis, 283
Coendou, 13, 92, 272–73
  bathygnathum, 272
  cascoense, 272
  cumberlandicum, *31*, 272–73
  stirtoni, 272
Compositae, 28
Condylura, 113–14
  cristata, 113–*114*
Condylurini, 113–14
Conepatus, 91–92, 164–65
  leuconotus, 60, *163*, 164–65
Corynorhinus, 124
Cosomys, 11, 15, 156, 254, 257
  primus, 254, *255*
Cratogeomys, 229
  bensoni, 229
  castanops, 229
Creccoides osborni, 19
Cricetidae, 209, 239–70, 360
Cricetinae, 92, 239–53
Cricetodipus, 234
Cricetus, 225

Crocuta, 166, 198
Cryptopterus, 13, 222
  tobieni, 222
  webbi, 222
Cryptotis, 103, 110, 111
  adamsi, 111
  meadensis, 111
  mexicana, 111
  parva, *110*, 111
Cudahyomys, 240
Cuon, 172
  alpinus, 74, 96, 172, 358
  priscus, 172
Cuvieronius, 10, 348
Cynomys, 25, 153, 210, 218–20
  gunnisoni, 85, 218, 219–20
  hibbardi, *211*, 218
  leucurus, 218, 219
  ludovicianus, 44, 64, 78, 218, 219
  meadensis, 218
  mexicanus, 218
  niobrarius, 78, 218–19
  parvidens, 218
  spispiza, 78, 219
  vetus, 218, 219
Cyonasua, 90, 91, 92
Cystophora, 207
  cristata, *206*, 207

Dama, 310
Dasypodidae, 91, 128–32, 360
Dasypodinae, 130–32
Dasypterus, 123
  floridanus, 123
Dasypus, 92, 130–32
  bellus, 56, 60, 70, 72, 79, 80, 84, 87, 128, *129*, 130–31, 132, 364
  novemcinctus, 128, 130, 131–32
Deinotherioidea, 343
Desmodontinae, 117–18
Desmodus, 56, 58, 117–18
  magnus, 117
  rotundus, 118
  stocki, 55, *74*, 117–18
Dibelodon, 346
Dicea, 278
Dicotyles, 295, 296
  fossilis, 296
  pennsylvanicus, 296
Dicrostonyx, 62, 95, 267, 268, 354
  hudsonius, 77, 267–68
  torquatus, *265*, 267
Didelphidae, 91, 101–2, 360
Didelphis, 92, 101–2
  marsupialis, 102
  virginiana, *82*, 101–2
Dinarctotherium merriami, 180
Dinobastis, 190
Dinofelis, 95, 186, 190–91
  abeli, 191
  cristata, 191
  diastemata, 191
  paleoonca, *189*, 190–91
Dinohippus, 285
Dipodomyinae, 230–32
Dipodomys, 5, 9, 22, 26, 27, 230, 231–32
  agilis, 232
  gidleyi, 231–32
  ingens, 232
  merriami, 218, *231*, 232
  ordii, 231, 232
  spectabilis, 232

INDEX TO LATIN NAMES

Dipoides, 15, 18, 235
   intermedius, 235, *237*
   rexroadensis, 235
   wilsoni, 235
Diprotodonta, 101
Dipus, 271
Dolichohippus, 3, 14, 22, 287
Dolomys, 256
Dugong, 340
Dugongidae, 340–41, 342, 358, 360
Duplicidentata, 275
Dusicyon, 91
Dusignathinae, 200
Dysopes, 126

Edentata, 128–45, 348, 359
Elephantidae, 343, 346, 348–54, 358, 360
Elephantinae, 348
Elephas, 95, 348, 349, 350, 351, 352, 353
   americanus, 344, 353
   eellsi, 352
   floridanus, 352
   jacksoni, 353
   maibeni, 351
   maximus, 349
   roosevelti, 352
   washingtoni, 352
Eligmodontia, 248
Enaliarctidae, 200
Enhydra, 52, 158, 160–61
   lutris, 159, 160
   macrodonta, 160
Enhydriodon, 158, 159–60, 161
   lluecai, 160, 161
   reevei, 160
   sivalensis, 160
Enhydrictis, 156
Eocastoroides lanei, 236
Eodipodomys, 231
Ephedra, 142
Eptesicus, 122, 124
   fuscus, 122
   grandis, 122
Equidae, 283–91, 360
Equus, 3, 9, 10, 11, 12, 15, 16, 18, 22, 25, 30, 35, 41, 45, 48, 54, 56, 57, 62, 67, 71, 72, 73, 78, 79, 80, 83, 84, 85, 87, 88, 91, 95, 283, 284, 285–91, 405, 406
   achates, 288
   altidens, 288
   bautistensis, 289, 290
   burchelli, 287
   caballus, 286, 287, *358*
   calobatus, 25, 32, *286,* 288
   complicatus, 285, 289, 290, 365
   conversidens, 44, 45, *285, 286,* 290, 291, 365
   crinidens, 289
   cumminsii, 286, 287–88
   excelsus, 291, 365
   francisci, 288
   fraternus, 290
   giganteus, 289
   grevyi, 287
   hatcheri, 290
   hemionus, 288, 364
   idahoensis, 285
   lambei, 291, 365
   laurentius, 287
   littoralis, 288
   mexicanus, 289

   midlandensis, 290
   minutus, 284
   niobrarensis, 290–91
   occidentalis, 285, 289–90, 365
   pacificus, 285, 289
   parastylidens, 286, 287
   pectinatus, 289
   proversus, 285
   quinni, 288
   scotti, 25, 28, 35, *285, 286,* 290, 291, 365
   semiplicatus, 291
   simplicidens, 14, 19, *285, 286,* 285–87
   tau, 285, 286, 288, 291
Eremotherium, 60, 78, 92, 138, 140–41
   carolinense, 140
   elense, 140
   rusconii, 140–41, 364
Erethizon, 62, 92, 272, 273
   dorsatum, 26, 63, 148, *273*
Erethizontidae, 91, 209, 272–73, 360
Erignathus, 206
   barbatus, 206
Erinaceomorpha, 103
Etadonomys, 230, 231
   tiheni, 231
Euarctos optimus, 183
Euceratheriini, 330
Euceratherium, 9, 22, 27, 50, 67, 73, 95, 330–31, 332
   bizzelli, 330
   collinum, 54, *330–31,* 365
Eumetopias, 203
   jubata, 203
Eumops, 126–27
   glaucinus, 126–27
   g. floridanus, 126–27
   perotis, 126
Euryboas, 199
Eutamias, 49, 220
   minimus, 220

Falco rusticolus, 87
Felidae, 146, 185–98, 360
Felinae, 185, 186, 191–98
Felis, 91, 186, 191, 192, 194–96, 197
   amnicola, 195
   bituminosa, 194
   concolor, 74, *193,* 194–95, *198*
   daggetti, 194
   eyra, 195
   hawveri, 194
   hillianus, 166
   imperialis, 191
   inexpectata, 26
   lacustris, 195
   longicrus, 194
   pardalis, 58, 195
   rexroadensis, 195
   veronis, 192
   wiedii, 196
   yagouaroundi, 58, 195–96
Feloidea, 146, 185–99
Ferinestrix, 155
   vorax, 155
Fiber, 266
   oregonus, 266
Fissipedia, 146–99
Fraxinus, 48

Galictis, 91, 156
   vittata, 156

Geochelone, 11, 12, 15, 16, 17, 18, 19, 20, 26, 33, 34, 35, 39, 65, 83
Geomyidae, 223–29, 230, 360
Geomys, 5, 9, 12, 16, 18, 26, 27, 66, 69, 79, 223, 225, 226, 227–29
   adamsi, 227
   bisulcatus, 228
   bursarius, 227, 228, 229
   b. texensis, 226
   garbanii, 224, 227–28
   jacobi, 224, 227
   parvidens, 227, 228
   personatus, 228
   pinetis, 228–29
   quinni, 227, 228
   tobinensis, 228
Gigantocamelus, 18, 29, 302
   fricki, 302
Glaucomys, 30, 57, 222–23
   sabrinus, 222–23
   volans, 77, 222, 223
Glossotherium, 9, 64, 67, 78, 92, 137, 138, 143–45
   chapadmalense, 143, 144
   harlani, 52, 66, 143–45, *144,* 364
   robustum, 143, 144
Glyptodon, 129, 135
   petaliferus, 135
Glyptodontidae, 91, 132–35, 358, 360
Glyptotherium, 9, 11, 25, 57, 67, 85, 92, *132,* 133–35
   arizonae, 133, 134, 135
   cylindricum, 135
   floridanum, 129, 133, 134, 135, 364
   mexicanum, 135
   texanum, 132–34, *133*
Gomphotheriidae, 343, 346–48, 358, 360
Gomphotherioidea, 343
Gomphotherium, 346
Gopherus, 19, 20, 83
Gramineae, 18, 28
Grammogale, 149
Grisonella cuja, 156
Grisoninae, 147, 155
Gulo, 30, 33, 95, 153–55
   gidleyi, 153
   gulo, 153–55, *154*
   luscus, 153, 155
   schlosseri, 153, 155

Halichoerus, 206–7
   grypus, 206–7
Hapalops, 138, 141
Haplomastodon, 348
Haplomylomys, 242–44
Hayoceros, 322
   falkenbachi, 322
Hemiauchenia, 18, 51, 57, 72, 83, 91, 301, 304, 306–8, *307,* 309
   blancoensis, 25, 302, 306–8, *307*
   macrocephala, 45, 56, *307,* 308, 365
   seymourensis, 308
   vera, 308
Hesperomys, 243, 245, 246, 247, 248, 250
Hesperoscalops, 20, 115
Heterogeomys, 229
   hispidus, 229
   onerosus, 229
Heteromyidae, 230–35, 360
Heteromyinae, 230, 234
Heteromys, 230, 234

Hexabelomeryx, 324
Hexameryx, 324
Hippidion, 91
Hippocamelus, 312, 313
  bisulcus, 313
Hippomorpha, 283
Hippotigris, 287
Histiotus, 122
  stocki, 122
Histriophoca, 204
Hodomys, 252
Holmeniscus, 304, 308
Holmesina, 67, 92, 128, 129–30
  septentrionalis, 29, 56, *58*, 128, *129–30*, 364
Hominidae, 355–56, 360
Homo, 37, 355–56
  erectus, 355
  sapiens, 96, 355–56
Homotheriini, 185
Homotherium, 19, 81, *82*, 95, 189, 190
  crenatidens, 190
  latidens, 190
  serum, *82*, *187*, *189*, 190
Hyaena, 166, 198, 199
Hyaenidae, 146, 165, 198–99, 358, 360
Hyaenognathus, 166
  dubius, 166
  pachyodon, 166
  solus, 165
Hydrochoeridae, 91, 92, 209, 274, 358, 360
Hydrochoerus, 92, 134, 236, 274, 358
  holmesi, *273*, 274
  hydrochoeris, 274
  isthmius, 274
Hydrodamalis, 340–41
  gigas, 340–*341*, 358
Hypolagus, 9, 10, 11, 12, 18, 20, 21, 27, 95, 156, 277–78
  arizonensis, 278
  browni, 278
  furlongi, 278
  limnetus, 277, 278
  regalis, 12, 277–78
  vetus, *277*, 278
Hypudaeus, 250, 262
Hystrix, 273

Ibex, 328
Ictidomys, 212
Insectivora, 103–15, 359
Ischnoglossa, 117
Ischyrosmilus, 11, 188–90, *189*, *193*
  crusafonti, 188
  idahoensis, 188
  ischyrus, 188–90
  johnstoni, 188

Jentinkia, 177
Juniperus osteosperma, 51

Kraglievichia, 92, 128–29, 130
  floridana, 128–29
  paranense, 28, 129

Lagomorpha, 275–82, 359
Lagomys, 276
Lagurus, 95, 263–64
  curtatus, 263–64
Lama, 301, 305, 306, 307, 309
  glama, 307, 308, 363

  guanicoe, 307
  pacos, 307
Lamini, 301, 305, 306–9
Lasionycteris, 121
  noctivagans, 121
Lasiurus, 122–24
  borealis, 122–23
  cinereus, *123*
  ega, 123
  fossilis, 123
  golliheri, 123
  intermedius, 123–24
Lemmiscus, 264
Lemmus, 23, 95, 267, 268
  sibiricus, 267
  trimucronatus, 267
Leporidae, 275, 276–82, 360
Leporinae, 276, 278–82
Leptobos, 95, 335, 339
Leptonycteris, 117
  nivalis, 117
Leptophoca, 204
Lepus, 3, 5, 9, 10, 11, 17, 22, 26, 27, 32, 95, 276, 277, 278, 279, 280, 281–82
  alleni, 281–82
  americanus, 77, 196, *277*, 282
  arcticus, 282
  benjamini, 281
  californicus, 282
  giganteus, 281
  othus, 282
  timidus, 282
  townsendii, 282
Leucocrossuromys, 218, 219
Libralces, 315, 316
  gallicus, 315, 316
Liomys, 230, 234–35
  irroratus, 234–35
Liops zuniensis, 332
Lontra, 159
Loxodonta, 348, 349
  africana, 349
Lupus niger, 169
Lutra, 91, 95, 158–59
  annectens, 159
  canadensis, 158–59
  iowa, 158, 159
  licenti, 159
  parvicuspidens, 158
  rhoadsii, 158
Lutravus, 155, 156
Lutreola, 149, 151
Lutrinae, 147, 158–61
Lycaon, 172
Lycyaena, 199
Lynx, 41, 95, 194, 195–98
  canadensis, 196, 197
  issiodorensis, 20, 195, 196, *197*
  i. kurteni, 196
  lynx, 196
  rufus, 195, 196, *197–198*
  r. calcaratus, 197
  r. koakudsi, 197

Machairodontinae, 185, 186–91
Machairodus ("*Machaerodus*"), 186, 188, 189, 190
  floridanus, 186
Macrorhinus, 207
Mammonteus, 350
Mammut, 57, 60, 67, 78, 79, 343, 344–46, 348, 404, 406

  americanum, 50, 56, *68*, 236–37, *344–346*, *347*, 365
  borsoni, 345
  progenium, 344
Mammuthus, 3, 5, 22, 27, 28, 41, 44, 45, 47, 57, 60, 67, 72, 78, 80, 84, 95, 203, 332, 345, 348, 349, 350–54, 404
  armeniacus, 350
  columbi, 23, 25, *61*, *344*, *347*, 350, 351–52
  imperator, 351, 352
  jeffersonii, *82*, *347*, 350, 351, 352–*353*, 354, 362, 365
  j. exilis, 49, 352, 365
  meridionalis, 23, 25, 28, *29*, 33, 34, 35, *347*, 350–51, 352
  m. nebrascensis, 350
  primigenius, *41*, 45, 96, *347*, 350, 352, 353–54, 362, 365
Mammutidae, 343–46, 358, 360
Mammutoidea, 343
Marmota, 30, 63, 210, 211–12
  arizonae, 212
  flaviventris, 212
  mexicana, 210
  minor, 211
  monax, *68*, 211–12
  nevadensis, 210
  oregonensis, 210
  sawrockensis, 210
  vetus, 211
Marsupialia, 101–2, 359
Marsupicarnivora, 101
Martes, 62, 147–49, 153
  americana, 75, 77, 96, 148, 149, 362
  a. caurina, 148, 149
  a. humboldtensis, 149
  diluviana, *31*, *147*–148
  martes, 149
  nobilis, 54, *147*, 148, 362, 364
  palaeosinensis, 147
  parapennanti, 147
  pennanti, 148, 149, 156, 273
  zibellina, 96, 149
Mastodon, 346
  giganteus, 344
  successor, 346
Megachiroptera, 116
Megalonychidae, 91, 92, 135–38, 358, 360
Megalonyx, 16, 18, 50, *61*, 63, 64, 67, 79, 128, 136–38, 141, 143, 144
  brachycephalus, 137
  californicus, 137
  curvidens, 136
  dissimilis, 137
  hogani, 137
  jeffersonii, 56, *61*, 74, 136, 137–38, *139*, 364
  laqueatus, 137
  leidyi, 137
  leptostomus, 19, 136, 137
  loxodon, 136
  mathisi, 136
  milleri, 137
  scalper, 136
  sierrensis, 137
  sphenodon, 136
  tortulus, 136
  wheatleyi, 33, 136–37
Megantereon, 9, 91, 95, 186, 187
  hesperus, 33, 186, *189*
  megantereon, 186
  orientalis, 186

INDEX TO LATIN NAMES

Megatheriidae, 91, 138–42, 358, 360
Megatherium, 137, 138, 140
   mirabile, 140
Megatylopus, 16, 301, 302–3
   cochrani, 302–3
   matthewi, 304
Meles, 157
Melinae, 147, 157–58
Mellivorinae, 146, 147, 155
Mephitinae, 147, 161–65
Mephitis, 162, 163, 164, 165
   fossidens, 164
   leptops, 164
   macroura, 164
   mephitis, *82*, 164
   mesoleucus, 164
   orthostichus, 164
Merycodontinae, 319
Metaxyomys, 269
Metaxytherium jordani, 341
Miacidae, 146
Microchiroptera, 116–27
Microdipodops, 230
Microsorex, 103, 109
   hoyi, 63, 106, 109, *110*
   minutus, 109
   pratensis, 109
Microtinae, 239, 253–70
Microtus, 23, 31, 34, 39, 95, 258–63, 264
   agrestis, 259
   aratai, *260*, 263
   arvalidens, 259
   arvalis, 259
   californicus, 260
   chrotorrhinus, 76, 261
   cumberlandensis, 258, 259, 262, 263
   deceitensis, 259
   dideltus, 262
   guildayi, 261–62
   hibbardi, 56, 263
   involutus, 261–62
   llanensis, 259, *260*, 262
   longicaudus, 260, 261
   ludovicianus, 262
   mcnowni, 263
   meadensis, 258, 259, *260*
   mexicanus, 260–61
   miurus, 41, 261
   montanus, 260
   ochrogaster, 262, 263
   oeconomus, 261
   operarius, 28
   paroperarius, 258, 259, *260*, 262
   parvulus, 263
   pennsylvanicus, 28, 29, 31, 32, 56, 80, 83, 86, 257, 259–*260*, 261
   pinetorum, 77, 262, 263
   pliocaenicus, 258–59
   speothen, 261, 262
   xanthognathus, 63, 76, 77, 86, 261
Mictomys, 269–70
Mimomys, 95, 254, 257
   monahani, 257
Miomephitis, 161
Miracinonyx, 192–94
Mirounga, 207–8
   angustirostris, 207–8
   leonina, 208
Moeritherioidea, 343
Molossidae, 125–27, 360
Molossides, 126
Molossus, 126
Monachinae, 204, 207–8

Monachus, 207
   tropicalis, 207, 358
Monotherium? wymani, 204
Mormoopidae, 116–17, 360
Mormoops, 116–17
   megaphylla, 58, 116–17
Morotherium leptonyx, 136
Mus, 222, 228, 240, 245, 252, 253, 259, 267
Musculus, 245
Mustela, 147, 148, 149–53, 158
   erminea, 80, 83, 95, 150, 151
   e. angustidens, 150
   eversmanni, 96, 152, 153
   e. beringiae, 152, 358
   e. michnoi, 152
   frenata, 149–50, 161
   gazini, 149
   macrodon, 151, 358
   meltoni, 151
   nigripes, 96, 151, 152–53
   nivalis, 150, 151
   putorius, 152
   reliquus, 149
   rexroadensis, 149, 150
   rixosa, 63, 96, 150–51
   stromeri, 152
   vision, 151, *152*, 153
   v. ingens, 151
Mustelidae, 146–65, 360
Mustelinae, 147–55
Mylodon, 143
   garmani, 143
   listai, 143
Mylodontidae, 91, 92, 142–45, 358, 360
Mylohyus, 13, 26, 31, 57, 79, 80, 81, 295, 296–97
   browni, 296
   exortivus, 296
   floridanus, 296
   gidleyi, 296
   nasutus, 77, 296–*297*, *298*, 300, 364
   tetragonus, 296
Myomorpha, 209
Myotis, 118–21, 124
   austroriparius, 121
   californicus, 119
   evotis, 119, 120
   grisescens, 56, 87, 120
   keenii, 86, 119
   leibii, 119
   lucifugus, 86, 119, 120–21
   magnamolaris, 83, 120, 364
   rectidentis, 83, 120, 364
   sodalis, 86, 119–20
   thysanodes, 118–19
   velifer, 120
   volans, 121
Myrmecophaga, 141

Nannippus, 3, 9, 10, 11, 15, 17, 22, 26, 284–85
   aztecus, 284
   beckensis, 284, 285
   minor, 284
   phlegon, 19, 284–*285*
Napaeozapus, 270, 271–72
   insignis, 271–72
Navahoceros, 73, 312–13
   fricki, 73, *74*, *311*, 312–*313*, *314*, 365
Nebraskomys, 254, 256, 257
   mcgrewi, 256
   rexroadensis, 256

Nekrolagus, 12, 21, 279, 280
   progressus, 277, 279
Neochoerus, 11, 57, 92, 134, 274
   pinckneyi, *58*, 274
Neofiber, 30, 31, 35, 57, 264–65
   alleni, 60, 264–65
   diluvianus, 264
   leonardi, 264
Neohipparion, 284
Neomeryx finni, 325
Neomyini, 103, 113
Neosorex, 107
   lacustris, 107
Neotoma, 18, 239, 243, 251–53, 404
   albigula, 252–53
   cinerea, 253
   floridana, 29, 252
   fossilis, 252
   fuscipes, 253
   lepida, 253
   magister, 252
   mexicana, 253
   micropus, *241*, 252
   quadriplicata, 251–52
   sawrockensis, 252
   spelaea, 252
   taylori, 252
Neotragocerus, 327
Nerterogeomys, 226–27
   minor, 226
   paenebursarius, 227
   persimilis, *224*, 227
   smithi, *224*, 226, 227
Neurotrichus, 113
Nothrotheriops, 27, 50, 51, 72, 92, 137, 138, 140, 141–42, 143, 144, 404, 405
   shastensis, 48, 52, 71, 141–*142*, 364
Nothrotherium, 141
   graciliceps, 141
   texanum, 141
Notiosorex, 65, 103, 113
   crawfordi, *110*, 113
   jacksoni, 111, 113
Notolagus, 278
   lepusculus, 278
   velox, 278
Notoungulata, 91
Nyctereutes, 173
Nycticeius, 124
   humeralis, 124
Nyctinomus, 126
Nyctophylinae, 125

Ochotona, 62, 87, 275–76
   collaris, 276
   complicidens, 275
   koslowi, 275
   princeps, 50, 276
   whartoni, 23, 275–76
Ochotonidae, 275–76, 360
Ochrotomys, 247
   nuttalli, 247
Odobenidae, 200–202, 360
Odobeninae, 200–202
Odobenus, 200–202
   rosmarus, 200–202, *201*
Odocoileinae, 309, 310–17
Odocoileus, 5, 9, 13, 16, 26, 27, 57, 60, 62, 64, 91, 310–12, *314*, 318, 406
   brachyodontus, 310, 312
   cascensis, 312
   cooki, 310
   halli, 312

Odocoileus (continued)
  hemionus, 310, 312, 313
  sheridanus, 310
  virginianus, 70, 296, 310–12, 313
Ogmodontomys, 16, 18, 253–54, 257, 265
  poaphagus, 253–54, *255*
  p. transitionalis, 254
  sawrockensis, 254
Onager altidens, 288
  arellanoi, 291
  hibbardi, 291
Ondatra, 3, 5, 9, 10, 13, 17, 22, 26, 27, 33, 74, 236, 264, 265–67
  annectens, 26, 29, 31, 32, 34, 264, *265*, 266
  hiatidens, 266
  idahoensis, 13, 17, 265–66
  kansasensis, 266
  nebracensis, 29, 32, *255*, 266
  triradicatus, 266
  zibethicus, 39, 56, 266–67
Onohippidium, 91
Onychomys, 18, 249–50
  bensoni, 249
  fossilis, *241*, 249, 250
  gidleyi, 249
  jinglebobensis, 249–50
  leucogaster, 249, 250
  pedroensis, 249
  torridus, 249, 250
Ophiomys, 15, 21, 156, 254–55
  fricki, *255*
  magilli, 255
  meadensis, *241*, 254, 255
  parvus, 13, 254, *255*
  taylori, 13, 254
Opuntia, 142
Oreamnos, 37, 87, 326, 327
  americanus, 327, 364
  harringtoni, 72, *74*, 327, *328*, 365
Oryctomys, 225
Oryzomys, 65, 239
  palustris, 239, *241*
  p. fossilis, 239
Osmotherium, 164
  spelaeum, 164
Osteoborus, 166
  crassipinaetus, 165
  cyonoides, 166
  hilli, 165
  progressus, 165
Otaria, 203
Otariidae, 202–4, 360
Otarioidea, 200–204
Otognosis, 234
Otospermophilus, 212, 213, 214
Ovibos, 96, 330, 332, 333, 334–35
  giganteus, 332
  moschatus, *330*, 332, 334–35, 364
  proximus, 334
  yukonensis, 334
Ovibovini, 326, 330–35
Ovis, 37, 51, 71, 328–330, 358
  canadensis, 328, 329–30, 364
  c. catclawensis, 329
  dalli, 328–29
  d. stonei, 329
  nivicola, 328

Pachyceros, 329
Pachygazella grangeri, 327
Paenemarmota, 199, 210–11
  barbouri, *14*, 210–*211*

Pagophilus, 204
Palaeolaginae, 276
Palaeolama, 50, 57, 92, 301, *307*, 308–9
  aequatorialis, 309
  mirifica, *307*, 308–9, 365
  weddelli, 309
Palaeomastodon, 343
Pampatherium, 129
Pannonictis, 156
Panthera, 22, 91, 190, 191–92, 194
  gombaszoegensis, 192
  leo, 39, 96, 191–92, 363, 404
  l. atrox, 40, 44, 45, 61, 67, 88, 191, 358, 364
  l. fossilis, 191
  l. spelaea, 191
  l. youngi, 191
  onca, 60, *74*, 192
  o. augusta, 56, 79, 192, 358, 364
Pantholops, 326
Pappogeomys, 74
Parabos, 338
Paracamelus, 301, 302
Paracryptotis, 103, 110–11
  gidleyi, 110–11
  rex, 110, 111
Paradipoides, 235
  stovalli, 235
Parahodomys, 251, 252
Parailurus, 21, 95, 175, 177–78
  anglicus, 177–78
  hungaricus, 177
Paramylodon, 143, 144
  nebrascensis, 143
Paraneotoma, 251–52
Paraonyx, 158
Parapliosaccomys, 229
Parascalopina, 114
Parascalops, 113, 114
  breweri, *114*
Parastylidequus, 287
Parelephas, 350
  progressus, 352
Paucituberculata, 101
Pecora, 295
Pediocetes phasianellus, 77
Pediolagus, 279
Pedomys, 258, 262
Pekania, 147–48
Pelynictis lobulatus, 164
Peramelina, 101
Perissodactyla, 283–94, 359
Perodipus, 232
Perognathinae, 230, 232–34
Perognathus, 20, 230, 232–34
  apache, 234
  californicus, 233, 234
  flavus, 233
  gidleyi, 232, 233
  hispidus, 29, 64, 233, 234
  inornatus, 233
  intermedius, 234
  longimembris, 234
  magnus, 233
  maldei, *231*, 233
  mclaughlini, 232, 233
  merriami, 234
  parvus, 233, 234
  pearlettensis, 233
  rexroadensis, 232–33
Peromyscus, 18, 26, 33, 39, 49, 242–47, 248, 249
  anyapahensis, 49, 243

  baumgartneri, 242–43
  berendsensis, 244
  boylii, 244, 246
  cochrani, 244–45
  cragini, *241*, 243, 244
  crinitus, 243
  cumberlandensis, 60, 244
  dentalis, 244
  difficilis, 246
  eremicus, 243
  floridanus, 247
  gossypinus, 244, 245
  hagermanensis, 244
  imperfectus, 243
  irvingtonensis, 243
  kansasensis, 242, 243
  leucopus, 244, 245, 246
  maniculatus, 244, 245, 246
  m. gambelli, 243
  nesodytes, 49, 243
  oklahomensis, 246–47
  pectoralis, 244, 246
  polionotus, 244, 245
  progressus, *241*, 244, 245, 246
  truei, 244, 246
Petauria, 222
Phenacomys, 30, 34, 87, 256, 258
  intermedius, 77, 258
Phiomia, 346
Phoca, 200, 202, 203, 204–6, 207
  fasciata, 205
  groenlandica, 75, 205–6
  hispida, 205
  pannonica, 205
  pontica, 205
  vitulina, 204–5
Phocidae, 204–8, 360
Phocinae, 204–7
Phocoidea, 200
Phyllostomatidae, 116, 117–18, 360
Phyllostomatinae, 117
Picea, 74, 86
Pika, 276
Pinaceae, 14
Pinnipedia, 200–208
Pinus, 18, 28, 74, 86
  aristata, 50–51
  flexilis, 51
Pipistrellus, 121–22
  hesperus, 121, 122
  subflavus, 121
Pitymys, 31, 258, 259, 262, 263
Planisorex, 18, 103, 109–10
  dixonensis, 109–*110*
Platycerabos, 338–39
  dodsoni, 338–39
Platygonus, 9, 10, 13, 31, 55, 56, 65, 72, 83, 91, 295, 296, 297–301, 404, 406
  bicalcaratus, 19, 27, 297, 298, 300
  compressus, 56, 66, 68, 69, 70, 79, 83, 89, 295, *297*, 298, *299*–301, 362, 364
  cumberlandensis, 298
  intermedius, 298
  leptorhinus, 299
  pearcei, 297–98, 299
  texanus, 298
  vetus, *31*, 32, 297, *298*–*299*, 300
Plecotus, 124–25
  alleganiensis, 124
  rafinesquii, 124
  tetralophodon, 124
  townsendii, 124–25
Plesictis, 175

INDEX TO LATIN NAMES

Plesiogulo, 153
Plesiothomomys, 224
Plesippus shoshonensis, 285
Pliauchenia, 301, 302, 306
Pliogeomys, 226
  parvus, 226
  transitionalis, 226
Pliohippus, 12, 285
  francescana, 285
Pliolemmus, 18, 258
  antiquus, 255, 258
Pliometanastes protistus, 136
Pliomys, 256
  deeringensis, 256
Plionarctos, 179
Pliophenacomys, 9, 18, 32, 254, 255–56, 258
  finneyi, 241, 255–56
  idahoensis, 254
  osborni, 256
  primaevus, 255, 256
Pliopotamys, 9, 15, 18, 21, 32, 156, 254, 264, 265
  meadensis, 265
  minor, 13, 265
Pliosaccomys, 227, 229
Pliotaxidea nevadensis, 158
Pliozapus, 270
Podomys, 242, 246–47
Poëphagus, 339
Poliocitellus, 212
Populus, 14, 74
Porthocyon, 166
  matthewi, 166
Potamotherium, 146
Praealces, 315
Praeovibos, 41, 334
  priscus, 334
Praesynaptomys, 268
Praotherium palatinum, 275
Pratilepus, 9, 278–79
  kansasensis, 279
  vagus, 277, 278–79
Predicrostonyx, 23, *265,* 267
  hopkinsi, 265, 267
Preptoceras, 330, 331, 334
  sinclairi, 330
Primates, 355–56, 359
Primelephas gomphotherioides, 348
Proantilocapra platycornea, 325
Proboscidea, 343–54, 359
Procamelus, 301, 302
Procastoroides, 18, 19, 21, 32, 235, 236
  idahoensis, 236
  sweeti, 235, 236, *237*
Procoileus edensis, 310
Procyon, 175–76
  cancrivorus, 175
  lotor, 175–176
  l. simus, 175, 176
  nanus, 175
  priscus, 175
  rexroadensis, 175
Procyonidae, 146, 175–78, 360
Procyoninae, 175–77
Prodipodomys, 12, 18, 20, 27, 230–31
  centralis, 230, *231*
  idahoensis, 230–31
  kansensis, 230
  rexroadensis, 230
Promephitis, 163
Proneofiber, 264, 265
  guildayi, 264

Proreithrodon, 91
Prosomys, 254
Prosopsis, 142
Prosthennops, 296
Protocyon, 91, 172
  scagliarum, 172
  texanus, 170, 172
Protolabidini, 301
Psammomys, 263
Pseudostoma, 229
Pteroneura, 158
  brasiliensis, 158
Puma, 194
Pusa, 204, 205
Putorius, 149.

Quercus, 74, 406

Ramapithecus, 355
Rangifer, 23, 25, 37, 41, 62, 71, 95, 309, 312, 314–15, 358
  muscatinensis, 314
  tarandus, 45, 76, 77, 314–15
  t. groenlandicus, 315
Reithrodon, 240, 242
Reithrodontomys, 240–42
  fulvescens, 242
  humulis, 29, 240*–241*
  megalotis, 240, 242
  montanus, 240, 241, 242
  moorei, 240, 241
  pratincola, 240
  rexroadensis, 240
  wetmorei, 240
Rhinocerotidae, 283
Rhynchotherium, 346
  falconeri, 346
  praecursor, 346
Rhytina, 340
Rodentia, 209–74, 275, 359
Romerolagus, 278
Ruminantia, 295
Rupicapra, 327
Rupicaprini, 326, 327

Saiga, 41, 326–27
  ricei, 326
  tatarica, 326–27, 358
Sangamona, 63, 79, 80, 313–14
  fugitiva, 313–14, 365
Satherium, 13, 18, 156, 158
  ingens, 158
  piscinarium, 158, *159*
Scalopina, 114–15
Scalopini, 113, 114–15
Scalopus, 18, 113, 114, 115
  aquaticus, 115
  blancoensis, 115
  rexroadi, 115
  sewardensis, 115
Scapanus, 113, 114–15
  latimanus, 114–15
Scaphoceros tyrelli, 332
Schistodelta sulcatus, 264
Sciuridae, 210–23, 360
Sciurus, 49, 57, 215, 217, 220–21, 222
  alleni, 221
  arizonensis, 221
  carolinensis, 221
  hudsonicus, 26
  niger, 220–21
  oculatus, 221
  vulgaris, 222

Scotophilus, 122
Serebelodon, 346
Serridentinus, 346
Sicistinae, 270
Sigmodon, 10, 12, 17, 20, 26, 56, 250–51, 256
  alleni, 251
  bakeri, 251
  curtisi, 251
  fulviventer, 251
  hilli, 250
  hispidus, 29, 251
  hudspethensis, 250–51
  intermedius, 250
  medius, 241, 250, 251
  minor, 250
  ochrognathus, 251
Simocyoninae, 165–67, 172
Simonycteris, 122
Sirenia, 340–42, 359
Smilodon, 9, 27, 53, 72, 79, 91, 186–88, 189, 190, 406
  californicus, 186
  fatalis, 23, 32, *52,* 56, 186–88, *187, 189,* 364
  gracilis, 23, 33, 186, 188, *189*
  nebraskensis, 186
  neogaeus, 188
  populator, 188
  trinitiensis, 186
Smilodontidion, 91, 186
  riggii, 186
Smilodontini, 185
Smilodontopsis
  conardi, 186
  mooreheadi, 194
  troglodytes, 186
Sminthosinus, 156
  bowleri, 156
Soergelia, 35, 95, *330,* 331–32
  elisabethae, 331, 332
  mayfieldi, 331–32
Sorex, 18, 20, 103–9, 110, 111, 112, 113, 115
  arcticus, 64, 76, 77, 104, 106, 107*–108*
  cinereus, 60, 74, 80, 104, *105*–106, 107
  c. meadensis, 105
  cudahyensis, 104
  dispar, 108
  franktownensis, 105
  fumeus, 26, 60, 107, 108
  hagermanensis, 104
  kansasensis, 106
  lacustris, 107
  leahyi, 104
  longirostris, 106
  megapalustris, 107
  meltoni, 104
  merriami, 104, 109
  obscurus, 104
  ornatus, 106
  palustris, 106, 107
  personatus, 105
  powersi, 104
  rexroadensis, 103–4, 111
  sandersi, 104
  saussurei, 109
  scottensis, 106
  taylori, 104, 111
  trowbridgei, 109
  vagrans, 106
Soricidae, 103–13, 360
Soricinae, 103–13

Soricini, 103–9
Soriciomorpha, 103
Sorisicus, 111
Speothos, 172
  venaticus, 172
Spermophilus, 18, 20, 26, 41, 210, 212–17, 219
  armatus, 215
  beecheyi, 213, 217
  bensoni, 213, 214
  boothi, 211, 213
  cochisei, 214
  columbianus, 213, 214, 215
  cragini, 214
  finlayensis, 214
  franklinii, 69, 213, 217
  howelli, 212–13
  johnsoni, 213
  lateralis, 48, 50, 217
  magheei, 213
  meadensis, 213, 214
  meltoni, 213
  mexicanus, 212, 213, 214, 217
  parryii, 213, 215, 218
  rexroadensis, 213
  richardsoni, 28, 214, 215
  spilosoma, 214, 216–17
  taylori, 214
  townsendii, 214, 215
  tridecemlineatus, 30, 66, 69, 76, 214, 215–*216,* 217
  tuitus, 214
  variegatus, 213, 217
Sphaeralcea, 142
Sphenophalos, 319
Spilogale, 161, 162–63
  gracilis, 163
  marylandensis, 162
  pedroensis, 162–63
  putorius, 162, 163
  pygmaea, 162
  rexroadi, 162
Stegodontidae, 343
Stegomastodon, 15, 16, 18, 22, 26, 27, 33, 346–48
  arizonae, 347
  mirificus, 346–48, *347*
  primitivus, 347
  rexroadensis, 347, 348
  texanus, 346
Stegotetrabelodontinae, 348
Stenocranius, 258, 261
Stockoceros, 321, 322–24
  conklingi, 74, 322, *323*–*24,* 365
  onusrosagris, 48, *320,* 322, 323–24, 365
  o. nebrascensis, 323
Subantilocapra, 325
Suina, 295–301
Sus scrofa, 297, 300, 301, 358
Sycium cloacinum, 266
Sylvilagus, 12, 27, 59, 279–81
  aquaticus, 277, 281
  audubonii, 280
  bachmani, 280–81
  floridanus, 277, 279–80
  leonensis, 281
  nuttalli, 277, 280
  palustrellus, 281
  palustris, 281
  transitionalis, 60, 280
Symbos, 32, 41, 63, 86, 332–33, 334, 404
  australis, 332
  cavifrons, 41, 76, 332–*333,* 365

  convexifrons, 332
  promptus, 332
Symmetrodontomys, 16, 249
  simplicidens, 241, 249
Synaptomys, 13, 30, 34, 95, 267, 268–70
  anzaensis, 269
  australis, 55, 56, 268, 269, 364
  borealis, 28, 60, 64, 76, 77, 268, 269, 270
  bunkeri, 268
  cooperi, 60, 77, 83, *265,* 268, 269
  europaeus, 268
  kansasensis, 269–70
  landesi, 269
  meltoni, 269, 270
  rinkeri, 15, 17, 268
  vetus, 265, 269

Tadarida, 58, 125–26
  brasiliensis, 120, 126
  b. mexicana, 125
  constantinei, 125–26
  femorosacca, 126
Talpidae, 113–15, 360
Talpinae, 113–15
Tamias, 30, 220
  aristus, 60, 220
  striatus, 83, 220
Tamiasciurus, 221–22
  douglasii, 222
  hudsonicus, 26, 77, 221
  h. tenuidens, 221
Tanupolama, 306, 308, 309
  hollomani, 308
  seymourensis, 306
  stevensi, 308
Tapiridae, 283, 292–94, 358, 360
Tapirus, 12, 13, 48, 54, 57, 64, 78, 79, 80, 91, 292–94, 358
  californicus, 292–93
  copei, 292, 293, 365
  excelsus, 293, 294
  haysii, 292, 293
  merriami, 292, 293
  tennesseae, 293
  terrestris, 292
  veroensis, 60, 83, *293*–94, 365
Tatu, 130
Taurotragus americanus, 330
Taxidea, 30, 69, 157–58
  marylandica, 157
  mexicana, 158
  robusta, 157
  sulcata, 157
  taxus, 9, 16, 66, *157*–58
  t. berlandieri, 158
  t. jeffersonii, 157
Tayassu, 296, 300
  peccari, 295, 296
  tajacu, 295, 301
Tayassuidae, 295–301, 360
Teonoma spelaea, 253
Teratornis merriami, 52, 53
Terrapene, 56, 57, 59, 65
  carolina, 55
  c. bauri, 55, 56
  c. major, 55
  c. putnami, 55, 56
Testudo, 15
Tetrabelodon shepardii, 346
Tetralophodon, 346
Tetrameryx, 9, 27, 73, 319, 321–22, 323, 324

  irvingtonensis, 321
  knoxensis, 29, 321
  mooseri, 321–22
  shuleri, 321, 322, 365
  tacubayensis, 322
Texoceras, 319
Thalassictis, 199
Thomomys, 58, 69, 74, 223–26
  bottae, 223, 225, 226
  gidleyi, 223, *224,* 225, 226
  mazama, 226
  microdon, 52, 226
  orientalis, 224
  potomacensis, 224
  talpoides, 63, 209, 223, 224, 225
  t. quadratus, 223
  townsendii, 225
  umbrinus, 225, 226
Tillandsia usneoides, 124
Tisisthenes, 153
  parvus, 153
Titanis walleri, 13
Titanotylopus, 16, 19, 23, 33, 301–2, 304
  merriami, 302
  nebraskensis, 301–2
  spatulus, 302
Tremarctinae, 178–82
Tremarctos, 9, 12, 178–80, 358
  floridanus, 56, *58,* 60, 74, 178–80, *179, 183,* 364
  mexicanus, 178
  ornatus, 179, 363
Trichechidae, 340, 341–42, 360
Trichechus, 341–42
  manatus, 341–42
Trigonictis, 9, 16, 21, 91, 95, 155–56
  cookii, 156
  idahoensis, 14, 155–56
  kansasensis, 155
Trochictis, 156
Trogontherium, 236
Trucifelis, 186
Tylopoda, 295

Uncia inexpectata, 192, 194
  mercerii, 186
Urmiabos, 339
Urocyon, 173, 174
  atwaterensis, 173
  cinereoargenteus, 170, 173
  littoralis, 173
  minicephalus, 173
  progressus, 173
Urotrichini, 113
Ursavus, 178
Ursidae, 146, 178–85, 360
Ursinae, 178, 182–85
Ursus, 9, 95, 153, 157, 175, 182–85
  abstrusus, 33, 182–83
  americanus, 33, 39, 74, 180, *182, 183*–84, 297, 301, 362
  a. amplidens, 183
  arctos, 39, 66, 85, 88, 96, 182, *183,* 184, *185,* 356, 362, 406
  a. dalli, 184
  a. horribilis, 184
  a. middendorffi, 184
  etruscus, 184
  haplodon, 180
  maritimus, 178, 185
  minimus, 182
  procerus, 184
  spelaeus, 179

INDEX TO LATIN NAMES

thibetanus, 183
vitabilis, 183

Vassalia, 129
Vesperimus, 246
Vespertilio, 119, 120, 121, 122, 123, 124, 125
  subulatus, 119
Vespertilionidae, 116, 118–25, 360
Vespertilioninae, 118–25
Vicugna, 301, 307
  vicugna, 307
Viverra, 162, 164
Vulpes, 173–74, 175

alopecoides, 174, 175
bengalensis, 174
corsac, 174
fulva, 174
macrotis, 174
palmaria, 174
velox, 173–74
vulpes, 87, 96, *170,* 174

Xenocyon, 172
Xenoglyptodon fredericensis, 134

Yucca, 142

Zalophus, 203–4
  californicanus, 203–204
Zapodidae, 209, 270–72, 360
Zapus, 17, 20, 270–71
  burti, 270, 217
  hudsonius, 270, 271
  h. adamsi, 271
  h. transitionalis, 271
  princeps, 270, 271
  rinkeri, 270–71
  sandersi, 34, *270,* 271
  s. rexroadensis, 271
  s. sandersi, 17
Zygogeomys, 228

Index to Common Names of Organisms

Alder, 238
Alga, 341
Alligator, 35
Alpaca, 307
Amphibian, 10, 15, 16, 17, 18, 26, 28, 30, 32, 35, 50, 55, 56, 57, 58, 59, 62, 64, 65, 68, 71, 77, 79, 81, 82, 83, 158, 174
Anteater, 91, 128
  giant, 141
Antelope, 13, 15, 326, 327
Antilocaprid, 21, 40, 319–25
Ape, 355
Aplodontid, 209–10
Archaeolagine, 276, 277–78
Armadillo, 35, 91, 128–32
  beautiful, 130–31
  nine-banded, 131–32
Arthropod, 77, 81
Artiodactyl, 41, 295–339, 357
Aspen, 238
Ass, 285, 287
  Cummins's, 287–88

Badger, 146, 157–58
  European, 157
  honey, 146
Bandicoot, 101
Bat, 12, 27, 56, 58, 76, 77, 83, 86, 116–27, 357
  Anza-Borrego, 125
  big brown, 122
  big-eared
    Allegany, 124
    four-lophed, 124
    Rafinesque's, 124
    Townsend's, 124–25
  evening, 124
  free-tailed, 125–27
    Brazilian, 126
    Constantine's, 125–26
  fruit, 116
  ghost-faced, 116–17
  Golliher's, 123
  hoary, 123
    extinct, 123
  leaf-chinned, 116–17
  leaf-nosed, 117
  mastiff, 126
    Florida, 126
    western, 126–27
  Mexican long-nosed, 117
  northern yellow, 123–24
  pallid, 125
  red, 122–23
  silver-haired, 121
  Stock's snub-nosed, 122
  vampire, 117–18
    Stock's, 117–18
  vespertilionid, 118–25
Bear, 31, 74, 154, 178–85
  Andean, 178
  black, 33, 178, 180, 182–84, 236, 297, 301
    primitive, 182–83

  brown, 182, 183, 184
  cave
    European, 179, 180
    Florida, 178–80
  grizzly, 182, 183, 184
  polar, 185, 205
  short-faced
    giant, 180–82
    lesser, 180
Beaver, 13, 14, 34, 71, 156, 235–38
  giant, 236–37, 238
  Hay's, 238
  Idaho, 236
  intermediate, 235
  Kellogg's, 237–38
  mountain, 209–10
  Rexroad, 235
  Stovall's, 235
  Sweet's, 236
Beetle, 17, 53, 125
  carrion, 53
Birch, 238
Bird, 10, 11, 13, 14, 15, 16, 17, 18, 19, 20, 26, 30, 32, 35, 36, 50, 51, 52, 53, 54, 55, 56, 58, 59, 62, 64, 65, 71, 72, 74, 77, 79, 81, 82, 86, 87, 89, 113, 149, 151, 154, 158, 174, 177, 217, 221
Bison, 43, 44, 47, 48, 51, 55, 56, 61, 62, 64, 65, 66, 84, 87, 88, 89, 326, 330, 335–38, 356, 358
  giant, 39, 40, 85, 336, 337
  steppe, 335–36
Bivalve, 18
Blowfly, 53
Boar, 300
Bobcat, 197–98, 273
Bovid, 27, 295, 326–39
Bradypus, 137
Buffalo, American, 326, 337–38
Bunomastodont, 346

Cacomistle, 177
Cactus, 48
Camel, 85, 87, 301–6
  Cochran's, 302–3
  furrow-toothed, 304
  Huerfano, 304
  Kansas, 303–4
  Meade's, 302
  Minidoka, 304–5
  Nebraska, 301–2
  spatulate-toothed, 302
  Traviswhite's, 304
  yesterday's, 305–6
Camelid, 15, 20, 23, 27, 40, 47, 51, 61, 80, 84, 91, 301–9
Cameline, 302
Camelopine, 301
Canid, 51, 53, 91, 165–75, 198
Capybara, 91, 134, 236, 274
  Holmes's, 274
  Pinckney's, 274
Caribou, 63, 76, 88, 154, 314–15, 356
Carnivore, 14, 17, 26, 31, 37, 41, 51, 53,
54, 59, 62, 73, 76, 80, 81, 88, 95, 137, 146–208, 357
  marsupial, 101
  procyonid, 90
Castorid, 235–38
Cat, 102, 131, 134, 164, 185–98, 209
  cheetahlike, 88
  lake, 195
  river, 195
  sabertooth, 145
  scimitar, 190
Cattle, 326
Cephalopod, 204
Cervid, 21, 35, 63, 91, 95, 309–18, 326
Chamois, 326, 327
Chaparral, 233, 234, 246, 281
Cheetah, 95, 192–94
  American, 194
  Studer's, 192–94
Chiropteran, 37, 116–27
Cingulate, 128
Clam, 84
Coendou, 272–74
  Appalachian, 273–74
  deep-jawed, 272
  Stirton's, 273
Composite (plant), 75
Condor, giant, 53, 72
Corn, 176
Cottontail, 177, 279–80
  desert, 280
  eastern, 279–80
  mountain, 280
  New England, 280
Cougar, 194, 273
Coyote, 95, 167–68, 171
  Johnston's, 167
  wolf, 168, 169
Crayfish, 82, 151, 158, 176
Cricetid, 33, 239–70
Cricetine, 239–53
Cricket, 125
Crocodile, 31, 60
Crustacean, 161, 202, 205
Cuttlefish, 208
Cypress, 49, 50, 53, 406

Dasypodid, 128–32
Dasypodine, 128, 130–32
Dasypus, 137
Deer, 33, 35, 58, 66, 295, 309–18
  brachydont, 310
  Cope's, 318
  fallow, 310
  fugitive, 313–14
  marsh, 312
    Florida, 312
  mountain, 312–13
  mule, 195, 198, 309, 312
  plesiometacarpalian, 309, 317
  roe, 309
  telemetacarpalian, 309
  white-tailed, 198, 296, 310–12
Dermestid, 53
Desmatophocid, 200

INDEX TO COMMON NAMES

Dhole, 172
Didelphoid, 101–2
Dinosaur, 357
Dipodomyine, 230–32
Dirktooth, 185, 186, 188
  western, 186
Diver (bird), 10
Dog, 165–75
  bone-eating, 166–67
  bush, 172
  Cedazo, 170–71
  domestic, 56, 59, 63, 171
  plundering, 165–66
  raccoon, 173
  simocyonine, 165–67
  Troxell's, 172
Donkey, 287, 306
Duck, 10, 72
Dugong, 340
Dusignathine, 200

Edentate, 59, 81, 90, 91, 128–45, 357
Elephant, 49, 342, 343, 345, 348–54
  African, 349
  Asiatic, 345, 349
Elk, American, 318
  false, 309–10
  Old World, 315
Elm, 63
Enaliarctid, 202
Equid, 15, 20, 40, 91, 283–91
Erethizontid, 272–73
Ermine, 150
Eutherian, 128

Felid, 91, 185–98
Ferret, 149, 152–53
  Beringian, 152
  black-footed, 152–53, 219
  European, 152
  steppe, 153
Fish, 13, 14, 15, 16, 17, 18, 20, 21, 29, 30, 32, 35, 51, 56, 60, 62, 64, 65, 68, 75, 76, 81, 82, 83, 89, 151, 158, 174, 202, 203, 204, 205, 206, 208
Fisher, 148, 156, 273
  diluvial, 147–48
Flesh-eater, voracious, 155
Fox, 165, 173–75
  arctic, 174–75
  flying, 116
  gray, 173
    island, 173
    progressive, 173
  kit, 173–74
  red, 87, 174
  swift, 174
Frog, 14, 20, 30, 75, 151

Gastropod, 18, 50, 52, 79
Gazelle-horse, 284–85
  Beck Ranch, 284
Geomyid, 223–29
Glyptodont, 26, 83, 91, 128, 130, 132–35, 362
  Gidley's, 134
  Mexican, 135
  Osborn's, 132–34
  Simpson's, 135
Goat, 326, 328
  mountain, 327
    Harrington's, 327
  wild, 328

Gomphothere, 343, 344, 345, 346–48, 349
  Cuvier's, 348
Gopher, pocket, 58, 223–29, 230
  Adams's, 227
  Benson, 229
  early plains, 227
  eastern, 224
  Garbani's, 227–28
  giant hispid, 229
  Gidley's, 223
  Hay's, 227
  Jacob's, 227
  mazama, 226
  Mexican, 229
  northern, 225–26
  plains, 228
  Potomac, 224
  pygmy, 225
  Quinn's, 227
  Sinclair's mazama, 226
  small, 226
  Smith's, 226
  southeastern, 228–29
  Tobin, 228
  Townsend's, 225
  valley, 225
  Zakrzewski's, 226
Grison, 155–56
  Cook's, 156
  Gazin's, 155–56
Groundhog, 211–12
Guanaco, 307

Half-ass, 287
Hamster, 239
Hare, 10, 276, 281–82
  arctic, 282
  Benjamin's, 281
  snowshoe, 196, 282
Hedgehog, 103
Hemione, 288
Herbivore, 19
Heteromyid, 230–35
Hickory, 63
Hippopotamus, 295
Hominid, 355–56
Horse, 11, 12, 17, 18, 19, 21, 47, 51, 55, 56, 61, 64, 70, 80, 84, 85, 88, 89, 91, 283–91, 356
  asslike, 26
  brother, 290
  complex-toothed, 290
  giant, 289
  Lambe's, 291
  Mexican, 291
  Mooser's, 287
  Niobrara, 290–91
  noble, 291
  Scott's, 290
  western, 289–90
Human, *see* Man
Hyaenid, 167, 198–99, 211
  cheetahlike, 211
Hyena, 165, 166, 171, 198–99
  American hunting, 199
Hyrax, 343

Ibis, 16
Iguana, 85
Iguanid, 74
Indian, *see* Man

Insect, 52, 53, 89, 106, 113, 116, 125, 130, 131, 154, 162, 164, 165, 174, 177, 178, 216, 217, 249, 270
Insectivore, 16, 103–15, 116, 357

Jaguar, 22, 79, 95, 186, 192
  Palearctic, 192
Jaguarundi, 195–96
Juniper, 51, 72, 168, 246, 280

Kangaroo, 101
Kiang, 288
Kulan, 288

Lagomorph, 198, 275–82
Larch, 68
Lemming, 23, 25, 88, 95, 239, 354
  bog, 268–70
    Anza, 269
    Bunker's, 268
    Florida, 269
    Kansas, 269–70
    Landes's, 269
    Melton's, 269
    northern, 270
    old, 269
    Rinker's, 268
    southern, 268
  boreal, 34
  brown, 267
  collared, 267
    Hudson Bay, 267–68
    Hopkins's, 267
  sagebrush, 263–64
  steppe, 263–64
Leporid, 276–82
Lichen, 315
Lion, 41, 145, 185, 186, 191–92
  American, 191–92
  cave, 191
  mountain, 194
Lizard, 20, 62, 82, 174, 249
Llama, 83, 91, 301, 306–9
  Blanco, 306–8
  large-headed, 308
  stout-legged, 308–9
Lupine, 88
Lynx, 95, 186, 195, 196–98
  Canada, 186–87
  Issoire, 186

Macroplankton, 206
Madrona, 221
Mammoth, 23, 33, 34, 40, 47, 48, 49, 50, 61, 62, 72, 73, 75, 76, 84, 85, 88, 89, 95, 166, 190, 344, 345, 349, 350–54, 356, 362, 363
  Columbian, 351–52
  Jefferson's, 190, 352–53
  southern, 350–51
  woolly, 40, 353–54
Man (artifacts, Paleo-Indians, etc.), 33, 41, 45, 47, 48, 50, 54, 55, 56, 57, 59, 61, 62, 63, 67, 71, 72, 73, 75, 79, 81, 83, 84, 87, 89, 145, 188, 190, 191, 195, 200, 201, 312, 315, 331, 338, 341, 345, 346, 348, 352, 353, 355–56, 362, 363
  modern, 355–56
  Neanderthal, 355
  Otavalo, 356
Manatee, 340, 341–42
Manzanita, 51

Margay, 195, 196
Marmot, 211–12
  Arizona, 212
  giant, 210–11
  yellow-bellied, 212
Marsupial, 90, 91, 101–2
Marten, 75, 146, 147–49, 154
  noble, 148
  pine, 149
Mastodont, 15, 18, 61, 63, 68, 70, 80, 83, 190, 236, 327
  American, 343–46
Megalonychid, 135–38
Megathere, 60, 138–41
Megatheriid, 138–42
Microchiropteran, 116–27
Microtine, 156, 253–70
Mink, 149, 151
  Melton's, 151
  sea, 151
Mole, 103, 113–15
  broad-footed, 114–15
  eastern, 115
  golden, 103
  hairy-tailed, 114
  Mt. Blanco, 115
  Rexroad, 115
  shrew, 113
  star-nosed, 113–14
Mollusk, 15, 16, 17, 28, 30, 34, 35, 53, 61, 63, 64, 65, 74, 75, 77, 80, 81, 82, 83, 84, 87, 89, 106, 161, 201, 205, 206, 405, 406
Monachine, 204
Monkey, 355
  New World, 90, 91
  Old World, 355
Moose, 37, 154, 309, 315, 316–17
  broad-fronted, 315–16
Mouse, 77, 164
  Anacapa, 243
  Baumgartner's, 242–43
  Benson, 248–49
    Arizona, 248
    Elias's, 248
    Meade, 249
  Berends, 244
  brush, 246
  cactus, 243
  canyon, 243
  Cochran's, 244
  cotton, 245–46
  Cragin's, 243
  cricetine, 239
  Cumberland, 244
  deer, 216, 245
  Florida, 247
  golden, 247
  grasshopper, 249–50
    Benson, 249
    fossil, 249
    Gidley's, 249
    Jinglebob, 249–50
    northern, 250
    San Pedro, 249
    southern, 250
  Hagerman, 244
  harvest, 240–42
    eastern, 240–41
    fulvous, 242
    meadow, 240
    Moore's, 240
    plains, 240
    Rexroad, 240
    western, 242
    Wetmore's, 240
    imperfect, 243
    Irvington, 243
  jumping, 270–73
    Burt's, 271
    meadow, 271
    Rinker's, 270–71
    Sanders's, 271
    western, 271
    woodland, 272
  Kansas, 242
  Oklahoma, 246–47
  Oldfield, 245
  piñon, 246
  progressive, 244
  pygmy, 247–48
    boreal, 247, 248
    Kolb's, 248
    least, 248
    northern, 248
    Rexroad, 247
    short-faced, 248
  rock, 246
  Santa Rosa, 243
  simple-toothed, 249
  white-ankled, 246
  white-footed, 242, 245
Mouse-eater, Bowler's, 156
Mushroom, 315
Muskox, 35, 37, 41, 70, 76, 86, 95, 326, 330, 331–35
  Harlan's, 334
  Staudinger's, 334
  woodland, 332–33
Muskrat, 30, 32, 34, 71, 151, 209, 239, 254, 264, 265–67
  Brown's, 266
  Cope's, 266
  Idaho, 265–66
  Meade, 265
  Nebraska, 266
  pygmy, 265
Mussel, 80, 81, 158
Mustelid, 91, 146–65, 204
  Cummins's, 156
  galictine, 156
Myotis, 118–21
  California, 119
  cave, 120
  fringed, 118
  gray, 120
  Indiana, 119
  Keen's, 119
  large, 120
  little brown, 120–21
  long-eared, 119
  long-legged, 121
  small-footed, 119
  southeastern, 121
  straight-toothed, 120

Nothrothere, 138, 141–42

Oak, 60, 221
  live, 49, 73, 233
Ocelot, 195
Ochotonid, 275–76
Odobenid, 200–202
Odobenine, 200–202
Onager, 288

  pygmy, 288
  stilt-legged, 288
Opossum, 91, 101–2
Opossum-rat, 101
Osage orange tree, 65
Ostracod, 17, 18, 64, 65, 82
Otariid, 52, 200, 202–4
Otter, 13, 146, 158–61, 204
  Blancan, 158
  river, 158–59
  sea, 158, 160–61
    ancestral, 159–60
    large-toothed, 160
  small-clawed, 158
Ovibovid, 79, 80
Ovibovine, 330–35
Owl, 26, 27, 58, 62, 85, 86
  barn, 58
Ox
  flat-horned, 326, 338–39
  Soergel's, 331–32

Paleo-Indian, see Man
Palmetto, 247
Paloverde tree, 48
Pampathere, 28, 29, 83, 84, 128–30
  Florida, 128–30
  northern, 129–30
Pampatheriine, 128–30
Panda
  English, 177–78
  lesser, 21
  red, 175, 177–78
Pangolin, 137
Peccary, 10, 13, 15, 21, 34, 79, 84, 295–301
  collared, 295, 301
  Cope's, 298
  flat-headed, 299–301
  Kinsey's, 296
  Leidy's, 298–99
  long-nosed, 296–97
  Pearce's, 197–98
  white-lipped, 295, 296
Pecoran, 295
Perissodactyl, 283–94, 357
Phalanger, 101
Phocid, 52, 204–8
Phocine, 204
Phocoid, 204
Pig, 295
Pika, 23, 275–76
  collared, 276
  Wharton's, 275–76
Pilosan, 128
Pine, 49, 51, 53, 65, 70, 73, 86, 221, 247, 258
  yellow, 72
Pinniped, 200–208
Piñon, 72, 246
Pipistrelle, 121–22
  eastern, 121
  western, 122
Polecat, 146
Poplar, 238
Porcupine, 91, 148, 272–73
Prairie dog, 30, 153, 218–20
  black-tailed, 219
  Gunnison's, 219–20
  Hibbard's, 218
  Niobrara, 218–19
  old, 218
  white-tailed, 219

INDEX TO COMMON NAMES

Primate, 355–56
  catarrhine, 355
Proboscidean, 22, 67, 81, 95, 340, 343–54, 357
Procyonid, 90, 175–78
Pronghorn, 295, 319–25
  Conkling's, 322
  diminutive, 320–21
  Hay's, 322
  Irvington, 321
  Knox, 321
  Matthew's, 319–20
  Mexican, 321
  Mooser's, 321–22
  Prentice's, 319
  Quentin's, 323–24
  Shuler's, 321
  Skinner's, 320
  Tacubaya, 322
Prosciurine, 209
Prosimian, 355
Puma, 186, 193, 194–95

Rabbit, 12, 59, 73, 156, 276–82, 357
  ancient, 277
  Benson, 279
  Brown's, 278
  brush, 280–81
  Furlong's, 278
  Hagerman, 277
  jack, 281–82
    antelope, 281–82
    black-tailed, 282
    white-tailed, 282
  Kansas, 279
  marsh, 281
    pygmy, 281
  plains, 278–79
  progressive, 279
  pygmy, 279
  royal, 277–78
  small, 278
  snowshoe, 282
  swamp, 281
  Tusker, 279
Raccoon, 175–77
  Rexroad, 175
Raptor, 53, 58, 86
Rat
  cotton, 250–51
    Baker's, 251
    Curtis, 251
    hispid, 251
    Hudspeth, 250–51
    intermediate, 250
    yellow-nosed, 251
  cricetine, 239
  kangaroo, 230–32
    agile, 232
    bannertail, 232
    central, 230
    giant, 232
    Gidley's, 231
    Idaho, 230–31
    Merriam's, 232
    Ord's, 232
    Tihen's, 231
  marsh rice, 239
  pack, 34, 251
  water, 239
    diluvial, 264
    Florida, 264–65

  Guilday's, 264
  Leonard's, 264
Redwood, 53
Reindeer, 95, 315
Reptile, 15, 16, 17, 18, 21, 26, 28, 30, 32, 35, 50, 54, 55, 56, 57, 58, 59, 64, 65, 68, 77, 79, 81, 83
Rhinoceros, woolly, 283
Rhynchothere, 346
  precursor, 346
Ringtail, 176–77
  Case's, 176
  Sonoita, 177
Rodent, 15, 16, 19, 26, 30, 37, 44, 54, 58, 59, 60, 64, 73, 75, 76, 81, 86, 149, 152, 157, 174, 177, 178, 195, 198, 209–74, 275, 357, 361
  caviomorph, 90, 91
  cricetid, 91
Ruminant, 295, 309

Sabertooth, 79, 85, 185, 186–88, 190
  false, 186, 190–91
  gracile, 186
  Idaho, 188–90
Sagebrush, 73, 234, 264, 280
Saguaro, 48
Saiga, 326–27
Salamander, 30
Scimitar-tooth, 185
Sciurid, 210–23
Sea-cow, 340–41
  Steller's, 44, 340–41
Seal, 146, 200–208
  bearded, 206
  elephant, 204, 207–8
    northern, 207–8
    southern, 207
  fur, 200, 202–3
    northern, 202–3
  gray, 206–7
  harbor, 204–5
  harp, 205–6
  hooded, 207
  monk, 207
    Caribbean, 207
  ringed, 205
Sea lion, 146, 200, 202
  California, 203–4
  Steller's, 203
Sea urchin, 161
Sedge, 251, 315, 316
Sewellel, 209–10
Shark, 13
Sheep, 326, 328–30
  bighorn, 329–30
  Dall, 328–29
  mountain, 329–30
  snow, 328
Shellfish, 84, 239; see also Crustacean
Shrew, 12, 15, 76, 77, 103–13
  Adams, 111
  arctic, 107–8
  Cudahy, 104
  desert, 113
  Dixon, 109–10
  Gidley's, 110–11
  Hagerman, 104
  Jackson's, 113
  Kansas, 106
  king, 110
  lake, 107
  Leahy, 104

  least, 111
  longtail, 108
  masked, 105–6
  Meade, 111
  meadow, 109
  Melton's, 104
  Merriam's, 109
  Mexican, 111
  minute, 109
  ornate, 106
  Powers's, 104
  pygmy, 109
  Rexroad, 103–4
  Sanders's, 104
  Saussure's, 109
  Scott, 106
  short-tailed, 112–13
    Ozark, 111–12
  smoky, 107
  southeastern, 106
  Taylor's, 104
  Trowbridge, 109
  vagrant, 106
  water, 107
    giant, 107
Shrub-ox, 166, 330–31
Sirenian, 340–42, 343
Sloth, 67, 71, 138
  ground, 34, 61, 67, 71, 81, 85, 91, 128, 135–45
    Chapadmalalan, 143
    Harlan's, 143–45
    Jefferson's, 137–38
    megalonychid, 135–38
    mylodont, 142–45
    narrow-mouthed, 136
    Rusconi's, 140–41
    shasta, 85, 141–42
    Wheatley's, 136–37
  tree, 128, 141
Snail, 51, 67
Snake, 14, 20, 30, 58, 62, 75
Solenodon, 103
Soricid, 28, 103–13
Spanish moss, 123, 247
Spider, 270
Squid, 202, 203
Squirrel, 210–23
  Allen's, 221
  antelope, 217–18
    white-tailed, 217–18
  Douglas, 222
  flying, 13, 210
    northern, 222–23
    southern, 222
    Webb's, 222
  fox, 220–21
  gray, 221
    Arizona, 221
  ground, 156, 210, 212–17
    Anita, 214
    Arctic, 215
    Benson, 213
    Booth's, 213
    California, 217
    Cochise, 214
    Columbian, 215
    Cragin's, 214
    Finlay, 214
    Franklin's, 214, 217
    Golden-mantled, 217
    Howell's, 212–13
    Johnson's, 213

Squirrel (*continued*)
    Maghee's, 213
    Meade, 214
    Melton's, 213
    Mexican, 217
    Rexroad, 213
    Richardson's, 214
    spotted, 216–17
    Taylor's, 214
    thirteen-lined, 215–16
    Townsend's, 215
    Uinta, 215
    red, 221–22
    rock, 217
    tree, 210, 220–21
Stag-moose, 317
Stegodont, 343, 349
Stegomastodont, 33, 91, 346–48
    wonderful, 346–48
Swallow, 58
Swine, 295

Talpid, 113–15
Tamarack, 63, 107
Tapir, 26, 33, 47, 83, 292–94
    California, 292–93
    Cope's, 293
    Merriam's, 292
    Vero, 293–94
Tenrec, 103
Tiger, 185
Tortoise, 83
    giant, 18, 33
    land, 19, 20
Toxodont, 91
Turkey, 85
Turtle, 13, 20, 30, 51, 69, 74, 75, 82
    box, 55, 56
        upland, 57
        lowland, 58
    giant, 16, 39
    pond, 20
    snapping, 71

Ungulate, 41, 44, 95
    even-toed, 295–339

    odd-toed, 283–94
    South American, 90, 91
Ursid, 91, 178–85, 200

Vespertilionid, 83, 116, 118–25
Vicuña, 307
Viverrid, 198
Vole, 25, 28, 34, 95, 239, 253–70
    ancient lemming, 258
    Balcones, 256–57
    boreal redback, 257
    California, 260
    Cape Deceit, 259
    Deering, 256
    field, 259
    Finney's, 255–56
    grass-eating furrowtooth, 253–54
    Guilday's, 261–62
    heather, 257, 258
    Hibbard's tundra, 259
    Kormos's steppe, 258–59
    Llano, 262
    long-tailed, 260
    Meade, 259
    meadow, 216, 259–60, 261
    Mexican, 260–61
    Monahan's, 257
    mountain, 260
    Nebraska, 256
        McGrew's, 256
        Rexroad, 256
    Osborn's, 256
    pine, 263
        Arata's, 263
        Cumberland, 263
        Hibbard's, 263
        McNown's, 263
    prairie, 262
    prime Coso, 254
    primeval, 256
    rock, 261
    Snake River, 254–55
        Frick's, 255
        Magill's, 255
        Meade, 255
        small, 254
        Taylor's, 254

    Trout Cave, 257
    yellow-cheeked, 261
Vulture, 58

Walrus, 146, 200–202
Wapiti, 66, 195, 312, 317–18
Waterfowl, 51
Weasel, 62, 146, 149–51
    least, 150–51
    long-tailed, 149–50
    Martin's, 153
    Rexroad, 149
    short-tailed, 150
Whale, killer, 205
Willow, 238
Wolf, 41, 167, 168–72
    Armbruster's, 169, 170
    dire, 64, 145, 170, 171–72, 188
    gray, 169–70, 171
    red, 167, 168, 169
    timber, 171
Wolf-coyote, 168, 169
Wolverine, 62, 153–55
    Schlosser's, 153
Woodchuck, 211–12
Woodpecker, 87
Woodrat, 10, 73, 86, 243, 251–53
    bushytail, 253
    cave, 252
    desert, 253
    dusky-footed, 253
    eastern, 252
    fossil, 252
    Mexican, 253
    Rexroad, 251–52
    southern plains, 252
    Taylor's, 252
    whitethroated, 252–53

Yak, 41, 326, 339

Zebra, 285, 289
    American, 285–87
    dolichohippine, 287
    Grévy's, 287
    lowland, 287

Index to Localities and Stratigraphic Terms

Adams, 25, 28, 39, 64
Aden Crater, 141, 404
Afton, 43, 75, 304
Aftonian Interglacial, 17, 25, 28, 33, 34, 180, 181, 264
Aguanga, 318
Aguascalientes, 272
Albiquiu, 330
Allen Cave, 58
Alton, 43, 63, 313, 330
Amchitka Island, 340
Ameca, 135
American Falls, 39, 60, 138, 191, 225, 238, 260, 282, 304, 337
Anacapa Island, 49, 243
Angus, 25, 31, 32, 167, 169, 174, 217, 218, 232, 238, 250, 257, 259, 262, 266, 271, 281, 288, 296, 304, 319, 350
Anita, 9, 10, 167, 168, 169, 199, 210, 212, 214, 278, 281
Anza Borrego Desert State Park, 6, 9, 11, 27, 272, 313
Anza Borrego Section, 7, 9, 22
Apollo Beach, 293
Arcata, 160
Arkalon, 168, 288
Arredondo, 43, 55, 118, 121, 124, 240, 245, 247, 259
Arroyo Seco, 6, 9, 11, 12, 27, 125, 220, 231, 232, 250, 277
Ashley River, 39, 78, 140, 169, 180
Asphalto, 9, 11, 12, 188, 289
Aucilla River IA, 43, 55, 195, 274, 341
Avery Island (Petite Anse Island), 43, 66, 67, 68

Back Creek Cave No. 2, 220
Baillie Islands, 326
Baker Bluff Cave, 43, 78, 107, 108, 109, 112, 113, 124, 131, 148, 149, 150, 157, 215, 220, 222, 258, 261, 271, 313, 314
Barrow Ice Cellar, 205
Bat Cave, 43, 68, 119, 120, 122, 148, 282, 300
Bautista Creek, 25, 26, 281, 290, 292, 312
Beartown Cave, 314
Beck Ranch, 162, 284, 288
Belding, 300
Bell Cave, 43, 87, 107, 120, 121, 148, 149, 217, 219, 225, 264, 267, 276, 327, 329
Bender, 9, 10, 12, 16, 250, 254, 258
Ben Franklin, 43, 80, 131, 217, 325, 364, 404
Benson, 9, 10, 12, 16, 25, 26, 113, 166, 213, 226, 229, 230, 233, 248, 249, 250, 252, 265, 277, 279, 347, 348
Berclair Terrace, 288
Berends, 25, 33, 112, 174, 234, 235, 238, 244, 259, 266
Beringia, 22, 39, 44, 92, 93, 95, 96, 153, 159, 171, 172, 190, 191, 192, 196, 222, 315, 316, 317, 326, 328, 329, 332, 334, 336, 338, 339, 350, 353, 355, 356, 358

Beringian (Alaska–Yukon) Refugium, 40, 41, 88, 182, 291, 316, 329
Berone Moore Cave, 70
Beverley Pit, 291
Big Bend, 118
Big Bone Cave, 43, 80
Big Bone Lick, 43, 66, 143, 334, 365, 404
Biharian Land Mammal Age, 95
Bindloss, 44, 47
Bishop Ash, 22
Blackwater Draw (Brown Sand Wedge), 43, 72, 131, 221, 246, 260, 320, 365, 404
Blackwater Draw (Gray Sand Unit), 43, 72, 352
Blancan–Irvingtonian boundary, 9, 22, 26, 37, 92, 95, 252
Blancan Land Mammal Age, xiii, xiv, 3, 5, 6, 7, 9, 10, 11, 12, 13, 14, 16, 17, 18, 19, 21, 22, 23, 25, 26, 27, 28, 32, 33, 34, 37, 91, 92, 95, 96, 99, 101, 103, 104, 105, 109, 110, 111, 113, 115, 123, 125, 128, 129, 130, 131, 134, 135, 136, 140, 143, 149, 150, 151, 155, 156, 157, 158, 160, 164, 165, 166, 167, 168, 169, 171, 173, 174, 175, 176, 177, 178, 179, 182, 186, 188, 190, 193, 195, 197, 199, 210, 212, 213, 214, 218, 222, 223, 225, 226, 227, 228, 229, 230, 232, 233, 235, 236, 238, 239, 240, 242, 243, 244, 247, 248, 249, 250, 251, 252, 254, 255, 256, 258, 265, 267, 269, 270, 271, 276, 277, 278, 279, 280, 281, 283, 284, 287, 288, 289, 290, 292, 293, 296, 297, 298, 299, 302, 303, 304, 307, 308, 309, 310, 319, 320, 327, 341, 344, 346, 347, 348, 357, 358, 360, 361
Blanco, 9, 12, 19, 20, 27, 104, 115, 126, 134, 136, 143, 156, 162, 167, 190, 195, 199, 210, 212, 230, 232, 233, 250, 277, 284, 287, 288, 297, 298, 302, 304, 308, 310, 346, 347
Blanco Ash, 19
Bliss Gravel Pit, 78
Bogoslof Hill, 185
Bonanza, 9, 10
Bonanza Creek, 329
Boney Spring, 43, 68, 212, 236, 271, 404
Boo Boo, 279
Bootlegger Sink, 43, 76, 107, 109, 113, 119, 124, 215, 222, 261, 271
Borchers, 9, 17, 104, 149, 162, 167, 173, 214, 231, 233, 240, 249, 250, 252, 265, 269, 271, 277, 280, 282, 289, 298, 304
Bradenton, 39, 56, 263, 274, 337
Bradford, 317
Brandon, 205
Broadwater, 9, 12, 15, 17, 18, 136, 143, 155, 156, 157, 158, 164, 167, 173, 186, 188, 210, 227, 232, 236, 270, 277, 280, 287, 298, 302, 304, 320, 347
Bruneau, 350
Brunhes–Matuyama boundary, 28, 37
Brunhes Normal Magnetic Epoch, 4, 22, 28
Brunswick Canal, 351

Brynjulfson Caves, 43, 69, 120, 131, 148, 163, 172, 217, 221, 225, 239, 273, 313, 314, 325, 364, 365, 404
Burnet Cave, 43, 73, 178, 217, 225, 229, 253, 260, 281, 282, 306, 312, 313, 323, 330, 331, 363, 365
Butler Spring, 39, 64, 65, 171, 219, 232, 242, 250
Buttonwillow, 254

Calabrian Marine Stage, 22
Calico, 356
California Wash, 9, 10, 230, 248, 250, 251, 348
Cape Blanco, 160, 204, 292
Cape Deceit, 23, 25, 34, 256, 259, 267, 275, 314, 318
Cape Hatteras, 200, 204
Carpinteria, 43, 49, 106, 109, 220
Carrier Quarry Cave, 43, 80
Carroll Cave, 70
Carter, 43, 74, 137, 148, 238, 365
Casper, 43, 87, 306, 338, 363, 404
Catclaw Cave, 329
Cave Bear Cave, 175
Cave North of Whitesburg, 43, 80
Cave on Lookout Mountain, 43, 80
Cavetown, 43, 67, 194, 221, 313
Cave Without a Name, 43, 80, 120, 131
Cedazo, 39, 40, 135, 170, 173, 180, 191, 192, 280, 287, 288, 291, 304, 321, 322
Centipede Cave, 47, 126, 164, 216, 217, 228, 251, 253
Champlain Sea, 75, 76, 201, 205, 206, 207
Channel Islands, 43, 49, 173
Channing, 188
Chapadmalalan Land Mammal Age, 13, 91, 92, 129, 143
Charleston, 274, 290
Chatanika, 276
Cherokee Cave, 43, 69, 131, 273, 300
Chimney Rock Animal Trap, 43, 54, 87, 148, 149, 150, 153, 217, 260, 276, 329, 404
China Lake, 43, 50, 404
Chipola River, 43, 59, 341
Choukoutien, 355
Cita Canyon, 9, 10, 12, 15, 19, 20, 134, 136, 143, 157, 166, 167, 175, 182, 188, 190, 193, 195, 197, 199, 230, 233, 271, 277, 287, 298, 302, 304, 308, 310, 347
Clamp Cave, 43, 80
Clark's Cave, 43, 85, 107, 108, 109, 119, 120, 124, 149, 215, 220, 222, 223, 261, 271, 280
Clear Creek, 43, 80, 113, 131, 219, 221, 244, 269, 404
Clovis, 43, 73, 306, 352, 356, 365, 404
Cochiti Magnetic Event, 11
Cochrane, 43, 44, 329, 404
Coleman IIA, 25, 27, 102, 121, 124, 130, 164, 169, 173, 175, 180, 192, 197, 221, 222, 228, 240, 247, 251, 263, 264, 273, 274, 281, 298, 309, 310
Columbus, 300

Comosi, 199
Conard Fissure, 25, 26, 31, 33, 106, 107, 109, 112, 115, 119, 122, 147, 149, 150, 151, 163, 164, 173, 183, 187, 192, 194, 197, 211, 215, 221, 228, 238, 259, 262, 263, 266, 273, 280, 281, 282, 296, 310, 318, 332
Coppell, 39, 81, 131, 251, 252, 304, 337
Corralillos Canyon, 292
Coso Mountains, 9, 11, 166, 254
Costeau Pit, 43, 50, 113, 217, 233, 241, 280
Cox Gravel Pit, 330
Cragin Quarry, 39, 55, 64, 65, 113, 123, 171, 174, 191, 192, 194, 242, 244, 245, 250, 288, 304, 319
Craighead Caverns, 43, 79
Crankshaft Cave (Pit), 43, 69, 107, 109, 112, 113, 114, 119, 120, 124, 131, 150, 163, 250, 269, 273, 282, 294
Crete, 350
Crevice Cave, 70
Cripple Creek Sump, 41, 334
Cromerian Interglacial, 4, 23, 95, 155, 170, 191, 192
Crypt Cave, 194
Crystal River Power Plant, 140, 348
Csarnotan, 93
Cudahy, 22, 25, 28, 29, 32, 35, 104, 105, 107, 109, 137, 150, 151, 169, 214, 215, 217, 225, 228, 240, 243, 249, 258, 259, 262, 266, 269, 271, 280, 298, 350
Cueva Las Cruces, 312
Cumberland Cave, 25, 30, 31, 33, 36, 39, 112, 119, 120, 122, 124, 147, 151, 153, 157, 158, 162, 163, 167, 169, 171, 173, 180, 183, 184, 192, 194, 212, 215, 220, 221, 222, 224, 235, 238, 244, 252, 257, 258, 259, 262, 263, 266, 268, 271, 273, 276, 282, 292, 296, 298, 310, 318, 330, 344
Curtis Ranch, 7, 10, 22, 25, 26, 27, 122, 134, 162, 168, 192, 194, 195, 214, 227, 231, 248, 249, 250, 251, 265, 278, 280, 282, 308, 310, 347

Dallas Sand Pits, 321
Dam (Hop-Strawn Pit), 43, 61, 62, 404
Damp Cave, 47, 216, 217
Deadman Pass Till, 10
Deering, 204
Deer Park, 9, 15, 19, 157, 167, 218, 227, 236, 254, 258, 287, 298, 304, 346, 347
Delight (Washtuckna Lake), 25, 35, 192, 318, 327, 344
Del Mar, 356
Delmont, 9, 19, 190, 236
Dent, 352
Denver, 300, 304
Devil's Den, 47, 56, 120, 121, 124, 172, 178, 188, 192, 221, 245, 247, 259, 269, 363, 364, 365
Dixon, 9, 15, 17, 20, 104, 109, 236, 255, 256, 258, 265, 267
Doby Springs, 39, 64, 65, 107, 112, 214, 244, 247, 250, 266, 271
Doeden, 39, 71
Domebo, 43, 75, 352, 365, 404
Dominion Creek, 88, 157
Donau Glaciation, 4
Donnelly Ranch, 9, 12, 55, 143, 167, 230, 250, 251, 287, 288, 293, 298, 302, 308
Dry Cave, 43, 73, 106, 109, 113, 123, 124, 164, 212, 225, 232, 243, 246, 253, 260, 264, 280, 282, 292, 329, 404
Dry Mountain, 320
Dubuque, 313
Duck Creek, 39, 64, 106, 257
Duck Point, 153
Durham Cave, 43, 76, 173
Dutchess Quarry Cave, 315
Dutton, 55

Eagle (Eagle Rock) Cave, 47, 86, 107, 149, 215, 222, 261
Early's Pit, 43, 86
Easley Ranch, 39, 81, 191, 216, 239, 240, 242, 245, 247, 250, 252, 269
Eastside Island, 10
Eden, 310
Edisto Beach, 207
Eemian Interglacial, 4, 37, 316, 328
El Casco, 136, 312
Elephant Point, 40
El Hatillo, 140
Elster Glaciation, 23, 37
Emery Borrow Pit, 43, 217, 253, 309, 325
Empress, 43, 44
Enon Sink, 43, 69, 294
Ensenadan Land Mammal Age, 91
Escapule, 352
Eschscholtz Bay, 40, 43, 316, 353
Evansville, 43, 63, 138, 364
Evil-Smelling Bluff, 45

Fairbanks I, 39, 40, 41, 314, 316, 318, 328, 334, 353
Fairbanks II, 40, 41, 43, 137, 150, 152, 153, 154, 157, 167, 174, 215, 238, 261, 267, 276, 291, 316, 326, 339, 353, 354, 365, 404, 405
Finch, 206
First American Bank Site, 43, 79, 188, 364
Fort Fisher, 341
Fort Qu'Appelle, 39, 78, 317
Fort Rock Cave, 356
Fossil Lake, 43, 76, 171, 225, 238, 260, 298
Fox Canyon, 9, 12, 15, 16, 103, 104, 110, 111, 123, 166, 210, 212, 213, 226, 227, 232, 233, 240, 242, 247, 248, 249, 254, 255, 256, 270, 271, 310
Frankstown Cave, 43, 76, 124, 163, 282, 313, 334
Friesenhahn Cave, 43, 81, 82, 113, 190, 219, 234, 243, 297
Fyllan Cave, 257, 258

Galena, 300
Galveston, 140
Gardner Cave, 43, 86
Gauss Normal Magnetic Epoch, 4, 6, 9, 10, 12, 14, 15, 16, 20
Gidley Locality, 25, 278
Gilbert Reversed Magnetic Epoch, 4, 6, 9, 11, 16, 287
Gilliland, 25, 33, 34, 35, 129, 134, 141, 168, 190, 194, 264, 280, 289, 290, 291, 293, 298, 308, 310, 321, 347, 348
Giltner, 317
Glass Cave, 43, 66
Glendale, 43, 71, 225, 234, 253, 260, 329
Glyptotherium Locality, 25
Gold Hill, 41, 275, 334
Gold Run Creek, 43, 88, 157, 288, 291, 316, 335, 353
Goleta, 199
Golf Course Locality, 57
Goodland, 300
Gordon, 25, 32, 180, 187, 192, 238, 266, 288, 298, 304
Grand View, 9, 13, 114, 136, 155, 156, 158, 166, 167, 178, 188, 232, 236, 238, 254, 265, 269, 272, 278, 287, 297, 302, 344, 347
Grapevine Cave, 180
Grassy Cove Saltpeter Cave, 80
Gray Point, 10
Grayson, 290
Green Saddle, 10
Green's Creek, 43, 75, 205
Guaje Ash, 19
Guanajuato, 284
Günz Glaciation, 4, 267, 334, 335
Guy Wilson Cave, 43, 80, 215, 258, 314, 405
Gypsum Cave, 43, 71, 329, 363, 365, 405

Hagerman, 6, 9, 13–15, 16, 20, 103, 104, 110, 114, 136, 149, 155, 156, 157, 158, 161, 166, 167, 178, 182, 186, 195, 199, 210, 212, 223, 226, 230, 233, 235, 238, 244, 247, 252, 254, 265, 277, 278, 287, 297, 302, 304, 319, 344
Haile VII A, 39, 56
Haile VIII A, 39, 56, 171, 173, 247, 251
Haile IX, 247
Haile XI B, 39, 56, 106, 118, 124, 240, 245, 247, 251
Haile XII A, 164
Haile XIV A, 43, 56
Haile XV A, 9, 13, 56, 111, 115, 129, 130, 143, 158, 186, 222, 238, 250, 280, 287, 292, 296, 308, 310
Haile XVI, 137, 140
Hand Hills, 43, 44, 219
Hartman's Cave, 173, 238, 297
Hawken, 338
Hawver Cave, 43, 50, 114, 209, 217, 246, 253, 330
Hay Springs, 25, 32, 37, 143, 167, 170, 180, 187, 218, 259, 266, 288, 304, 310, 319, 322
Hemphillian–Blancan boundary, 16, 93
Hemphillian Land Mammal Age, xiii, 6, 9, 11, 12, 16, 91, 110, 111, 113, 115, 136, 137, 158, 166, 167, 168, 173, 174, 179, 193, 196, 199, 210, 211, 222, 226, 227, 229, 232, 234, 248, 252, 253, 254, 275, 284, 285, 287, 302, 304, 308, 325, 327
Herculaneum, 43, 70, 247, 281, 282
Hickman, 300
Hill-Shuler, 43, 81, 131, 221, 304, 321
Historic Bluff, 40
Holloman, 25, 33, 34, 35, 134, 219, 288, 289, 290, 298, 302, 347, 350
Holocene Interglacial, 4, 39, 44, 45, 56, 59, 62, 63, 69, 70, 71, 72, 73, 75, 80, 83, 84, 86, 87, 89, 102, 115, 120, 122, 126, 131, 148, 150, 153, 154, 157, 163, 164, 169, 175, 178, 184, 192, 196, 200, 205, 228, 229, 245, 247, 251, 253, 258, 261, 264, 288, 291, 296, 297, 303, 312, 316, 317, 318, 328, 329, 338, 343, 345, 349, 357, 362, 363

INDEX TO LOCALITIES

Holsteinian Interglacial, 4, 37
Horned Owl Cave, 43, 87, 217, 219, 220, 264, 276, 327, 329
Hornsby Springs, 47, 57, 130, 145, 364, 365
Horse Quarry, 14, 287
Hosterman's Pit, 47, 77, 405
Houston, 130
Howard Ranch, 43, 82, 107, 112, 217, 225, 232, 239, 242, 405
Huayquerian Land Mammal Age, 91
Hudspeth, 9, 20, 134, 136, 166, 213, 214, 227, 250, 287, 288, 293, 302, 310
Hull, 205
Hunker Creek, 291, 353
Hydro, 330

Ichetucknee River, 43, 57, 195, 238, 247, 274
Illinoian Glaciation, 4, 5, 26, 27, 28, 30, 31, 32, 33, 35, 37, 39, 40, 41, 55, 56, 61, 64, 65, 75, 78, 106, 107, 123, 130, 131, 137, 148, 151, 152, 169, 170, 171, 214, 232, 239, 240, 241, 242, 244, 247, 249, 250, 251, 252, 260, 261, 264, 266, 269, 271, 299, 314, 316, 318, 323, 326, 328, 329, 332, 334, 335, 337
Ingleside, 39, 82–83, 178, 219, 294, 309, 320, 348
Inglis I A, 25, 27, 112, 115, 129, 130, 134, 136, 140, 143, 164, 186, 190, 199, 221, 274, 281, 292, 298, 310, 320, 344
Irvington, 7, 22, 25, 26, 167, 169, 180, 187, 190, 192, 213, 225, 243, 248, 304, 312, 321, 330
Irvingtonian Land Mammal Age, xiii, 3, 5, 6, 7, 10, 11, 12, 22, 23, 25, 26, 27, 28, 30, 31, 32, 33, 34, 35, 37, 39, 64, 91, 92, 95, 96, 102, 104, 107, 109, 111, 112, 113, 114, 115, 119, 120, 121, 122, 124, 128, 129, 130, 134, 135, 136, 137, 140, 141, 142, 143, 147, 150, 151, 153, 158, 159, 160, 163, 164, 166, 167, 168, 169, 170, 171, 172, 173, 174, 175, 178, 180, 181, 183, 186, 187, 190, 192, 193, 194, 195, 196, 197, 199, 211, 213, 215, 216, 218, 219, 220, 221, 224, 225, 226, 227, 228, 230, 231, 232, 234, 236, 238, 240, 243, 244, 248, 249, 250, 251, 252, 256, 257, 258, 259, 261, 262, 263, 264, 265, 266, 267, 268, 269, 271, 272, 273, 274, 275, 276, 280, 281, 282, 288, 289, 290, 291, 292, 293, 296, 297, 298, 302, 304, 308, 309, 310, 312, 314, 318, 320, 321, 326, 330, 331, 332, 333, 339, 343, 344, 347, 348, 349, 350, 351, 357
Irvingtonian–Rancholabrean transition, 37
Island Bluff, 45
Isleta Cave, 43, 73, 119, 125, 153, 212, 216, 219, 225, 232, 233, 234, 250, 252, 264, 329

Jackass Butte, 236, 272
Jaguar Cave, 43, 55, 61, 148, 153, 171, 184, 191, 215, 217, 220, 225, 263, 264, 267, 276, 279, 280, 282, 291, 329, 356, 363, 364, 365, 405
Java, 25, 34, 153, 231, 238, 258, 266, 271, 280, 281, 298
Jinglebob, 39, 65, 112, 232, 239, 240, 244, 249, 262, 269, 271

Jinks Hollow, 334
Jones, 43, 65, 217
Joshua Creek, 43, 59
Jupiter Inlet, 43, 59

Kaena Reversed Magnetozone, 10
Kanopolis, 25, 29, 129, 243, 244, 252, 259, 260, 264, 293
Kansan Glaciation, 4, 25, 26, 27, 28, 30, 32, 33, 34, 36, 45, 69, 105, 107, 112, 153, 169, 180, 214, 240, 243, 244, 251, 257, 259, 266, 269, 302, 317, 332
Kansas River, 287
Keefe Canyon, 9, 16, 136, 166, 167, 235, 249, 298, 302
Kendrick I A, 43, 57, 121
Kentuck, 25, 29, 30, 39, 219, 228, 251, 258, 259, 262, 266, 269
Kettleman Hills, 160, 238
Kimmswick, 43, 70
Kincaid Shelter, 43, 83, 131
Klamath River, 330
Klein Cave, 47, 83, 113, 119, 120, 122, 150, 225, 232, 240, 242, 246
Kokolik River, 201
Kokoweef Cave, 43, 50, 329, 330
Kougechuck Creek, 202
Kyle, 164, 221, 325
Kyle Quarry Cave, 43, 79

Ladds, 43, 60, 107, 120, 131, 148, 164, 178, 194, 195, 220, 244, 264, 268, 269, 280
Laguna Beach, 356
Lake Simcoe, 184
La Mirada, 43, 51, 172, 365, 405
Laubach Cave, 43, 83, 120, 135, 300, 364, 405
Layer Cake, 6, 9, 11, 27, 250, 277, 279
Lecanto Cave, 58
Lehner Ranch, 43, 47, 48, 292, 352, 356, 365, 405
Leikem, 352
Levi Shelter, 43, 83, 405
Lewisville, 81, 245
Lincoln Bight, 203
Lindenmeier, 365
Lisco, 9, 17, 18, 136, 166, 167, 171, 188, 227, 277, 287
Little Box Elder Cave, 43, 55, 87, 107, 109, 119, 121, 148, 153, 157, 174, 182, 184, 217, 219, 220, 260, 264, 267, 276, 282, 312, 313, 327, 329, 363
Little Kettle Creek, 43, 60
Little Salt River Cave, 43, 80
Livermore, 305
Longhorn Cavern, 43, 83, 113, 234, 240
Los Angeles River, 356
Lost Chicken Creek, 43, 192, 291, 328, 353, 364, 365, 405
Loup Fork, Nebraska, 347, 351
Loup Fork, Oregon, 318
Lower Carp Lake, 137
Lower Cleary Creek, 41, 334
Lubbock Lake, 43, 83, 182, 363, 405
Lujanian Land Mammal Age, 91

Mabel, 136
McGee Till, 10
McKittrick, 43, 51, 106, 125, 217, 232, 233, 253, 280, 320, 330, 405
McLeod, 137, 169, 264
McPherson *Equus* beds, 30

Magdalenian, 171
Malaspina, 205
Mammoth Cave, 43, 66, 126
Mammoth Magnetic Event, 10
Manix, 43, 51
Maricopa, 43, 51, 325
Matuyama Reversed Magnetic Epoch, 4, 6, 9, 10, 11, 13, 17, 19, 20, 22, 26, 27
Mazama Ash, 62
Meade County, Kansas, xiii, 6, 64, 75
Meadowview Cave, 43, 86
Medford Cave, 131
Medicine Hat (late Nebraskan–early Kansan, Units D and E), 25, 34, 141, 238, 288, 290, 304, 308, 350
Medicine Hat (Yarmouthian, Unit G), 25, 219, 290
Medicine Hat (Sangamonian, Unit K), 39, 45, 137, 152, 174, 196, 219, 225, 280, 282, 318, 329, 337, 356
Medicine Hat (early Wisconsinan, Unit M), 43, 45, 219
Medicine Hat (middle Wisconsinan, Unit O), 43, 45, 47, 162, 282
Medicine Hat (late Wisconsinan, Unit R), 43, 45
Medicine Hat (early Holocene, Unit T), 45, 354
Melbourne, 43, 57, 59, 126, 195, 207, 274, 281, 293
Mendevil, 10
Merritt Island, 57, 195
Mesa De Maya, 12, 39, 55, 219, 243, 244
Meyer Cave, 47, 63, 112
Miami, 352
Middle Butte Cave, 47, 62, 154, 167, 279
Mifflin, 298
Miller Creek, 88
Miller's Cave, 47, 84, 120, 131, 364
Mináca Mesa, 168
Mindel Glaciation, 37, 334
Minidoka, 304
Mitchell Bluff, 45, 405
Mockingbird Gap, 352
Monterey Bay, 340
Moonshiner Cave, 47, 62, 148, 149, 150, 153, 154, 157, 167, 174, 184, 215, 217, 220, 234, 245, 260, 264, 279
Moonstone Beach, 160
Moore Pit, 43, 81, 153, 293, 304, 321, 337, 364, 365
Mosherville, 301, 364
Mt. Eden, 302
Mt. Scott, 39, 64, 65, 106, 107, 112, 123, 235, 238, 239, 244, 252, 266, 269, 271
Muaco, 348
Mullen I, 25, 32, 196, 236, 238, 256, 257, 259, 265, 266, 281, 298, 347
Mullen II, 25, 32, 115, 167, 170, 194, 218, 228, 238, 280, 282, 293, 302, 304, 323, 332, 344
Murray Springs, 43, 47, 292, 352, 356, 365

Naco, 43, 48, 352
Natchez, 43, 67, 293
National City, 292
Natural Chimneys, 43, 86, 107, 109, 112, 114, 119, 120, 122, 148, 149, 150, 215, 221, 222, 261, 271, 282
Natural Trap Cave, 43, 88, 153, 194, 329, 364, 405
Nebraskan Glaciation, 4, 28, 136

Nehawka, 339
New Cave, 126
New Paris No. 4, 43, 77, 86, 107, 108, 109, 112, 113, 119, 120, 148, 149, 150, 215, 221, 222, 223, 245, 257, 261, 263, 267, 271, 272, 282, 296, 364, 405
Newport Bay Mesa, Loc. 1066, 39, 51, 52, 54, 125, 160, 207, 217, 225, 292, 337
Newport Bay Mesa, Loc. 1067, 43, 51, 52, 106, 113, 114, 232, 233, 242, 243, 253, 280
Newport Lagoon, 203
Nichol's Hammock, 126, 169
Nome, 334
North Havana Road, 43, 59
North Point, 203
Nuglungnugtuk Estuary, 205

Old Crow River, Loc. 11 A, 43, 88, 148, 153, 171, 332, 336
Old Crow River, Loc. 14N, 43, 88, 89, 153, 172, 275, 282, 315, 356
Old Crow River, Loc. 44, 39, 88, 89, 261, 276, 282
Old Crow River (miscellaneous localities), 137, 152, 153, 174, 238, 316, 334, 353
Olduvai Magnetic Event, 9, 10, 13, 22, 34, 350
Olsen-Chubbuck, 405
111 Ranch, 11, 197
Organ-Hedricks Cave, 43, 87, 119, 120, 122, 131
Orr Cave, 152
Ottawa, 43, 75
Overpeck, 184
Owl Cave, 62

Palos Verdes Sand, 203, 204
Papago Springs Cave, 43, 48, 119, 120, 124, 125, 126, 176, 212, 217, 234, 253, 260, 319, 323
Parsnip River, 329
Pashley, 291
Patton Cave, 120
Payne's Prairie, 39, 57
Pearlette Ash, 30, 32, 33, 35
Pearlette Ash Type "B", 9, 22
Pearlette Ash Type "O", 22, 28
Pearlette Ash Type "S", 22, 32
Peccary Cave, 43, 49, 107, 109, 112, 113, 114, 131, 148, 150, 157, 215, 221, 223, 258, 300, 313
Perkins Cave, 70
Pikimachay Cave, 356
Plattsburgh, 207
Playa Del Ray, 204
Pleistocene, early, 32, 129, 132, 160, 161, 162, 168, 169, 172, 204, 238, 256, 258, 275, 308, 343, 346, 348, 349, 355, 360
Pleistocene, middle, 23, 44, 95, 152, 153, 172, 184, 185, 190, 205, 206, 211, 259, 275, 315, 316, 318, 348, 350, 351, 354
Pleistocene, late, 41, 44, 48, 51, 54, 60, 61, 67, 70, 71, 72, 73, 76, 77, 78, 79, 80, 85, 86, 87, 91, 95, 102, 108, 109, 119, 124, 126, 131, 132, 135, 136, 148, 150, 151, 152, 164, 170, 174, 176, 177, 178, 182, 183, 185, 203, 204, 228, 233, 234, 244, 253, 268, 283, 288, 289, 295, 300, 301, 319, 321, 325, 328, 329, 348, 351, 357, 358, 360, 361, 362

Pleistocene Epoch, xiii, xiv, 3, 5, 22, 28, 30, 52, 59, 60, 80, 81, 84, 85, 88, 90, 92, 95, 101, 102, 103, 106, 109, 111, 113, 114, 116, 117, 118, 120, 121, 122, 124, 126, 127, 128, 129, 130, 131, 135, 137, 141, 143, 146, 147, 153, 161, 162, 165, 167, 168, 170, 171, 174, 175, 176, 178, 179, 180, 182, 184, 185, 186, 189, 190, 191, 192, 197, 200, 202, 203, 205, 207, 208, 210, 211, 212, 215, 217, 220, 221, 232, 233, 234, 235, 236, 237, 238, 239, 247, 257, 258, 259, 269, 270, 274, 275, 281, 283, 284, 285, 287, 288, 290, 291, 292, 295, 301, 307, 309, 310, 312, 315, 319, 324, 325, 326, 328, 339, 340, 341, 343, 344, 355, 357, 358, 360, 365
Pleistocene faunas, xiii, xv, 39, 50, 52, 59, 60, 66, 68, 70, 73, 75, 81, 85, 87, 95, 112, 130, 151, 164, 180, 191, 219, 220, 221, 222, 319, 344, 365
Pleistocene mammals, xiii, xiv, xv, 40, 43, 99, 118, 131, 132, 150, 164, 167, 168, 171, 182, 184, 185, 186, 187, 189, 190, 191, 209, 275, 281, 283, 284, 287, 290, 291, 292, 293, 295, 296, 300, 301, 307, 308, 309, 310, 312, 314, 319, 323, 324, 344, 358, 361, 365
Pleistocene megafauna, 57, 59, 61, 63 71, 76, 78, 172, 314, 357, 358, 362, 363
Pliocene, xiii, 93, 102, 103, 111, 118, 129, 136, 140, 153, 156, 158, 161, 166, 174, 182, 188, 190, 200, 207, 222, 230, 235, 253, 254, 256, 274, 277, 278, 279, 283, 296, 308, 309, 310, 319, 324, 325, 327, 339, 340, 341, 348, 349, 350, 355, 361
Pliocene, late, xiii, 103, 113, 132, 150, 157, 159, 160, 161, 162, 172, 218, 222, 238, 247, 272, 285, 301, 319, 340, 341, 342, 348, 360
Pliocene–Pleistocene boundary, 5
Point Barrow, 160, 161
Polecat Creek, 43, 63, 184
Pontian, 147, 159
Pool Branch, 25, 28
Porcupine River, 353
Port Kennedy Cave, 25, 31, 33, 112, 136, 137, 147, 153, 157, 158, 163, 164, 168, 169, 173, 180, 183, 186, 192, 194, 195, 197, 238, 262, 264, 266, 271, 272, 275, 280, 289, 293, 296, 310, 344
Post Ranch, 10
Potter Creek Cave, 43, 50, 52, 114, 118, 125, 141, 148, 181, 184, 209, 217, 220, 222, 223, 225, 226, 327, 330
Powder Mill Creek Cave, 70, 405
Pre-Nebraskan Alpine Glaciation, 12, 15, 17, 55, 149, 162, 214, 218
Proctor Pits, 136
Prospect, 26
Puebla De Valverde, 169
Punta Gorda, 25, 28, 129, 197, 290, 310, 341, 348, 350
Püspökfürdö (= Betfia), 153

Quesnel Forks, 137, 327
Quitaque, 43, 84

Rainbow Beach, 43, 61, 62, 219, 225, 279, 337, 365, 405
Rampart Cave, 43, 48, 142, 212, 327, 329, 364, 365, 405
Rancho La Brea, 43, 51, 52–54, 106, 137, 143, 144, 145, 167, 170, 172, 184, 187, 188, 191, 194, 217, 232, 233, 243, 250, 280, 289, 290, 292, 305, 320, 321, 325, 337, 406
Rancholabrean Land Mammal Age, 3, 5, 12, 22, 28, 33, 37, 39, 49, 50, 51, 52, 57, 59, 64, 71, 75, 78, 80, 91, 92, 95, 96, 99, 107, 113, 115, 122, 128, 129, 130, 131, 134, 135, 137, 140, 143, 154, 160, 162, 167, 168, 169, 170, 171, 178, 180, 181, 182, 183, 190, 191, 192, 194, 195, 197, 214, 217, 222, 224, 225, 228, 229, 238, 240, 243, 245, 247, 250, 251, 253, 257, 259, 260, 261, 262, 263, 264, 266, 267, 271, 276, 279, 280, 281, 282, 287, 288, 291, 292, 293, 296, 297, 298, 304, 305, 308, 310, 312, 313, 318, 319, 321, 330, 335, 339, 344, 348, 350, 351, 353, 355, 357, 360
Rapps Cave, 276
Red Corral, 10, 12, 167, 236, 250
Reddick, 39, 57, 118, 121, 122, 124, 126, 164, 171, 173, 184, 194, 195, 228, 239, 240, 245, 247, 251, 263, 274, 281
Red Light, 9, 20, 134, 136, 143, 157, 166, 173, 227, 230, 233, 249, 284, 288, 290, 298, 302, 304, 310
Red Willow, 43, 71, 329
Renick, 298
Rexroad, 9, 10, 12, 14, 16, 19, 104, 113, 115, 136, 149, 151, 155, 157, 158, 161, 162, 166, 167, 173, 175, 176, 186, 195, 199, 226, 227, 230, 233, 236, 242, 247, 248, 249, 250, 252, 254, 256, 277, 278, 279, 287, 298, 308, 310, 347, 348
Rexroad 3, 212, 213, 235, 271
Rezabek, 25, 30, 259, 262, 264, 280
Ricardo, 188
Riddell, 78
Rincon, 158
Riss Glaciation, 329, 335
Riverside Cave, 43, 80
Robert, 39, 43, 64, 107, 112, 406
Robinson Cave, 43, 79, 107, 108, 109, 112, 119, 120, 124, 131, 148, 149, 150, 215, 221, 222, 223, 271, 313
Rock Creek, 25, 34, 35, 134, 167, 169, 172, 180, 181, 288, 290, 291, 298, 304, 332
Rock Springs, 39, 58, 116, 195, 341
Rome Beds, 192
Roosevelt Lake, 43, 78
Ruscinian Land Mammal Age, 93, 153, 173, 177, 182, 191, 196, 199
Rushville, 25, 32, 187, 218, 238, 266, 281, 288, 298, 304, 310

Saale Glaciation, 37
Sabertooth Cave, 43, 58, 224, 274, 293, 312
Sacramento, 114
Saltpeter Cave, 43, 80, 178
Saltville, 43, 86, 406
Samwel Cave, 43, 50, 54, 148, 184, 209, 217, 222, 223, 225, 226, 282, 327, 330
Sandahl, 25, 30, 214, 219, 234, 238, 240, 244, 271, 302
Sand Draw, 9, 12, 15, 17, 18, 19, 104, 109, 115, 155, 156, 157, 158, 162, 166, 167, 188, 213, 227, 230, 235, 236, 242, 249, 250, 254, 255, 256, 258, 277, 287, 288, 298, 302, 304, 347
Sanders, 9, 12, 15, 20, 104, 228, 231, 233, 249, 250, 255, 258, 265, 271, 284, 287

INDEX TO LOCALITIES 437

Sand Hollow, 318
Sandia, 306
Sand Point, 238
Sangamonian Interglacial, 4, 31, 35, 37, 39, 40, 45, 55, 56, 58, 59, 61, 64, 65, 66, 67, 70, 74, 76, 78, 80, 81, 83, 85, 89, 106, 113, 116, 118, 122, 123, 124, 125, 126, 131, 152, 170, 171, 173, 174, 191, 194, 195, 196, 197, 201, 203, 205, 220, 224, 225, 228, 229, 239, 240, 241, 242, 244, 245, 247, 249, 251, 252, 253, 260, 263, 269, 271, 280, 281, 282, 293, 300, 312, 316, 317, 318, 327, 328, 329, 335, 336, 337, 338
San Josecito Cave, 43, 73, 74, 109, 111, 117, 118, 123, 124, 141, 164, 172, 176, 178, 183, 192, 194, 195, 212, 216, 221, 225, 229, 235, 242, 246, 252, 260, 268, 272, 281, 312, 319, 322, 327
San Miguel Island, 49, 352
San Nicholas Island, 204
San Pedro, 39, 54, 160, 253, 280, 337
San Pedro Valley, 6, 7, 10, 22, 25, 27, 47, 285
Santa Cruz Island, 49
Santa Fe River I B, 9, 13, 101, 129, 130, 134, 136, 143, 155, 167, 186, 236, 250, 287, 292, 296, 298, 308, 310, 320, 341, 344, 346
Santa Fe River II A, 59, 192, 238, 274, 341
Santa Rosa Island, 49, 50, 160, 243, 352, 365, 406
San Timoteo, 9, 12
Sappa, 25, 32, 269, 271
Saskatoon Area, 39, 78
Savage Cave, 43, 66, 221
Sawrock Canyon, 16, 115, 232, 235, 252, 254
Scharbauer, 406
Schuiling Cave, 43, 54, 320
Schulze Cave, 43, 84, 106, 112, 113, 119, 120, 123, 150, 164, 184, 195, 216, 217, 225, 232, 234, 239, 240, 242, 246, 248, 271, 282, 406
Sebastian Canal, 43, 59, 180
Seger, 9, 17, 287, 347
Selby, 55
Seminole Field, 43, 59, 135, 228, 273, 274, 341
Shelter Cave, 322
Shishimaref Inlet, 203
Silver Creek, 39, 85, 107, 150, 196, 215, 220, 225, 279
Sims Bayou, 43, 85, 269
Sinton, 140, 274, 348
Skiddaway Island, 140
Slaton, 25, 35, 106, 123, 130, 131, 171, 216, 232, 234, 242, 244, 249, 252, 264, 282, 288, 291, 304, 310, 319, 321
Slaughter Canyon Cave, 312
Smith Creek Cave, 43, 72, 148, 276, 327, 365, 406

Snake Creek, 166
Spring Creek, 15
Stanton Cave, 327, 329
Sullivan Gravel Pit, 330
Surprise Bluff, 45
Süssenborn, 315, 335
Sutherland Gravel Pit, 78

Taber Child Site, 43, 47, 356, 406
Talara, 187, 191
Taunton, 9, 21, 177
Taylor Bayou, 43, 85
Tehama, 9, 12
Tehuacán, 363
Tequixquiac, 39, 67, 135, 180, 320, 321, 330
Thurman, 112
Timm's Point, 160
Tlapacoya, 312, 356
Tofty Placer District, 43, 44
Toronto, 39, 76, 317
Trail Creek Cave, 356
Trollinger Spring, 43, 68, 70, 406
Trout Cave, 25, 35, 107, 114, 121, 122, 124, 183, 197, 212, 215, 221, 222, 224, 238, 244, 252, 257, 258, 261, 264, 266, 268, 273, 276
Tule Springs, 43, 72, 260, 279, 306
Turlin, 330
Tusker, 9, 11, 134, 232, 274, 278, 281, 348

Union Pacific Mammoth Kill Site, 352, 365
Upper Cleary Creek, 333
Upper Etchegoin, 310
Upper Hunker Creek, 43, 88
Upper Porcupine River, 334
Uquian Land Mammal Age, 91, 172

Vallecito Creek, 6, 9, 11, 12, 22, 25, 27, 113, 137, 141, 167, 168, 178, 186, 225, 227, 231, 232, 249, 251, 269, 272, 280, 281, 298, 302, 308, 330
Valley of Mexico, 288, 289, 291
Valsequillo, 186, 356
Ventana Cave, 43, 48, 217, 219, 293, 322, 356, 365, 406
Vera, 25, 34, 35, 259, 262, 266
Vera Cruz, 135
Vero, 43, 59, 121, 122, 123, 228, 240, 245, 274, 281
Villafranchian Land Mammal Age, 93, 95, 167, 168, 169, 170, 174, 175, 177, 182, 184, 190, 191, 192, 193, 196, 199, 211, 237, 257, 263, 267, 315, 317, 318, 328, 334, 335, 345

Waccasassa River, 39, 59, 140, 180, 195, 259, 273, 341

Wakulla Springs, 43, 59
Warm Springs I, 43, 71, 264
Wasden, 43, 62, 150, 215, 220, 225, 232, 234, 264, 279, 280
Wathena, 258, 259, 262
Watkin's Quarry, 43, 60, 140, 364
Weichsalian Glaciation, 185
Wellsch Valley, 22, 25, 34, 166, 214, 259, 298, 330, 350
Welsh Cave, 43, 66, 107, 109, 157, 184, 215, 228, 229, 282, 300, 406
Wendell Fox, 235
White Bluffs, 9, 21, 114, 136, 143, 155, 166, 167, 182, 210, 223, 238, 252, 277, 279, 287, 297, 302, 309, 344
White Rock, 9, 17, 104, 105, 115, 136, 213, 218, 230, 233, 236, 238, 240, 243, 249, 250, 252, 255, 256, 265, 267, 271, 287, 298, 302, 308
Whitesburg, 43, 80, 221, 313
Wichman, 9, 18
Wilber Creek, 276
Williams (Kansas), 106
Williams Cave, Texas, 43, 85, 217, 219, 229
Williston III A, 39, 164, 251, 263
Wilson Butte Cave, 43, 63, 148, 355, 406
Winding Stairs Cave, 43, 86
Windy Mouth Cave, 120
Wisconsinan Glaciation, 4, 34, 36, 37, 39, 40, 41, 44, 45, 47, 49, 50, 51, 52, 56, 57, 59, 60, 61, 62, 63, 65, 66, 67, 69, 70, 71, 72, 74, 75, 76, 78, 79, 80, 81, 83, 84, 85, 87, 89, 93, 95, 101, 109, 114, 117, 118, 119, 120, 121, 124, 128, 131, 135, 136, 137, 138, 148, 149, 150, 151, 152, 153, 154, 157, 160, 162, 163, 164, 168, 171, 173, 174, 175, 177, 180, 182, 183, 184, 190, 191, 192, 196, 201, 205, 206, 209, 212, 215, 219, 221, 222, 225, 226, 228, 229, 236, 239, 242, 244, 245, 246, 253, 258, 260, 261, 264, 267, 268, 269, 270, 271, 272, 280, 281, 282, 288, 290, 291, 296, 297, 300, 303, 308, 309, 313, 314, 315, 316, 317, 318, 326, 327, 329, 330, 332, 333, 334, 335, 336, 337, 338, 344, 352, 356, 357, 358, 363
Withlacoochee River, 39, 59, 259, 341
Wolfe City, 135
Wolf Ranch, 249
Womack Gravel Pit, 207
Woodbridge, 354
Woodpecker Bluff, 47
Woolper Creek Deposit, 66

Yarmouthian Interglacial, 4, 17, 29, 30, 31, 33, 35, 137, 160, 170, 244, 264, 266, 335
Yuba City, 356

Zoo Cave, 43, 70, 131, 300
Zuma Creek, 43, 54, 292

INDEX TO AUTHORS

Adam, E. K., 62, 87, 226, 260, 264, 276, 280, 375
Adams, D. B., xvi, 88, 193, 194, 195, 330, 364, 367, 387, 405
Adams, R. McC., 70, 367
Aguirre, E., 350, 367
Akersten, W. A., xvi, xvii, 21, 53, 54, 173, 196, 223, 226, 227, 229, 288, 367, 372
Alexander, H. L., Jr., 83, 367
Alekseev, M. N., 95, 367
Allen, G. M., 126, 127, 367
Allen, G. T., 57, 371
Allen, J. A., 313, 314, 332, 333, 334, 367
Allison, I. S., 76, 225, 367
Alt, D., 55, 367
Alvarez, T., 313, 367
Anderson, E., 43, 61, 62, 88, 148, 149, 151, 152, 153, 154, 155, 157, 158, 171, 174, 182, 260, 291, 315, 337, 364, 365, 367, 385, 387, 405
Anderson, S., xvi
Arata, A. A., 176, 367
Armstrong, D. M., 109, 113, 367
Armstrong, R. L., 9, 10, 14, 15, 367, 371
Auffenberg, W., 55, 56, 57, 58, 59, 131, 367
Ayer, M. Y., 85, 367
Azzaroli, A., 315, 316, 368

Bada, J. L., 4, 356, 368, 398
Bader, R. S., 55, 57, 130, 368
Baird, D., xvi
Bandy, O. L., xiii, 5, 22, 23, 368
Banfield, A. W. F., 314, 315, 368
Bannikov, A. G., 327, 368
Barbour, E. H., 18, 325, 335, 339, 352, 368
Barbour, R. W., 117, 119, 120, 121, 122, 123, 124, 125, 126, 368
Barnes, L. G., 204, 205, 208, 368
Barton, J. B., 61, 368, 404
Baskin, J. A., 248
Bateman, G. C., 116, 117, 399
Bayrock, L. A., 291, 337, 373
Beaumont, G. de, 199, 368
Beebe, B., xvi, 89, 171
Bender, M. S., 375
Beninde, J., 318, 368
Benninghoff, W. S., 86, 392
Bensley, B. A., 368
Berger, R., 51, 54, 365, 368, 381, 405, 406
Berggren, W. A., xiii, 4, 5, 22, 23, 377
Bjork, P. R., xvii, 13, 15, 104, 110, 111, 149, 151, 155, 156, 158, 161, 166, 167, 180, 183, 195, 368, 381
Black, C. C., xvi, 84, 211, 212, 219, 368, 405
Blake, W. P., 368
Bode, F. D., 73, 397
Bogan, A. E., 79, 391
Bojanus, L. H., 368
Bonifay, M.-F., 167, 368
Bonnichsen, R., 61, 382
Borchardt, G. A., 19, 382

Bossert, W. H., 171, 385
Bovard, J. F., 368
Brantley, A. G., 369
Breyer, J., 302, 308, 369
Brodkorb, P., 13, 15, 55, 57, 58, 369
Brooks, H. K., 55, 56, 367
Brothwell, D., 356, 369
Brown, B., 26, 109, 135, 143, 145, 148, 164, 228, 282, 369
Bryan, A. L., 356, 369
Bryant, V. M., Jr., 363, 365, 369
Burleigh, R., 356, 369
Burt, W. H., xvi, 111, 215, 217, 220, 221, 222, 223, 232, 233, 234, 245, 369
Butler, B. R., 338, 369
Buwalda, J. P., 51, 369

Cahn, A. R., 130, 369
Cain, C. H., 190
Carr, W. J., 61, 369
Carson, L. C., 261, 377
Carter, D. C., 112, 113, 383
Chaffee, R. G., 319, 321, 324, 370
Chamberlain, T. C., 3, 369
Chandler, A. C., 321, 325, 369
Chantell, C. J., 15, 369
Choate, J. R., 83, 120, 369
Christian, W. G., 198, 199, 397
Churcher, C. S., xvi, xvii, 25, 34, 44, 47, 89, 167, 172, 174, 184, 190, 280, 291, 318, 336, 337, 350, 356, 369, 370, 396, 398, 404
Clark, D., 4, 373
Clark, T. W., 219, 370
Clemens, W. A., Jr., xvi
Clulow, F. V., 88, 182, 288, 291, 317, 336, 377
Clutton-Brock, J., 172, 173, 174, 370
Cockerell, T. D. A., 144, 145, 370
Colbert, E. H., 319, 321, 324, 370
Coleman, D. C., 370, 404
Compton, L. V., 106, 113, 370
Condie, K. C., 330, 397
Cook, H. J., 33, 370, 378, 379
Cooke, H. B. S., 4, 37, 39, 370
Cooper, B. N., 86, 392
Cope, E. D., xiii, 19, 33, 35, 136, 137, 148, 153, 155, 156, 159, 164, 167, 169, 186, 195, 262, 275, 298, 370
Corbet, G. B., 172, 173, 174, 370
Corner, R. G., 71, 370
Courtright, M., xvi
Cox, A., 10, 370
Crompton, A. W., xvi
Crook, W. W., 81, 370
Crusafont, M., 169, 385
Cuffey, R. J., 260, 370
Cummins, W. F., 19
Curry, R. P., 10, 370
Cushing, J. E., 229, 281, 370

Dalquest, W. W., xvi, xvii, 19, 33, 34, 35, 40, 81, 82, 84, 112, 115, 126, 131, 132, 156, 162, 167, 171, 173, 174, 196, 217, 218, 219, 220, 225, 232, 234, 239, 240,
242, 245, 246, 247, 248, 250, 252, 253, 260, 264, 271, 280, 282, 284, 285, 287, 288, 289, 290, 291, 302, 304, 308, 321, 322, 346, 370, 371, 381, 390, 405, 406
Dalrymple, G. B., 9, 10, 370, 371
Davies, D. M., 356, 371
Davis, E. M., 83, 364, 365, 398, 404, 405, 406
Davis, L. C., 49, 371
Davis, W. B., 264, 371
Davis, W. H., 117, 119, 120, 121, 122, 123, 124, 125, 126, 368
Dawkins, W. B., 178, 371
Dawson, M. R., xvi, xvii, 278, 280
DeGeer, G., 4, 371
Del Campana, D., 371
Denny, C. S., 301, 364, 392
Denton, G. H., 9, 10, 371
Dice, L. R., 243, 277, 371
Dietrich, W. O., 371
Doell, R. R., 10, 370
Dolan, E. M., 57, 371
Domning, D. P., xvii, 67, 68, 340, 341, 371
Donovan, J. J., 284, 371
Dort, W., 62, 371
Doutt, J. K., 267, 375
Downey, J. S., 11, 278, 279, 371
Downs, T., xvi, 12, 27, 54, 65, 227, 228, 229, 248, 278, 371, 372, 400
Dreimanis, A., 344, 346, 372

Eames, A. J., 142, 372
Eddleman, C. D., 196, 372
Edmund, G., xvi, xvii, 129, 130, 131, 140, 141, 142, 145
Einsohn, S. D., 259, 372
Eldredge, N., xiii, 93, 372
Elftman, H. O., 76, 225, 372
Emry, H. L., xvi
Erdbrink, D. P., 184, 372
Erickson, B. R., 237, 372
Eshelman, R. E., xvi, 16, 17, 104, 105, 106, 213, 218, 219, 220, 249, 250, 252, 256, 301, 332, 372
Etheridge, R., 64, 372
Evans, G. L., 19, 81, 190, 372
Evenson, E. B., 301, 372
Evernden, J. F., 4, 15, 372
Ewer, R. F., 151, 155, 162, 163, 165, 176, 372

Fahlbusch, V., 93, 372
Falconer, H., 351, 352, 372
Faro, J. B., 161, 393
Feduccia, J. A., 15, 372
Figgins, J. D., 372
Finch, W. C., 300, 301, 372
Findley, J. S., 73, 111, 233, 234, 250, 372, 378
Fleischer, R. L., 4, 372
Flerov, C., 333, 336, 338, 372, 373
Flint, R. F., 2, 37, 39, 373
Foley, R. L., 68, 70, 172, 300, 301, 378
Fortsch, D. E., xvi, 50, 61, 373, 382, 404

INDEX TO AUTHORS

Foster, H. L., 44, 192, 400
Frankforter, W. D., 37, 39, 317, 337, 393, 394
Frazier, M. K., xvi, xvii, 264, 265, 273, 373
Freudenberg, W., 373
Frick, C., 12, 26, 290, 292, 312, 316, 319, 320, 322, 326, 327, 330, 339, 373
Friedman, I., 4, 373
Frison, G. C., 87, 335, 336, 338, 356, 373, 404
Fry, W. E., 20, 35, 310, 318, 327, 373
Frye, J. C., 373
Fuller, W. A., 291, 337, 373
Funderberg, J. B., 341, 342, 373
Furlong, E. L., 50, 54, 319, 320, 321, 322, 330, 373, 395, 397

Gabunia, L. K., 95, 373
Galbreath, E. C., 63, 70, 172, 373, 405
Galusha, T., xvi
Garbani, H. J., 27
Gard, L. M., Jr., 44, 340, 341
Gardner, A. L., 102, 373
Gazin, C. L., 11, 15, 26, 31, 33, 57, 124, 140, 141, 148, 153, 155, 156, 158, 159, 164, 169, 184, 186, 195, 213, 214, 224, 226, 227, 229, 248, 249, 252, 262, 268, 272, 276, 277, 278, 279, 281, 287, 298, 299, 319, 331, 373, 374
Geist, V., 318, 329, 374
Genoways, H. H., 112, 113, 383
Getz, L., 163, 173, 374
Gidley, J. W., xiii, 10, 26, 31, 33, 35, 57, 124, 134, 148, 153, 155, 158, 159, 164, 169, 176, 184, 224, 248, 252, 262, 268, 272, 276, 289, 290, 297, 298, 299, 331, 332, 333, 348, 374
Gilbert, B. M., 88, 194, 330, 364, 387, 405
Gildersleeve, H., 369
Giles, E., 167, 168, 374
Gillette, D. D., xvi, xvii, 56, 81, 132, 134, 135, 195, 374
Goldman, E. A., 159, 375
Goodrich, C., 65, 375
Gould, S. J., xiii, 93, 372
Graham, R. W., 26, 39, 55, 112, 375
Green, M., 78, 219, 375
Gregory, J. T., xvi
Griggs, C. D., 297, 395
Grigorescu, D. 202, 204, 205, 207, 208, 393
Gromova, V. I., 99, 336, 339, 375
Grossenheider, R. P., 111, 215, 217, 220, 221, 222, 223, 232, 233, 234, 245, 369
Gruhn, R., 63, 375
Guilday, J. E., xvi, xvii, 31, 33, 36, 62, 63, 66, 69, 74, 76, 77, 79, 80, 86, 87, 102, 106, 107, 108, 109, 111, 112, 114, 115, 119, 120, 121, 122, 123, 124, 174, 180, 184, 188, 192, 212, 215, 216, 220, 221, 222, 223, 226, 228, 229, 242, 244, 245, 257, 258, 260, 261, 262, 263, 264, 267, 268, 270, 272, 276, 280, 282, 296, 297, 299, 300, 301, 306, 312, 315, 318, 348, 362, 364, 365, 375, 376, 387, 389, 391, 405, 406
Gunter, H., 59, 376
Gustafson, E. P., 21, 35, 177, 178, 310, 318, 327, 373, 376, 398
Gut, H. J., 58, 118, 124, 126, 127, 165, 195, 376, 392

Guthrie, R. D., xvi, 25, 41, 43, 215, 256, 259, 261, 267, 276, 318, 328, 329, 336, 338, 354, 376

Hager, M. W., xvi, 12, 55, 376, 404
Hall, E. R., xvi, xvii, 83, 99, 119, 120, 124, 148, 149, 150, 151, 153, 155, 158, 159, 162, 164, 165, 212, 213, 221, 225, 234, 280, 282, 369, 376, 377
Hall, E. T., 4, 377
Hall, J. S., 66, 126, 382
Hallberg, G. R., 261, 377
Hamilton, H. W., xvi, 66, 76, 79, 86, 87, 107, 108, 109, 111, 112, 120, 184, 212, 216, 220, 222, 223, 228, 229, 270, 300, 301, 315, 375, 376, 406
Hamon, J. H., 58, 377
Hancock, G. A., 52
Handley, C. O., Jr., 87, 120, 124, 244, 375, 377
Haq, B. U., xiii, 4, 5, 22, 377
Harington, C. R., xvi, xvii, 34, 40, 43, 44, 75, 76, 78, 88, 89, 138, 172, 174, 175, 182, 192, 201, 202, 205, 206, 207, 261, 276, 282, 288, 291, 312, 315, 316, 317, 326, 327, 329, 330, 331, 332, 333, 334, 335, 336, 338, 339, 353, 364, 365, 377, 382, 405
Harksen, J. C., 19, 388
Harlan, R., 378
Harlow, R. F., 184, 378
Harris, A. H., xvi, 73, 212, 226, 233, 234, 243, 246, 250, 329, 378, 404
Harris, R. K., 81, 370
Haury, E. W., 47, 49, 378
Hausman, L. A., 142, 378
Hawksley, O., xvi, 68, 70, 172, 300, 301, 378, 382
Hay, O. P., xiii, 10, 33, 34, 59, 63, 64, 67, 68, 75, 78, 80, 81, 198, 199, 212, 214, 218, 219, 274, 278, 281, 294, 312, 314, 317, 318, 346, 378, 379
Haynes, C. V., Jr., 47, 48, 352, 353, 356, 365, 379, 404, 405, 406
Hays, J. D., 4, 37, 39, 379
Helfman, P. M., 356, 368
Hemmer, H., 191, 192, 379
Hemmings, E. T., 352, 353, 379
Hendey, Q. B., xv, xvi
Hendry, C., xvi
Heppenstall, C. A., xvi
Hershkovitz, P., 92, 250, 379
Hesse, C. J., 325, 379
Hester, J. J., 73, 145, 313, 363, 364, 365, 379
Hibbard, C. W., xiii, xvi, 15, 16, 17, 18, 20, 28, 29, 30, 32, 33, 34, 35, 37, 39, 64, 65, 67, 104, 107, 109, 110, 111, 112, 113, 115, 123, 130, 149, 156, 157, 158, 161, 162, 166, 167, 175, 176, 213, 214, 218, 226, 227, 230, 231, 233, 239, 240, 242, 243, 244, 245, 247, 248, 249, 250, 252, 254, 255, 256, 258, 259, 260, 262, 263, 264, 265, 266, 268, 271, 277, 278, 279, 284, 285, 287, 288, 289, 290, 291, 294, 301, 302, 303, 308, 319, 320, 321, 329, 330, 332, 333, 339, 372, 373, 379, 380, 381, 395
Hibbard, E. A., 317, 381
Hill, J. E., 215, 381
Hillerud, J. W., 22, 23, 37, 39, 337, 338, 381, 393
Hills, M., 172, 173, 174, 370

Hinds, F. J., 332, 333, 381
Hirschfeld, S. E., 136, 138, 381
Ho, T. Y., 54, 381
Hoffmann, R. S., xvi, 219, 220, 370, 391
Hoffstetter, R., 143, 145, 381
Holland, W. J., 77, 381
Holman, J. A., 15, 59, 263, 382
Hood, C. H., 70, 300, 301, 382
Hoover, R., 80, 260, 262, 364, 396, 404
Hopkins, D. M., xvi, 44, 92, 257, 318, 382, 391, 392
Hopkins, M. L., xvi, 61, 304, 382
Howard, E. B., 73, 253, 260, 281, 312, 313, 331, 393
Howard, H., 54, 72, 382
Hubbs, C. L., 203, 393
Hughes, G. T., 291, 371
Hutchison, J. H., xvi, xvii, 113, 114, 115, 118, 176, 367, 382
Huxley, J. S., xiii, 382

Ikeya, M., 4, 382
Imbrie, J., 4, 37, 39, 379
Irving, D. C., 80, 180, 376
Irving, W. N., 89, 315, 382
Ives, P. C., 365, 382, 404
Izett, G. A., 19, 382

Jackson, C. G., 382
Jakway, G. E., 32, 74, 221, 246, 333, 382
James, G. T., 130, 382
Jammot, D., 105, 106, 382
Järvi, A., xvi, 363, 365, 382
Jefferson, G. T., xvi, xvii, 50, 51, 54
Jefferson, T., 66, 137, 138, 334, 382
Jegla, T. C., 66, 126, 382
Jelinek, A. J., 382
Jenkins, F. A., xvi
Jepsen, G. L., 346, 382
Johnson, D. L., 49, 50, 382
Johnson, E., xvi, 84
Johnson, G. H., 260, 370
Johnson, N. M., 5, 6, 7, 10, 11, 15, 16, 17, 19, 20, 22, 23, 26, 27, 28, 29, 39, 285, 350, 351, 383, 386
Johnston, C. S., 19, 20, 167, 183, 184, 297, 383, 398
Johnston, R., 383
Jones, J. K., Jr., xvi, 109, 112, 113, 118, 124, 150, 151, 367, 383
Jones, R. E., 340, 341, 383
Jorgensen, C. D., 253, 400
Judd, F., 84, 174, 196, 217, 225, 232, 234, 239, 240, 242, 246, 248, 250, 252, 253, 271, 280, 282, 371, 406

Kahlke, H. D., 95, 316, 317, 334, 335, 383
Kaisen, O. C., 335, 336, 337, 338, 396
Kapp, R. O., 28, 383
Kaspar, T. C., 85, 383
Kaye, C. A., 238, 383
Kaye, J. M., 382, 383
Kellogg, L., 226, 238, 253, 383
Kellogg, R., 204, 383
Kelsall, J. P., 315, 383
Kelson, K., 99, 119, 124, 162, 164, 165, 212, 213, 221, 225, 234, 280, 377
Kennedy, G. E., 356, 383
Kennerly, T. E., Jr., 234, 384
Khan, E., 78, 317, 384
Kilmer, F. H., 160, 384
King, J. E., 69, 70, 389, 404, 406

Kinsey, P. E., 296, 384
Kirby-Smith, H. T., 80, 387
Kitts, D. B., 335, 384
Klein, J., 28, 384
Klingener, D., 271, 384
Koch, A. K., 70
Kolb, K. K., 64, 384
Kontrimavichus, B. L., 92, 384
Köppen, W., 4
Koopman, K. F., xvi, xvii, 118, 122, 126, 127, 384
Kormos, T., 153, 384
Korobitsyna, K. U., 328, 329, 330, 384
Kowalski, K., 268, 384
Kraglievich, J. L., 172, 384
Kretzoi, M., 93, 384
Krochak, G., xvi
Kukla, G. J., 384
Kurtén, B., xiv, 13, 20, 34, 55, 56, 57, 58, 59, 62, 80, 88, 148, 151, 153, 155, 166, 167, 168, 169, 170, 171, 172, 173, 180, 182, 184, 185, 186, 190, 191, 192, 194, 195, 196, 198, 199, 282, 291, 313, 314, 334, 336, 354, 364, 365, 384, 385

Lambe, L. M., 385
Lammers, G. E., 385
Lance, J. F., 274, 385
Langenwalter, P. E., xvi, 50, 54, 309, 385
Langguth, A., 385
Laudermilk, J. D., 72, 142, 385
Lawrence, B., xvi, 62, 126, 171, 385
LeConte, J., 299
Leeman, W. P., 14, 15, 367
Leffler, S. R., 161, 385
Leidy, J., 59, 67, 68, 289, 299, 351, 385, 386
Leonard, A. B., 29, 373, 386
Leonhardy, F. C., 75, 365, 386, 404
Lewis, A. P., xvi
Lewis, G. E., xvi, 44, 301, 340, 341, 373, 386
Libby, W. F., 4, 51, 54, 365, 368, 386, 405, 406
Ligon, J. D., 57, 386
LiLiu, C., 370, 404
Lillegraven, J. A., xvi, 138, 386
Lindsay, E. H., xvi, 5, 6, 7, 10, 11, 13, 14, 15, 16, 17, 19, 20, 22, 23, 26, 27, 28, 29, 39, 49, 69, 70, 210, 211, 249, 285, 287, 350, 351, 383, 386, 389, 404, 406
Lipps, L., xvi, 60, 273, 386, 392
Long, A., 142, 386
Long, C. A., 157, 158, 386
Loomis, F. B., 57, 386
Lull, R. S., 142, 386
Lundelius, E. L., Jr., xvi, xvii, 73, 80, 81, 83, 176, 228, 246, 251, 288, 289, 290, 291, 294, 297, 348, 386, 401
Lydekker, R., 386
Lyell, C., 5, 387
Lynch, T. F., 356, 387
Lyon, G. M., 138, 387

McAnulty, W. N., 138, 398
McClung, C. E., 387
McClure, W. L., 85, 383, 396
McCoy, J. J., 57, 387
McCrady, A. D., xvi, 66, 76, 77, 79, 87, 106, 107, 108, 109, 111, 112, 119, 120, 121, 122, 184, 212, 216, 221, 222, 223, 228, 229, 245, 257, 258, 261, 263, 268, 270, 272, 282, 297, 300, 301, 315, 364, 376, 405, 406
McCrady, E., 80, 387
McDonald, H. G., xvi, xvii, 34, 61, 62, 136, 137, 138, 337, 365, 387, 405
Macdonald, J. R., 15, 18, 51, 387
McGinnis, H., xvi, 80, 192, 376
McGowan, J., 378
McGrew, P. O., xiii, xvi, 287, 302, 387
Mack, C. W., xvi, 151, 399
McKenna, M. C., xvi, 103, 128, 387
McMullen, T. L., 64, 106, 387
Macneish, R. S., 356, 362, 363, 365, 387
Macpherson, A. H., 267, 387
McPike, M. E., xvi
Madden, C. T., 50
Madsen, J. H., xvi
Maglio, V. J., xvi, 343, 348, 349, 350, 351, 352, 353, 354, 387
Maher, W. J., 335, 387
Malde, H. E., 14, 15, 350, 351, 367, 387
Marcus, L. F., 53, 54, 168, 170, 172, 381, 387, 401
Marsh, O. C., 387
Martin, H. T., 387
Martin, L. D., xvi, xvii, 18, 32, 39, 88, 186, 190, 194, 196, 218, 256, 257, 259, 265, 269, 329, 330, 337, 364, 387, 394, 406
Martin, P. S., xvi, 47, 48, 77, 106, 107, 108, 112, 119, 121, 122, 142, 212, 216, 221, 222, 223, 245, 257, 258, 261, 263, 268, 270, 272, 282, 297, 306, 348, 356, 358, 362, 363, 364, 365, 375, 386, 387, 390, 404, 405, 406
Martin, R. A., 13, 19, 27, 34, 55, 56, 57, 59, 78, 153, 169, 173, 229, 235, 237, 241, 243, 250, 251, 259, 263, 264, 281, 363, 364, 365, 387, 388
Matthew, W. D., 32, 35, 166, 167, 226, 287, 388
Matthews, J. V., Jr., 25, 89, 256, 259, 267, 276, 354, 376, 388
Mawby, J. E., xvii, 20, 72, 190, 279, 388
Maxfield, J., 64, 257, 401
Mayr, E., xiii, 388
Mead, R. A., 162, 163, 388
Meade, G. E., 19, 33, 81, 134, 167, 190, 191, 319, 320, 348, 372, 388
Mech, L. D., 170, 388
Mehl, M. G., 70, 388
Mehringer, P. J., Jr., 47, 48, 49, 69, 70, 72, 388, 389, 404, 406
Melton, W. G., Jr., 134, 389
Merriam, J. C., xiii, 11, 168, 172, 182, 188, 190, 192, 196, 198, 389
Meyer, K. J., 4, 389
Milankovitch, M., 4
Miller, B. B., 332, 333, 395
Miller, G. J., xvi, 172, 188, 313, 389
Miller, G. S., 184, 389
Miller, L. H., 74, 389
Miller, S. J., xvi, 63, 72
Miller, W. E., xvi, xvii, 50, 51, 52, 54, 85, 114, 138, 145, 161, 208, 215, 217, 225, 233, 241, 242, 243, 260, 280, 281, 321, 337, 365, 389, 405
Mills, R. S., xvi, 74, 365, 389
Milstead, W. M., 81, 390
Mitchell, E. D., 161, 204, 205, 208, 368, 390

Moore, J., 237, 390
Mooser, O., 40, 171, 173, 280, 284, 287, 288, 289, 291, 304, 321, 322, 390
Morgan, A. V., 184, 369
Mosimann, J. E., 362, 365, 390
Mott, R. J., 353, 377
Müller, H., 4, 390
Mundel, P., 329, 378
Munz, P. A., 72, 385
Musil, R., 171, 390
Myers, T. P., 71, 370

Nadler, C. F., 219, 370
Neff, N. A., 29, 390
Neill, W. T., 57, 390
Nelson, M. E., 64, 384
Nelson, R. S., 32, 265, 266, 267, 390
Nevo, E., 209, 398
Nichols, R. H., xvi
Nowak, R. M., xvii, 74, 167, 168, 169, 170, 172, 390

Oakley, K. P., 354, 390
Odano, M. J., xvi
Oelrich, T. M., 175, 310, 390
Oesch, R. D., 69, 70, 114, 164, 226, 242, 280, 314, 364, 365, 390, 391, 404
Ognev, S. I., 267, 390
Olsen, S. J., xvi, 58, 118, 126, 127, 165, 195, 269, 345, 346, 354, 390, 392
Olson, E. C., xvi, 70, 281, 390
Opdyke, N. D., 5, 6, 7, 10, 11, 12, 14, 15, 16, 17, 19, 20, 22, 23, 26, 27, 28, 29, 39, 285, 331, 350, 351, 383, 386, 390
Orr, P. C., 50, 194, 390
Osborn, H. F., 134, 343, 346, 348, 350, 351, 352, 353, 354, 390
Osgood, W. H., 333, 391
Ostrom, J. H., xvi
Owen, R., 391

Packard, R., 248, 391
Palmer, H. A., 328, 391, 392
Palmquist, J. C., 334, 335, 392
Paradiso, J. L., xvi
Parmalee, P. W., xvi, 63, 69, 79, 86, 114, 164, 220, 222, 223, 226, 242, 266, 280, 314, 315, 364, 365, 376, 391, 404
Parris, D. C., xvi
Pascual, R., 90, 91, 92, 136, 140, 391
Patterson, B., xv, xvi, 90, 91, 92, 132, 136, 140, 391
Patton, T. H., 84, 132, 252, 257, 262, 263, 282, 364, 391
Paula Couto, C. de, 130, 135, 136, 138, 141, 142, 391
Paulson, G. R., 29, 107, 112, 113, 228, 240, 243, 259, 262, 269, 391
Peterson, O. A., 77, 282, 314, 391
Peterson, R. L., 206, 207, 391
Peterson, R. S., 203, 393
Péwé, T. L., 41, 43, 44, 215, 316, 318, 327, 334, 354, 365, 391, 404, 405
Pizzimenti, J. J., 219, 220, 391
Portis, A., 315, 316, 391
Powell, L. H., 237, 391
Powers, H. A., 350, 351, 387
Prentiss, D. W., 151, 391
Price, P. B., 4, 372
Purdy, R. W., xvi

INDEX TO AUTHORS

Quackenbush, L. S., 40, 391
Quick, H. F., xvi
Quinn, J. II., 288, 290, 291, 391

Rasmussen, D. L., 71, 238, 260, 370, 391
Rausch, R. L., 155, 261, 267, 385, 391, 392
Rausch, V. R., 267, 392
Ray, C. E., xvi, 44, 57, 58, 60, 76, 124, 126, 127, 151, 165, 185, 195, 200, 202, 204, 205, 207, 208, 220, 245, 268, 269, 273, 280, 281, 294, 301, 318, 334, 335, 364, 376, 386, 392, 393, 399
Reed, C. A., 328, 392
Reed, K. M., 115, 392
Reichstein, H., 150, 151, 392
Reinhart, R. H., 342, 392
Rensberger, J. M., xvii, 210, 392
Repenning, C. A., xvi, xvii, 44, 103, 109, 110, 111, 113, 159, 160, 161, 199, 200, 201, 202, 203, 204, 205, 206, 207, 208, 211, 212, 257, 392, 393
Reynolds, J. E., 68, 70, 172, 300, 301, 378
Reynolds, R., xvi
Reynolds, R. E., xvi, 51
Rhoads, S. N., 393
Richards, H. G., xvi, 397
Richardson, J., 393
Richey, K. A., xvi
Ridenour, J., 60, 61, 393
Riggs, E. S., 16, 167, 302, 303, 381
Rinker, G. C., 381
Ritchie, R., 80, 81, 245, 396
Robertson, J. S., Jr., 13, 56, 57, 59, 129, 143, 222, 338, 393
Robinson, P., xvi, 88
Romer, A. S., xv, xvi, 99, 142, 301, 306, 393
Roth, E. L., 83, 84, 174, 196, 217, 225, 232, 234, 239, 240, 242, 246, 248, 250, 252, 253, 271, 280, 282, 371, 393, 406
Rubin, M., 44, 257, 301, 364, 392
Russell, R. D., 12, 393
Russell, R. J., 74, 224, 229, 393

Sabels, B. E., 48, 142, 387
Sadek-Kooros, H., xvi, 62, 393, 405
Saunders, J. J., 69, 70, 237, 393
Savage, D. E., xvi, 3, 11, 19, 20, 22, 23, 26, 27, 37, 39, 168, 183, 194, 195, 196, 198, 243, 284, 287, 289, 290, 291, 304, 305, 306, 312, 331, 348, 383, 393
Sawyer, G. J., xvi
Saylor, E. B., 47, 378
Schaub, S., 199, 332, 393
Scheffer, V. B., 202, 393
Schneider, D., 176, 393
Schneider, K. B., 161, 393
Schultz, C. B., xvi, 18, 22, 23, 32, 37, 39, 66, 73, 132, 186, 190, 196, 253, 260, 269, 281, 312, 313, 317, 325, 331, 337, 339, 368, 393, 394
Schultz, G. E., xvi, 19, 20, 28, 39, 64, 65, 106, 107, 242, 262, 319, 320, 394, 406
Schultz, J. R., 11, 51, 184, 253, 287, 394
Schütt, G., 172, 394
Scott, W. B., 317, 394
Sellards, E. H., 33, 59, 81, 130, 281, 294, 394
Semken, H. A., Jr., xvi, xvii, 30, 32, 39, 49, 112, 113, 214, 218, 219, 234, 240, 242, 244, 251, 260, 261, 262, 265, 266, 267, 269, 297, 332, 333, 375, 377, 390, 394, 395
Sergeant, D. E., 205, 377
Shackleton, D. M., 34, 377
Shackleton, N. J., 4, 37, 39, 379
Sher, A. V., xvi, 316, 327, 332, 395
Shotwell, J. A., 13, 236, 238, 287, 395
Shutler, D., 48, 72, 142, 387
Sikes, S. K., 350, 395
Simons, E. L., xvi
Simpson, G. G., xiii, xvi, 56, 57, 59, 69, 70, 79, 92, 130, 131, 135, 143, 145, 176, 188, 192, 195, 224, 269, 274, 283, 292, 293, 294, 297, 301, 309, 312, 326, 327, 331, 395
Sims, J. D., 300, 301, 372
Sinclair, W. J., 52, 226, 331, 373, 395
Skeels, M. A., 346, 353, 395
Skinner, M. F., xvi, xvii, 15, 16, 18, 20, 48, 104, 110, 156, 158, 162, 177, 213, 227, 230, 242, 248, 249, 250, 254, 255, 256, 258, 284, 287, 288, 291, 320, 322, 323, 324, 335, 336, 337, 338, 395, 396
Slaughter, B. H., xvi, xvii, 73, 80, 81, 83, 131, 168, 188, 221, 245, 246, 260, 262, 290, 294, 301, 312, 361, 364, 365, 386, 396, 404
Smith, R. L., 4, 373
Soiset, J. M., xvi, xvii, 51, 337
Sokolov, I. I., 159, 396
Sommers, J., 369
Stalker, A. MacS., xvi, 25, 34, 44, 45, 47, 167, 280, 291, 318, 336, 337, 356, 370, 396, 398, 405, 406
Starrett, A., 33, 266, 267, 396
Steller, G., 340
Stephens, J. J., 75, 112, 196, 245, 247, 266, 267, 396
Stevens, J. B., 332, 333, 395
Stevens, M. S., 173, 288, 386, 396
Stirton, R. A., 11, 27, 122, 166, 167, 198, 199, 237, 283, 285, 321, 325, 388, 396, 397
Stock, A. D., 330, 397
Stock, C., xiii, 49, 50, 52, 54, 72, 73, 74, 138, 142, 143, 145, 166, 170, 173, 180, 182, 188, 192, 196, 198, 253, 290, 308, 319, 321, 322, 327, 353, 389, 397
Stokes, W. L., 330, 397
Storer, J. E., 44, 219, 397
Stout, T. M., 18, 32, 37, 39, 394
Stovall, J. W., 138, 184, 398
Strain, W. S., 20, 213, 214, 227, 288, 398
Studer, F. V., 20
Szabo, B. J., 47, 398

Tanner, L. G., xvi, 32, 394
Taylor, B., xvii
Taylor, D. W., 15, 16, 28, 33, 61, 64, 65, 123, 244, 245, 319, 320, 321, 381, 398
Taylor, W. P., 321, 398
Tedford, R. H., xv, xvi, xvii, 5, 21, 169, 177, 178, 200, 201, 202, 392, 398
Templeton, H., 80, 387
Tener, J. S., 335, 398
Tessman, N., xvi, 10, 11, 26, 49, 71, 210, 211, 249, 386
Thaeler, C. S., 209, 398
Thenius, E., 172, 398
Tikhomirov, B. A., 354, 398
Tipper, H. W., 353, 377
Tobien, H., 93, 95, 153, 346, 398
Toohey, L., 317, 394
Torre, D., 167, 398
Trapp, G. R., 177, 398
Trever, J. E., xvi
Trimble, D. E., 61, 369
Troxell, E. L., 35, 172, 288, 290, 332, 398
Turekian, K. K., 4, 398
Turnbull, W. D., xvi

Valastro, S., Jr., 83, 364, 365, 398, 404, 405, 406
Van Couvering, J., xvi
Van Couvering, J. A., xiii, 4, 5, 22, 23, 377
VanderHoof, V. L., 12, 166, 167, 337, 393, 397, 398
Van Der Meulen, A. J., 31, 258, 259, 262, 263, 398, 399
Van Gelder, R. G., xvi, 162, 163, 165, 194, 195, 328, 335, 399
Vangengeim, E. A., 95, 399
Van Valen, L., xvi, 113, 357, 365, 399
Van Zyll de Jong, C. G., 159, 399
Varela, A. G., 83, 364, 365, 398, 404, 405, 406
Vaughan, T. A., 99, 116, 117, 118, 126, 132, 204, 213, 218, 232, 238, 325, 399
Vereshchagin, N. K., 192, 399
Voorhies, M. R., xvi, 60, 140, 141, 231, 232, 364, 399

Walker, D. N., xvi, 87, 399, 401
Walker, E. P., 107, 109, 118, 151, 155, 161, 177, 178, 202, 203, 204, 205, 206, 207, 208, 210, 218, 221, 238, 267, 273, 276, 301, 315, 325, 327, 328, 335, 342, 399
Walker, M. V., xvi
Walker, R. M., 4, 372
Warren, J. C., 345, 346, 399
Warter, J. K., 53, 54, 399
Wasley, W. W., 47, 378
Waters, J. H., 151, 399
Webb, S. D., xvi, xvii, 13, 27, 28, 55, 56, 57, 58, 59, 90, 91, 92, 129, 131, 132, 136, 138, 180, 188, 245, 246, 247, 263, 264, 281, 301, 302, 303, 304, 306, 307, 308, 309, 312, 324, 325, 348, 360, 363, 364, 365, 381, 388, 399
Wegener, A., 4
Weigel, R. D., 59, 229, 399
Wells, P. V., 253, 399, 400
Werdelin, L., xvi, 195, 196
Wetzel, R. M., 296, 400
Wheat, J. B., xvi, xvii, 338, 400, 405
Whistler, D. P., xvi, xvii
Whitaker, J. O., Jr., 111, 400
White, J. A., xvi, xvii, 12, 27, 61, 62, 125, 225, 227, 228, 229, 243, 248, 272, 273, 278, 279, 280, 281, 315, 367, 371, 400
Whitmore, F. C., Jr., xvi, xvii, 44, 192, 284, 300, 301, 340, 341, 346, 350, 372, 373, 400
Wilcox, R. W., 19, 382
Wilcoxon, J. A., 368
Williams-Dean, G., 363, 365, 369
Williston, S. W., 301, 400
Willoughby, D. P., 290, 400
Wills, D. L., 334, 335, 392
Wilson, D. J., 338, 373
Wilson, J. A., xvii

Wilson, J. W., III, xvii, 353, 361, 365
Wilson, M. V., xvii, 71, 78, 335, 336, 337, 338, 362, 365, 373, 400
Wilson, R. L., 317, 400
Wilson, R. W., 49, 223, 254, 265, 266, 272, 327, 400
Windham, S., xvii
Wistar, C., 66, 299
Woldstedt, P., 37, 39, 401
Wood, A. E., 232, 401

Wood, H. E., 3, 401
Woodard, G. D., 53, 54, 401
Woodburne, M. O., 235, 236, 295, 297, 298, 348, 401
Woods, C. A., 273, 401
Wright, B. A., 329, 330, 381
Wright, N. E., xvii
Wright, T., 176, 401

Young, S. P., 198, 401

Zakrzewski, R. J., xiii, xvii, 15, 16, 29, 31, 36, 64, 65, 156, 210, 211, 212, 213, 223, 224, 226, 228, 231, 233, 235, 236, 237, 238, 243, 244, 247, 254, 255, 256, 257, 259, 260, 264, 265, 269, 273, 274, 381, 384, 401
Zazhigin, V. S., 95, 399
Zeimens, G., 87, 401
Zeuner, F. E., 4, 401

FIGURE CREDITS

Barbro Elgert: 1.1, 2.1, 8.1–4, 9.1, 10.3–4, 10.6–7, 11.1–7, 11.9–15, 11.17, 11.22–25, 12.1–14, 13.1, 14.1–3, 15.4–7, 15.18, 16.1, 17.2. Maps 1–5.

Erica Hansen: 10.5.

Margaret Lambert: 2.2, 3.1–2, 4.1–2, 4.4, 4.9, 10.1–2, 11.8, 11.16, 11.19, 14.4, 15.2, 15.9–10, 15.12–14, 15.16, 17.3.

Margaret Lambert and Dawn B. Adams: 2.2, 4.6–7.

Margaret Lambert and Barbro Elgert: 11.21, 15.3, 15.17.

Riggert Munsterhjelm: 11.18, 15.8, 15.15.

Hubert Pepper: 11.20.

James D. Senior: 4.3, 4.5, 4.8, 15.1, 15.11, 17.1.